AF323825

Optical Processes in
Microparticles and **Nanostructures**

A Festschrift Dedicated to **Richard Kounai Chang**
on His Retirement from Yale University

ADVANCED SERIES IN APPLIED PHYSICS

Editors-in-Charge: R. K. Chang (*Yale Univ.*) &
A. J. Campillo (*Naval Research Lab.*)

Published

Advanced
Series in
Applied
Physics
Volume 6

Optical Processes in
Microparticles
and
Nanostructures

A Festschrift Dedicated to
Richard Kounai Chang
on His Retirement from Yale University

Editors

Ali Serpengüzel
Koç University, Turkey

Andrew W Poon
The Hong Kong University of Science & Technology, China

World Scientific

NEW JERSEY · LONDON · SINGAPORE · BEIJING · SHANGHAI · HONG KONG · TAIPEI · CHENNAI

Published by

World Scientific Publishing Co. Pte. Ltd.

5 Toh Tuck Link, Singapore 596224

USA office: 27 Warren Street, Suite 401-402, Hackensack, NJ 07601

UK office: 57 Shelton Street, Covent Garden, London WC2H 9HE

British Library Cataloguing-in-Publication Data
A catalogue record for this book is available from the British Library.

Advanced Series in Applied Physics — Vol. 6
OPTICAL PROCESSES IN MICROPARTICLES AND NANOSTRUCTURES
A Festschrift Dedicated to Richard Kounai Chang on His Retirement from Yale University

ISBN-13 978-981-4295-77-2
ISBN-10 981-4295-77-9

Printed in Singapore by B & Jo Enterprise Pte Ltd

CONTENTS

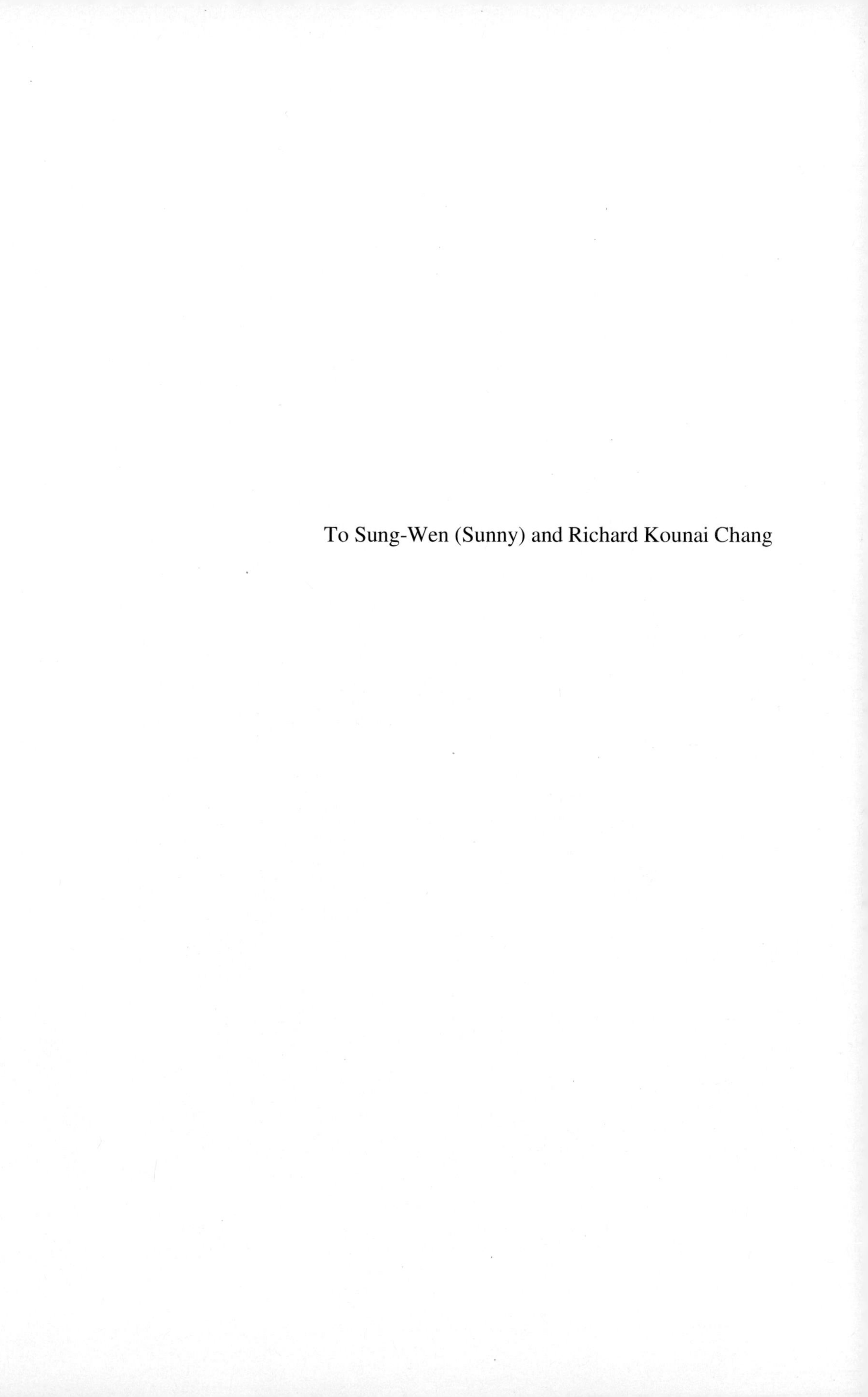

To Sung-Wen (Sunny) and Richard Kounai Chang

Richard Kounai Chang (Chinese New Year 2010)

FOREWORD

Richard Kounai Chang has made a lasting contribution to the development of quantum optoelectronics during the past fifty years. This volume contains a large variety of papers written by his former students and colleagues. They provide testimony to the global influence of the teaching and research of professor Chang.

I had the good fortune to have him as a graduate student in the early sixties. He did research towards a doctor of philosophy degree in Applied Physics at Harvard University. He was a co-author on a half-dozen papers concerned with the fundamentals of second harmonic generation of light in a variety of materials. They verified the basic laws of nonlinear reflection and refraction.

Richard's promise as a graduate student has been amply fulfilled by his subsequent as a professor of applied physics at Yale University. This volume represents a well-deserved testimony to his scientific legacy.

Nicolaas Bloembergen
University of Arizona, Tucson, Arizona, USA

PREFACE

This Festschrift is a collection of articles contributed by former students and colleagues of an eminent scholar on a special occasion: Richard Kounai Chang on his retirement from Yale University on June 12, 2008. Richard's extensive experimental and theoretical research spanning four decades defined the frontiers of what can be learnt using linear and nonlinear optical techniques in surface science, electrochemistry and colloids, microparticles and bioaerosols, microcavities and nanostructures.

Chang's approach to multidisciplinary research is reflected in the Yale University Center for Laser Diagnostics (CLD), which he founded in 1976. CLD is a testament to the importance and cross fertilization of interdisciplinary research with faculty members from Applied Physics, Physics, Electrical Engineering, Mechanical Engineering and Chemical Engineering. Chang himself was a professor of Applied Physics, Physics and Electrical Engineering. For his distinguished academic achievements, Yale University honored him with the Henry Ford II Chair Professor for Applied Physics. For his lasting contributions to optics, laser diagnostics, and chemistry of aerosols, droplets and microparticles, The American Association for Aerosol Research (AAAR) honored him with the David Sinclair Award. Colleagues have noted that, professor Chang's work is inventive, seminal, and very often beautiful.

We, his former students, are most indebted for the inspiration this great scholar gave us. Richard, our deeply respected teacher and a close friend, lit the path for us and also travels with us. We are most fortunate to see first-hand how this incredible gentleman brings out the very best in academic research and in life. This volume is only in a crystallizing glimpse of Richard Kounai Chang's scientific legacy, and his global influence to the people, whom he has come to interact with in both scientific and personal ways.

Chang started his academic life working with Nobel Laureate Professor Nicolaas Bloembergen during the early sixties. His thesis work on second harmonic generation (SHG) from surfaces of centrosymmetric media verified the basic laws of nonlinear reflection and refraction, and provided basis for later development of important nonlinear spectroscopic tools for surface and interface studies.

In 1966, he joined the Department of Applied Physics at Yale University, where he continued his nonlinear optics research in a variety of materials. His early work on nonlinear optical processes in surfaces led to his later role in spearheading the development of such technologically relevant nonlinear spectroscopic tools using surface-enhanced Raman scattering (SERS) and other nonlinear processes for electrochemistry and colloidal science. His early work on SERS is seminal for what we now call nanoplasmonics.

Chang pioneered the study of nonlinear optics in micrometer-sized droplets, which elegantly confine light in the droplet circumference as sharp morphology-dependent resonances (MDRs) or what are now commonly known as whispering-gallery modes (WGMs). Chang's group was the first to observe in the WGMs of microdroplets optical processes such as lasing, stimulated Raman scattering, sum-frequency generation and stimulated Brillouin scattering. Chang's group was also among the first to explore a large variety of linear and nonlinear effects in microdroplets and microcylinders including elastic scattering, fluorescence, spontaneous Raman scattering, coherent Raman mixing, coherent anti-Stokes Raman scattering (CARS), multiorder Stokes emission, laser induced-plasma breakdown spectroscopy (LIBS), explosive vaporization and two-photon pumped lasing. These early work formed the basis of using WGMs for sensitive laser-based microparticle sizing and diagnostic of chemical and biochemical compositions in microparticles, and also stimulated much research interest in optical processes in microcavities.

In the last decade, Chang's group has worked intensively on developing linear and nonlinear spectroscopy for microparticles. His unparalleled insights into linear and nonlinear optical processes in microparticles have led to development of techniques to rapidly interrogate and sort on the fly, particle by particle, biological agents such as anthrax from other ambient respirable airborne particles. Richard initiated this research well before the September 11 attacks, which now forms the basis for some of the most promising defenses against biological weapons.

Chang's group has also been studying lasing micropillars of different cavity shapes in polymers and compound semiconductors. This work has opened new avenues for microcavities, namely controlling the optical coupling via cavity shape deformations and optimizing the optical and electrical pumping configurations by selectively exciting the WGMs. Recently, Chang's group introduced the microspiral resonators, which demonstrated unidirectional lasing in InGaN quantum-well (QW) microlasers, thus breaking the clockwise and counter-clockwise symmetry. Colleagues have noted how beautifully simple the microspiral resonator concept is, and how important that microstructure will be in the nascent development of electrophotonic integrated circuits (EPICs).

The Festschrift has contributions from academia, national laboratories and industries, across disciplines and countries. We include work from Richard's native Hong Kong China, to his blooming in the United States, as well as Canada, Germany, Israel, Korea, Russia, Switzerland, Taiwan and Turkey. The contributors span departments and schools of Applied Physics, Biomedical Engineering, Biophotonics, Chemistry, Chemical Engineering, Chemical Physics, Electrical Engineering, Electronic and Computer

Engineering, Materials Science, Mechanical Engineering, Information Technology, Optics, Photonics, Physics and Physical Chemistry. In the light of the broad variety and global coverage of Richard's influence, including a chapter by everyone is a huge, if not an impossible, task. We therefore seek to reflect a representative set of Richard's many research contributions in this collection, while recognizing this might only do partial justice to his scientific legacy and the many people, who were part of this journey.

Part I *Nonlinear Optics & Spectroscopy* includes chapters 1 - 4.

Chapter 1 by Y. Ron Shen of the University of California, Berkeley, on the topic of *Nonlinear optical spectroscopy for interfaces* highlights earlier pioneering work of Richard Chang on laying the foundation for development of second-harmonic generation and sum-frequency generation as surface-specific spectroscopic tools.

Chapter 2 by Andreas Otto of Heinrich-Heine-Universität Düsseldorf on the topic of *Surface enhanced Raman scattering (SERS) of carbon dioxide on cold-deposited copper films* discusses the SERS and infrared reflection absorption spectroscopy spectra of carbon dioxide on cold-deposited copper films, and reveals an electronic effect at a minority of surface sites.

Chapter 3 by Alfred Leipertz and Thomas Seeger of Friedrich-Alexander Universität Erlangen-Nürnberg, on the topic of *Combustion diagnostics by pure rotational coherent anti-stokes Raman scattering (CARS)* reviews pure rotational CARS fundamentals and their technological applications to combustion diagnostics, flame research and practical combustion systems.

Chapter 4 by Marshall Long of Yale University on the topic of *Imaging flames* reviews the use of laser light scattering mechanisms in multiparameter imaging experiments for quantitative study of flames and comparison with combustion models.

Part II *Linear & Nonlinear Optics in Microparticles* includes chapters 5 - 8.

Chapter 5 by E. James Davis of University of Washington, Seattle, on the topic of *Elastic and inelastic light scattering from levitated microparticles* reviews numerous applications of laser light scattering measurements from single levitated solid-state microspheres or microdroplets to basic studies of physical, chemical, and optical properties in microparticles and droplets. Morphology-dependent resonances (MDRs) are applied to study the size and refractive index of droplets, microspheres and coated spheres. The application of inelastic scattering to biological particles detection is also discussed.

Chapter 6 by Claudiu Dem, Michael Schmitt, Wolfgang Kiefer and Jürgen Popp of Friedrich-Schiller University Jena, University Würzburg and Institute of Photonic Technology, Jena, on the topic of *Physical chemistry and biophysics of single trapped microparticles* reviews work on combined inelastic / elastic (Raman-Mie) light scattering studies of single microparticles trapped by optical / electrodynamical forces and their applications to several microchemical reactions and on elastic scattering on a femtosecond timescale. The use of MDRs-induced peaks in the Raman spectra as a diagnostic probe of the microparticles physical properties and dynamics of chemical reactions is discussed. A few pharmaceutical applications are highlighted.

Chapter 7 by Alfred S. Kwok of Pomona College on the topic of *Cavity-enhanced emission in fluorescent microspheres* revisits some of Richard Chang's first MDR experiments. MDRs in the lasing spectra of dye-coated polystyrene microspheres and the fluorescence spectra in quantum dot-coated microspheres are discussed.

Chapter 8 by Mikhail V. Jouravlev of Pohang University of Science and Technology and Gershon Kurizki of Weizmann Institute of Science on the topic of *Theory of Raman amplification in microspheres* presents the unified theory of spontaneous and stimulated Raman scattering in microspherical cavities. The reduction of the threshold intensity and the increase of the gain upon input and output resonances near MDRs are analyzed based on the coupling of the internal partial waves in nonlinear cavities.

Part III *Linear & Nonlinear Spectroscopy of Bioaerosols* includes chapters 9 - 11.

Chapter 9 by Hermes C. Huang of Yale University and Yong-Le Pan, Steven C. Hill and Ronald G. Pinnick of the U.S. Army Research Laboratory on the topic of *Fluorescence-based classification with selective collection and identification of individual airborne bioaerosol particles* reviews the development of techniques for bioaerosol detection and characterization spearheaded by Richard Chang's group and collaborators at the U.S. Army Research Laboratory over the last decade. The chapter focuses on the development of the Single-Particle Fluorescence Spectrometer (SPFS) for real-time, in-*situ* monitoring, classification, sorting and collection of bioaerosols. The chapter highlights the investigations of ultraviolet-laser-induced fluorescence of aerosols, the evolution of the SPFS technology, and the application of the SPFS to characterization of ambient aerosols at three different geographic locations.

Chapter 10 by Stephen Holler of Thermo Fisher Scientific Incorporation and Kevin B. Aptowicz of West Chester University on the topic of *Discerning single particle morphology from two-dimensional light scattering patterns* reviews the development of two-dimensional angular optical scattering (TAOS) technique pioneered by Richard Chang's group and collaborators at University of Hertfordshire, the U.S. Army Research Laboratory, and Porton Down over the past decade. The chapter presents measured and numerically simulated TAOS patterns of non-spherical particles (smooth spheroids or corrugated aggregates) and data analysis techniques for extracting single particle morphology from the information-rich TAOS patterns.

Chapter 11 by Jean-Pierre Wolf of University of Geneva on the topic of *Femtosecond spectroscopy for biosensing* discusses a new femtosecond spectroscopy technique for bioaerosol sensing. The induced nonlinear light emission from the particles is backward-enhanced, which is favorable for remote detection. The technique also features quantum control schemes which enable discrimination of bioaerosols from ambient organic particles.

Part IV *Optical Microcavities & Nanostructures* includes chapters 12 - 18.

Chapter 12 by Hui Cao of Yale University on the topic of *Lasing in random media* provides an overview of the random laser which represents a class of non-conventional laser whose feedback is facilitated by random fluctuation of the dielectric constant in space. The chapter details the random laser fundamentals, key characteristics and potential applications.

Chapter 13 by Wen-Feng Hsieh, Hsu-Cheng Hsu, Wan-Jiun Liao, Hsin-Ming Cheng, Kuo-Feng Lin, Wei-Tze Hsu and Chin-Jiu Pan of National Chaio Tung University on the topic of *Optical properties of zinc oxide quantum dots* presents work on size-dependence of efficient ultraviolet photoluminescence (PL) and absorption spectra of various sizes of zinc oxide (ZnO) quantum dots (QDs), and reveals evidence for quantum confinement effect. Temperature-dependent and power-dependent PL spectra of ZnO QDs are analyzed.

Chapter 14 by Martin Claassen and Hakan Türeci of Eidenössische Technische Hochschule Zürich on the topic of *Constant Flux states and their applications* reviews constant flux (CF) states that describe the steady-state response of a photonic medium with an arbitrary, possibly frequency-dependent index of refraction to a harmonically oscillating source.

Chapter 15 by Vladimir Ilchenko, Andrey Matsko, Anatolily Savchenkov and Lute Maleki of OEwaves Incorporation on the topic of *Electro-optical applications of high-Q crystalline WGM resonators* presents whispering-gallery mode resonators in functional radio-frequency signal processing devices operating at frequencies in microwave and millimeter-wave band.

Chapter 16 by El-Hang Lee of INHA University on the topic of *VLSI Photonics* reviews optical microcavity microspheres, microdisks, and microrings and microcylinders, which found their applications in photonic integration and VLSI photonics of today. It then discusses on the outlook of the VLSI photonics in the future.

Chapter 17 by Ali Serpengüzel of Koç University on the topic of *Integrating Spheres* proposes the multidimensional integration of spherical microresonators coupled with optical waveguides will usher in a new age in integrated photonics.

Chapter 18 by Xianshu Luo and Andrew W. Poon of The Hong Kong University of Science and Technology on the topic of *Microspiral and double-notch-shaped resonators for integrated photonics* highlights the development of the microspiral resonator and its derivative for integrated photonics. The chapter reviews how the microspiral resonator invented by Richard Chang's group for the application of unidirectional-emission microcavity lasers has been finding new applications in silicon-based passive photonic devices, featuring gapless direct coupling between the microdisk and the integrated waveguides and also gapless inter-cavity coupling between adjacent microdisks.

Part V *Photoscapes or Multidisciplinary Applications* **in biomedical engineering, telecommunications and astronomy includes chapters 19 - 22.**

Chapter 19 by Grace D. Metcalfe, Paul H. Shen and Michael Wraback of the U.S. Army Research Laboratory on the topic of *Terahertz radiation from nitride semiconductors* discusses the generation of broadband terahertz (THz) radiation from nitride semiconductors by ultrafast optical pulses. The THz regime is technologically important for example in biomedical imaging. The chapter emphasizes on the generation mechanisms related to the photo-Dember effect and acceleration of photogenerated charges in built-in electric fields in-plane and normal to the sample surface.

Chapter 20 by Noelle R. B. Stiles, Benjamin P. McIntosh, Patrick J. Nasiatka, Michelle C. Hauer, James D. Weiland, Mark S. Humayun and Armand R. Tanguay, Jr. of the University of Southern California on the topic of *An intraocular camera for retinal prostheses* reviews the implantation of an intraocular retinal prosthesis representing one possible approach to the restoration of sight in those with minimal light perception due to photoreceptor degenerating diseases such as retinitis pigmentosa and age-related macular degeneration.

Chapter 21 by Dipak Chowdhury and Michal Mlejnek of Corning Incorporation and Shiva Kumar of McMaster University on the topic of *Model for optical propagation in randomly coupled birefringent fiber and its implementation* summarizes the model for birefringent propagation in optical fibers, and describes its detail numerical implementation.

Chapter 22 by Pui Tang Leung of The Chinese University of Hong Kong and Jun Wu of University of Pittsburgh on the topic of *Applications of quasi-normal mode expansion* reviews the application of quasi-normal mode expansion and the properties of quasi-normal modes for various open wave systems from microdroplets to black holes and neutron stars.

This work would have been impossible without the help of dear friends and colleagues Nejat Bilkay Tulgar, Sinan Utku, Xianshu Luo, Shaoqi Feng and Ting Lei. We would like to thank Mike Tsoupko-Sitnikov for image processing of the front cover. The professional assistance of Alvin Chong of World Scientific is most gratefully acknowledged.

May 2010

ALI SERPENGÜZEL
Koç University, Istanbul, Turkey

ANDREW W. POON
Hong Kong University of Science & Technology,
Hong Kong China

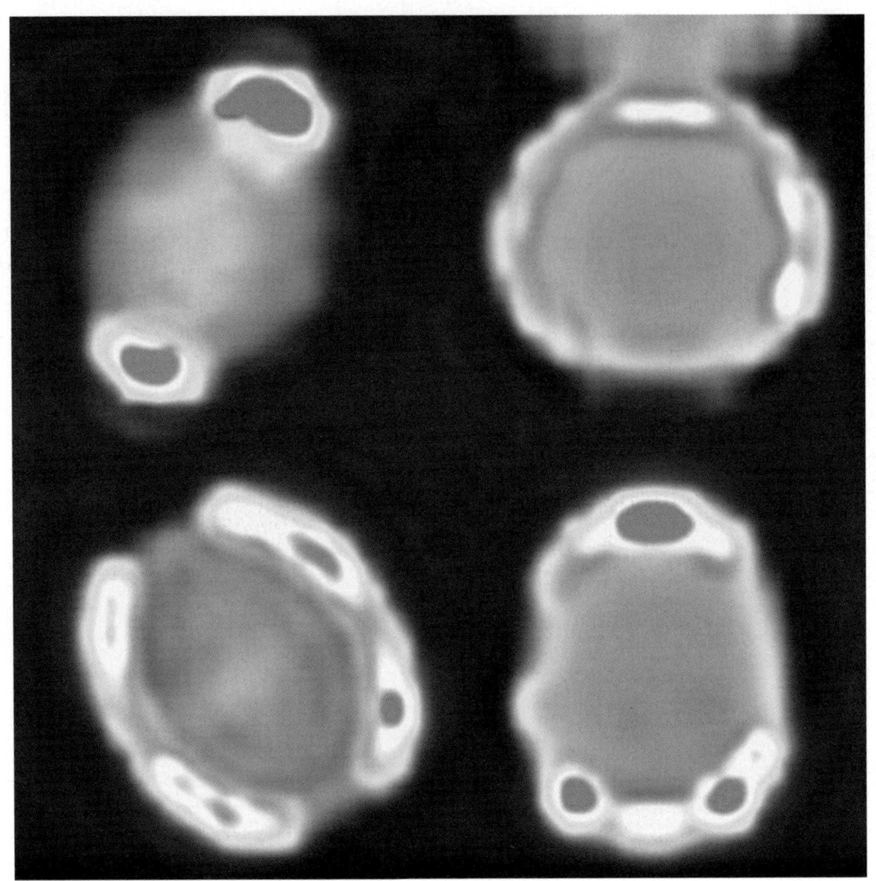

Front Cover

Experimental lasing images of broadly pumped prolate and oblate microdroplets, from Seongsik Chang, Richard K. Chang, A. Douglas Stone and Jens U. Nöckel, *Journal of Optical Society of America B* 17 1828 (2000). Reproduced with permission.

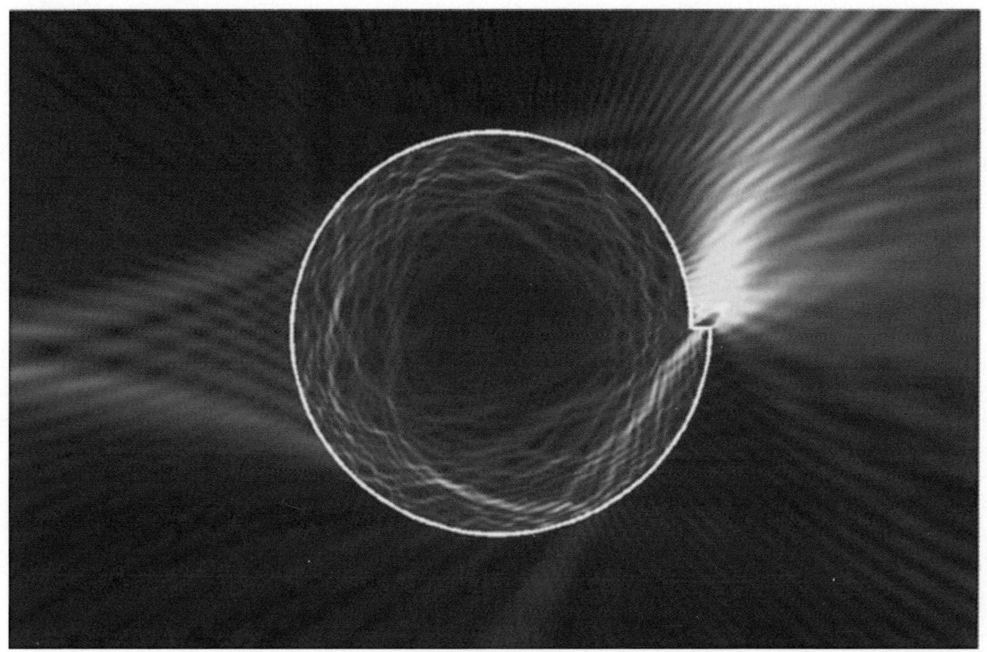

Back Cover

Real-space false-color plot of the modulus of the electric field for a calculated quasibound state in a microspiral, from G. D. Chern, H. E. Türeci, A. Douglas Stone, R. K. Chang, M. Kneissl and N. M. Johnson, *Applied Physics Letters* **83**, 1710 (2003) Reproduced with permission.

Fundamental Physical Constants

Quantity	Symbol	Value	Exponent of 10	Unit
Avogadro	N_A	6.022 141 99(47)	+23	mol^{-1}
Boltzmann	k	1.380 6503(24)	−23	$J\,K^{-1}$
Planck	h	6.626 068 76(52)	−34	$J\,s$
electric	ε_o	8.854 187 817	−12	$F\,m^{-1}$
magnetic	μ_o	4π	−07	$N\,A^{-2}$
speed of light in vacuum	c	299 792 458	00	$m\,s^{-1}$
elementary charge	e	1.602 176 462(63)	−19	C
electron mass	m_e	9.109 381 88(72)	−31	kg
electron magnetic moment	μ_e	−928.476 362(37)	−26	$J\,T^{-1}$
proton mass	m_p	1.672 621 58(13)	−27	kg
proton magnetic moment	μ_p	1.410 606 633(58)	−26	$J\,T^{-1}$
neutron mass	m_n	1.674 927 16(13)	−27	kg
neutron magnetic moment	μ_n	−0.966 236 40(23)	−26	$J\,T^{-1}$

Source: Peter J. Mohr and Barry N. Taylor, CODATA Recommended Values of the Fundamental Physical Constants: 1998, Journal of Physical and Chemical Reference Data, Vol. 28, No. 6, 1999 and Reviews of Modern Physics, Vol. 72, No. 2, 2000.

PART I

NONLINEAR OPTICS & SPECTROSCOPY

CHAPTER 1

NONLINEAR OPTICAL SPECTROSCOPY FOR INTERFACES

Y. RON SHEN

Physics Department, University of California
Berkeley, California 94720 USA
yrshen@berkeley.edu

Earlier pioneering work of Richard Chang set the foundation for the development of second-harmonic generation and sum-frequency generation as surface-specific spectroscopic tools that have created many new opportunities for research in surface science.

Nonlinear optical spectroscopy has created many exciting new opportunities for material studies. One of the areas it has had strong impact on is surface science. We discuss here the development of second-harmonic and sum-frequency generation as surface analytical tools.[1] They have become in recent years the most powerful and versatile spectroscopic probes for surface and interface studies. As second-order nonlinear optical processes, they are forbidden in media with inversion symmetry under the electric-dipole approximation, but at the surface or interface, the inversion symmetry is broken and the processes become allowed. Dominated by surface contribution, the processes can then be used as surface-specific probes. More generally, surface and bulk have different structural symmetry, and it is possible to have the two contributions to the nonlinear optical processes deduced separately.

There are a number of advantages using second harmonic generation (SHG) and sum-frequency generation (SFG) as surface probes.[1] They are surface-specific, and have submonolayer sensitivity. Being coherent optical processes, they generate an output that is highly directional and easily distinguishable from the non-coherent background such as fluorescence, and can be used for *in situ*, remote sensing studies. The techniques are non-detrimental as long as the input laser intensities are below the laser damage threshold. As laser spectroscopy, they are capable of having very high spatial, temporal, and spectral resolutions. Most important of all is that the techniques are applicable to all interfaces as long as the interfaces are accessible by both input and output light. The last point makes the techniques most versatile in comparison with other surface probes. As a result, SHG and SFG have found many unique applications. Among them are studies of buried interfaces, surface structures of neat materials such as polymers and liquids, molecular adsorption and reactions under ambient conditions, surface dynamics, and surface microscopy. All these are important areas of surface science, but have hardly been explored because of limitation of conventional surface probes.

Investigation of SHG and SFG from surfaces can be traced back to the earlier work of Richard Chang as a graduate student in Professor Bloembergen's lab. His work[2] on SHG in reflection from Si and Ge surfaces, and Brown and coworkers' work[3] on SHG in reflection from silver surfaces constitute the first study of SHG from media with inversion symmetry. They had in mind the use of SHG to probe surface nonlinearity,[4] but unfortunately their results were complicated by surface contamination and lack of good understanding of surface contributions to SHG at the time. In a subsequent paper, however, Bloembergen, Chang, et al.[5] developed the macroscopic theory of SHG from a medium with inversion symmetry. It assumes a three-layer model with the surface or interface treated as a thin layer with optical constants different from the bulk. This is still the model we follow today to understand surface SHG and SFG.

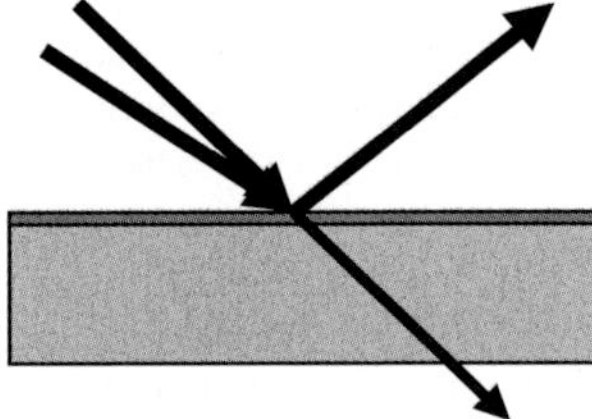

Fig. 1.   Schematic of the beam geometry with respect to the sample in a sum-frequency spectroscopic measurement.

From the three-layer model with the beam geometry sketched in Fig. 1, one can show that the sum-frequency output of photons per pulse generated from the system by the overlapping input beams at ω_1 and ω_2 is given by[6]

$$S_i(\omega = \omega_1 + \omega_2) \sim \frac{8\pi^3\omega}{\hbar c^3} \mid F_{ii}(\omega)\chi^{(2)}_{eff,ijk}F_{jj}(\omega_1)F_{kk}(\omega_2)\mid^2 I_1(\omega_1)I_2(\omega_2)AT \qquad (1)$$

with

$$\ddot{\chi}^{(2)}_{eff} = \ddot{\chi}^{(2)}_S + i\ddot{\chi}^{(2)}_B / \Delta k \qquad (2)$$

Here, $I_1(\omega_1)$ and $I_2(\omega_2)$ are intensities of input beams at ω_1 and ω_2, respectively, from medium 1, $F_{\alpha\alpha}(\omega_i)$ is the transmission Fresnel coefficient at ω_i, A and T the beam overlapping area and time at the interface, $\ddot{\chi}^{(2)}_S$ and $\ddot{\chi}^{(2)}_B$ are the surface and bulk susceptibilities of the interfacial layer and medium 2, respectively, assuming the bulk nonlinear susceptibility of medium 1 is negligible, and Δk is the phase mismatch for the SFG process in medium 2. Note that SHG is a special case of SFG with $\omega_1=\omega_2$. From Eqs. (1) and (2), it is seen that because $\Delta k \sim 10^5$/cm for SFG in reflection (much smaller in transmission) and $\mid\chi^{(2)}_S\mid \sim 10^{-7}\mid\chi^{(2)}_B\mid$ in esu units, if $\mid\chi^{(2)}_B\mid$ is allowed, the bulk contribution often dominates, but if $\mid\chi^{(2)}_B\mid$ is not allowed, then $\mid\chi^{(2)}_S\mid$ could be dominant and SFG become surface-specific. This is often the case for bulk media with inversion symmetry in which second-order nonlinear optical processes are forbidden by symmetry.[1]

A crude estimate using typical values of $|\chi_S^{(2)}| \sim 10^{-15}\,esu$ for a surface monolayer, and $I_1 \sim I_2 \sim 10$ GW/cm^2 for 10-ps inputs of 100 J/pulse focused to an overlapping area of $A \sim 0.1$ mm^2 in Eq. (1) yields a surface SFG signal of 10^4 photons/pulse, which should be readily detectable. This shows that SFG can be very sensitive as a surface analytical tool. Moreover, the existence of several independent, nonvanishing tensor elements of $\chi_{S,ijk}^{(2)}$ that can be separately deduced from SFG measurements with different input/output polarization combinations allows us to obtain information on interfacial structure and molecular arrangement. Here, we use the neat water/vapor interface as an example to illustrate how SFG can serve as a powerful surface spectroscopic tool.

Water interfaces play key roles in many relevant physical, chemical, and biological processes. Understanding of these processes requires knowledge of water interfacial structures at the molecular level.[7] Orientations and bonding arrangement of water molecules at an interface are expected to be very different than in the bulk, and they determine the interfacial properties and hence the interfacial processes. Such information can be obtained from the vibrational spectra of water interfaces. To date, however, SFG is the only technique that can provide such spectra. Vapor-water interfaces have attracted much interest not only because they are important in environmental science,[8] but also because they serve as standard references for all water interfaces.

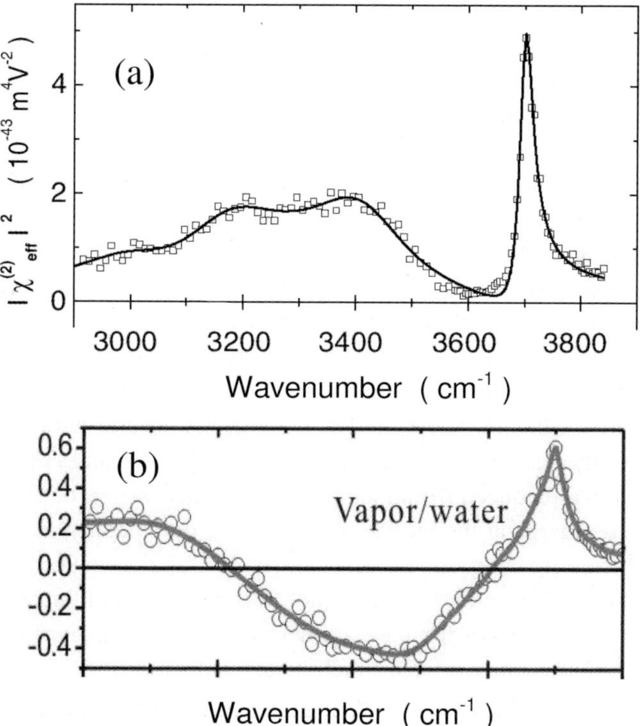

Fig. 2. Sum-frequency vibrational spectra of water/vapor interface. (a) $|\chi_{S,eff}^{(2)}|^2$ and (b)Im $\chi_S^{(2)}$ versus input IR frequency. (Reproduced from Refs. 10 and 13).

What we can learn from the spectrum in Fig. 2a and other spectra with different input/output polarization combinations for the same water interface is that the interfacial water structure appears as a highly disordered, strongly fluctuating, largely tetrahedral H-bonding network with some molecules having local ice-like bonding structure. The surface has an overall structure more ordered than the bulk, as expected from surface tension. One would, however, like to learn more about the interface from the spectrum such as which water species contribute to which spectral region and how they react to ions or adsorbates appearing at the interface. This would require more detailed characterization of surface resonances contributing to the spectrum. In analogy to Im $\varepsilon(\omega)$ that describes an absorption or emission spectrum, we need to have the spectrum of Im $\chi_S^{(2)}$ (instead of $|\chi_S^{(2)}|^2$) in order to directly characterize surface resonances. For water interfaces, we have

$$|\chi_S^{(2)}|^2 = |\chi_{NR}^{(2)} + \int \frac{A_q(\omega_q)\rho(\omega_q)d\omega_q}{\omega_2 - \omega_q + i\Gamma_q}|^2$$

$$\text{Im}\,\chi_S^{(2)} = \text{Im}\int \frac{A_q(\omega_q)\rho(\omega_q)d\omega_q}{\omega_2 - \omega_q + i\Gamma_q} \propto A_q(\omega_q)\rho(\omega_q)$$

(3)

As seen from the above equations, it is not possible to obtain Im $\chi_S^{(2)}$ from $|\chi_S^{(2)}|^2$. Measurements of both $|\chi_S^{(2)}|^2$ and the phase ϕ of $\chi_S^{(2)}$ are necessary to acquire the complete information on $\chi_S^{(2)}$ and deduce Im $\chi_S^{(2)}$. This is generally true for all coherent nonlinear wave-mixing processes as first recognized by Chang et al.[12] They set up the first nonlinear interferometric experiment to measure the phase of a nonlinear susceptibility for SHG from a nonlinear crystal.

Phase-sensitive sum-frequency vibrational spectroscopic measurement on neat-water/vapor interface was recently carried out.[13] It provided the Im $\chi_S^{(2)}$ spectrum for the interface displayed in Fig. 2b. Knowing that the dangling OH peak describes O→H pointing out of the water, we can easily identify the positive ice-like band comes mainly from DDAA water species straddling DDA and DAA molecules in the topmost layer with upward ice-like donor H-bonds. The more loosely bonded DDAA species with net polar orientations of O→H pointing toward the water bulk contributes to the negative liquid-like band up to ~3500 cm^{-1}, and the DDA and DAA species with O→H pointing toward water are mainly responsible for the part of the negative band above 3500 cm^{-1}. The latter assignment was deduced from the observation that this part of the spectrum did not change when ions emerge at the interface; the surface field created by the ions could reorient the more loosely bonded DDAA molecules and change their spectrum accordingly, but not the DDA and DAA molecules because they could not experience the strong surface field. Extension of the measurement to water/vapor interfaces of various acid, base, and salt solutions has yielded much better understanding of how ions would appear at a water interface and affect the water interfacial structure.[14]

SF spectroscopy with ultrashort pulses in a pump/probe scheme also allows us to investigate *in situ* ultrafast surface dynamics. Here again, we use water/vapor interfaces as an example.[15] There is strong interest in the science community in learning how vibrational excitation energy at a water interface decays away. Obviously, we need a surface-specific probe to monitor the decay of the surface excitation. In the reported experiment,[14] an infrared femtosecond pump pulse was directed onto a water/silica interface. It excited a narrow range of OH stretch vibrations of water molecules both at the interface and in the bulk. Significant population transfer from the ground state to the vibrational excited state of the pumped molecules could occur with a sufficiently strong pump, and created a spectral hole in the vibrational spectrum of the water molecules. Evolution of the spectral hole in time provided information on the relaxation of the excitation. Since the interest was on surface vibrational dynamics, time-dependent SF vibrational spectroscopy with femtosecond pulses was employed as a probe to discriminate surface against the bulk. It monitored only how the OH stretch spectrum of the water interface changed with time after the pump pulse even though bulk excitation was also present. The result showed that spectral diffusion first broadened the spectral hole appreciably in less than 100 fs after the pump, vibrational relaxation from the excited state then happened with a time constant of ~300 fs, and finally the deposited energy appeared as heat to increase the local water temperature in ~700 ps. This relaxation pathway is qualitatively similar to that of vibrational excitation in bulk water,[16] and is probably the generic characteristic of vibrational excitation/relaxation processes in a hydrogen-bonding network.

Applications of SHG and SFG as surface analytical tools are not restricted to bulk media with inversion symmetry. Because surface and bulk generally have different structural symmetry, it is possible to use specific geometry and input/output polarization combinations to suppress the bulk contribution and make the surface contribution dominant.[17] We take crystalline α–quartz as an example.[18] SHG and SFG are allowed in quartz.[18] The nonvanishing nonlinear susceptibility elements of the bulk are $\chi^{(2)}_{B,aaa} = -\chi^{(2)}_{B,abb} = -\chi^{(2)}_{B,bba} = -\chi^{(2)}_{B,bab}$, $\chi^{(2)}_{B,abc} = -\chi^{(2)}_{B,bac}$, $\chi^{(2)}_{B,acb} = -\chi^{(2)}_{B,bca}$ and $\chi^{(2)}_{B,cab} = -\chi^{(2)}_{B,cba}$, where $(\hat{a}, \hat{b}, \hat{c})$ refer to the crystalline coordinates with $\hat{c}$ along the 3-fold axis, $\hat{a}$ along the 2-fold axis, and $\hat{a}, \hat{b} \perp \hat{c}$. For the (0001) surface (with $\hat{c}$ along the surface normal), the nonvanishing elements of the surface nonlinear susceptibility are $\chi^{(2)}_{S,ccc}$, $\chi^{(2)}_{S,aac} = \chi^{(2)}_{S,bbc}$, $\chi^{(2)}_{S,aca} = \chi^{(2)}_{S,bcb} \approx \chi^{(2)}_{S,caa} = \chi^{(2)}_{S,cbb}$. If the input beams are in an incident plane rotated by Φ away from the a axis, then it can be readily seen from Eq. (2) that for SSP (denoting S-, S-, and P-polarizations for SF, visible, and infrared fields, respectively) and PPP polarization combinations, $\chi^{(2)}_{eff}$ has the form

$$\chi^{(2)}_{SSP,eff} = L_{yy}(\omega_{SF})L_{yy}(\omega_{vis})[L_{zz}(\omega_{IR})\sin\theta_{IR}\chi^{(2)}_{S,aac} + L_{xx}(\omega_{IR})\cos\theta_{IR}i\chi^{(2)}_{B,aaa}\cos 3\Phi / \Delta k]$$

$$\chi^{(2)}_{PPP,eff} \approx -L_{xx}(\omega_{SF})L_{xx}(\omega_{vis})L_{zz}(\omega_{IR})\cos\theta_{SF}\cos\theta_{vis}\sin\theta_{IR}\chi^{(2)}_{S,aac}$$

$$+L_{zz}(\omega_{SF})L_{zz}(\omega_{vis})L_{zz}(\omega_{IR})\sin\theta_{SF}\sin\theta_{vis}\sin\theta_{IR}\chi^{(2)}_{S,ccc} \qquad (4)$$

$$+L_{xx}(\omega_{SF})L_{xx}(\omega_{vis})L_{xx}(\omega_{IR})\cos\theta_{SF}\cos\theta_{vis}\cos\theta_{IR}i\chi^{(2)}_{B,aaa}\cos 3\Phi / \Delta k$$

If Φ is set to make $\cos3\Phi$ vanish, then only $\tilde{\chi}^{(2)}_{S}$ contributes to $\chi^{(2)}_{eff}$. Shown in Fig. 3 are the SSP spectra of the α–quartz (0001) surface. It is seen that the bulk phonon mode at

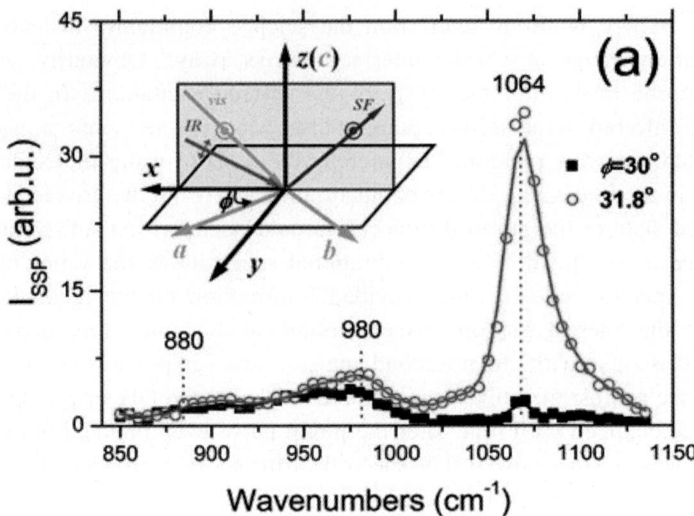

Fig. 3. SSP sum-frequency vibrational spectra of α-quartz(0001) taken at $\phi = 30°$ and 31.8° (curves are guides to eyes). (Reproduced from Ref. 18).

1064 cm^{-1} is strongly suppressed when Φ changes from 31.8^0 to 30^0, while the surface phonon modes at 880 cm^{-1} (Si-O-Si) and 980 cm^{-1} (Si-OH) remain unchanged. This illustrates how the surface spectra of a non-centrosymmetric crystal can be obtained.

There are many areas of research surface-specific second-harmonic and sum-frequency spectroscopy have hardly explored. Bio-interfaces, electrochemical interfaces, nanostructure interfaces, and interfacial catalytic reactions are among the most important ones. Further development of the techniques is also anticipated as their applications become more wide spread.

Acknowledgments

This work was supported by the NSF Science and Technology Center of Advanced Materials for Purification of Water with Systems (Water CAMPWS; CTS-0120978), and by the Department of Energy under Contract No. DE-AC03-76SF00098.

References

1. Y. R. Shen, Nature 337, 519 (1989).
2. R. K. Chang, C. H. Lee, and N. Bloembergen, Phys. Rev. Lett. 16, 986 (1966).
3. F. Brown and M. Matsuoka, Phys. Rev. Lett. 18, 985 (1969).
4. N. Bloembergen and R. K. Chang, in "Physics of Quantum Electronics", edited by B. Lax and P. M. Kelley (McGraw Hill, NY, 1965), p.80.
5. N. Bloembergen, R. K. Chang, S. S. Jha, and C. H. Lee, Phys. Rev. 174, 813 (1968).
6. Y.R.Shen, in "Frontiers in Laser Spectroscopy", Proceedings of the International School of Physics, Enrico Fermi, Course CXX (North-Holland, Amsterdam, 1994), p.139.
7. See, for example, P. Jungwirth, Chem. Rev. 106, 1137 (2006).
8. See, for example, J. G. Irwin and M. L. Williams, Environ Pollut 50, 29 (1988).
9. Q. Du, R. Superfine, E. Freysz, and Y. R. Shen, Phys. Rev. Lett. 70, 2313 (1993).

10. X. Wei and Y. R. Shen, Phys. Rev. Lett. 86, 4799 (2001).

11. U. Buck, I. Ettischer, M. Melzer, V. Buch, and J. Sadlej, Phys. Rev .Lett. 80, 2578 (1998).

12. R. K. Chang, J. Ducuing, and N. Bloembergen, Phys. Rev. Lett. 15, 6 (1965).

13. N. Ji, V. Ostroverkhov, C. S. Tian, and Y. R. Shen, Phys. Re . Lett.100, 096102 (2008).

14. C. S. Tian, N. Ji, G. A. Waychunas, and Y. R. Shen. J. Am. Chem. Soc. 130, 13033 (2008).

15. J. A. McGuire and Y. R. Shen, Science 313, 1945 (2006).

16. M. L. Cowan et al., Nature 434, 199 (2005); A. J. Lock and H. J. Bakker, J. Chem. Phys. 117, 1708 (2002);. A. J. Lock, S. Woutersen, and H. J. Bakker, J. Phys. Chem. A105, 1238 (2001).

17. T. Stehlin, M. Feller, P. Guyot-Sionnest, and Y. R. Shen, Optical Lett. 13, 389 (1988).

18. W.-T. Liu and Y. R. Shen, Phys. Rev. Lett. 101, 016101 (2008).

CHAPTER 2

SURFACE-ENHANCED RAMAN SCATTERING (SERS) OF CARBON DIOXIDE ON COLD-DEPOSITED COPPER FILMS: AN ELECTRONIC EFFECT AT A MINORITY OF SURFACE SITES

ANDREAS OTTO

Institut für Physik der kondensierten Materie, Heinrich-Heine-Universität Düsseldorf, Germany
otto@uni-duesseldorf.de

After some special points from the long history of surface-enhanced Raman scattering (SERS) and a discussion of silver (Ag) films on nano-spheres and their relation to the "chemical effect" the focus of this chapter is on SERS and infrared reflection absorption spectroscopy (IRRAS) of carbon dioxide (CO_2) on cold-deposited copper films. The SERS spectra of CO_2 on copper films deposited at 40 K display neutral species at SERS active sites with bands not observed by Raman spectroscopy of CO_2 gas, but identical to the loss bands of gaseous CO_2 in electron energy loss spectroscopy. The absence of one component of the Fermi doublet of CO_2 in SERS proves that the local electromagnetic field enhancement at SERS active sites cannot deliver signals above the noise level. The activated anionic CO_2^- is observed by transient electron transfer from the anionic molecule to the copper metal at a subgroup of SERS active sites, which are annealed below 200 K. The IRRAS spectra show only the expected infrared (IR) active modes of neutral CO_2 representing the "majority species" of adsorbed CO_2.

1. Introduction

Surface-enhanced Raman scattering (SERS) was first introduced to a wider public by a book edited by Richard K. Chang and Thomas E. Furtak[1] and a review article by Richard K. Chang.[2] SERS has not developed in a straightforward way. To give two examples, from the early days and from present days: I remember that I attended the Conference on the optical properties of thin films, Southampton, 24.-26.9. 1973. A chairman allowed someone not belonging to the conference before the session started to point out an important new discovery. As it was in Southampton it is probable that the strange intruder was Martin Fleischmann. I cannot recall what the topic was (maybe the contents of Ref. 3?). Only after D.L. Jeanmaire and R.P. Van Duyne published their famous paper[4] (received by the publisher 7. Oct.1976) the first "SERS rush" started.

Triggered by two papers by Katrin Kneipp and her collaborators[5,6] and by Ref. 7 the "Single molecule SERS rush" started. The state of the field in 2007 has been described in Ref. 8. The fantastic high enhancement factors quoted have been questioned recently.[9,10]

The resonance Raman effect is no prerequisite of single molecule SERS[11,12] Single molecule SERS on single colloidal particles has been reported in Ref. 13. Tip enhanced Raman spectroscopy of adsorbates at single crystal noble metal surfaces in ultra high vacuum does now reach single molecule sensitivity.[14]

SERS has also come a long way to discuss the mechanisms involved. Here I cite Martin Moskovits:[15] "The electromagnetic theory of surface-enhanced Raman spectroscopy (SERS), despite its simplicity, can account for all major SERS observations. However, the electromagnetic model does not account for all that is learned through SERS. Molecular resonances, charge-transfer transitions and other processes such as ballistic electrons transiently probing the region where the molecule resides and then modulating electronic processes of the metal as a result certainly contribute to the rich information SERS reports; and by virtue of the fact that these contributions will vary from molecule to molecule, they will constitute the most interesting aspects reported by SERS."

The most interesting developments beyond classical metal optics come from the theory of small clusters. The low laser frequency (chemical), transient charge transfer (CT) and electromagnetic (EM) local field enhancement of the vibrational bands of pyridine at vertex-site and in the middle of a (111) facet (S-complex) of an Ag_{20} cluster have been calculated by Schatz and coworkers.[16] At laser frequencies fitting to the charge transfer energies, surprising high enhancements have been found. A somewhat different calculation for pyridine in an S-complex of a Ag_{20} cluster comes to the same conclusion.[17] Zhao et al.[18] presented a detailed time-dependent density functional theory (TDDFT) investigation of the absorption and Raman spectra of a pyrazine molecule located at the junction between two Ag_{20} clusters. Surface-enhanced Raman scattering enhancements of the order of 10^6 have been found for the junction system, which are similar to enhancements of 10^5 found for individual silver nanoparticles. Surprisingly, the chemical enhancement was found to account for as much as 10^5, suggesting that this mechanism might be more important than previously believed, in particular for nanoparticle aggregates. Moreover, TDDFT calculations suggested that unlike larger nanoparticles, the junction between small Ag_{20} tetrahedral clusters does not provide an electromagnetic "hot spot." DJ Wu et al[19] employed density functional theory (DFT) to obtain information about surface bonding and adsorption by calculating and analyzing the relative intensity of SERS spectra of pyridine on noble and transition metals. The transient charge transfer model has been described recently in Ref. 20, the shift of the resonance frequencies and the selection rules of SERS at electrodes as function electrode potential for instance in Ref. 21. Based on the standard Kramers-Heisenberg equation and the Born-Oppenheimer approximation, Ref. 22 shows that SERS can be considered as a single effect, drawing on up to three resonances, plasmon, charge transfer, and molecular resonance.

In spite of this, the analytical practitioner of SERS understandably longs for the pure EM-enhancer, with the same sensitivity for all different adsorbents and a good EM theory as a guide for the nano-fabrication of the substrates. A very promising candidate for reproducible EM nano-enhancers are Metal Films Over uniform and densely packed

Nano-spheres (MFON, M=Ag, Au), developed in the Van Duyne-group. In the next section I will point out their relation with the "chemical effects."

2. AgFON and their relation to the "chemical effect"

2.1. *AgFON in ultra high vacuum (UHV) (Ref. 23)*

This paper Ref. 23 concludes: "Room temperature annealing does not irreversibly destroy the SERS enhancement capability of this surface, thereby permitting for repeated use in UHV experiments. The AgFON surface morphology and localized surface plasmon resonance frequencies, as monitored by UV-VIS extinction, change as the AgFON surface temperatures increases from 300 to 548 K, and the SERS activity corresponds with these changes. Because the AgFON surface is thermally stable at room temperature and retains high SERS-activity following temperature annealing to 573K, it is unlikely that adatoms or adatom clusters play a significant role as adsorption sites supporting the chemical enhancement mechanism. Rather, one can conclude that the electromagnetic enhancement mechanism is the most likely origin of the SER spectra from benzene, pyridine, and C60 adsorbed on AgFON surfaces."

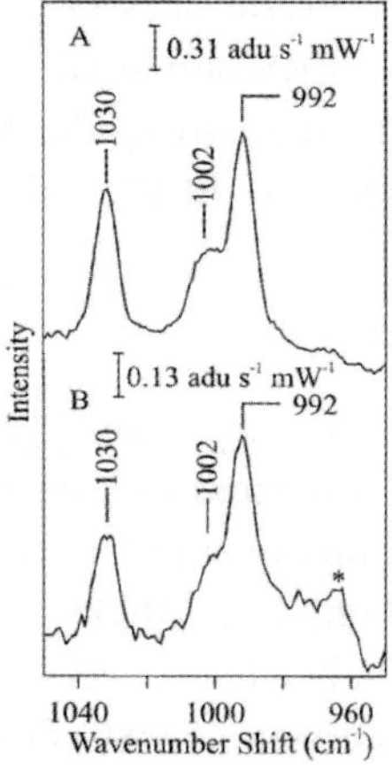

Fig. 1. Regeneration of the AgFON surface. (A) SER spectra of 5L of pyridine on a new AgFON surface. (B) AgFON surface which has been Ar+-ion sputtered for 3h at 1keV, annealed to 573K for 2 h, and then re-deposited with 120-nm of silver. Spectra were taken with 18mW of λex = 514.5nm for 200s.

From Fig. 2B in Litorja et al.'s article[23] follows an EM enhancement G of about 21375 (log G = 4.32) for the breathing vibration of pyridine on AgFON. This is the same order of magnitude as the EM enhancement of the Raman scattering of the benzene breathing vibration of one monolayer of benzene adsorbed on well isolated silver islands on randomly distributed SiO_2 posts ("stochastic post structure") on large areas, produced by an early and efficient nano-technique.[24] These silver islands, probably prolate spheroids, were deposited at room temperature and yielded log G_{EM} about 4.0, see Fig. 10 in Ref. 25 or Fig.10 in Ref. 26. The isolated prolate silver islands still show an extra chemical enhancement of about log G_{chem} = 0.5. This extra enhancement was assigned to adsorption at sites of atomic roughness, so-called SERS active sites.[27] The remnants of the chemical effect at surface defects (SERS active sites) are observed in Fig. 1 (which is Fig. 11 in Ref. 23) as a "non-annealable" shoulder at 1002 cm⁻¹ assigned to defects by Litorja et al.

The number of surface defects has not been estimated in Ref. 23. If one assumes surface defect coverage of for instance 5%, the true enhancement of the 1002 cm⁻¹ band with respect to the majority band at 992 cm⁻¹ would be about a factor of 10.

One should mention that the results of the absolute values of the enhancement are exact because the condensed film thickness is much smaller than the focalisation length of the sample illumination. The sticking coefficient at low temperatures is one, as investigated in Ref. 28.

The stability of SERS active sites is very different on different Cu samples. On cold-deposited Cu films they are annealed at about 250 K[27] and on 80-nm Cu films on rough CaF_2 films they are annealed between 300 K and 350 K.[29] When Cu island films were prepared on 2-3-nm aluminum oxide films on top of bulk polished Cu supports the SERS active sites were still present after annealing up to 400 K.[29] At higher temperatures the aluminum-oxide films broke down.

In summary the results on AgFON surfaces in UHV[23] are in line with our previous reports on the electronic effect at SERS active sites. These remarkable results of Van Duyne group just come from a surface, where the SERS signal is dominated by EM enhancement, because the concentration of SERS active sites is low.

2.2. *AgFON electrodes*

Ref. 30 states: "Tremendous stability to extremely negative potential excursions is observed for MFON electrodes as compared to standard metal oxidation reduction cycle (MORC) roughened electrodes. Consequently, irreversible loss of SERS intensity at negative potentials is not observed on these MFON electrodes. We conclude that MFON electrodes present a significant advantage over MORC electrodes because SERS enhancement is not lost upon excursion to extremely negative potentials. This work demonstrates that the MFON substrate, while easily prepared and temporally stable, offers unprecedented stability and reproducibility for electrochemical SERS experiments. Furthermore, one can conclude that irreversible loss is not a distinguishing characteristic of electrochemical SERS and consequently cannot be used as evidence to support the chemical enhancement mechanism." Nevertheless a chemical effect is observed in the spectra from AgFON sample, see Fig. 2 (which is Fig. 9 from Ref. 30).

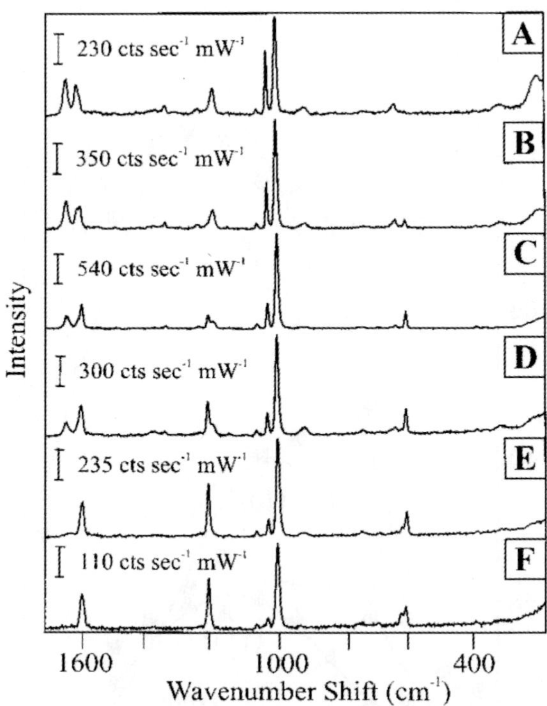

Fig. 2. Potential dependence of SER spectra of 50-mV pyridine in 0.1MKCl on AgFON. (A) -0.3 V, (B) –0.5 V, (C) –0.7 V, (D) -0.9 V, (E) –1.1 V, and (F) –1.3 V vs. Ag/AgCl. Laser excitation was 4-mW 632.8 nm, scan rate 1 Angstrom/s, 1-s dwell time.

At first sight one may believe that the intensity of the breathing vibration of pyridine at about 1006 cm^{-1} does not depend on the potential of the AgFON electrode, but of course it does when one considers the scale bars. A "chemical effect" by dynamic charge transfer in the case of pyridine on copper electrodes was demonstrated in Ref. 31. It was also observed for pyridine at Cu single crystal electrodes with an attenuated total reflection (ATR) method. Smooth Cu single crystal electrodes were exposed to an "external EM enhancement,"[32] given by the surface plasmon polariton resonance in so called Otto-attenuated total reflection (ATR) configuration.[33] There is a strong and reversible dependence of the SERS signal on the electrode potential (see Fig. 3).

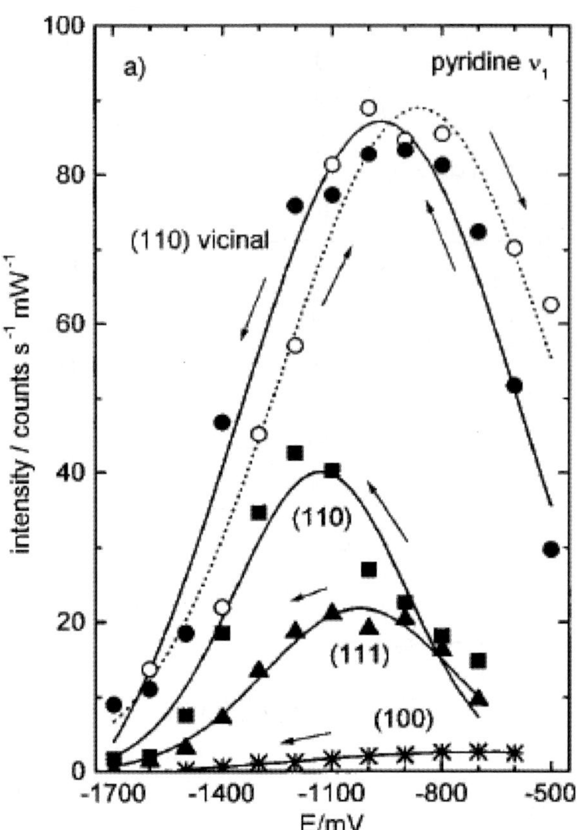

Fig. 3.   Intensity potential profiles of the Raman intensities of Cu in 0.01-M pyridine +0.1M KClO$_4$ for four different crystallographic orientations. Given are the peak intensities of the breathing vibration of pyridine (v1 at ca 1010 cm^{-1}). The lines are Gaussian fits as a guide to the eye. The arrows give direction of the scan. From Ref. 32.

The enhancement values for the Cu(100) crystalline samples with respect to a monolayer of pyridine on an ideal flat Cu sample with only the "external enhancement" by the ATR–coupling is about 3. Accordingly, the overall enhancement for the Cu(110) vicinal face at −1100 mV is of the order of 50. The change of pyridine coverage in the potential range displayed in Fig. 3 is small as inferred from electrochemical capacity measurements. Consequently one must assign the intensity variation as function of the potential in Fig. 3 to the electronic enhancement by transient charge transfer. There is no irreversible "cathodic quenching." Maybe there are also constant EM enhancements, depending on the orientation of the samples. The Raman bands on the Cu single crystals were tentatively assigned to defects (SERS active sites).[32]

For CO$_2$ however, the vibrational selectivity of SERS provides clear evidence for an electronic mechanism. Here we discuss previous SERS results[34,35] in a new way using recent electron energy loss[36,37] and infrared spectroscopic results. A short version of these results has appeared in Ref. 38.

3. SERS of CO_2 at cold-deposited Cu films

At first glance one would expect nothing exciting for the neutral linear molecule O-C-O. For gas phase CO_2, infrared spectroscopy yields the double degenerate O-C-O bending mode at 667 cm^{-1} and the antisymmetric CO stretch vibration at 2349 cm^{-1} Raman spectroscopy yields only the Fermi resonance doublet[39] composed by the symmetric C-O stretch vibration and the second harmonic O-C-O bending mode.[40] The lower frequency component (in gas phase at 1286 cm^{-1}) will be denoted FRI, the higher frequency component (in gas phase at 1388 cm^{-1}) FRII, see Fig. 4.

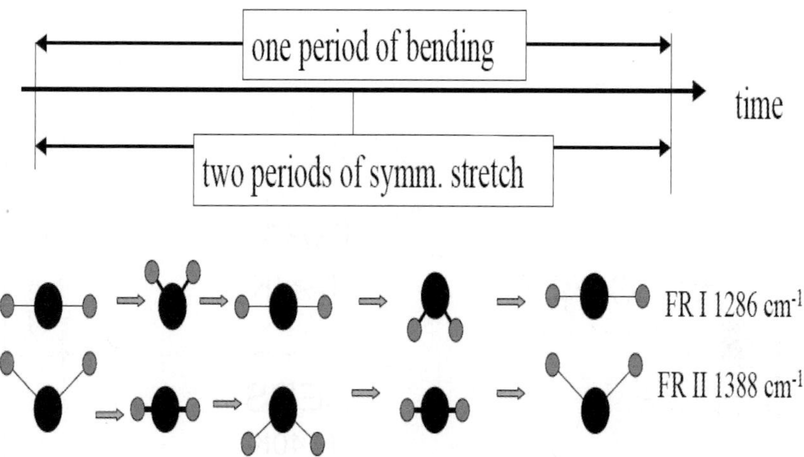

Fig. 4. Qualitative picture of the Fermi resonance in CO_2, following the quantum mechanical wave functions in Ref. 40.

The linear neutral CO_2 can be transformed under special circumstances into a metastable bent anionic form,[41] see Fig. 5.

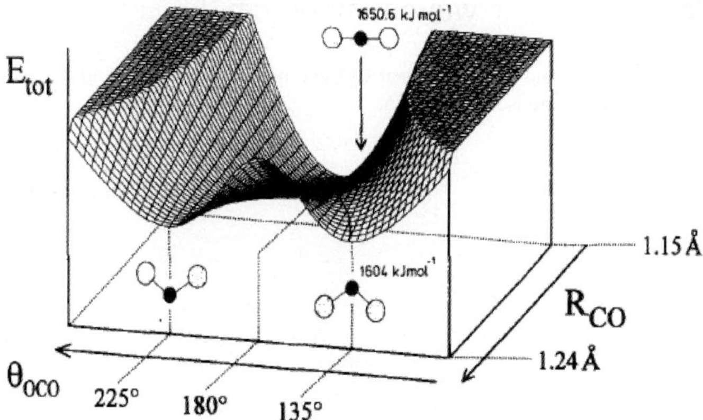

Fig. 5. Schematic energy diagram connecting linear neutral CO_2 with bent anionic CO_2^-. Experimentally determined energies are 1650.6 kJmol-1 for CO_2 and 1604 kJmol-1 for CO_2^-. Reproduced from Ref. 41.

This is the radical form, which allows for a CO_2 – chemistry, for instance the formation of formate HCO_2 on copper films with a potassium coverage, as observed by SERS.[42]

The infrared reflection absorption spectroscopy (IRRAS) spectrum of CO_2 adsorbed on cold-deposited Cu films (taken with the UHV IR equipment described in Refs. 43 and 44) is displayed in the upper part of Fig. 6. Only the IR active modes of CO_2 are observed. The double degenerate bending mode is split and the antisymmetric CO-stretch mode has a non-symmetric line shape depending on exposure, probably caused by the disordered Cu surface. The broad structure near 2100 cm^{-1} is caused by CO contamination.

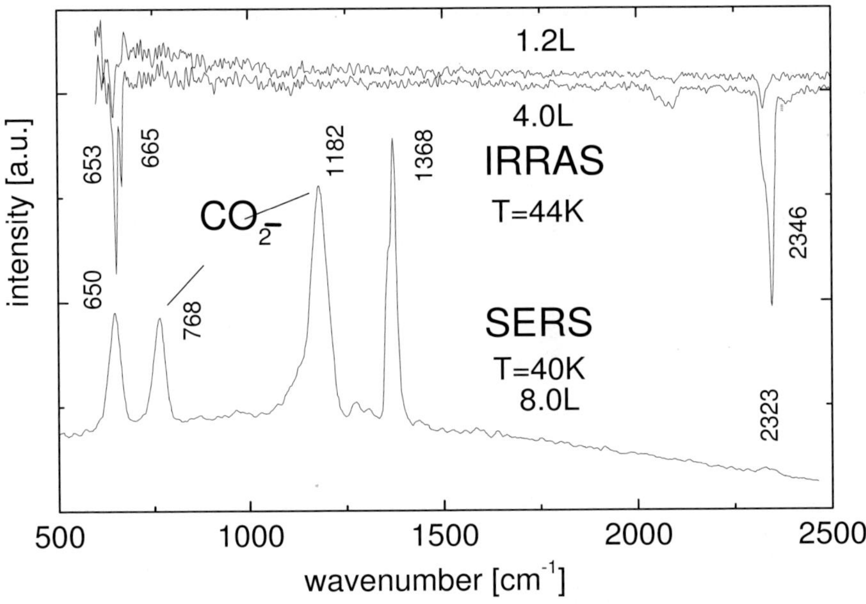

Fig. 6.   Comparison of IRRAS and SERS of linear CO_2 and activated CO_2^- on cold-deposited Cu films (40 K). The SERS spectrum at 8L exposure is from Ref. 35.

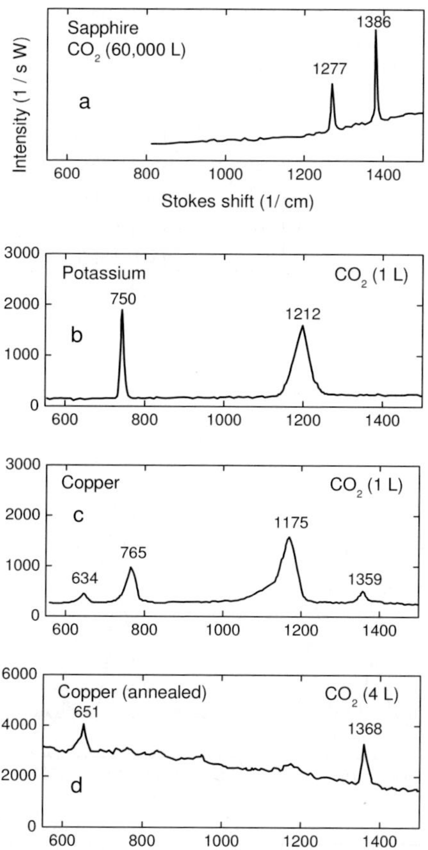

Fig. 7.   a) Raman spectrum of CO_2 condensed on sapphire showing the Fermi resonance components FRI (1277 cm^{-1}) and FRII (1386 cm^{-1}). b) 1L CO_2 on cold-deposited potassium. c) 1L CO_2 on cold deposited (40 K) Cu. d) Cold-deposited Cu, annealed at 200 K, exposed at 40 K to 4L CO_2. The excitation was provided by an Ar$^+$-laser with 514.5 nm emission in the case of the K films and by a Kr$^+$-laser with 647.1 nm emission in the case of the Cu films. All peak positions are given in Table 1.

The SERS results in Fig. 6 are explained on the basis of the results in Fig. 7.[34]

The Fermi dyad is seen in the Raman spectrum of condensed solid CO_2 in Fig. 7a. It is conspicuous, that only the higher wave number component FRII contributes to the SERS spectra of CO_2 on cold-deposited Cu (Figs. 6 and 7c,d). For CO_2 on cold-deposited K, see Fig. 7b and Table 1, only the anionic species is observed, because the low work function of potassium favors the permanent electron transfer to CO_2. The anionic species is also observed on cold-deposited Cu (Fig. 7c), but not when the Cu film was first annealed at 200 K before exposure at 40 K (Fig. 7d). For CO_2 on cold-deposited Ag,[35] one observes the bending mode and FRII, but no bands of the anionic CO_2 species. The appearance of the bending mode of linear, neutral CO_2 can be understood by using group theory and the model in,[22] whereas all the other observations cannot be explained in this way (J R Lombardi, private communications).

Table 1. Compilation of the band positions of carbon dioxide in gas and solid phase and in SERS from cold-deposited silver, copper and potassium. δ: bending mode, v_s and v_{as} : symmetric and antisymmetric C-O stretch mode, Δ: frequency shifts with respect to the solid phase. They are smaller than the shift of the C-C vibration of ethene on cold-deposited copper, with $\Delta(v_{C-C})$ = -70 cm^{-1}.[47] vw: very weak,, NA: Not Applicable.

CO_2 frequencies/cm^{-1}	δ	$\Delta(\delta)$	FR:Fermi resonance	Δ(FRII)	v_{as}
gas phase[45]	667	-7	FR I 1286		2349
			FR II 1388	-2	
solid phase[34,46]	674	0	FR I 1277	0	2374
			FRII 1386		
SERS on silver, 1L[34]	649	-25	FR I –		vw
			FR II 1368	-18	
SERS on copper, 1L[34, 35]	634	-40	FR I –		vw 2323
			FR II 1359	-27	
anionic CO_2- frequencies/ cm^{-1}	δ		vs		
SERS on copper, 1L[34]	765	NA	1175	NA	NA
SERS on potassium, 1L[34]	750	NA	1212	NA	NA

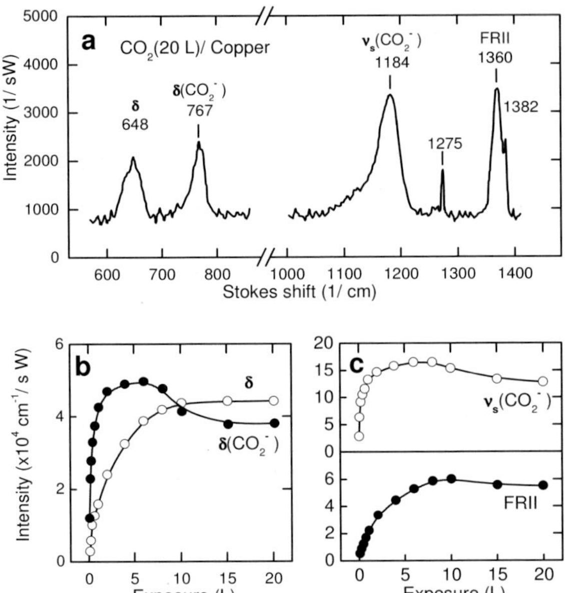

Fig. 8. a) SERS spectra at 20L exposure in the range of the symmetric stretch and bending vibrations. Note the Fermi doublet of CO_2 not adsorbed at SERS active sites with narrow line width at1275 cm^{-1} and 1382 cm^{-1}. b) Integrated intensities of the bending band δ of CO_2 and $\delta(CO_2^-)$ versus exposure. c) Integrated bands of FRII and the symmetric stretch mode $v_s(CO_2^-)$.

At the time of the publications[34,35] it was not clear, why only the higher component FRII of the Fermi resonance appeared strongly and why the Raman forbidden bending mode was observed.

Now a consistent electronic enhancement mechanism is indicated by comparison with the inelastic cross sections of electron scattering by gaseous CO_2 in the energy range of

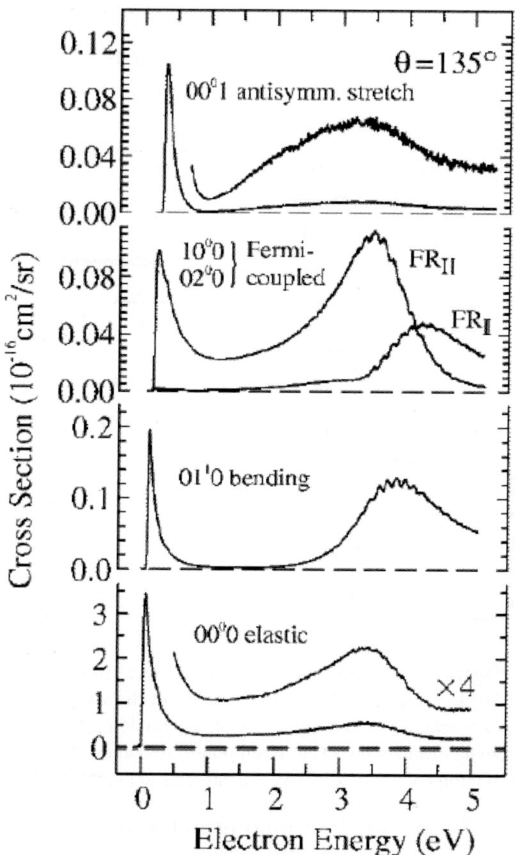

Fig. 9.　Overview of the elastic (no loss) and vibrational inelastic electron scattering cross sections in gaseous CO_2 at a scattering angle of 135°. (100) signifies the symmetric stretch, (010) the bending mode and (001) the antisymmetric stretch. The energy scale refers to the primary energy.

0-5 eV, including the $^2\Pi_u$ shape resonance at 2 - 5 eV[36,37] and the so-called virtual state $^2\Sigma$ of anionic CO_2^- below 1 eV[48] in Fig. 9 (from Ref. 37).

The cross sections of the Fermi doublet components below 1 eV display at decreasing electron energy an increase for FRII and a negligible contribution of FRI. This experimental result has been corroborated by theory.[49,48] The selection rules in the $2\Pi_u$ shape resonance and the virtual state $^2\Sigma$ forbid the excitation of the antisymmetric stretch mode. The relatively strong cross section of the antisymmetric stretch mode (see Fig. 9) at low electron energies is probably caused by a comparatively long-range electron interaction with the dynamic electric dipole of the antisymmetric stretch mode. This interaction is shielded at the surface. The SERS spectra can easily be explained with the electronic energy scheme in Fig. 10.

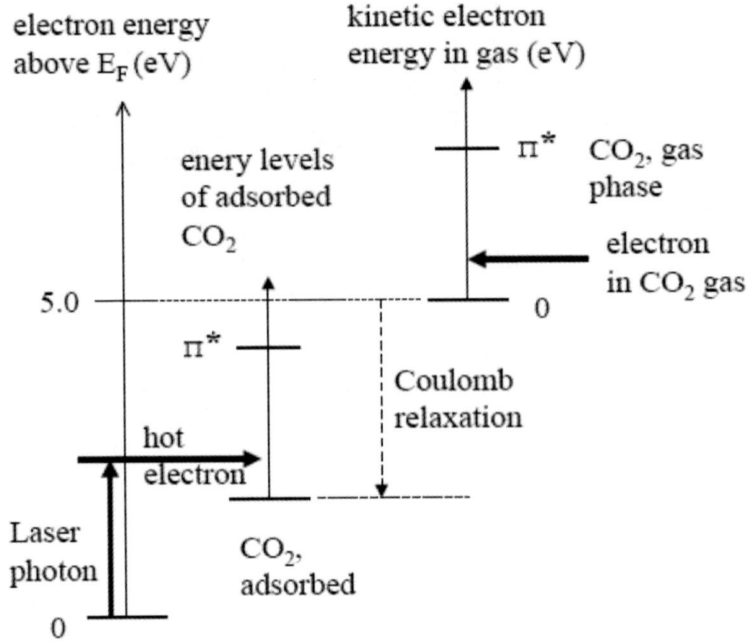

Fig. 10. Electron energy diagram above the Fermi level of Cu. The work function of Cu(111) is 5.0 eV, fixing the vacuum level, e.g. the level of zero kinetic electron energy in the CO_2 gas phase. The anionic electronic levels of the adsorbed CO_2 are relaxed by the attraction of the shielding hole at the metal surface. The most likely energy of the intermediate electron involved in temporary transfer to CO_2 (indicated as "hot electron") is the Fermi energy plus laser photon energy of 1.92 eV. All energies are up to scale.

The value of about 3.5 eV of the Coulomb relaxation in Fig. 10 has been inferred from two-photon-emission spectroscopy[50] of CO on Cu(111) and the resonant low energy electron scattering in CO gas.[51]

Within the coherent Raman process the "hot electron" created by laser photon annihilation will temporarily occupy the virtual state $^2\Sigma$ of the directly adsorbed CO_2 species and excite the various vibrational bands according to the cross sections in Fig. 9 at energies below 0.5 eV. These are the bending mode and the higher frequency mode FRII. This explains the SERS spectrum of CO_2 on Cu in Fig. 6, especially the observation of the bending mode which is not Raman active for free CO_2 and of only the higher frequency band FRII of the Fermi dyad. Strong EM enhancement should show both FRI and FRII. But the near absence of the FRI component in SERS proves that the local EM field enhancement cannot deliver signals above the noise level. The exclusive contribution of EM enhancement may cause the weak appearance of the narrow line Fermi doublet in Fig. 8a at high exposure. But even this is not granted, as demonstrated by a special electronic effect for the modes of C_2H_4 adsorbed in first layer on Cu(111) facets.[52]

The SERS spectrum of CO_2 on Cu films cold-condensed at 40 K (Figs. 6, 7c) contains a second couple of lines, which coincide with the positions of the CO_2^- bands observed on potassium, see Fig. 7b. These bands indicate the existence of the bent form of anionic CO_2^-, now permanently charged with one electron, see Fig. 5. The bending and symmetric stretching mode of the anionic form are observed by the transient transfer of the extra electron from CO_2^- to the metal, because these two nuclear motions are involved in the configuration change towards the linear molecular form of neutral CO_2. This direction of photo-induced charge transfer is similar to the mechanism found in SERS of CN^- on silver electrodes.[53,54]

When a cold-deposited Cu film, cryocondensed at 40 K, annealed at 200 K, re-cooled to 40 K, is exposed to CO_2, then the anionic CO_2^- is no more observed, only the bending modes and FRII of neutral CO_2 persist, see Fig. 7d. Apparently, the adsorption sites on Cu where CO_2 is activated to its anionic shape are annealed below 200 K. Apparently, these "chemically active sites" form a subgroup of the "SERS active sites."[38]
The fact, that the infrared active bending and symmetric stretch modes of anionic CO_2^- is not observed in IRRAS (see Fig. 6) corroborates the assignment of the bonding of these species to a minority of surface sites. This resembles the case of NO exposure of cold-deposited Cu:[43] SERS provides high spectroscopic sensitivity and is focusing on the "chemical active sites" covered with N_2; $N_2^{-\delta}$, N_2O, $O_{ads.}$, but shows no NO, whereas with IRRAS only adsorbed NO was observed.

The model of transient charge transfer at sites of atomic scale roughness is supported by the observation of Raman active bands in IRRAS of ethene on cold-deposited copper films. Recently, it was demonstrated, that submonolayers of Cu on Cu(111) yielded the Raman bands of ethane, the IR bands did not appear at the low exposures employed.[55,56] The advantage of IRRAS with respect to Raman spectroscopy is the relatively easy observation of molecular monolayers on smooth single crystalline surfaces without any enhancement. Further progress on understanding the atomic structure of SERS-active sites may be expected by a combination of IRRAS and scanning tunneling microscopy.

4. Conclusions

Silver films over Nanospheres (AgFON), produced for instance in Van Duyne group are stable electromagnetic (EM) enhancers both in UHV and in electrolytes. Nevertheless they show some characteristics of the "chemical effect." The structure of cold-deposited copper films is not well known, but these films display a wealth of effects, which are beyond an explanation by local metal optics ("EM enhancement"). Adsorption sites of atomic scale roughness with dynamic charge transfer between metal and an adsorbed molecule, so called "SERS active sites" on copper islands are stable up to at least 400 K. The comparison of infrared and Raman spectra of the same cold-deposited metal-adsorbate system confirm the existence of SERS active sites. If infrared reflection-absorption spectroscopy of smooth Cu single crystal surfaces with a cold-deposited submonolayer of Cu would be combined with scanning tunneling spectroscopy, the structure of SERS active sites might be unraveled. CO_2 on Cu is a very good system to study the SERS mechanisms. All spectral SERS features of neutral CO_2 on SERS active sites of cold-deposited Cu involve transient electron transfer from Cu to a temporary virtual $^2\Sigma$ state of anionic CO_2. This can be proven by a comparison of the SERS spectra with low electron energy loss spectra of gaseous CO_2. The EM enhancement alone is not strong enough to deliver a Raman band of the Fermi doublet component FRI from CO_2 species not adsorbed at SERS active sites. On Cu films deposited at 40 K also a stable form of anionic CO_2^- is observed by SERS, involving transient electron transfer from anionic CO_2^- to Cu. The adsorption sites of the stable anionic CO_2 are lost by annealing the film to 200 K, whereas SERS of neutral CO_2 is still observed after re-cooling to 40 K. IRRAS observes only the infrared active modes of the majority species of neutral CO_2 adsorbed on cold-deposited Cu, whereas SERS is focusing on a minority of surface sites on the atomically rough Cu surface.

Acknowledgments

I thank W. Akemann for digitizing analogue spectra and A. Pucci for valuable discussions.

References

1. R. K. Chang, and B. L. Laube, "Surface-Enhanced Raman-Scattering and Nonlinear Optics Applied to Electrochemistry," CRC Critical Reviews in Solid State and Materials Sciences **12**, 1-73 (1984).
2. M. Fleischmann, P. Hendra, and A. McQuillan, "Raman-Spectra of Pyridine Adsorbed at a Silver Electrode," Chemical Physics Letters **26**, 163-166 (1974).
3. D. L. Jeanmaire, and R. P. Van Duyne, "Surface Raman Spectroelectrochemistry .1. Heterocyclic, Aromatic, and Aliphatic-Amines Adsorbed on Anodized Silver Electrode," Journal of Electroanalytical Chemistry **84**, 1-20 (1977).
4. K. Kneipp, Y. Wang, R. R. Dasari, and M. S. Feld, "Approach to Single-Molecule Detection Using Surface-Enhanced Resonance Raman-Scattering (SERS) - a Study Using Rhodamine 6G on Colloidal Silver," Appl. Spectrosc. **49**, 780-784 (1995).
5. K. Kneipp, Y. Wang, H. Kneipp, L. T. Perelman, I. Itzkan, R. Dasari, and M. S. Feld, "Single molecule detection using surface-enhanced Raman scattering (SERS)," Phys. Rev. Lett. **78**, 1667-1670 (1997).
6. S. M. Nie, and S. R. Emory, "Probing single molecules and single nanoparticles by surface-enhanced Raman scattering," Science **275**, 1102-1106 (1997).
7. J. A. Dieringer, R. B. Lettan, K. A. Scheidt, and R. P. Van Duyne, "A frequency domain existence proof of single-molecule surface-enhanced Raman Spectroscopy," Journal of the American Chemical Society **129**, 16249-16256 (2007).
8. A. Otto, "On the significance of Shalaev's hot spots in ensemble and single-molecule SERS by adsorbates on metallic films at the percolation threshold," J. Raman Spectrosc. **37**, 937 (2006).
9. P. G. Etchegoin, E. C. Le Ru, M. Meyer, and H. Kneipp, "SERS assertion addressed," in *Physics Today* (2008), pp. August , 13-18.
10. K. Kneipp, H. Kneipp, V. B. Kartha, R. Manoharan, G. Deinum, I. Itzkan, R. R. Dasari, and M. S. Feld, "Detection and identification of a single DNA base molecule using surface-enhanced Raman scattering (SERS)," Phys. Rev. E **57**, R6281-R6284 (1998).
11. P. Paul, J. Goulet, and R. Aroca, "Distinguishing Individual Vibrational Fingerprints: Single-Molecule Surface-Enhanced Resonance Raman Scattering from One-to-One Binary Mixtures in Langmuir-Blodgett Monolayers," Anal. Chem. **79**, 2728-2734 (2007).
12. C. Eggeling, J. Schaffer, C. A. M. Seidel, J. Korte, G. Brehm, S. Schneider, and W. Schrof, "Homogeneity, transport, and signal properties of single Ag particles studied by single-molecule

surface-enhanced resonance Raman scattering," Journal of Physical Chemistry A **105**, 3673-3679 (2001).

13. J. Steidtner, and B. Pettinger, "Tip-enhanced Raman spectroscopy and microscopy on single dye molecules with 15 nm resolution," Phys. Rev. Lett. **100**, 236101 (2008).

14. M. Moskovits, "Surface-enhanced Raman spectroscopy: a brief perspective," J. Raman Spectrosc. **36**, 485-496 (2006).

15. L. L. Zhao, L. Jensen, and G. C. Schatz, "Pyridine-Ag-20 cluster: A model system for studying surface-enhanced Raman scattering," Journal of the American Chemical Society **128**, 2911-2919 (2006).

16. M. T. Sun, S. Liu, M. Chen, and H. Xu, "Direct visual evidence for the chemical mechanism of surface-enhanced resonance Raman scattering via charge transfer," J. Raman Spectrosc. **40**, 137–143 (2009).

17. L. L. Zhao, L. Jensen, and G. C. Schatz, "Surface-enhanced Raman scattering of pyrazine at the junction between two Ag-20 nanoclusters," Nano Letters **6**, 1229-1234 (2006).

18. D. Y. Wu, X. M. Liu, S. Duan, X. Xu, B. Ren, S. H. Lin, and Z. Q. Tian, "Chemical enhancement effects in SERS spectra: A quantum chemical study of pyridine interacting with copper, silver, gold and platinum metals," Journal of Physical Chemistry C **112**, 4195-4204 (2008).

19. D. Y. Wu, J. F. Li, B. Ren, and Z. Q. Tian, "Electrochemical surface-enhanced Raman spectroscopy of nanostructures," Chemical Society Reviews **37**, 1025-1041 (2008).

20. S. P. Centeno, I. Lopez-Tocon, J. F. Arenas, J. Soto, and J. C. Otero, "Selection rules of the charge transfer mechanism of surface-enhanced Raman scattering: The effect of the adsorption on the relative intensities of pyrimidine bonded to silver nanoclusters," Journal of Physical Chemistry B **110**, 14916-14922 (2006).

21. J. R. Lombardi, and R. L. Birke, "A unified Approach to surface enhanced Raman spectroscopy," J. Phys. Chem. C **112**, 5605-5617 (2008).

22. M. Litorja, C. L. Haynes, A. J. Haes, T. R. Jensen, and R. P. Van Duyne, "Surface-enhanced Raman scattering detected temperature programmed desorption: Optical properties, nanostructure, and stability of silver film over SiO2 nanosphere surfaces," Journal of Physical Chemistry B **105**, 6907-6915 (2001).

23. G. P. Goudonnet, T. Inakaki, T. L. Ferell, R. J. Warmack, M. C. Buncick, and E. T. Arakawa, "Enhanced Raman scattering from benzoic acid on silver and gold prolate spheroids on large and transparent patterned areas," Chemical Physics **106**, 225-232 (1986).

24. I. Mrozek, and A. Otto, "Quantitative Separation of the Classical Electromagnetic and the Chemical Contribution to Surface Enhanced Raman-Scattering," Journal of Electron Spectroscopy and Related Phenomena **54**, 895-911 (1990).

25. A. Otto, I. Mrozek, H. Grabhorn, and W. Akemann, "Surface-Enhanced Raman-Scattering," Journal of Physics-Condensed Matter **4**, 1143-1212 (1992).

26. M. Hein, P. Dumas, M. Sinther, A. Priebe, A. Bruckbauer, P. Lilie, A. Pucci, and A. Otto, "Relation between surface resistance, infrared-, surface enhanced infrared- and Raman-spectroscopies of CO and C_2H_4 on copper," Surf. Sci. **600**, 1017-1925 (2006).

27. C. Pettenkofer, I. Mrozek, T. Bornemann, and A. Otto, "On the Contribution of Classical Electromagnetic-Field Enhancement to Raman-Scattering from Adsorbates on Coldly Deposited Silver Films," Surface Science **188**, 519-556 (1987).

28. C. Siemes, A. Bruckbauer, A. Goussev, A. Otto, M. Sinther, and A. Pucci, "SERS-active sites on various copper substrates," Journal of Raman Spectroscopy **32**, 231-239 (2001).

29. L. A. Dick, A. D. McFarland, C. L. Haynes, and R. P. Van Duyne, "Metal film over nanosphere (MFON) electrodes for surface-enhanced Raman spectroscopy (SERS): Improvements in surface nanostructure stability and suppression of irreversible loss," Journal of Physical Chemistry B **106**, 853-860 (2002).

30. J. C. Ingram, and J. E. Pemberton, "Comparison of Charge-Transfer Enhancement in the Surface Enhanced Raman-Scattering of Pyridine on Copper and Silver Electrodes," Langmuir **8**, 2034-2039 (1992).

31. A. Bruckbauer, and A. Otto, "Raman spectroscopy of pyridine adsorbed on single crystal copper electrodes," Journal of Raman Spectroscopy **29**, 665-672 (1998).

32. A. Otto, "Excitation of Nonradiative Surface Plasma Waves in Silver by Method of Frustrated Total Reflection," Zeitschrift für Physik **216**, 398-410 (1968).

33. W. Akemann, and A. Otto, "Roughness induced reactions of N_2 and CO_2 on noble and alkali-metals," Surface Science **272**, 211-219 (1992).

34. W. Akemann, and A. Otto, "The effect of atomic scale surface disorder on bonding and activation of adsorbates: Vibrational properties of CO and CO_2 on copper," Surf. Sci. **287/288**, 104-109 (1993).

35. M. Allan, "Selectivity in the excitation of Fermi-coupled vibrations in CO_2 by impact of slow electrons," Phys. Rev. Lett. **87**, 33201 (2001).

36. M. Allan, "Vibrational structures in electron-CO_2 scattering below the (2)Pi(u) shape resonance," Journal of Physics B-Atomic Molecular and Optical Physics **35**, L387-L395 (2002).

37. A. Lust, A. Pucci, W. Akemann, and A. Otto, "SERS of CO_2 on cold-deposited Cu: An electronic effect at a minority of surface sites," Journal of Physical Chemistry C **112**, 11075-11077 (2008).

38. E. Fermi, "Über den Ramaneffekt des Kohlendioxyds," Zeitschrift für Physik **71**, 250-259 (1931).

39. D. T. Colbert, and E. L. Silbert III, "Variable Curvature coordinates for molecular vibrations," J. Chem. Phys **91**, 350 (1989).

40. H. J. Freund, and M. W. Roberts, "Surface chemistry of carbon dioxide," Surface Science Reports **25**, 225-273 (1996).

41. M. Pohl, and A. Otto, "Adsorption and reaction of carbon dioxide on pure and alkali promoted cold-deposited copper films," Surf. Sci. **406**, 125-137 (1998).

42. M. Lust, A. Pucci, and A. Otto, "SERS and infrared reflection-absorption spectroscopy of NO on cold-deposited Cu," Journal of Raman Spectroscopy **37**, 166-174 (2006).

43. O. Krauth, G. Fahsold, and A. Pucci, "Asymmetric line shapes and surface enhanced infrared absorption of CO adsorbed on thin iron films on MgO(001)," Journal of Chemical Physics **110**, 3113-3117 (1999).

44. R. P. Madden, "A High-Resolution Study of CO_2 Absorption Spectra between 15 and 18 Microns," J. Chem. Phys. **35**, 2083 (1961).

45. M. Lust, "Infrarot- und SERS-Spektroskopie zur Untersuchung katalytischer Reaktionen auf rauhen Kupferoberflächen," Dissertation Heidelberg (2004).

46. J. Grewe, U. Ertürk, and A. Otto, "Raman scattering of C_2H_4 on copper films, absorbed at (111) terraces and "annealable sites"," Langmuir **14**, 696-707 (1998).

47. W. Vanroose, Z. Y. Zhang, C. W. McCurdy, and T. N. Rescigno, "Threshold Vibrational Excitation of CO_2 by slow electrons," Phys. Rev. Letters **92**, 053201 (2004).

48. W. McCurdy, W. Isaacs, H.-D. Meyer, and T. Rescigno, "Resonant vibrational excitation of CO_2 by electron impact: Nuclear dynamics on the coupled components of the 2Pu resonance," Phys. Rev. C**67**, 42708 (2003).

49. M. Wolf, A. Hotzel, E. Knoesel, and D. Velic, "Direct and indirect excitation mechanisms in two-photon photoemission spectroscopy of Cu(111) and CO/Cu(111)," Physical Review B **59**, 5926-5935 (1999).

50. H. Ehrhardt, L. Langhans, and H. Taylor, "Resonance scattering of slow electrons from H_2 and CO " Phys. Rev. **173**, 222-230 (1968).

51. A. Otto, W. Akemann, and A. Pucci, "Normal bands in Surface-Enhanced Raman Scattering (SERS) and Their Relation to the Electron-Hole Excitation Background in SERS," Israel Journal of Chemistry **46**, 307-315 (2006).

52. T. E. Furtak, and S. H. Macomber, "Voltage-Induced Shifting of Charge-Transfer Excitations and Their Role in Surface-Enhanced Raman-Scattering," Chemical Physics Letters **95**, 328-332 (1983).

53. J. Billmann, and A. Otto, "Charge-transfer between adsorbed cyanide and silver probed by SERS," Surface Science **138**, 1-25 (1984).

54. O. Skibbe, M. Binder, A. Otto, and A. Pucci, "Electronic contributions to infrared spectra of adsorbate molecules on metal surfaces: Ethene on Cu(111)," Journal of Chemical Physics **128**, 194703 (2008).

55. O. Skibbe, M. Binder, A. Otto, and A. Pucci, "Erratum: "Electronic contributions to infrared spectra of adsorbate molecules on metal surfaces: Ethene on Cu(111)" [J. Chem. Phys. 128, 194703 (2008)]," Journal of Chemical Physics **129**, 149901 (2008).

CHAPTER 3

COMBUSTION DIAGNOSTICS BY PURE ROTATIONAL COHERENT ANTI-STOKES RAMAN SCATTERING

ALFRED LEIPERTZ and THOMAS SEEGER

Lehrstuhl für Technische Thermodynamik
and
Erlangen Graduate School in Advanced Optical Technologies
Friedrich-Alexander Universität Erlangen-Nürnberg
Erlangen, Germany
Alfred.Leipertz@ltt.uni-erlangen.de

Since its first use in Richard Chang's laboratory in 1982 in a comparative study with vibrational coherent anti-Stokes Raman scattering (VCARS) in a flame, pure rotational coherent anti-Stokes Raman scattering (RCARS) has gained tremendous importance for gas temperature and relative species concentration measurements in combustion diagnostics. The field of application covers basic studies on diagnostics development and on flame research as well as its use in technical combustion systems, e.g., for the determination of the gas-phase temperature in the vaporizing spray of a gasoline direct injection (GDI) injector or for the simultaneous measurement of gas temperature and exhaust-gas-recirculation rate (EGR rate) in a homogeneous charge compression ignition (HCCI) engine. An overview is given on the fundamentals of the technique and on its most important technical applications.

1. Introduction

Laser diagnostics forms a powerful tool for the characterization of combustion processes. There are several different options to find access to the required measurement information, e.g., by using laser-induced fluorescence (LIF) for temperature and minor species concentration measurements[1-5] or by linear Raman scattering for the determination of temperature and major species concentrations[6-10] to mention just a few of them. These techniques are mostly restricted to applications in a relatively clean combustion environment. For real technical combustion processes the signals are often disturbed or even covered by other sources of radiation, e.g., soot emissions or different kinds of background radiation. Here the application of nonlinear optical methods is advantageous as they provide signals with higher intensities which are emitted laser-like into one particular room direction.[11] The most often used nonlinear spectroscopic technique is coherent anti-Stokes Raman scattering (CARS) where three laser beams are directed into the region of interest inside the combustion field by forming a particular angle configuration between the beams generating by their interaction a fourth beam,

     Optical Processes in Microparticles and Nanostructures

which is emitted into a fixed direction due to momentum conservation between the four beams involved.[12] Depending on the molecular energy levels probed pure rotational CARS (RCARS) or rotational-vibrational CARS (VCARS) signals are generated when probing Raman transitions between rotational energy levels within one vibrational state or between rotational energy levels of different vibrational states, respectively. These signals contain the required measurement information.

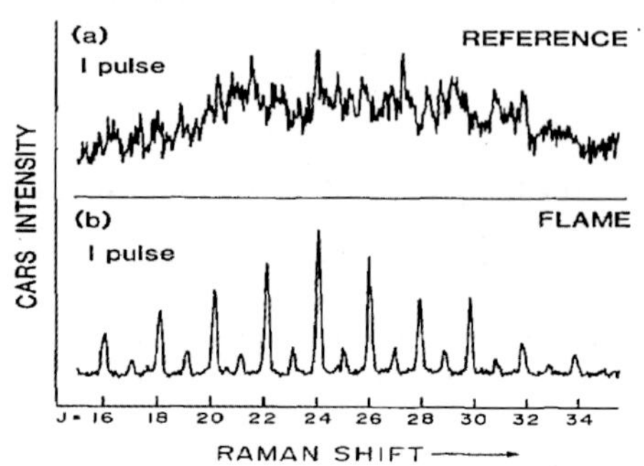

Fig. 1. Single-pulse RCARS spectrum taken in a 10-ns laser pulse: (a) the non-resonant signal of the Ar reference and (b) the normalized RCARS spectrum of N_2 in a flame (copyright permission obtained).[13]

VCARS is a well established technique for temperature and concentration measurements[14-24] and is advantageous to RCARS at higher temperatures.[25,26] RCARS provides more accurate results at lower temperatures[25-28] and was for the first time used for temperature measurements at room temperature in 1981 at Yale University[29] and in flames some time later in the same laboratory (see Fig. 1 where a single-shot high-temperature RCARS spectrum is shown[13]). When used simultaneously with coherent Stokes Raman scattering (CSRS) which is possible for RCARS and RCSRS, the same measurement information is contained in both signals and can be extracted from both signals simultaneously (see Fig. 2).[30] Since these days forming the starting point of RCARS investigations several different developments have been done improving the signal separation,[31,32] the Raman linewidth information,[33,34] the measurement accuracy[35,36] and the stray light suppression.[37-40] Due to these improvements measurements can nowadays also be executed in technical environments which have not been accessible before, e.g., in highly sooting flames,[41-43] inside evaporating engine sprays[38,44,45] and inside internal combustion engines.[40,46,47] After a brief treatment of the theoretical background of CARS, with particular emphasis on polarization resolved RCARS,[38,48] its application in these not so easily accessible combustion systems is described.

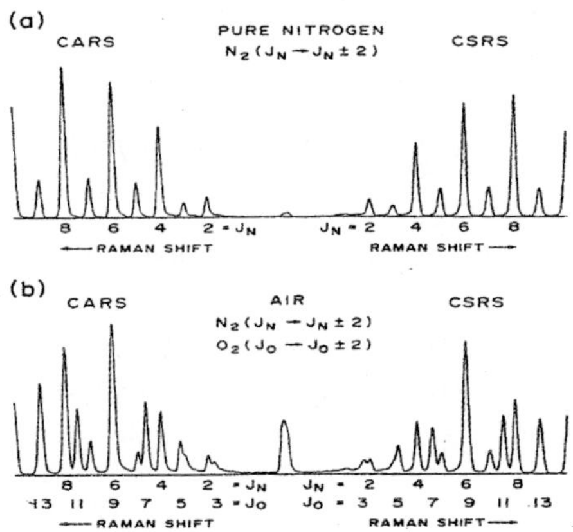

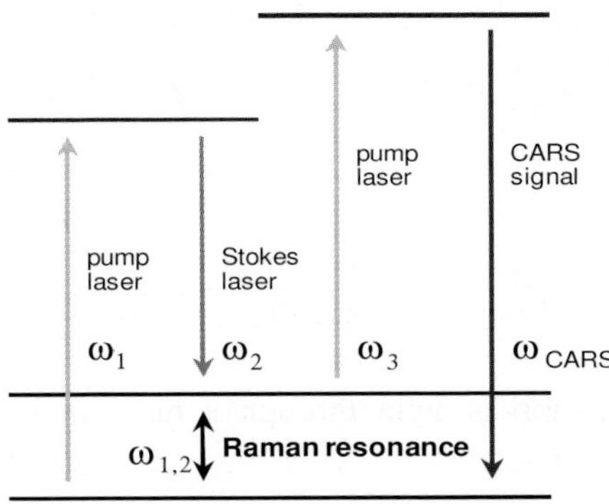

Fig. 2. Simultaneous detection of RCARS and RCSRS: (a) pure nitrogen and (b) air (copyright permission obtained).[30]

2. The CARS process

2.1. *Theoretical background*

The CARS signal is generated as the result of the interaction between generally three focused laser beams of frequency ω_1, ω_2 and ω_3 and the molecules in the sample. In their common overlap region two monochromatic laser beams coherently excite a Raman transition $\omega_{1,2}=\omega_1-\omega_2$ and the ω_3 beam is mixed in, resulting in an oscillating polarization of frequency $\omega_{CARS}=\omega_1-\omega_2+\omega_3$, the CARS signal. This process is shown in the energy level diagram displayed in Fig. 3.

Fig. 3. Energy level diagram of the CARS process.

In order to gain quantitative information out of a CARS spectrum, it is in most cases necessary to calculate accurately the details of the spectrum. The theoretical basis is described in several review papers and books,[49-56] therefore only a short description is given here. The interaction of an incident electromagnetic field with an ensemble of molecules can be described by the induced polarization $\vec{P}$, which is expressed by a Taylor series in powers of the electric field $\vec{E}$:

$$\vec{P} = \chi^1 \vec{E} + \chi^2 \vec{E}\vec{E} + \chi^3 \vec{E}\vec{E}\vec{E} + \dots \tag{1}$$

The susceptibility χ is a tensor quantity. The first term, the linear susceptibility χ^1 describes two photon processes like Rayleigh scattering or spontaneous Raman scattering. For gases, which are isotropic, and thereby display inversion symmetry, the even-order polarization terms vanish in Eqn. (1). The first non-vanishing nonlinear term is the third-order susceptibility χ^3, which is responsible for four-wave mixing (FWM) processes like CARS:

$$\vec{P}_\mu^3(\vec{r},t) = \varepsilon_0 \int\limits_{-\infty}^{\infty}\int\limits_{-\infty}^{\infty}\int\limits_{-\infty}^{\infty} \chi^3(-\omega_\mu,\omega_1,\omega_2,\omega_3) \cdot \vec{E}(\vec{r},\omega_1) \cdot \vec{E}(\vec{r},\omega_2) \cdot \vec{E}(\vec{r},\omega_3) \cdot$$
$$\cdot \exp(-i\omega_\mu t)\, d\omega_1 d\omega_2 d\omega_3 \tag{2}$$

The third-order susceptibility χ^3 is a fourth-rank tensor which has 81 separate elements. For an isotropic medium 21 of these are non-zero, and for arbitrary frequencies of the incoming lasers three elements are independent. There are four types of non-zero tensor elements:

$$\left.\begin{aligned}
\chi_{xxxx}^3 &= \chi_{yyyy}^3 = \chi_{zzzz}^3 \\
\chi_{yyzz}^3 &= \chi_{zzyy}^3 = \chi_{zzxx}^3 = \chi_{xxzz}^3 = \chi_{xxyy}^3 = \chi_{yyxx}^3 \\
\chi_{yzyz}^3 &= \chi_{zyzy}^3 = \chi_{zxzx}^3 = \chi_{xzxz}^3 = \chi_{xyxy}^3 = \chi_{yxyx}^3 \\
\chi_{yzzy}^3 &= \chi_{zyyz}^3 = \chi_{zxxz}^3 = \chi_{xzzx}^3 = \chi_{xyyx}^3 = \chi_{yxxy}^3
\end{aligned}\right\} , \tag{3}$$

which are related by

$$\chi_{xxxx}^3 = \chi_{xxyy}^3 + \chi_{xyxy}^3 + \chi_{xyyx}^3 \tag{4}$$

Beside this the propagation of light through a medium is described by the wave equation:

$$\nabla \times \nabla \vec{E}(\vec{r},\omega) - \frac{\omega^2}{c^2}\vec{E}(\vec{r},\omega) = \mu_0 \cdot \omega^2 \cdot \vec{P}(\vec{r},\omega) \tag{5}$$

which can be derived from Maxwell´s equations, where c is the speed of light and μ_o the permeability in vacuum. Solving this equation by using Eqn. (2), the intensity of the generated CARS signal can be calculated. Assuming for simplicity a CARS process with the polarization of the laser beams being all parallel and the incident electromagnetic waves as plane waves, then the intensity of the CARS signal can be expressed by:

$$I(\omega_{CARS}) = \frac{n_3 \cdot c \cdot \varepsilon_0}{2} \cdot \left| E(\omega_{CARS}) \right|^2$$

$$= \frac{\omega_{CARS}^2}{n_1^2 \cdot n_2 \cdot n_{CARS} \cdot c^4} \cdot \left| 3 \cdot \chi_{xxxx}^3 \right|^2 \cdot I_1 \cdot I_2 \cdot I_3 \cdot L^2 \cdot \left(\frac{\sin(\Delta k \cdot L/2)}{\Delta k \cdot L/2} \right)^2 \tag{6}$$

From this equation it can be seen that the intensity is proportional to the intensities of the generating fields and to the square of the interaction length L. The CARS signal intensity also depends on the degree of phase mismatch:

$$\Delta \vec{k} = \vec{k}_1 - \vec{k}_2 + \vec{k}_3 - \vec{k}_{CARS} \tag{7}$$

The wave vectors of the interacting four waves are denoted with k_i. Perfect phase matching is achieved for $\left| \Delta k \right| = 0$. There are different perfect phase-matching schemes which can be used in a typical RCARS experiment: planar and folded BOXCARS. Both configurations are explained in Fig. 4.

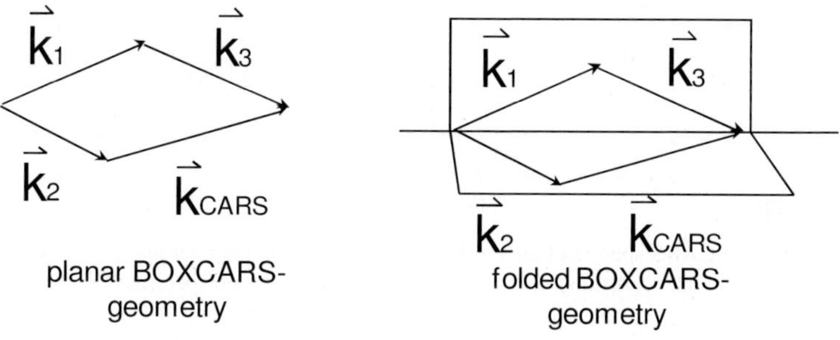

Fig. 4. Phase-matching geometries used for RCARS.

Neglecting one- and two-photon resonances and using the isolated line approximation the third-order non-linear susceptibility is given by the sum of a non-resonant (χ^3_{NR}) and a resonant part (χ^3_R):

$$\chi_{xxxx}^3 = \chi_{NR}^3 + \chi_R^3$$

$$\chi_{xxxx}^3 = \chi_{NR}^3 + \sum_{if} \frac{c^4}{4 \cdot \hbar \cdot \omega_2^4} \cdot \frac{\Delta N_{if}}{(\omega_{if} - \omega_1 + \omega_2 - i \cdot \Gamma_{if})} \cdot \left(\frac{d\sigma_{if}}{d\Omega} \right), \tag{8}$$

where Γ_{if} represents the collisional Raman linewidth and $(d\sigma_{if}/d\Omega)$ the differential Raman cross section. The temperature and concentration dependence is mainly contained in the population difference between two molecular states ΔN_{if}. In thermodynamic equilibrium the number of molecules in a particular vibrational and rotational level is given by the Boltzmann distribution. Typical RCARS spectra of air in dependence of temperature and pressure can be seen in Fig. 5.

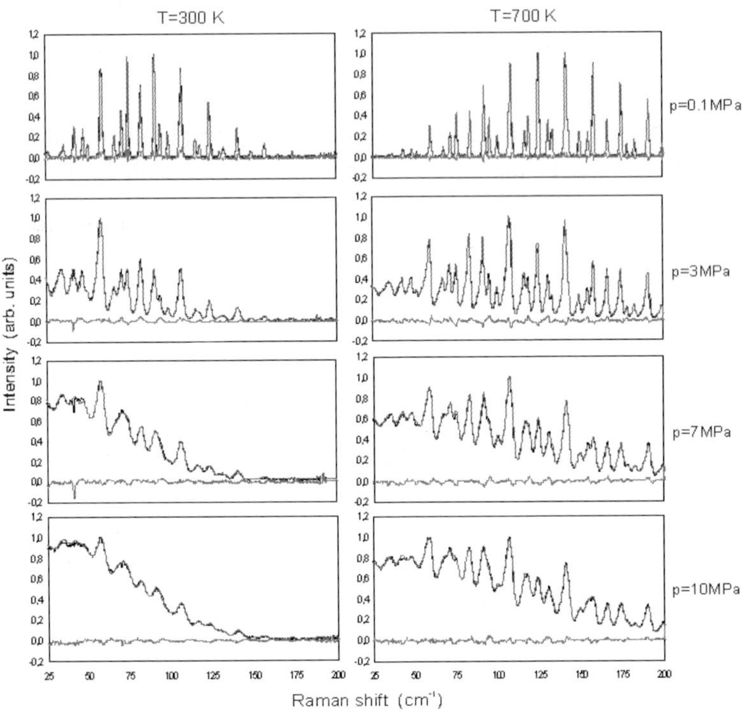

Fig. 5.   Pure rotational CARS spectra of air for different pressures and temperatures. The deviation between the best fit and experimental data is displayed as the curve located below each spectrum.[28]

2.2. *Rotational CARS*

There exist two variants of pure rotational CARS: conventional rotational CARS (C-RCARS) and dual-broadband rotational CARS (DB-RCARS). The principal approaches for both cases are shown in Fig. 6. In both approaches the probed molecules are coherently excited by pairs of photons of frequency ω_1 and ω_2 with a frequency difference exactly equal to the rotational Raman frequency shift $\omega_{1,2}$ of the transition. A third photon of frequency ω_3 is then scattered off the resonance to generate the CARS signal at the frequency ω_{CARS}. In C-RCARS a narrowband laser, typically a frequency doubled Nd:YAG laser (λ=532nm) is used for the transitions corresponding to the frequencies ω_1 and ω_3. The Stokes transitions (ω_2) are driven by a broadband dye-laser operating with the dye Coumarin. In DB-RCARS two broadband dye lasers, e. g., being operated with the dyes DCM or Rhodamine, are used for the transitions corresponding to the frequencies ω_1 and ω_2, whereas a narrowband frequency doubled Nd:YAG laser is usually used for the transition corresponding to the frequency ω_3. In both cases the rotational CARS signal contains all anti-Stokes frequencies since broadband dye lasers are used. This offers the possibility to achieve time resolution in the measurements being limited by the pulse width of the lasers which is typically a few ns.

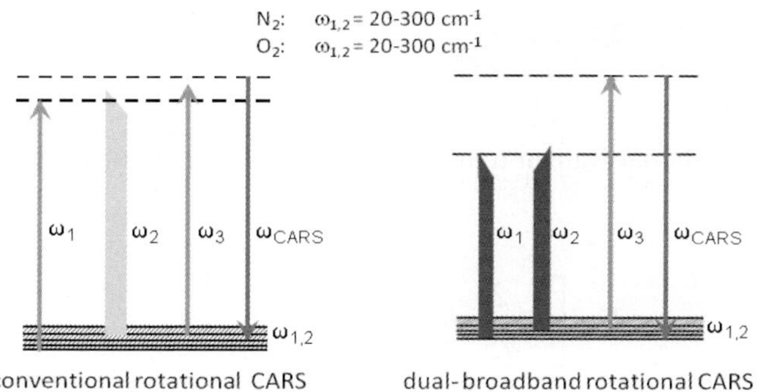

Fig. 6.   Energy level diagram of the conventional RCARS and the DB-RCARS process.

In DB-CARS spectra a spectral averaging effect occurs because each rotational Raman resonance is driven by multiple photon pairs. Therefore the influence of the statistical mode fluctuations on the shape of the DB-CARS spectrum is significantly reduced which leads to an improved precision[25,27,57] in comparison to C-RCARS where only one photon pair drives each resonance. Beside this for DB-RCARS the dye can be arbitrarily chosen. Usually DCM is used which is a stable dye with a wavelength centered at 630 nm and spectrally well separated from the DB-RCARS signal. As a result a planar BOXCARS configuration can be used, because in this case the CARS signal can be separated from the dye laser by a spectral filter. This is not possible for C-RCARS since the wavelength of all three laser beams are located spectrally very close to the rotational CARS signal. Then a more complex folded BOXCARS geometry is necessary in order to avoid stray light in the detection system.

2.3. *Polarization CARS*

The most efficient generation of the CARS signal is achieved with the polarization of the three laser beams being oriented parallel to each other. Nevertheless it can be necessary to use different polarizations for the laser beams. This can be motivated by a more efficient stray light reduction or a total suppression of the non-resonant background.[38,48] As shown in Eqn. (1) and (6) the CARS intensity scales with the square of the third-order susceptibility. Owyoung has demonstrated that the elements of this tensor can be expressed in terms of two nuclear response functions and a non-resonant contribution σ.[58] In the case of DB-RCARS applied to the gas phase in combustion processes the tensor elements can be reduced to:[38,48]

$$\chi^{3}_{xxyy} = \frac{1}{24} \cdot (\sigma - 3 \cdot a_{(\omega 1 - \omega 2)})$$

$$\chi^{3}_{xyxy} = \frac{1}{24} (\sigma + 2 \cdot a_{(\omega 1 - \omega 2)})$$

$$\chi^{3}_{xyyx} = \frac{1}{24} (\sigma - 3 \cdot a_{(\omega 1 - \omega 2)})$$

$$\chi^{3}_{1111} = \frac{1}{24} (3 \cdot \sigma - 4 \cdot a_{(\omega 1 - \omega 2)})$$

$$(9)$$

For a single rotational Raman transition the response function a can be written in the form:

$$a_{(\omega k - \omega 2)} = K_{JJ'} \left[-\frac{2}{45} \cdot b_{JJ'} \cdot (\gamma')^2 \right]$$

$$K_{JJ'} = \frac{1}{\hbar} \left[N_J - \frac{(2J+1)}{(2J'+1)} \cdot N_{J'} \right] \cdot (\omega_{AS} - \omega_k + \omega_2 - i\Gamma)^{-1}$$

$$(10)$$

where, γ' is the anisotropic part of the Raman polarizability, and $b_{JJ'}$: the Plazek-Teller coefficient.

Typical polarization arrangements keep the dye laser (ω_1, ω_2) polarization unchanged and only the Nd:YAG laser polarization (ω_3, θ_3) is changed. This leads to the fact, that the polarization of the resonant (θ_R) and the non-resonant part (θ_{NR}) of the CARS signal can be determined only by the polarization of the Nd: YAG laser (θ_3):

$$\tan \theta_{NR} = \frac{1}{3} \tan \theta_3$$

$$\tan \theta_R = -\frac{1}{2} \tan \theta_3$$

$$(11)$$

Then, in order to suppress most of the elastically scattered stray light a polarization analyzer in the signal path has to be perpendicular to the Nd: YAG laser polarization. In this case minimal loss of the resonant CARS signal is obtained for the Nd: YAG laser polarization θ_3 set to 45° with the analyzer set to -45°.[38]

A total suppression of the non-resonant part of the CARS signal can be achieved by a polarization analyzer perpendicular to χ_{NR}. Then maximum signal detection is achieved for the Nd:YAG laser polarization θ_3 set to 60° with the analyzer set to -60°.[48]

3. Measurements in sooting flames

Fuel combustion, which is one of the major energy conversion processes, is connected with the production of pollutants such as soot and nitric oxides. Soot formation and oxidation, which are in the focus of the investigation reported here, are sensitive to the local combustion conditions, since the preparticle soot inception chemistry is relatively slow.[59] The local temperature, residence time, and mixture fraction have been identified as important variables in soot production, in addition to the chemical structure of the fuel. Therefore quantitative information about the phenomenology of soot formation i.e. the dependence of soot formation on fuel, mixture composition, temperature, pressure, additives, or turbulence, requires experiments over a large range of the operation parameters under well defined conditions. Generally, for nitrogen vibrational CARS thermometry a frequency-doubled Nd:YAG laser emitting at 532 nm is used along with a broadband dye laser, resulting in a CARS signal wavelength of 473 nm. Unfortunately the emission of laser produced C_2 radicals occurs in the same spectral region and disturbs the signal by interferences especially in fuel-rich, sooting flames,[60] as can be seen in Fig. 7. These interferences can be a severe problem resulting in erroneous temperature evaluation. One possibility to solve this problem is to use a dual-pump vibrational CARS setup, but for strongly sooting flames, it is still difficult to find interference-free spectral regions.[61-63] In such an environment it may be useful to apply the rotational CARS technique where the signal is generated around 532 nm to avoid spectral distortion from soot fragments.[31,42]

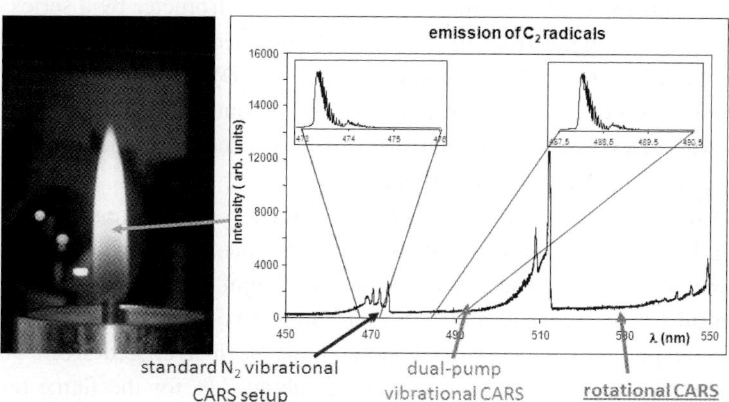

Fig. 7. Emission spectrum of C_2 taken from a sooting flame. In addition the spectral regions of the different CARS approaches are marked.

As an example measurements taken in a non-premixed laminar bluff-body flame operated with methane at atmospheric pressure are shown. The burner consists of two coaxial tubes of 18 and 56 mm diameter and 820 mm length. The inner tube supplies the combustion gases and the outer tube is used to stabilize the flame. A photo of this flame is shown in Fig. 8a where also the frame is displayed for the presentation of the results in Fig. 9.

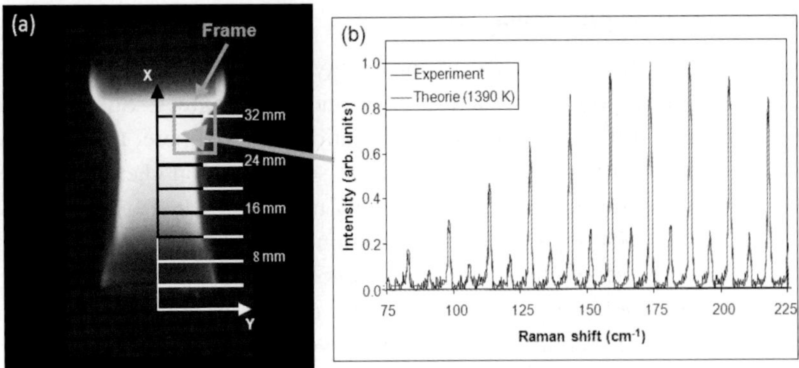

Fig. 8.   Single-shot DB-RCARS spectrum taken from a non-premixed sooting methane diffusion flame with a bluff body.

For the dual-broadband RCARS setup a frequency-doubled Nd:YAG laser is used as pump source for a broadband dye laser and simultaneously as the narrow-band source for the CARS process by splitting of a fraction of 30 mJ. A solution of DCM in methanol was used in the dye laser. The dye laser beam was split into equal beams of 25 mJ each, with a bandwidth of about 200 cm^{-1}, centered around 635 nm. The three beams were arranged in a planar BOXCARS configuration with a 300-mm focusing lens and a probe volume of diameter 0.1 mm and length 1.5 mm was achieved. The signal is re-collimated by another 300-mm focusing lens and directed to the spectrometer by a series of dichroic mirrors separating the signal from the broadband laser beam. The spectra were recorded by a back-illuminated 16-bit charge coupled device (CCD) camera mounted to a double-monochromator to cover a spectral region of about 300 cm^{-1}. The temperature information was determined by a comparison of the measured spectra to a library of calculated spectra using a contour fit procedure.

A typical disturbance-free CARS spectrum from the sooting region of the flame is shown in Fig. 8b together with the evaluation. RCARS has been used at 40 points along 5 horizontal lines of 8 points each at a distance of 1 mm. The lines were separated by 2 mm. The results have been interpolated to provide a two-dimensional temperature field (Fig. 9). The temperature data for each point have been averaged from 50 evaluated single-shot spectra with an average rms-value of about 7 % for the flame temperatures. The maximum temperature values can be found near the reaction zone about 6-8 mm from the centerline at 26 mm downstream the burner´s exit diverging with increasing height. A maximum temperature of 1950 K has been determined and no interference effects have been observed.

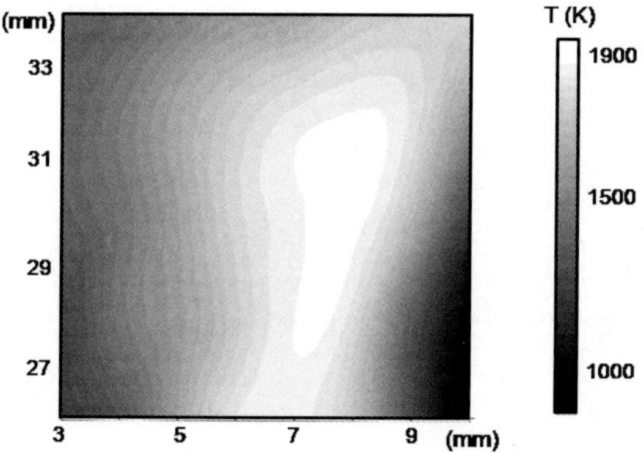

Fig. 9. Enlarged scale representation of the sooting region (Frame in Fig 8a) of the flame providing information of the two-dimensional distribution of the gas temperature.

4. Evaporating sprays

The development of stratified-charge gasoline direct injection (GDI) engines is one of the most promising strategies for the reduction of fuel consumption and pollutant emission in spark-ignition (SI) engine combustion. A necessary requirement for the successful application of the desired stratified-charge concept over a wide range of load and engine speed is the fast and effective atomization and vaporization of the fuel before ignition. Thus, the design of new efficient injectors and adjustment of parameters like fuel temperature and injection pressure is required. This can only be done by thorough experimental characterization of the injection spray. Beside properties like droplet temperature and size, the surrounding gas phase temperature has to be determined to assess the vaporization of the fuel droplets. The RCARS technique is a useful tool for the local determination of this important quantity inside the vaporizing spray of a GDI injector.

For the first time RCARS was applied in an evaporating GDI spray by Beyrau et al.[38] The laser and detection system was nearly the same as described in Sec. 3. A heated injection chamber was used with three rectangular fused silica windows providing optical access.[64] A multihole GDI injector with an injection pressure of 80 bar with isooctane fuel heated up to 90°C was used. The chamber was heated by a constant air flow with the pressure set to 3 bar. The CARS probe volume was positioned at a distance of 70 mm from the injector nozzle orifice in the centre of the spray cone. In Fig. 10 the spray and the location of the measurement position are displayed.

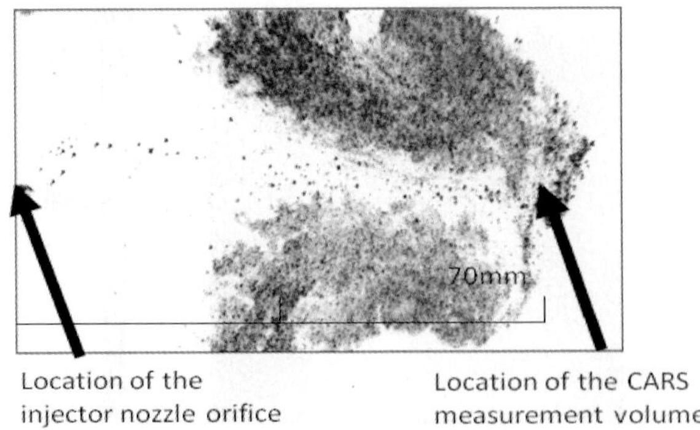

Fig. 10.   Location of the measurement position within the spray.[38]

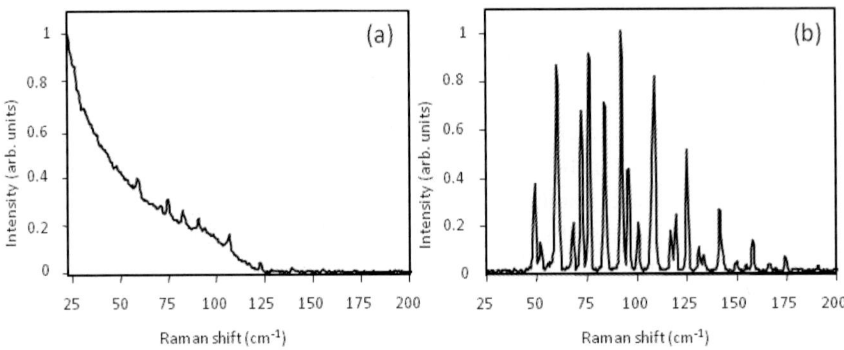

Fig. 11.   Experimental spectrum taken at 4 ms after start of injection (SOI) without (a) and with (b) stray light suppression.[38]

In the presence of droplets, particles, or more general surfaces, near the probe volume or cell windows in the beam path, light from the narrowband laser can be scattered or reflected in the direction of the signal beam. This elastically scattered light "fills" the spectrometer and leads to an interference with the rotational CARS signal over a wide spectral range, see Fig. 11a. Therefore a polarization scheme was used in order to suppress most of the elastically scattered stray light. As described in Sec. 2.3 the dye laser polarization was kept unchanged and only the Nd:YAG laser polarization is changed. Thus, maximum signal detection is achieved for the Nd:YAG laser polarization set to 45° with the analyzer set to -45°. Then only depolarized light from multiple scattering from the droplets and birefringence in the optics and chamber windows is transmitted through the analyzer. This greatly reduced the stray light problem, but was not sufficient for the spray experiments. Thus, in addition a knife edge was placed at the exit plane of the first part of the double-monochromator, where the light is already spectrally dispersed. The knife edge acts as a very sharp spectral filter with a slope width of a few wavenumbers. A single-shot spectrum taken inside the spray with this setup is shown in Fig. 11b.

Fig. 12 shows the temporal behavior of the gas phase temperature in the spray for the two different injection pressures of 40 bar and 100 bar at a distance of 70 mm downstream the injector nozzle. For each measurement point 100 single-shot CARS spectra have been acquired at various times after start of injection within the injected fuel sprays. For better comparison, each profile is shifted to zero milliseconds when the spray arrives at the probe volume to compensate for the different spray velocities at different injection pressures. A sharp decrease of the temperature can be observed when the spray front enters the probe volume due to vaporization cooling of the fuel, where the required vaporization enthalpy is taken from the surrounding gas phase. At this particular time, the gas phase is cooled down for about 30 K in both measurement series, independent of the injection pressure. The identical temperature drop for both pressures indicates that fuel vapor saturation has been reached in the gas phase in this region of the spray front due to the large amount of droplets available. This can be considered as a barrier condition for further fuel vaporization. At later times, the temperature increases slowly to its initial value but even at 20 ms after spray arrival at the measurement position the temperature has not reached its initial value. The temperature increase is faster for higher pressures. This probably is caused by the larger amount of droplets and thus fuel mass available inside the probe volume for the case of a lower injection pressure after the minimum temperature has been reached.

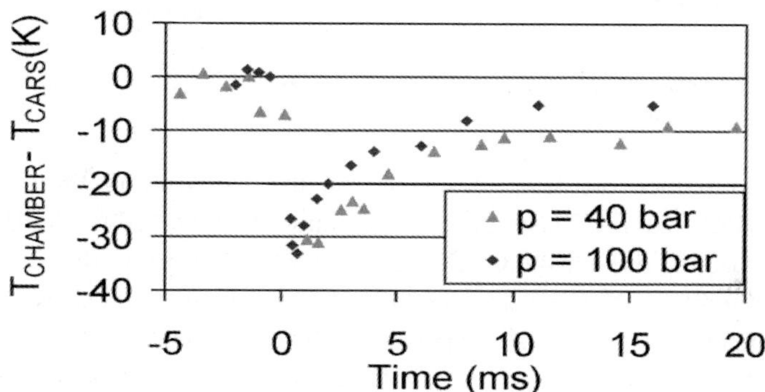

Fig. 12.   Gas phase temperature results for injection pressures of 40 bar and 100 bar and fuel temperature of 363 K as a function of time after the spray front arrives at the probe volume (t = 0 ms).[44]

5. IC engine

During the last decades optical techniques have been of great importance for the understanding of processes governing the combustion in internal combustion (IC) engines. For an understanding and improvement of such processes often high spatial and temporal resolution is necessary, which can be achieved by laser-based techniques. Important thermodynamic parameters are the temperature and the exhaust-gas-recirculation rate since they govern the chemical reactions leading to ignition and to formation of pollutants. Since RCARS has a high spectral sensitivity in the temperature and pressure region corresponding to conditions, prior to ignition RCARS has some advantages for the application at IC engines and was first demonstrated by Bengtsson et al.[46] Since then RCARS was mainly used to characterize knocking conditions in SI engines[65-68]. Based on these experiences limitations of this technique, e.g., due to possible laser induced damages of the windows, the high unknown nonresonant susceptibility and a necessary stray light suppression are reported.[69] Recently Weikl et al. solved these problems and demonstrated DB-RCARS measurements for the simultaneous determination of temperature and exhaust-gas recirculation in a homogeneous charge-compression ignition (HCCI) engine using a nearly production-line engine fired with standard fuel (RON 95).[70,71]

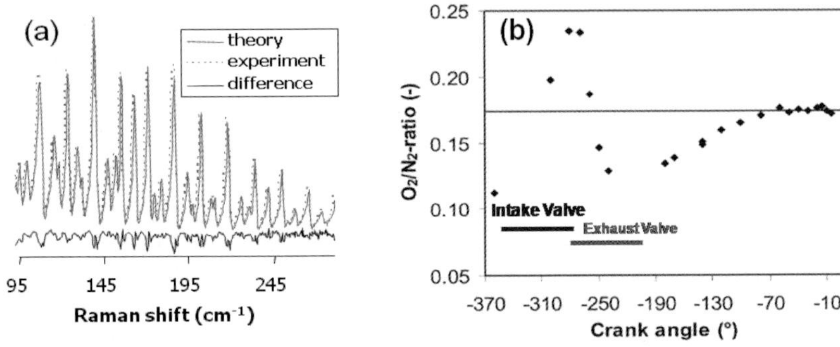

Fig. 13.   (a): Single-pulse spectrum taken at -10° crank angle (CA).   (b): Evaluation of the O_2/N_2 ratio in the intake and compression stroke.[70]

In order to reduce the laser irradiance on the windows in this work the initial frequency-doubled Nd:YAG laser pulse was temporally stretched in a pulse stretcher and split into two beams, one for the CARS process and one to pump the dye laser. The rest of the setup is similar to the one described in Sec. 3 and again a planar BOXCARS phase-matching geometry was used. A proper stray light rejection was achieved by using a knife edge at the exit plane of the first part of the double-monochromator and a polarization scheme for the incoming laser beams[38] as already mentioned in Sec. 4. Also in this case maximum signal detection is achieved for the Nd:YAG laser polarization set to 45° with the analyzer set to -45°. As an example Fig 13a shows a typical single-shot spectrum taken at -10° crank angle (CA). In Fig. 13b the O_2/N_2 ratio in the intake and compression ratio evaluated from RCARS measurements can be seen. At first a very low oxygen concentration is present that is enhanced by the intake of fresh air. When the exhaust valve is opened again the concentration ratio decreases strongly due to the presence of exhaust gas, which is subsequently mixed with fresh air. This results at least in an O_2/N_2 ratio of 0.174, which corresponds to an exhaust-gas-recirculation (EGR) ratio of 45.5%.

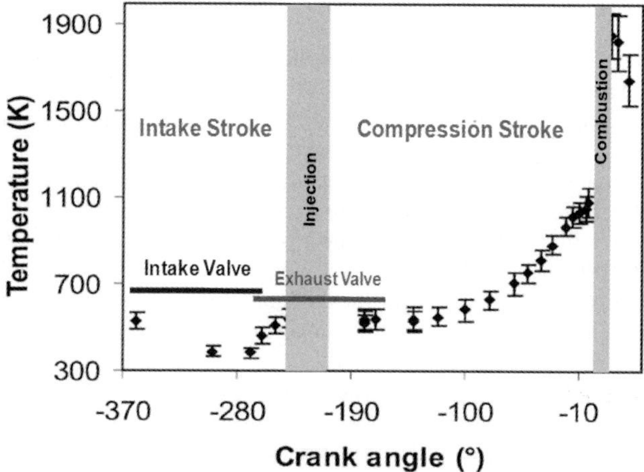

Fig. 14.   CARS temperatures measured before autoignition and after combustion at various crank angles (CAs[70]).

To demonstrate the potential of DB-RCARS for making measurements in a broad range of different crank angles a measured temperature profile is shown in Fig. 14. The temperatures shown indicate mean values with the standard deviation of the single-pulse measurements represented by error bars. When the intake valve opens, fresh air was sucked in and the temperature was decreasing. When the exhaust gas valve was opened again for the EGR, the temperature in the probe volume was rising due to the hot exhaust gas. At -180°CA the valves were closed and the mixture is compressed to top dead center. During injection a too-high droplet density and during combustion beam steering effects made measurements impossible.

6. Conclusions

The development of pure rotational coherent anti-Stokes Raman scattering (RCARS) for combustion diagnostics over now nearly three decades has shown a tremendous improvement. It has turned out that RCARS has important advantages for an application to technical combustion systems. It can be applied, e.g. to sooting flames, sprays or IC engines for precise temperature and concentration determination. This is possible due to important improvements of this technique like the development of the DB-CARS to increase the accuracy, the introduction of a polarization scheme to eliminate the stray light problem and the application of a pulse stretcher to reduce the window damage risk. Therefore, in the future, RCARS will probably be applied to even more practical combustion systems.

Acknowledgments

The authors gratefully acknowledge financial support for parts of the work by the German Science Foundation (DFG), which also funds the Erlangen Graduate School in Advanced Optical Technologies in the framework of the German excellence initiative.

References

1. N. M. Laurendeau, Temperature measurements by light-scattering methods Prog. Energy Combust. Sci. **14**, pp 147-170 (1988)
2. D. R. Crosley, Semiquantitative laser-induced fluorescence in flames Combust. Flames **78** , pp 153-167 (1989)
3. K. Kohse-Höinghaus, Laser techniques for the quantitative detection of reactive intermediates in combustion systems, Prog. Energy Combust. Sci. **20**, pp 203-279 (1994)
4. J. W. Daily, Laser induced fluorescence spectroscopy in flames, Prog. Energy Combust. Sci. **23**, pp 133-199 (1997)
5. C. Schulz, V. Sick, Tracer-LIF diagnostics: quantitative measurement of fuel concentration, temperature and fuel/air ratio in practical combustion systems Prog. Energy Combust. Sci. **31**, pp 75-121 (2004)
6. M. C. Drake and G. M. Rosenblatt, Flame temperatures from Raman scattering Chem. Phys. Lett. **44**, pp 313-316 (1976)
7. S. Ledermann, The use of laser Raman diagnostics in flow fields and combustion Prog. Energy Combust. Sci. **3**, pp 1-34 (1977)
8. J. Haumann and A. Leipertz, Giant-pulsed laser Raman oxygen measurements in a premixed laminar methane-air flame, Appl. Opt. **24**, pp 4509-4515 (1985)
9. A. Leipertz, J. Haumann and M. Fiebig, Contact-free measurements of oxygen concentration in industrial flames by Raman scattering, Chem. Eng. Technol. **10**, pp 190-203 (1987)
10. R. W. Dibble, A. R. Masri, and R. W. Bilger, The spontaneous Raman scattering technique applied to nonpremixed flames of methane Combust. and Flame **67**, pp 189-206 (1987)
11. W. Kiefer, Nonlinear Raman spectroscopy in "Infrared and Raman Spectroscopy", ed.: B. Schrader, Weinheim, pp 162-188 (1995)
12. P. D. Marker and R. W. Terhune, Study of optical effects due to an induced polarization third order in the electric field strength Phys. Rev. **137**, pp A801-A818 (1965)
13. B. Zheng, J. B. Snow, D. V. Murphy, A. Leipertz, R. K. Chang, R. L. Farrow Experimental comparison of broadband rotational coherent anti-Stokes Raman scattering (CARS) and broadband vibrational CARS in a flame Opt. Lett. **9**, pp 341-343 (1984)
14. D. Klick, K. A. Marko, L. Rimai, Broadband single-pulse CARS spectra in a fired internal combustion engine, Appl. Opt. **20**, pp 1178-1181 (1981)
15. K. Aron, L. E. Harris, J. Fendell, N_2 and CO vibrational CARS and H_2 rotational CARS spectroscopy of CH_4/N_2O flames, Appl. Opt. **22**, pp 3604-3611 (1983)
16. C. Eckbreth, T. J. Anderson, Dual broadband CARS for simultaneous, multiple species measurements, Appl. Opt. **24**, pp 2731-2736 (1985)
17. W. A. England, J. M. Milne, S. N. Jenny, D. A. Greenhalgh, Application of CARS to an operating chemical reactor, Appl. Spect. **38**, pp 867-876 (1984)
18. Y. Yueh, E. J. Beiting, Simultaneous N_2, CO, and H_2 multiplex CARS measurements in combustion environments using a single dye laser, Appl. Opt. **27**, pp 3233-3243 (1988)
19. A. C. Eckbreth, T. J. Anderson, G. M. Dobbs, Multi-color CARS for hydrogen-fueled scramjet applications, Appl. Phys. B **45**, pp 215-223 (1988)
20. K. W. Boyack, P.O. Hedman, Dual-Stokes CARS system for simultaneous measurement of temperature and multi species in turbulent flames, Proc. of the Combust. Inst. **23**, pp 1893-1899 (1990)
21. S. Kampmann, T. Seeger, A. Leipertz, Simultaneous CARS and 2D laser Rayleigh thermometry in a contained technical swirl combuster, Appl. Opt. **34,** pp 2780-2786 (1995)
22. S. Roy, T. R. Meyer, R. P. Lucht, V. M. Belovich, E. Corporan, and J. R. Gord Temperature and CO_2-concentration measurements in the exhaust stream of a liquid-fueled combustor using dual-pump coherent anti-Stokes Raman scattering Combust. and Flame **138**, pp 273-284 (2004)

23. A. Datta, F. Beyrau, T.Seeger, and A. Leipertz, Temperature and CO concentration measurements in a partially premixed CH_4/Air CO-flowing jet flame using coherent anti-Stokes Raman scattering, Combust. Sci. Technol. **176**, pp 19765-1984 (2004)

24. A. Braeuer, F. Beyrau, T. Seeger, J. Kiefer, A. Leipertz, A. Holzwarth, A. Soika Investigation of the combustion process in an auxiliary heating system using dual-Pump coherent anti-Stokes Raman scattering (CARS): evaluation of temperature and oxygen/nitrogen ratio, J. Raman Spectrosc. **37**, pp 633-640 (2006)

25. T. Seeger, A. Leipertz, Experimental comparison of single-shot broadband vibrational and dual-broadband pure rotational coherent anti-Stokes Raman scattering in hot air, Appl. Opt. **35**, pp 2665-2671 (1996)

26. Thumann, M. Schenk, J. Jonuscheit, T. Seeger, A. Leipertz Simultaneous temperature and relative nitrogen-oxygen concentration measurements in air with pure rotational coherent anti-Stokes Raman scattering for temperatures to as high as 2050 K, Appl. Opt. **36,** pp 3500-3505 (1997)

27. M. Aldén, P.-E. Bengtsson, H. Edner, S. Kröll, D. Nilsson, Rotational CARS: a comparison of different techniques with emphasis on accuracy in temperature determination, Appl. Opt. **28**, pp 3206-3219, (1989)

28. T. Seeger, F. Beyrau, A. Braeuer, and A. Leipertz, High pressure pure rotational CARS: comparison of temperature measurements with O_2, N_2 and synthetic air, J. Raman Spectrosc. **34**, pp 932-939 (2003)

29. D. V. Murphy, R. K. Chang, Single-pulse broadband rotational coherent anti-stokes Raman-scattering thermometry of cold N_2 gas, Opt. Lett. **6**, pp 233-235 (1981)

30. J. B. Zheng, A. Leipertz, J. B. Snow, R. K. Chang, Simultaneous observation of rotational coherent Stokes Raman scattering and coherent anti-Stokes Raman scattering in air and nitrogen, Opt. Lett. **8**, pp 350-352 (1983)

31. M. Aldén, P.-E. Bengtsson, H. Edner, Rotational CARS generation through a multiple four-color interaction, Appl. Opt. **25**, pp 4493-4500 (1986)

32. A. C. Eckbreth, T.J. Anderson, Simultaneous rotational coherent anti-Stokes Raman spectroscopy and coherent Stokes Raman spectroscopy with arbitrary pump-Stokes spectral separation, Opt. Lett. **11**, pp 496-498 (1986)

33. G. Millot, R. Saint-Loup, J. Santos, R. Chaux, H. Berger, J. Bonamy Collisional effects in the stimulated Raman Q branch of O_2 and O_2-N_2 J. Chem. Phys. **96**, pp 961-971 (1992)

34. L. Martinsson, P-E. Bengtsson, M. Aldén, S. Kröll, J. Bonamy, A test of different rotational Raman linewidth models: accuracy of rotational coherent anti-Stokes Raman scattering thermometry in nitrogen from 295 to 1850 K, J. Chem. Phys. **99**, pp 2466-2477 (1993)

35. F. Beyrau M. C. Weikl, T. Seeger, A. Leipertz, Application of an optical pulse stretcher to coherent anti-Stokes Raman spectroscopy, Opt. Lett. **29**, pp 2381-2383 (2004)

36. M. Afzelius, P.-E. Bengtsson, Precision of single-shot dual-broadband rotational CARS thermometry with single-mode and multi-mode Nd:YAG lasers, J. Raman Spectrosc. **34**, pp 940-945 (2003)

37. M. C. Weikl, Y. Cong, T. Seeger, A. Leipertz, Development of a simplified dual-pump dual-broadband CARS system, Appl. Opt. 48, **B 43-B50** (2009)

38. F. Beyrau, A. Braeuer, T. Seeger and A. Leipertz, Gas-phase temperature measurement in the vaporizing spray of a gasoline direct injection injector using pure rotational coherent anti-Stokes Raman scattering, Opt. Lett. **29**, pp 247-249 (2004)

39. J. Bood, P.-E. Bengtsson, M. Aldén, Stray light rejection in rotational coherent anti-stokes Raman spectroscopy by use of a sodium-seeded flame, Appl. Opt. **37**, pp 8392-8396 (1989)

40. J. Bood, P.-E. Bengtsson, F. Mauss, K. Burgdorf, and I. Denbratt, Knock in spark-ignition engines: end-gas temperature measurements using rotational CARS and detailed kinetic calculations of the autoignition process, SAE Technical Paper 971669 (1997)

41. P.-E. Bengtsson, L. Martinson, M. Aldén, S. Kröll, Rotational CARS thermometry in sooting flames, Comb. Sci. and Tech. **81**, pp 129-140 (1992)

42. C. Brackmann, J. Bood, P.-E. Bengtsson, T. Seeger, M. Schenk, A. Leipertz Simultaneous vibrational and pure rotational coherent anti-Stokes Raman spectroscopy for temperature and multi-species concentration measurements demonstrated in sooting flames Appl. Opt. **41**, pp 564-572 (2002)

43. T. Seeger, J. Egermann, S. Dankers, F. Beyrau, and A. Leipertz Comprehensive characterization of a laminar sooting methane diffusion flame by applying a combination of different laser techniques Chem. Eng. Tech. **27**, pp 1150-1156 (2004)

44. F. Beyrau. M. Weikl, I. Schmitz, T. Seeger and A. Leipertz Locally resolved investigation of the vaporization of GDI sprays applying different laser techniques Atomization and Sprays **16**, pp 319-330 (2006)

45. S. Hildenbrand, S. Staudacher, D. Brüggemann, F. Beyrau, M. C. Weikl, T. Seeger and A. Leipertz Numerical and experimental study of the vaporization cooling in gasoline direct injection sprays Proc. of the Combust. Inst. **31**, pp 3067-3073 (2007)

46. P.-E. Bengtsson, L. Martinsson, M. Aldén, B. Johansson, B. Lassesson, K. Marforio, G. Lundholm Dual-broadband rotational CARS measurements in an IC engine Proc. of the Combust. Inst. **25**, pp 1735-1742 (1994)

47. M. C. Weikl, F. Beyrau, A. Leipertz Simultaneous temperature and exhaust-gas recirculation-measurements in a homogeneous charge-compression ignition engine by use of pure rotational coherent anti-Stokes Raman spectroscopy Appl. Opt. **45**, pp 3646-3651 (2006)

48. F. Vestin, M. Afzelius, P.-E. Bengtsson Development of rotational CARS for combustion diagnostics using a polarization approach Proc. of the Combust. Inst. **31**, pp 833-840 (2007)

49. S. Kröll, D. Sandell Influence of laser-mode statistics on noise in nonlinear-optical processes-application to single-shot broadband coherent anti-Stokes Raman scattering thermometry J. Opt. Soc. Am. B **5**, pp 1910-1926 (1988)

50. A. Owyoung The origin of the nonlinear refractive indices of liquids and gasses Ph. D. Dissertation, California Institute of Technology (1971)

51. I. Glassman Soot formation in combustion processes Proc. Combust. Inst. **22**, pp 295-311 (1988)

52. P.-E. Bengtsson, M. Aldén, S. Kröll and D. Nilsson, Vibrational CARS thermometry in sooty flames: Quantitative Evaluation of C2 Absorption Interference Combust. Flame **82**, pp 199-210 (1990)

53. F. Beyrau, A. Datta, T. Seeger and A. Leipertz Dual-pump CARS for the simultaneous detection of N_2, O_2 and CO in CH_4-flames J. Raman Spectrosc. **33**, pp 919-924 (2002)

54. M.C. Weikl, T. Seeger, M. Wendler, R. Sommer, F. Beyrau and A. Leipertz Validation experiments for spatially resolved one-dimensional emission spectroscopy temperature measurements by dual-pump CARS in a sooting flame Proc. of the Combust. Inst. **32**, pp 745-752 (2009)

55. S. P. Kearney, M. N. Jackson Dual-pump coherent anti-Stokes Raman scattering thermometry in heavily sooting flames AIAA Journal **45**, pp 2947-2956 (2007)

56. I. Schmitz, W. Ipp, and A. Leipertz, Flash boiling effects on the development of gasoline-direction engine sprays SAE Technical Paper 2002-01-2661 (2002)

57. J. Bood, P.-E. Bengtsson, F. Mauss, K. Burgdorf and I. Denbratt Knock in spark-ignition engines: end-gas temperature measurements using rotational CARS and detailed kinetic calculations of the autoignition process SAE Technical Paper Series No. 971669 (1997)

58. B. Grandin, I. Denbratt, J. Bood, C. Brackmann and P.-E. Bengtsson The effect of knock on the heat transfer in an SI engine: thermal boundary layer investigation using CARS temperature measurements and heat flux measurements SAE Technical Paper Series No. 2000-01-2831 (2000)

59. B. Grandin, I. Denbratt, J. Bood, C. Brackmann, P.-E. Bengtsson, A. Gogan, F. Mauss und B. Sundén Heat release in the end-gas prior to knock in lean, rich and stoichiometric mixtures with and without EGR SAE Technical Paper Series No. 2002-01-0239 (2002)

60. B. Grandin, I. Denbratt, J. Bood, C. Brackmann and P.-E. Bengtsson A study of the influence of exhaust gas recirculation and stoichiometry on the heat release in the end-gas prior to knock using rotational coherent anti-Stokes-Raman spectroscopy thermometry Int. J. Engine Research **3**, pp 209-221 (2002)

61. C. Brackmann, J. Bood, M. Afzelius and P.-E. Bengtsson Thermometry in internal combustion engines via dual-broadband rotational coherent anti-Stokes Raman spectroscopy Meas. Sci. Technol. **15**, pp R13-R25 (2004)

62. M.C. Weikl, F. Beyrau and A. Leipertz Simultaneous temperature and exhaust-gas recirculation-measurement in a homogeneous charge-compression ignition engine using pure rotational coherent anti-Stokes Raman spectroscopy Appl. Opt. **45**, pp 1-10 (2006)

63. M. C. Weikl, F. Beyrau, A. Leipertz, A. Loch, C. Jelitto and J. Willand Locally Resolved Measurement of Gas-Phase Temperature and EGR-Ratio in an HCCI-Engine and Their Influence on Combustion Timing SAE Technical Paper Series 2007-01-0182 (2007)

64. I. Schmitz, W. Ipp, and A. Leipertz, Flash boiling effects on the development of gasoline-direction engine sprays SAE Technical Paper 2002-01-2661 (2002)

65. J. Bood, P.-E. Bengtsson, F. Mauss, K. Burgdorf and I. Denbratt Knock in spark-ignition engines: end-gas temperature measurements using rotational CARS and detailed kinetic calculations of the autoignition process SAE Technical Paper Series No. 971669 (1997)

66. B. Grandin, I. Denbratt, J. Bood, C. Brackmann and P.-E. Bengtsson The effect of knock on the heat transfer in an SI engine: thermal boundary layer investigation using CARS temperature measurements and heat flux measurements SAE Technical Paper Series No. 2000-01-2831 (2000)

67. B. Grandin, I. Denbratt, J. Bood, C. Brackmann, P.-E. Bengtsson, A. Gogan, F. Mauss und B. Sundén Heat release in the end-gas prior to knock in lean, rich and stoichiometric mixtures with and without EGR SAE Technical Paper Series No. 2002-01-0239 (2002)

68. B. Grandin, I. Denbratt, J. Bood, C. Brackmann and P.-E. Bengtsson A study of the influence of exhaust gas recirculation and stoichiometry on the heat release in the end-gas prior to knock using rotational coherent anti-Stokes-Raman spectroscopy thermometry Int. J. Engine Research **3**, pp 209-221 (2002)

69. C. Brackmann, J. Bood, M. Afzelius and P.-E. Bengtsson Thermometry in internal combustion engines via dual-broadband rotational coherent anti-Stokes Raman spectroscopy Meas. Sci. Technol. **15**, pp R13-R25 (2004)

70. M.C. Weikl, F. Beyrau and A. Leipertz Simultaneous temperature and exhaust-gas recirculation-measurement in a homogeneous charge-compression ignition engine using pure rotational coherent anti-Stokes Raman spectroscopy Appl. Opt. **45**, pp 1-10 (2006)

71. M. C. Weikl, F. Beyrau, A. Leipertz, A. Loch, C. Jelitto and J. Willand Locally Resolved Measurement of Gas-Phase Temperature and EGR-Ratio in an HCCI-Engine and Their Influence on Combustion Timing SAE Technical Paper Series 2007-01-0182 (2007)

<h1 style="text-align:center">CHAPTER 4</h1>

IMAGING FLAMES: FROM ADVANCED LASER DIAGNOSTICS TO SNAPSHOTS

MARSHALL LONG

Department of Mechanical Engineering, Yale University
New Haven, Connecticut 06520 USA
marshall.long@yale.edu

Laser imaging techniques have become productive tools for studying combustion. A variety of laser light scattering mechanisms can be used to measure temperature, species concentrations, and velocities in flames. In addition to giving valuable insight through visualization of the flow, imaging experiments can provide quantitative data that can be used to guide and verify combustion models. Several different approaches will be considered, ranging from a set of multiparameter imaging experiments in a turbulent nonpremixed flame designed to measure reaction rate and scalar dissipation, to measurements of soot in laminar flames. The first set of experiments required simultaneous imaging of four different quantities, using four lasers, while the second was carried out using a consumer digital camera. Data from both types of measurements are valuable for quantitative comparisons with model predictions.

1. Introduction

One important role for experiments in combustion research is to guide and verify computational models. The advances in lasers, detectors, and computers that occurred in the 1970's and 80's led to the development of many new techniques for gathering data in the harsh combustion environment. Laser-based planar imaging was among the first of the new techniques developed and the approach has been in use now for more than thirty years. The basic idea is quite simple – a laser is formed into a thin sheet intersecting the flame and scattered light (e.g., from Rayleigh, Raman, Lorenz/Mie or fluorescence) is imaged onto a two-dimensional detector. Some of the first work of this kind was done in Richard Chang's laboratories[1-3] and took advantage of image-intensified detectors that Professor Chang and his group developed before such systems became available commercially.[4]

Combustion represents one of the most challenging measurement environments. The early work using optical diagnostics understandably concentrated on those quantities that were easiest (and therefore possible) to measure. Each new spectroscopic technique or measured species provided an additional piece of the puzzle. Interpretation of the information was often difficult, however, due to the complex nature of the combustion environment and the sensitivity of the techniques to that environment. The enabling technologies for these measurements (lasers, detectors, and computers) have continued to improve, as has our understanding of the physics of the processes that create the signals. The current state of the art allows quantitative measurement of some of the quantities of most interest to modelers. In the following sections, examples of recent quantitative imaging experiments will be given. First, a set of measurements will be described in which mixture fraction and reaction rate were imaged in a turbulent nonpremixed flame. The goal of this work was to add spatial information to the database of single-point measurements already available for this flame, which is being modeled by several groups worldwide. Following that, a much simpler experiment will be described in which the sooting characteristics of a laminar flame are quantified using a consumer digital camera and compared with state-of-the-art modeling results.

2. Multi-scalar imaging in turbulent flames

The ability to predict the behavior of turbulent nonpremixed flames would be of tremendous value for the design and optimization of a broad range of practical devices including diesel engines, gas turbines, boilers, furnaces and incinerators. This remains an extremely difficult problem and a general "design code" that is applicable to a broad range of problems is still not available. However, in the past decade or so, considerable progress has been made. The focus of some of this progress has been a series of "Turbulent Nonpremixed Flame (TNF) Workshops" (http://www.ca.sandia.gov/TNF/). These workshops have brought together an international group of collaborators, both experimentalists and combustion modelers, to work at advancing the state of the art.

The group has identified a number of "target flames" with well-defined boundary conditions, spanning a range of flow and chemistry conditions. Independent groups have performed measurements in these flames and the results have been made available on the internet. Modelers are thus provided with detailed boundary conditions and a rich database of measured quantities including major species, velocities, and temperature, as well as selected radicals and pollutants. A shortcoming of the database, however, is the lack of image data on the target flames. Such data are needed, not only for the insight that would be provided into the flow through visualization, but also for the spatial gradient information (i.e., scalar dissipation) that is important in the modeling of the flames.

The goal of a recent set of experiments[5,6] was to add simultaneous images of temperature, mixture fraction, reaction rate, and scalar dissipation to the database of one of the target flames. One of the flames selected for the work was a piloted methane-air jet flame, consisting of a mixture of 25% methane and 75% air (by volume) issuing from a 7.2 mm diameter inner nozzle. An annular premixed pilot flame (18.2 mm diameter) surrounds the main jet and helps anchor the main flame. The flame database covers conditions from laminar to near blowout conditions, designated Flames A-F in the database.

2.1. *Mixture fraction imaging*

The mixture fraction, defined as the atomic mass fraction of all species originating in the fuel stream, is a key parameter used in modeling turbulent nonpremixed combustion. For the single-point measurements that make up the TNF database, the mixture fraction is obtained by tracking all of the major species, which are determined simultaneously by Raman scattering or (for CO, NO, and OH) by laser-induced fluorescence.[7] Extending this same approach to imaging experiments is infeasible, due to the extremely weak nature of the Raman signal as well as the practical problems associated with the simultaneous imaging of seven or more quantities.

Over a decade ago, Stårner *et al.*,[8] showed that for methane flames of the type being considered here, a reasonable measurement of the mixture fraction could be obtained by imaging only two quantities: the Rayleigh scattering, which depends on the overall number density, and the fuel concentration. Measurements were undertaken by a number of groups,[9-14] but the results suffered from several shortcomings, some related to the problems of accurate measurement of the fuel concentration and others more inherent in the two-scalar approach itself.

In early work, an attempt was made to determine the fuel concentration using fluorescence, either from the fuel or from a fuel tag.[12] Because of the favorable fluorescence scattering cross section, the images had high signal/noise, which is of particular importance for deriving the scalar dissipation. The scalar dissipation derived from mixture fraction gradients will be accurate only if the spatial resolution is commensurate with the smallest flow features and if the noise level is sufficiently low. The main difficulty associated with the fluorescence-based fuel concentration measurements lies in an inherent assumption of the two-scalar technique that the chemistry is simple. Molecules that are commonly used as fuel tags, (e.g., acetone, acetaldehyde, or biacetyl) tend to break down well before the stoichiometric contour, and the mixture fraction derived from these measurements has obvious errors related to the premature disappearance of the tracer (and therefore the fluorescence signal). Simpler molecules, such as methane, are not readily accessible to fluorescence measurements above the vacuum ultraviolet. A slightly different approach for using fluorescence to determine mixture fraction was developed by Sutton and Driscoll.[15] In this work, relatively high levels of NO were seeded into the fuel stream of a carbon monoxide/air flame, with the assumption that the NO behaved as a nonreactive scalar and could thus be used to determine the mixture fraction.

Other work employing the two-scalar approach relied on Raman scattering to determine the fuel concentration. While this allows measurements of relatively simple fuels (e.g., H_2 or CH_4), the extremely weak Raman signal produces noisy images, even when innovative smoothing techniques are used.[16] A technique capable of providing fuel images from simple fuels with improved signal/noise could be expected to improve the accuracy and confidence in the mixture fraction and scalar dissipation images obtained. With this requirement in mind, a new technique for determining the methane concentration with an order of magnitude larger signal was developed and will be briefly described in the next section. Details are given in Fielding *et al.*[17]

2.2. *Polarized/depolarized rayleigh scattering*

In the standard Rayleigh imaging experiment, a vertically polarized laser beam is formed into a sheet that intersects the region of interest in the flow or flame. The elastically-scattered light is imaged onto a detector at a 90-degree angle with respect to the illumination sheet. The vast majority of the scattered light (typically ~98%) has the same vertical polarization as the exciting laser. However, the classical treatment of Rayleigh scattering [e.g., Woodward, 1967 in Ref. 18] predicts that some fraction of polarized incident radiation will become depolarized for scattering objects that are not isotropic (i.e., spherically symmetric). The depolarization ratio, ρ_p, is defined as the ratio of the radiant intensities scattered with polarizations perpendicular ($I_\perp$) and parallel ($I_\parallel$) relative to the polarization of the source (the subscript p denotes a polarized source). The depolarization ratio for a given molecular species is related to the mean value of the polarizability tensor, α, and the anisotropy, γ, as follows:

$$\rho_p = \frac{3\gamma^2}{45\alpha^2 + 4\gamma^2} \tag{1}$$

For spherically symmetric atoms or molecules, the anisotropy is zero – examples include noble gases, or, of interest here, methane. For most other gases important in combustion, depolarization ratios are on the order of a few percent (actual values are dependent on the excitation wavelength, see Fielding *et al.*[17]).

The fact that methane has a significantly different depolarization ratio than the other species present in a flame suggests that measurement of the depolarized Rayleigh signal can be used to gain information about the concentration of methane in the flame (necessary for determining the mixture fraction). While the depolarized Rayleigh signal is roughly a factor of 50 smaller than the polarized component, this is still more than a factor of 10 larger than the Raman scattering signal. (Raman cross sections are typically a factor of 1000 smaller than Rayleigh cross sections.) In a simple two-stream mixing case, if one stream contains methane and the other stream consists of a gas mixture with a non-zero depolarization (e.g., air), then a properly normalized linear combination of polarized and depolarized Rayleigh signals has been shown to be proportional to the methane concentration.[17]

2.3. *Reaction rate imaging*

The ability to image reaction rates within the flame is beneficial for the study of turbulence/chemistry interactions. Several different techniques have been proposed that allow imaging different reaction rates in a flame.[19-21] In the context of the current experiment, it has been shown that a pixel-by-pixel product of appropriately measured OH and CO fluorescence images will yield a scalar field proportional to the forward rate of reaction for $CO + OH \Rightarrow CO_2 + H$.[5] This is the main reaction pathway for the production of CO_2 in flames.

The OH radical is perhaps the most often measured intermediate species in flames and its photophysics are well understood (e.g., quenching rates, saturation behavior, energy transfer mechanisms). OH fluorescence is also quite easily excited and detected using readily available lasers and detectors. CO, on the other hand, presents more of a challenge. The main electronic absorption lines of CO fall in the vacuum ultraviolet. In order to access these transitions in flames, two-photon techniques are most often used. The use of nonlinear two-photon processes in imaging experiments presents several experimental challenges. The excitation scheme requires high intensity in the laser sheet. In addition, the nonlinear relationship of the fluorescence signal to the laser sheet intensity necessitates careful corrections of the images for shot-to-shot intensity fluctuations.

2.4. *Experimental details*

The experimental configuration for simultaneous imaging of temperature and fuel concentration (via polarized/depolarized Rayleigh imaging) and reaction rate (via OH and CO LIF) is shown schematically in Fig. 1. The experiments were performed in the Advanced Imaging Laboratory at Sandia National Laboratory's Combustion Research Facility in Livermore, California. The experiments required five cameras and four lasers.

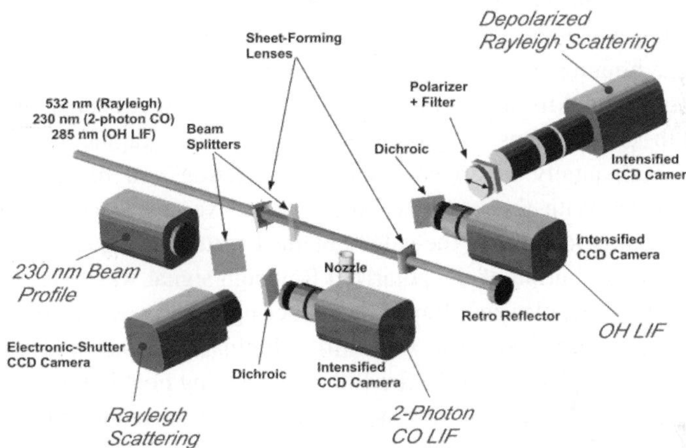

Fig. 1.   Experimental configuration for simultaneous imaging of polarized and depolarized Rayleigh scattering, OH LIF, and two-photon CO LIF.

Rayleigh scattering was excited using the second harmonic from two separate Nd:YAG lasers providing a combined energy of 0.92 J per 10 ns pulse at 532 nm. The beams were spatially overlapped and a cylindrical lens focused the beams into a sheet that intersected the flame. The beams were re-collimated with a second cylindrical lens and reflected back through the flame to provide additional intensity in the imaging sheet. For OH fluorescence, the frequency-doubled output of a Nd:YAG-pumped dye laser (0.25 mJ per 10 ns pulse) was tuned near 285 nm to pump the $Q_1(12)$ transition of the A-X(1,0) band. Fluorescence from CO was probed using a two-photon process. Excitation was provided by a Nd:YAG-pumped optical parametric oscillator (OPO) that was frequency-doubled, generating 22 mJ per 5 ns pulse near 230.1 nm. This produced a two-photon excitation of overlapping transitions in the B-X(0,0) Hopefield-Birge system of CO near the bandhead. The laser was tuned to maximize the CO LIF signal in a laminar flame with the same fuel composition as in the TNF workshop flames (25% methane, 75% air). Both fluorescence lasers were spatially overlapped with the 532 nm Rayleigh beams and focused into a sheet using the same cylindrical lens. Because the wavelengths used to form the illumination sheets differed significantly and because of different divergence characteristics of the lasers, separate beam telescopes were required in the beams before they were combined to ensure that all sheets focused at the same point within the flow. The fluorescence lasers were not retro reflected. The four lasers were fired sequentially over a period of <1 μs to minimize the chance for crosstalk between the four measurements.

The Rayleigh and depolarized Rayleigh signals were detected with separate CCD cameras located on opposite sides of the illumination sheets. The polarized Rayleigh signal was imaged onto an unintensified interline transfer CCD camera (Sensicam, 640 x 512 pixels after 2 x 2 binning) with a projected pixel dimension of 53 x 53 μm^2 and an exposure period of 600 ns bracketing both Nd:YAG lasers. Rayleigh scattering was transmitted through a dichroic beamsplitter and collected with an f/1.4 50-mm focal length camera lens. On the opposite side of the burner, the horizontal component of Rayleigh scattering was imaged onto an intensified CCD camera (Sensicam, 320 x 240 pixels after 2 x 2 binning) with a projected pixel dimension of 90 x 90 μm^2. Rayleigh scattering was transmitted through a dichroic beamsplitter and collected with an f/1.2 85-mm focal length camera lens. A polarizer (B+W photographic circular polarizer) transmitted the horizontally polarized Rayleigh scattering and blocked the vertically polarized component. With this polarizer, the leakage was measured to be $\ell = 0.04\%$, which was significantly lower than the value obtained with a high-quality linear polarizer. Thus the parasitic contribution of the polarized Rayleigh signal was reduced to about 4% of the expected true depolarized Rayleigh scattering, and was neglected. A narrow bandpass filter ($\lambda_{center} = 532$ nm, $\Delta\lambda = 10$ nm), eliminated broadband laser-generated interference. The intensifier was gated for 400 ns bracketing both Nd:YAG laser pulses.

As shown in Fig. 1, both fluorescence signals were separated from the Rayleigh scattering using dichroic beam splitters. The OH fluorescence from the (0,0) and (1,1) bands was reflected by a dichroic beam splitter with a reflective coating from 300-350 nm and imaged onto an intensified CCD camera (Andor Technology, 512 x 512 pixels) with an f/1.8 Cerco quartz camera lens. The projected pixel size was 67 x 67 μm^2. The image intensifier was gated for 400 ns bracketing the dye laser pulse, eliminating any interference from the other lasers. The OH LIF images were corrected for spatial variations in the laser sheet. For this purpose, the beam profile of the dye laser was recorded using acetone LIF excited at 285 nm, which was the same wavelength used for OH LIF.

The CO fluorescence was reflected by a dichroic beam splitter and imaged onto an intensified CCD camera (Andor Technology, 512 x 512 pixels). The imaging system included an f/1.2 camera lens and a narrow bandpass interference filter ($\lambda_{center} = 484$ nm, $\Delta\lambda = 10$ nm), which transmitted fluorescence from the B-A(0,1) transition at 483.5 nm and blocked laser-generated Swan Band emission from C_2^*. The projected pixel size was 71 x 71 μm^2. The image intensifier for the CO LIF camera was gated for 400 ns bracketing the OPO pulse. The average laser beam profile was measured using CO LIF from a dilute mixture of CO in N_2 (0.3% CO by volume). Shot-to-shot fluctuations in the light sheet profile were recorded on an unintensified CCD camera. The OPO beam profile was sampled by a fused silica wedge positioned after the sheet forming optics. Filters placed in front of the beam profile camera transmitted the 230 nm beam while blocking light from the other lasers. Corrections for shot-to-shot fluctuations were important for the CO LIF measurements because of the sensitivity of two-photon CO LIF to variations in beam intensity.

2.5. *Experimental results*

Simultaneous measurements of CO and OH LIF combined with measurements of polarized and depolarized Rayleigh scattering were performed in both a laminar flame ("Flame A", Re = 1100) and a turbulent flame ("Flame D", Re = 22,400) stabilized on the burner described above. For Flame A, detailed single-point measurements[7,22] allow mapping of the two Rayleigh scattering signals and CO LIF onto mixture fraction. The polarized Rayleigh scattering has the best signal/noise and spatial resolution but is a dual-valued function of mixture fraction and varies only weakly near the stoichiometric value. In the current three-scalar approach the CO and difference-Rayleigh measurement were then used to improve on these shortcomings. Fuel-rich regions with values of the mixture fraction greater that 0.5 were identified using the difference-Rayleigh signal. For mixture fraction values between 0.5 and 0.2, CO LIF was able to provide improved sensitivity, while regions with mixture fraction less than 0.2 could be identified by a lack of signal from both difference-Rayleigh and CO LIF.

Images from two separate single-shot measurements are shown in Fig. 2. OH and CO LIF are displayed without corrections for quenching or Boltzmann fraction variations according to their use in obtaining the reaction rate through direct pixel-by-pixel multiplication. Small-scale turbulence is seen near the jet centerline on the right side of the mixture-fraction image. As expected, the reaction zone is a rather thin strip in the overlap of OH and CO near the stoichiometric contour and at the location of highest temperature.

One of the overall goals of these experiments was to add statistical information on scalar dissipation to the turbulent flame database of the TNF workshop. The scalar dissipation, χ, is related to the mixture fraction by $\chi \equiv 2D\, \nabla \xi \cdot \nabla \xi$, where D is the diffusivity. It is clear that the full determination of $\nabla \xi$ (and therefore χ) would require three-dimensional measurements. In a homogeneous flow, a single measured component of $\nabla \xi$ would be sufficient to characterize χ statistically, but the applicability of that assumption to the current case is not clear. Our two-dimensional measurements of ξ allow the comparison of the contributions of two different components of $\nabla \xi$ to the scalar dissipation. Figure 3 shows the average of χ conditioned on ξ obtained from 70 images of "Flame D" at 15 diameters downstream. The contribution of radial gradients is seen to be consistently larger than that of the axial gradients. In addition, an analysis of the correlation coefficient between the radial and axial scalar dissipation ranges from 0.2 – 0.4, depending on the mixture fraction. This makes it clear that for this case, a single component of $\nabla \xi$ is not sufficient to fully characterize χ.

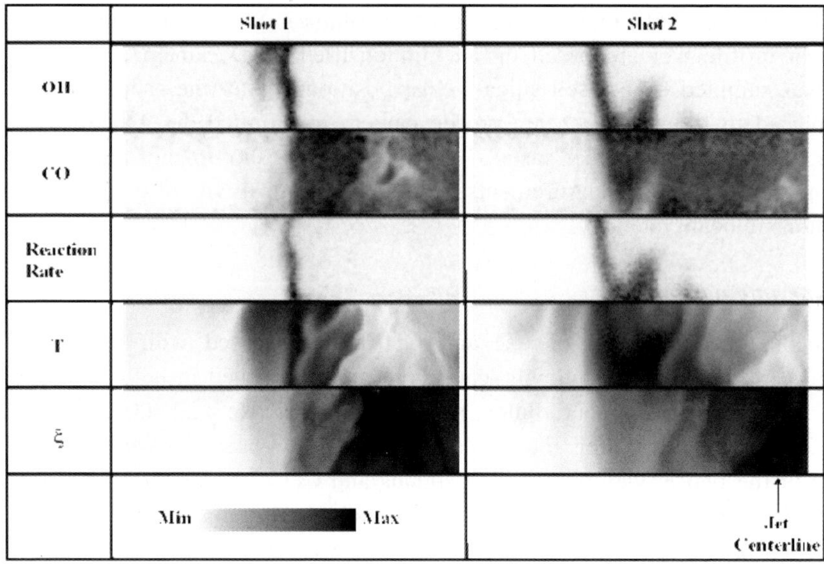

Fig. 2. Simultaneous single-shot measurements of CO LIF, OH LIF, reaction-rate (RR), temperature (T), and mixture fraction (ξ) in a turbulent CH_4/air jet flame. Images are centered at x/d = 15.

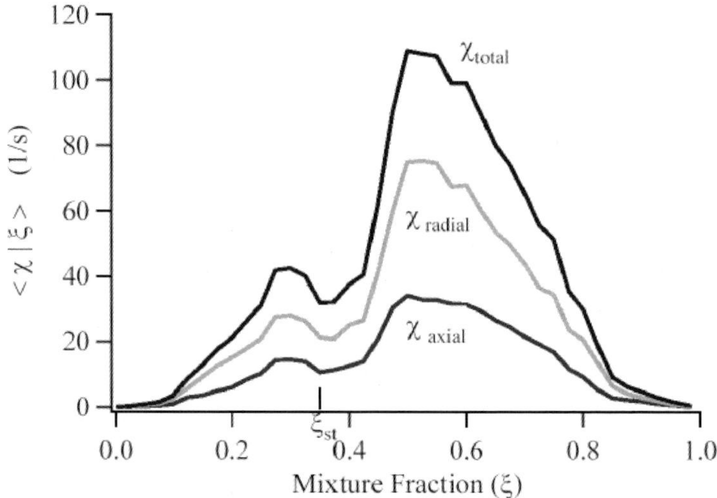

Fig. 3.   Mean value of the axial, radial and total scalar dissipation, conditioned on mixture fraction. Results are from a set of 70 images at x/d = 15.

Although the first mixture fraction imaging work (based on the two scalar approach) was done some time ago, and statistics of the scalar dissipation were published for a few flames, it has been hard to assess the reliability of those measurements since there were no independent checks of the results. That situation has changed with a set of experiments by Barlow and Karpetis.[23] In their work, they have extended the single-point

Raman measurements (that have formed the core of the TNF mixture fraction data set) to allow measurements of all major species along a line. From these results, the mixture fraction can be determined using a more complete formulation. While the spatial resolution for the Raman measurements is currently not as good as for the imaging work and only one component of the gradient is obtained, the availability of their data allows an important check of the imaging results. The magnitude of the conditional scalar dissipation from the imaging work is nearly a factor of two greater than from the line measurements. However, if the pixel volume of the imaging measurements is degraded to more nearly match those of the line measurements, the agreement is remarkable.

The multiscalar imaging work described here is an example of experiments in which laser imaging techniques have provided quantitative data on temperature, mixture fraction, scalar dissipation, and reaction rate in a well-characterized turbulent flame. These quantities are predicted directly by models of turbulent nonpremixed combustion. The experiments required simultaneous measurements of polarized and depolarized Rayleigh scattering as well as laser-induced fluorescence from OH and CO. In order to extend the use of the data beyond flow visualization, the underlying photophysics of the scattering mechanisms must be understood and care must be taken to perform the necessary corrections and calibrations. Performing measurements in the target flames established by the recent set of Turbulent Nonpremixed Flame Workshops allows comparison of the imaging results with those obtained by other single-point or line-imaging techniques.

3. Reconsidering the interaction of experiments and computations

As mentioned previously, early diagnostics work often concentrated on development of the technique. New diagnostic techniques were themselves of interest, and the relevance of the measured quantities to the theoretical and computational models under development at the same time was of secondary importance. However, as laser diagnostic techniques matured in the 1990's, the focus shifted to measuring fundamental quantities that were of more significance to the modelers. As seen in the experiments described above, achieving this important goal can be difficult, often involving simultaneous measurement of many quantities (each with their own noise, uncertainties, and interferences) to get the fundamental quantity of interest to compare with simulations. Errors are often very difficult to estimate in these cases. In the meantime, computational models have become more sophisticated, more quantitative, and more complete. The availability of more complete information allows the possibility of using simulation results to derive predictions of measured signals rather than measuring many quantities to derive a single fundamental quantity. In some cases, comparison of computed and measured signals may be more informative and reliable than the comparison of computed and measured mole fractions, temperatures, mixture fractions and scalar dissipations.

A recent paper by Connelly *et al.*,[24] explores several cases in which the comparison of measured and computed signals is advantageous. While the comparison of measured and computed signals has been done in the past, oftentimes, this is seen as a last resort – when direct measurements of fundamental quantities are impossible due to the complexity of the flow environment or the limited availability of diagnostics. The main premise of the "paradigm shift" suggested by Connelly *et al.*,[24] is that, in designing experiments and choosing diagnostics, experimentalists should consider the accurate measurement of signals that can be calculated with little uncertainty as well as the more conventional approach of making direct comparison with computed results. It may be that deriving predictions of signals from numerical results is better than measuring the fundamental quantities that are normally output by simulations. Quantitative comparisons can be obtained by using appropriate calibration on both sides of the comparison. This approach clearly necessitates that modelers and experimentalists work together. Further, a detailed understanding of signal generation, as well as such effects as quenching, signal interferences, detector characteristics, and spatial resolution on signals is required. In the next section an example of comparing measured and computed signals is presented in which the signals are luminosity images of sooting diffusion flames.

4.　Testing the soot model in computations of laminar flames

A new computational model for predicting soot formation in laminar diffusion flames was introduced by Smooke *et al.*[25] A conventional means of testing the soot submodel involves comparing soot volume fractions – an output of the simulations – with measured soot volume fractions. The latter have been determined via multiple experimental techniques, providing a consistency check and helping improve confidence in each individual measurement. In one experiment, laser extinction measurements were coupled with laser-induced incandescence (LII) measurements to obtain calibrated two-dimensional soot volume fraction images.[25] Results from the comparison are shown for two different flames in Fig. 4. From the figure a number of observations can be made regarding differences in the measured and computed soot levels. Fig. 4(a) shows the computed and measured soot volume fractions for an 80% C_2H_4 flame. In the computed 80% C_2H_4 flame, the soot region extends farther down than in the measurements, and the shape at the tip of the flame is seen to be different. The calculated soot volume fraction is only slightly higher than that measured experimentally and in both computation and experiment, the soot peaks on the wings. In Fig. 4(b), the computed and measured soot volume fractions for a 40% C_2H_4 flame are shown. For this flame, the calculated soot region is again larger than in the experiment, but now, the calculated soot volume fraction is higher than the measurement by a factor of ~2.4. Furthermore, the calculated soot peaks on the wings of the flame, while the measured soot peaks on the centerline. Reconciling these differences will be important in the next generation of the soot modeling efforts.

Another, much simpler, way to compare measurements and computations of sooting flames is by using line-of-sight emission as the point of comparison. Since the flame simulations include detailed information on the temperature and soot volume fraction, this can be combined with Planck's law to get an expected intensity distribution as a function of wavelength as shown in Fig. 5. The calculated two-dimensional intensity profiles can be rotated about the symmetry axis to get a three-dimensional intensity distribution as a function of wavelength. This intensity distribution is then convolved with the measured color filter array profiles for the red, green, and blue filters used on a digital single lens reflex camera, producing a computed soot luminosity image. In addition to the soot luminosity, chemiluminescence from excited-state CH (denoted CH*) is an important contributor to the overall visible flame emission. Since the chemical mechanism used in Smooke *et al.*,[25] does not include CH* chemistry, the spatial distribution of CH is used as a surrogate for CH* chemiluminescence. While these are entirely different chemical species, it has been shown that, at the base of the flame where the CH* luminosity is greatest, the two species are spatially coincident.[26]

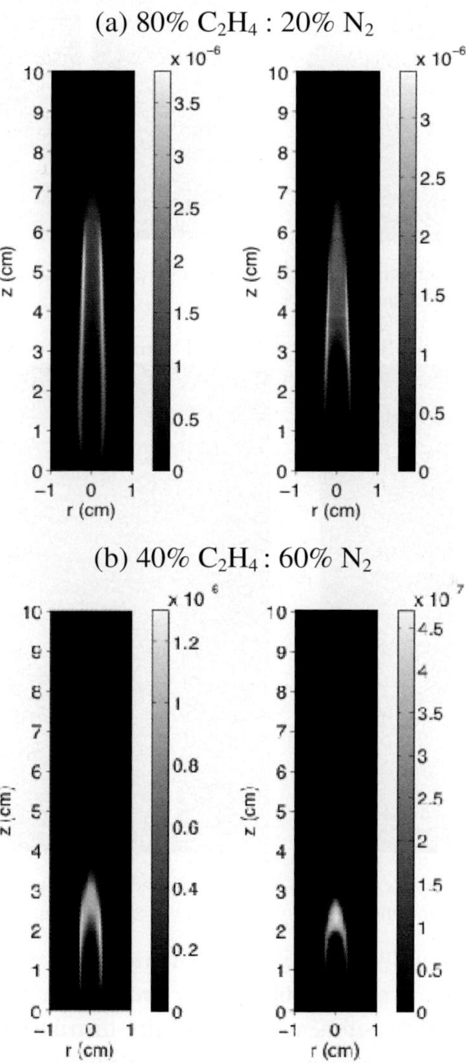

Fig. 4. Computed and (left) and measured (right) soot volume fractions in N_2 diluted C_2H_4 flames. Results for 80% and 40% cases are shown in (a) and (b), respectively.

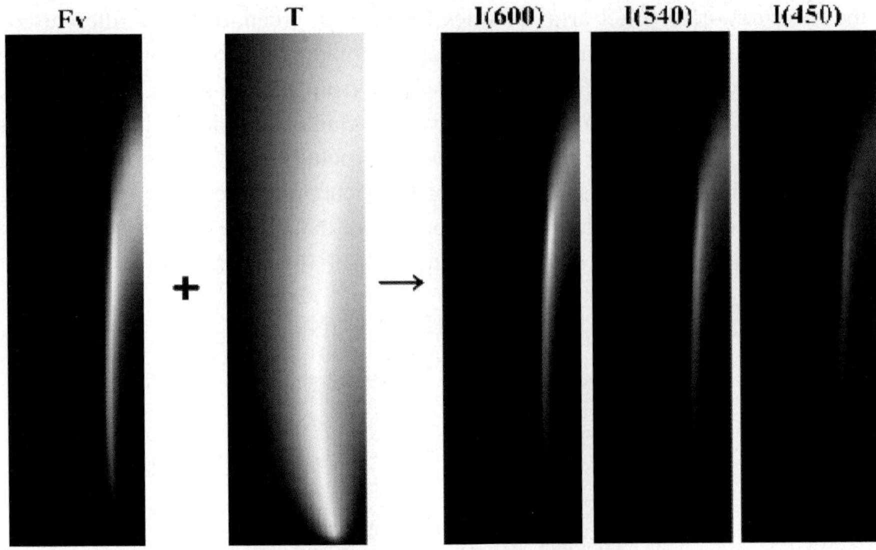

Fig. 5.   The computed soot volume fraction (Fv) and temperature (T) are used with Planck's law to determine the intensity of soot luminosity as a function of wavelength.

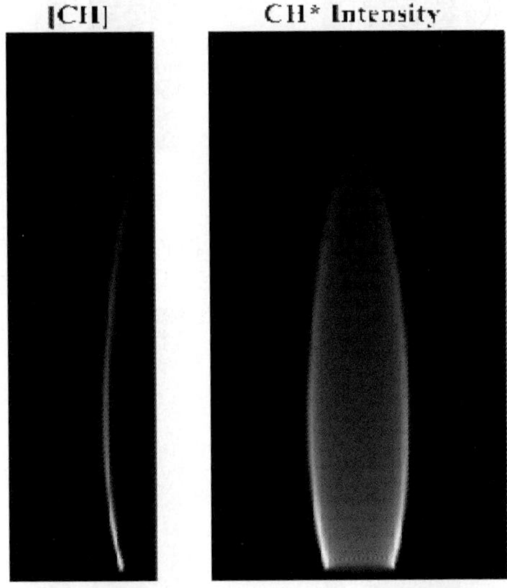

Fig. 6.   The computed CH concentration (left) is rotated around the symmetry axis and combined with the geometric characteristics of the imaging system to get the simulated CH* image (right).

Fig. 6 shows the spatial distribution of the CH radical as well as the simulated CH* chemiluminescence image, which is obtained by rotating the CH distribution about the symmetry axis to get the three-dimensional intensity distribution. The geometry of the imaging system is also taken into account so that the final image includes the effects of the finite distance between the camera and the flame. The simulated CH* luminosity is added to the blue image channel using an empirical scaling constant and the computed

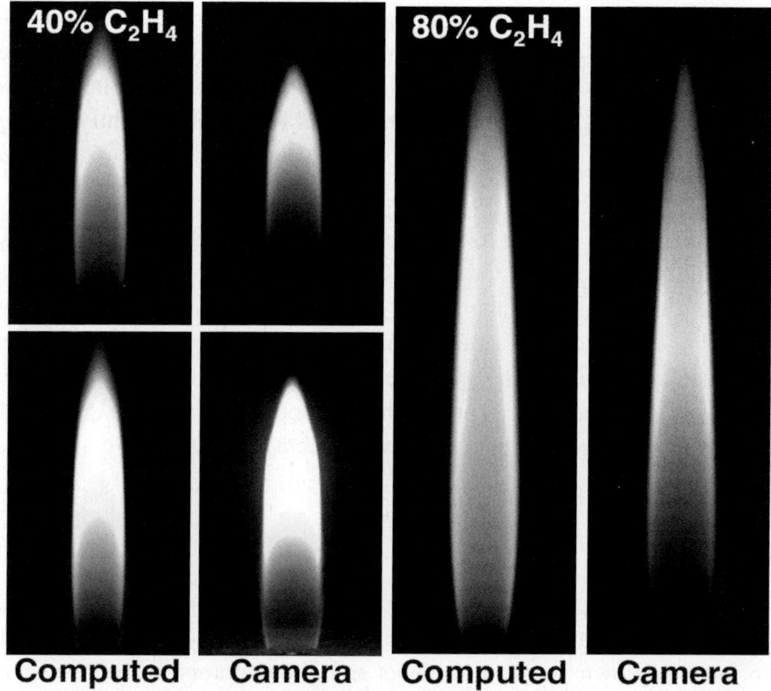

Fig. 7.　Comparison of the computed and measured (camera) flame luminosity images for the moderately sooting flame (40% C_2H_4/60% N_2) and the heavily sooting flame (80% C_2H_4/20% N_2). In the upper leftmost and rightmost pairs of images, the red and green components are partially saturated to look more like standard photographs, which often saturate the most intense parts of an image to stimulate the response of the eye.　In the lower leftmost pair of images, the red and green components are heavily saturated.

luminosity images can then be compared with images taken with the digital camera. If the appropriate reactions are added to the computations, CH* can be computed directly, eliminating the need for the arbitrary scale factor.

Results comparing computed and measured (i.e., photographed) luminosity from the 40% and 80% C_2H_4 flames are shown in Fig. 7. The experimental images were obtained with a Nikon D70 camera. In displaying these images, the red and green components are partially saturated in the display to look more like standard photographs, which often saturate the most intense parts of an image to simulate human visual response. The comparison of the 40% C_2H_4 flame in the lower left of the figure is even more heavily saturated, resulting in images that more closely resemble what is observed when viewing this flame by eye. By comparing the digital camera photos (the output of a very simple diagnostic technique) with the computed luminosity images, it is possible to draw nearly all of the same conclusions noted in the comparisons with the more difficult LII measurements. The computed 40% C_2H_4 flame shows a significantly taller soot-containing region than the camera image, with the distribution of the most intense luminosity more heavily located on the wings of the flame. The computed 80% C_2H_4 flame also shows a larger area of soot that extends further down than the camera image, with the more intense regions covering a larger portion of the flame. A more quantitative look at the intensities in the computed and measured images shows that the relative intensity of the 40% and 80% cases is larger in the computations than in the experiments, once again indicating an over-prediction of the model for the more lightly sooting case.

5. Conclusions

Imaging experiments have proven to be valuable tools for understanding the combustion process. In addition to the insight that comes from visualizing the flow, quantitative data can be provided if the physics of the signal generation process is understood and care is taken to perform the necessary calibrations and corrections. A broad range of light scattering mechanisms can be employed and, as seen from the examples, experimental complexity can vary a great deal. Whether the diagnostics are complex or simple, proper experimental design can provide data that are valuable for quantitative comparisons with model predictions.

Acknowledgments

I would like to express my sincere gratitude to Richard Chang, who, as my Ph.D. thesis advisor, started me along a path to a rewarding career in research and academia. I would also like to thank my own graduate students and collaborators, who are responsible for the work summarized here. The DOE Office of Basic Energy Sciences, the National Science Foundation, NASA, and the AFOSR have provided research support.

References

1. Webber, B. F., Long, M. B. and Chang, R. K. (1979), "2-Dimensional Average Concentration Measurements in a Jet Flow by Raman-Scattering," Applied Physics Letters 35(2): 119-121.
2. Long, M. B., Webber, B. F. and Chang, R. K. (1979), "Instantaneous 2-Dimensional Concentration Measurements in a Jet Flow by Mie Scattering," Applied Physics Letters 34(1): 22-24.
3. Chang, R. K. and Long, M. B. (1982), "Optical Multichannel Detection," Topics in Applied Physics 50: 179-205.
4. Black, P. C. (1977). Development of a parallel-channel spectrographic detector: applications to Raman scattering. Ph.D. Thesis, Yale University.
5. Frank, J. H., Kaiser, S. A. and Long, M. B. (2003), "Reaction-rate, mixture-fraction, and temperature imaging in turbulent methane/air jet flames," Proc. Combust. Inst. 29: 2687-2694.
6. Frank, J. H., Kaiser, S. A. and Long, M. B. (2005), "Multiscalar imaging in partially premixed jet flames with argon dilution," Combust. Flame 143(4): 507-523.
7. Barlow, R. S. and Frank, J. H. (1998), "Effects of Turbulence on Species Mass Fractions in Methane/Air Jet Flames," Proc. Combust. Inst. 27: 1087-1095.
8. Stårner, S. H., Bilger, R. W., Dibble, R. W. and Barlow, R. S. (1992), "Measurements of Conserved Scalars in Turbulent Diffusion Flames," Combustion Science and Technology 86: 223-236.
9. Long, M. B., Frank, J. H., Lyons, K. M., Marran, D. F. and Stårner, S. H. (1993), "A Technique for Mixture Fraction Imaging in Turbulent Nonpremixed Flames," Berichte Der Bunsen-Gesellschaft-Physical Chemistry Chemical Physics 97(12): 1555-1559.

10. Stårner, S. H., Bilger, R. W., Lyons, K. M., Frank, J. H. and Long, M. B. (1994), "Conserved scalar measurements in turbulent diffusion flames by a Raman and Rayleigh ribbon imaging method," Combust. Flame 99(2): 347-54.
11. Stårner, S. H., Bilger, R. W., Long, M. B., Frank, J. H. and Marran, D. F. (1997), "Scalar dissipation measurements in turbulent jet diffusion flames of air diluted methane and hydrogen," Combustion Science and Technology 129(1-6): 141-163.
12. Frank, J. H., Lyons, K. M., Marran, D. F., Long, M. B., Stårner, S. H. and Bilger, R. W. (1994), "Mixture Fraction Imaging in Turbulent Nonpremixed Hydrocarbon Flames," Proc. Combust. Inst. 25: 1159-1166.
13. Kelman, J. B., Stårner, S. H. and Masri, A. R. (1994), "Wide-Field Conserved Scalar Imaging by a Raman and Rayleigh Method," Proc. Combust. Inst. 25: 1141-1147.
14. Kelman, J. B. and Masri, A. R. (1997), "Quantitative technique for imaging mixture fraction, temperature, and the hydroxyl radical in turbulent diffusion flames," Appl. Opt. 36(15): 3506-3514.
15. Sutton, J. A. and Driscoll, J. F. (2006), "A method to simultaneously image two-dimensional mixture fraction, scalar dissipation rate, temperature and fuel consumption rate fields in a turbulent non-premixed jet flame," Experiments in Fluids 41(4): 603-627.
16. Stårner, S. H., Bilger, R. W. and Long, M. B. (1995), "A method for contour-aligned smoothing of joint 2D scalar images in turbulent flames," Combustion Science and Technology 107(1-3): 195-203.
17. Fielding, J., Frank, J. H., Kaiser, S. A., Smooke, M. D. and Long, M. B. (2003), "Polarized/depolarized Rayleigh scattering for determining fuel concentrations in flames," Proc. Combust. Inst. 29: 2703-2709.
18. Woodward, L. A. (1967). Raman Spectroscopy: Theory and Practice. H. A. Szymanski. New York, Plenum Press.
19. Rehm, J. and Paul, P. H. (2000), "Reaction rate imaging," Proc. Combust. Inst. 28.
20. Paul, P. H. and Najm, H. N. (1998), "Planar laser-induced fluorescence imaging of flame heat release rate," Proc. Combust. Inst. 27: 43-50.
21. Böckle, S., Kazenwadel, J., Kunzelmann, T., Shin, D-I., Schulz, C., and Wolfrum, J. (2000), "Simultaneous Single-Shot Laser-Based Imaging of Formaldehyde, OH, and Temperature in Turbulent Flames," Proc. Combust. Inst. 28: 279-286.
22. Frank, J. H. and Barlow, R. S. (1998), "Simultaneous Rayleigh, Raman, and LIF measurements in turbulent premixed methane-air flames," Proc. Combust. Inst. 27: 759-766.
23. Barlow, R. S. and Karpetis, A. N. (2004), "Measurements of scalar variance, scalar dissipation, and length scales in turbulent piloted methane/air jet flames," Flow Turbulence and Combustion 72(2-4): 427-448.
24. Connelly, B. C., Bennett, B. A. V., Smooke, M. D. and Long, M. B. (2009), "A paradigm shift in the interaction of experiments and computations in combustion research," Proc. Combust. Inst. 32(1): 879-886.
25. Smooke, M. D., Long, M. B., Connelly, B. C., Colket, M. B. and Hall, R. J. (2005), "Soot formation in laminar diffusion flames," Combust. Flame 143(4): 613-628.
26. Walsh, K.T., Long, M.B., Tanoff, M.A., and Smooke, M.D. (1998), "Experimental and computational study of CH, CH*, and OH* in an axisymmetric laminar diffusion flame," Proc. Combust. Inst. 27: 615-623.

PART II

LINEAR & NONLINEAR OPTICS IN MICROPARTICLES

CHAPTER 5

ELASTIC AND INELASTIC LIGHT SCATTERING FROM LEVITATED MICROPARTICLES

E. JAMES DAVIS

Department of Chemical Engineering
University of Washington,
Seattle, WA 98195-1750 USA
davis@cheme.washington.edu

Laser light scattering measurements from single levitated solid particles, or microdroplets, have been used to study a wide variety of physical, chemical, and optical properties and processes. A brief overview of the electrodynamic levitation is provided, and numerous applications of light scattering are reviewed. The interpretation of elastic scattering data based on Mie theory has made it possible to measure precisely the size and refractive index of microspheres and to apply such measurements to analyze droplet evaporation and condensation processes for pure components and multicomponent droplets. In addition, the Rayleigh limit of charge on a droplet has been explored using elastic scattering to determine the size and charge corresponding to droplet explosions, when a droplet reaches the Rayleigh limit. Phase function measurements (intensity versus scattering angle) and data on morphology-dependent resonances (MDRs) provide alternative methods to determine the size and refractive index of droplets and microspheres, and the characteristics of coated spheres. The absorption of electromagnetic radiation leads to particle heating, and for volatile droplets this increases the evaporation rate. A review of theory and applications of microsphere heating is provided. Inelastic scattering (Raman and fluorescence scattering) has been used to measure the chemical composition of microparticles and to study gas/particle and gas/droplet chemical reactions such as the uptake of carbon dioxide by reaction with sorbent particles. Inelastic scattering has also been used to follow polymerization of monomer microparticles. It is shown that the temperature of a microparticle can be determined by measuring the ratio of the anti-Stokes to Stokes scattering intensities. Finally, the application of inelastic scattering to the detection of biological particles such as pollen and bacteria is explored.

1. Introduction

The ability to isolate single particles or nanoparticles in the path of one or more laser beams makes it possible to perform light scattering measurements without the complication of the effects of other particles that may be of different sizes and morphologies and different chemical and/or optical properties. This can be achieved by trapping a charged microparticle using an electrodynamic balance (EDB) or ion trap, by optical levitation, or by acoustic levitation.   Of particular interest here is electrodynamic levitation. The EDB is an outgrowth of the quadrupole electric mass filter of Paul and Räther,[1] and the bihyperboloidal electrode configuration introduced by Wuerker et al.[2] has been widely used. One version of the bihyperboloidal EDB is shown in Fig. 1.   As discussed by Davis and Schweiger,[3] many other configurations have been applied to single microparticle studies.

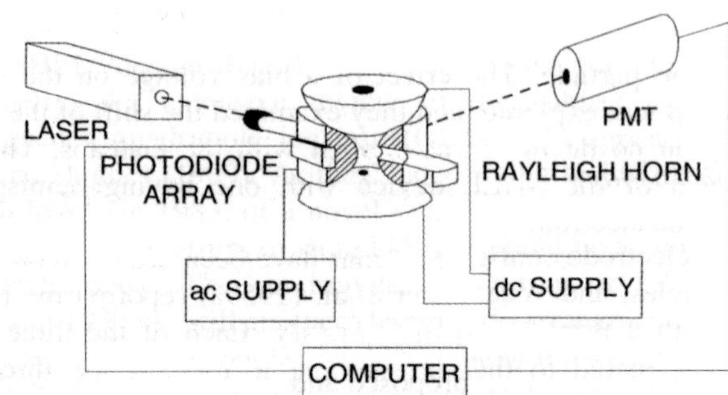

Fig. 1.   The bihyperboloidal EDB equipped for light scattering measurements from the author's laboratory.

In the conventional operation of the EDB shown in Fig. 1 an ac potential is applied to the ring electrode, and a dc field is generated by applying equal but opposite polarity dc potentials to the endcap electrodes. A photodiode array is mounted on the ring electrode shown in the figure to measure the irradiance of the scattered light as a function of angle, and a photomultiplier tube (PMT) is located at right angles to the incident laser beam to record the irradiance at a single angle as a function of time.

The ac field is used to trap the particle in an oscillatory mode, and the dc field serves of the purpose of the Millikan condenser to balance the gravitational force and any other vertical forces on the particle. The appropriate ac trapping frequency depends on the particle mass and drag force on the particle. When the vertical forces are balanced by means of the dc field the particle can be maintained at the midpoint of the EDB provided that the ac potential is not too large. If the ac field is too large, the particle is expelled from the balance chamber. This effect is used to isolate single particles by varying the ac potential and frequency to eliminate undesired particles. The principles and stability characteristics of the EDB were first analyzed by Frickel et al.,[4] and other aspects of the EDB are discussed by Davis and Schweiger.[3]

The Paul trap also led to the quadrupole mass spectrometer, and in 1989 Paul and Dehmelt shared the Nobel Prize in Physics for the development of the ion trap. Dehmelt[5] used radiofrequencies to store atomic ions with a quadrupole trap, and for particles of order 1 μm frequencies of order 100 Hz are appropriate. The dc field strength depends on the charge-to-mass ratio of the particle.

A particularly simple and effective electrode configuration is the double-ring balance shown in Fig. 2. In this case the ac and dc potentials are superposed on the two electrodes to generate the electric fields used to stably trap the particle. One advantage of the double-ring device is that optical access to the particle is quite unrestricted.   In the device shown in Fig. 2 the upper and lower rings are segmented to permit forces in the X,

Y, and Z directions to be exerted on the levitated particle. This is accomplished by applying different dc potentials to the various segments. Zheng[6] called this device the octopole EDB.

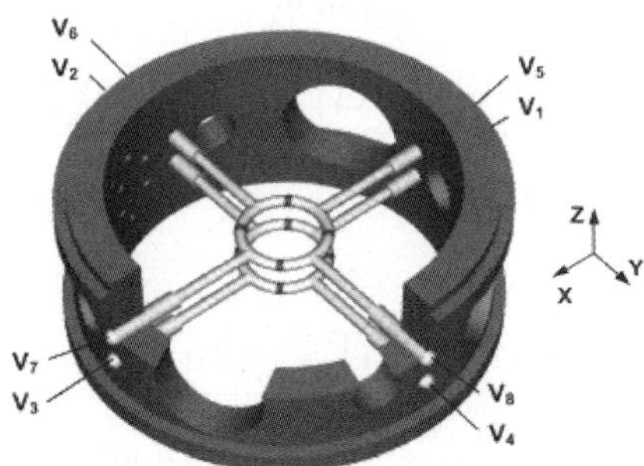

Fig. 2.　The octopole EDB of Zheng[6] reproduced with the permission of Dr. Feng Zheng.

Forces that can be encountered on a small particle include the gravitational force, radiometric forces exerted by a laser beam, aerodynamic drag due to a flowing gas, the thermophoretic force, and the photophoretic force.

2.　Elastic scattering

There has developed an extensive literature dealing with elastic scattering since Lord Rayleigh[7] published his analysis of light scattering by small particles in 1871 and Mie[8] applied electromagnetic theory to describe scattering from spheres in 1908. Mie theory, or Lorenz-Mie-Debye theory as some prefer to call it, is featured in the often-cited treatises of van de Hulst,[9] Kerker,[10] and Bohren and Huffman.[11] Many of the optical effects of small particles of special interest here are found in the book edited by Barber and Chang,[12] so it is not necessary to provide a detailed review of the relevant theory, except where theory is needed to interpret experimental results.

Two aspects of light scattering have been particularly useful for measuring the optical properties of small spheres. The first is the angular dependence of the scattered irradiance, sometimes called the phase function. The second is the existence of morphology-dependent resonances (MDRs) or whispering gallery modes (WGMs). Using the notation of Bohren and Huffman, we can interpret phase functions and MDRs in terms of the scattering functions S_1' and S_2' defined by

$$S_1' = \sum_n \frac{2n+1}{n(n+1)}\left[a_n \pi_n (\cos\theta) + b_n \tau_n (\cos\theta) \right],$$

$$(1)$$

and

$$S_2' = \sum_n \frac{2n+1}{n(n+1)}\left[a_n \tau_n (\cos\theta) + b_n \pi_n (\cos\theta) \right], \qquad (2)$$

where the functions π_n and τ_n are related to the associated Legendre polynomials, $P_n^1 (\cos\theta)$, as follows

$$\pi_n(\cos\theta) = \frac{P_n^1(\cos\theta)}{\sin\theta}, \text{ and } \tau_n(\cos\theta) = \frac{d}{d\theta}\left[P_n^1(\cos\theta)\right]. \tag{3}$$

The scattering coefficients a_n and b_n are given by

$$a_n = \frac{m^2 j_n(mx)\left[xj_n(x)\right]' - j_n(x)\left[mxj_n(mx)\right]'}{m^2 j_n(mx)\left[xh_n^{(1)}(x)\right]' - h_n^{(1)}(x)\left[mxj_n(mx)\right]'}, \tag{4}$$

and

$$b_n = \frac{j_n(mx)\left[xj_n(x)\right]' - j_n(x)\left[mxj_n(mx)\right]'}{j_n(mx)\left[xh_n^{(1)}(x)\right]' - h_n^{(1)}(x)\left[mxj_n(mx)\right]'}. \tag{5}$$

Here, $j_n(x)$ and $h_n^{(1)}$ are spherical Bessel functions of the first and third kind, respectively, $m = N_1/N$, N_1 is the refractive index of the sphere, N is the refractive index of the surrounding medium ($N\approx1$ for air), and x is the dimensionless light scattering size (or size parameter) defined by $x = 2\pi aN/\lambda$, in which λ is the wavelength of the light, a is the particle radius.   The prime denotes differentiation with respect to the variable in parentheses.

The scattered irradiance depends on the polarization of the incident beam.   For incident light polarized parallel to the scattering plane (the lab bench) and perpendicular to the scattering plane the results are, respectively,

$$S_P^{sca} = \frac{1}{k^2 r^2}\left|S_2'\right|^2 S_{inc}, \tag{6}$$

and

$$S_\perp^{sca} = \frac{1}{k^2 r^2}\left|S_1'\right|^2 S_{inc}, \tag{7}$$

in which $k = 2\pi N/\lambda$, r is the distance between the particle and the observer (detector), and S_{inc} is the incident beam irradiance.

The basis of phase function measurements is the angular dependence associated with functions π_n and τ_n, and MDRs correspond to zeros in the denominators of scattering coefficients a_n and b_n.   The resonance modes corresponding to the zeros of the denominator of b_n are called transverse electric or TE modes because there is no radial component of the electric field in this case.   The resonance modes associated with the zeros of the denominator of a_n are called transverse magnetic or TM modes because there is no radial component of the magnetic field.

Using an EDB of the type shown in Fig. 1, Ray et al.[13] recorded the output of the photodiode array as a levitated droplet of 1,8-dibromooctane evaporated.   The unpolarized laser had a wavelength of 632.8 nm, and $m = 1.4977$. By comparing the data with Mie theory, the dimensionless light scattering sizes shown in the figure were obtained.   These correspond to droplet radii of 30.73 µm, 24.54 µm, and 21.39 µm, respectively. In another set of experiments using a rotating photomultiplier tube and a double-ring EDB, they obtained the results presented in Fig. 4 for a dioctyl phthalate (DOP) droplet (m = 1.4860). There is excellent agreement between theory and experiment, and by varying the size and refractive index in the theoretical computations it was determined that the size and refractive index can be determined to 2 parts in 10^4.

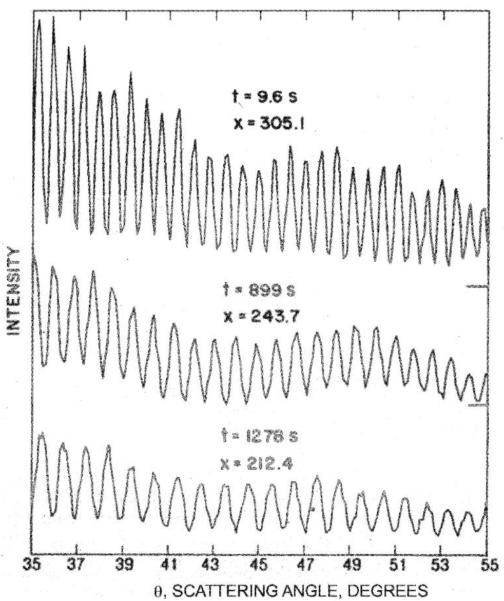

Fig. 3. Phase functions for an evaporating droplet of 1,8-dibromooctane in air reported by Ray et al.[13] (Reproduced with permission of the Optical Society of America).

Numerous applications of MDRs have been explored, particularly for the study of droplet evaporation and/or growth. Richardson et al.[14] recorded the scattered intensity at $\theta = 90°$ as a function of time for single levitated droplets of sulfuric acid evaporating in a vacuum to determine the evaporation rate from which the vapor pressure could be established. For a spherical droplet they used the kinetic theory of gases result to relate the time derivative da/dt to the vapor pressure, that is,

$$\rho \frac{da}{dt} = -0.0583\sqrt{M_i / T} p_i^o , \qquad (8)$$

where ρ is the liquid density (in g/cm^3 using the numerical constant in the equation), M_i is the molecular weight of the evaporating species, p_i^o is its vapor pressure, and T is the droplet temperature.

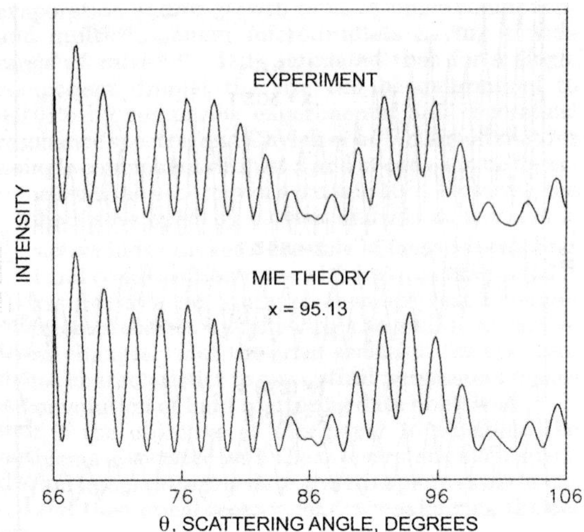

Fig. 4. Experimental and theoretical phase functions for a levitated droplet of dioctyl phthalate from Ray et al.[13] (Reproduced with permission of the Optical Society of America).

Taflin and Davis[15] recorded MDRs to explore droplet evaporation in a carrier gas. For diffusion-controlled, quasi-steady, droplet evaporation in a stagnant gas the change in droplet radius is related to the gas phase binary diffusion coefficient, D_{ij}, by[3]

$$\frac{da^2}{dt} = \frac{2D_{ij}p_i^o(T_s)M_i}{\rho RT} , \tag{9}$$

where i denotes the diffusing species, j denotes the surrounding gas, T_s is the liquid surface temperature, and R is the ideal gas constant. When the surrounding gas is not stagnant but flows around the droplet, the mass rate is increased. Taflin and Davis varied the gas flow rate through the EDB chamber to determine the effects of flow on the evaporation rate.

Allen et al.[16] extended the work of Taflin and Davis using MDR measurements to study binary droplet evaporation. From the measurements of the evaporation rates of several pairs of organic compounds they were able to determine thermodynamic activity coefficients. Additional work on the determination of activity coefficients by means of droplet measurements was reported by Tu et al.[17,18].

Taflin et al.[19] also applied the properties of MDRs to study the Rayleigh limit of charge on droplets. As a charged droplet evaporates in the EDB a point is reached at which the repulsive force of the surface charges overcomes the surface tension force. At this point the droplet ruptures, losing mass and charge in the process. Rayleigh[20] determined that a droplet becomes unstable when its charge reaches a critical value given by

$$q_R = 8\pi\sqrt{\varepsilon_0\gamma a^3} , \tag{10}$$

where ε_0 is the permittivity of the vacuum and γ is the surface tension.

More recently, Li et al.[21] made careful measurements of MDRs, droplet mass, and droplet charge to explore the Rayleigh limit. Figure 5 shows their MDR measurements and computations based on Mie theory for the evaporation of a diethyl phthalate (DEP) droplet. The TE mode spectrum was obtained at scattering angle $\theta = 95.75°$. Not shown in the figure are the simultaneous measurements of the EDB balance voltage that was used to determine the charge loss. There is excellent agreement between the observed and calculated intensities except for the size parameter region $130.153 \leq x \leq 131.4503$. At $t = 4961.5$ s a sudden discontinuity in the MDR record occurred due to the droplet breakup. After breakup the droplet was re-balanced by changing the dc voltage. The levitation voltages before and after breakup were $V^-_{dc,0} = 3.34$ V and $V^+_{dc,0} = 4.22$ V, respectively, and the droplet radii before and after breakup were $a^- = 13.234$ µm and $a^+ = 13.108$ µm, respectively. Based on the known droplet density and the measured radii, the mass loss was 2.86%.

If the only vertical forces on the droplet are the gravitational force and the electrical force due to the dc field of the EDB, a force balance yields

$$C_0 q \frac{V_{dc,0}}{z_0} = -mg ,\qquad(11)$$

in which q is the charge on the droplet having mass m, $2z_0$ is the distance between the dc electrodes, $V_{dc,0}$ is the potential difference between the electrodes in the absence of forces other than gravity, g is the acceleration of gravity, and C_0 is a geometrical constant that accounts for the distortion of the dc field from that of flat plate electrodes. In this case Li et al. reported $C_0 = 0.70\pm0.06$ for their double ring EDB.

Using Eq. (11), the charges before and after breakup were found to be $q^- = 6.825\times10^{-13}$ C and $q^+ = 5.249\times10^{-13}$ C, respectively. Consequently, the charge loss was 23.1%. Comparing the charge at breakup with Eq. (9), the breakup corresponds to 99.8% of the Rayleigh limit.

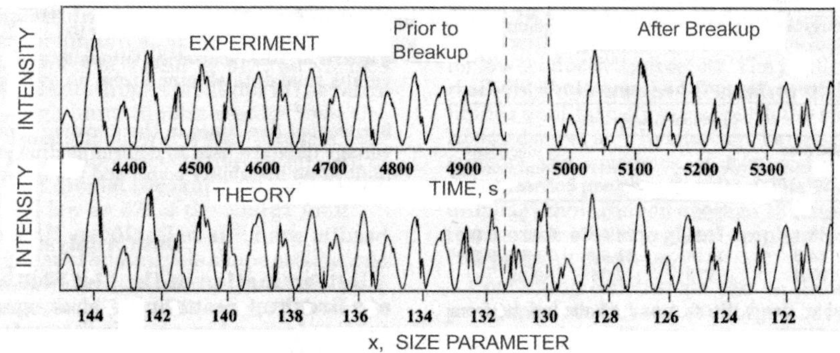

Fig. 5. Experimental and computed resonance spectra obtained by Li et al.[21] for an evaporating DEP droplet for TE mode resonances. (Reproduced with permission of the American Chemical Society).

Aden and Kerker[22] extended Mie theory to concentric spheres, and Ray and his coworkers[23,24] applied the theory to analyze light scattering data from levitated layered droplets. Figure 6 presents some of the results of Tu and Ray[24] for a glycerol droplet coated with dimethyl phthalate (DMP) as the DMP layer evaporated. There is very good agreement between theoretical and experimental results for the TE mode, but less satisfactory agreement was found for the TM mode even when the system parameters were varied to obtain the best agreement. For this system $m_{DMP} = 1.510$ and $m_{glycerol} = 1.4706$.

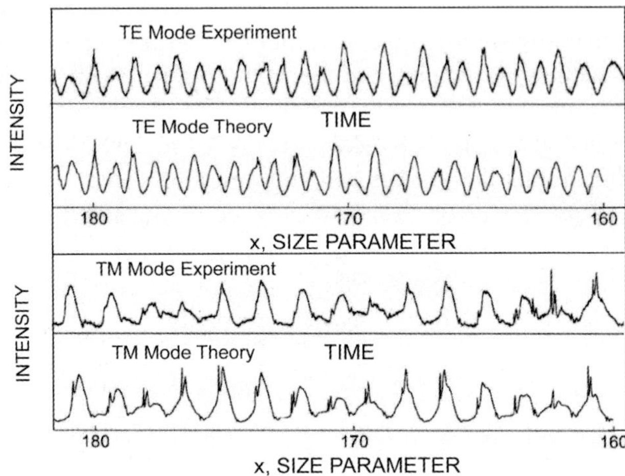

Fig. 6.   Experimental and theoretical resonance spectra for a glycerol sphere having a core radius of 16.10 μm coated with DMP from Tu and Ray.[24] (Reproduced with permission of the Optical Society of America).

3. Radiometric effects

Although the properties of the scattered light are germane to the experiments discussed above, there are several other aspects of the optics of small particles of interest. These include the radiation pressure on a sphere, and the internal electric and magnetic fields. Ashkin and Dziedzic[25] demonstrated optical levitation of small spheres in 1971, and in 1977 they[26] reported observations of optical resonances in the radiation pressure. For a sphere with projected area πa^2 the radiation pressure is given by

$$p_r = \frac{F_r}{\pi a^2} = Q_{pr} \frac{N}{c_o} S_{inc} \, , \tag{12}$$

in which F_r is the force on the sphere due to the radiation, c_o is the velocity of light in a vacuum, and Q_{pr} is the efficiency for radiation pressure related to the extinction and scattering efficiencies, Q_{ext} and Q_{sca}, by

$$Q_{pr} = Q_{ext} - Q_{sca} \langle \cos\theta \rangle , \tag{13}$$

where

$$Q_{ext} = \frac{2}{x^2} \sum_n (2n+1) \, \mathrm{Re}\{a_n + b_n\} , \tag{14}$$

and

$$Q_{sca}\langle\cos\theta\rangle = \frac{4}{x^2}\left[\begin{array}{l}\sum_n \frac{n(n+2)}{n+1}Re\left\{a_n a_{n+1}^* + b_n b_{n+1}^*\right\} \\ +\sum_n \frac{(2n+1)}{n(n+1)}Re\left\{a_n b_n^*\right\}\end{array}\right]. \tag{15}$$

The asterisk denotes the complex conjugate of a scattering coefficient. As MDRs are associated with the coefficients a_n and b_n, it is to be expected that resonances should be observed in the radiation pressure.

Direct measurements of the radiation pressure force on a sphere were made by Allen et al.[27] using an EDB. The light beam was directed upward, so the dc field was used to maintain a droplet at the midplane of the device. In the absence of the laser illumination the force balance on the droplet is given by Eq. (11). When the droplet is illuminated from below the force balance becomes

$$C_0 q \frac{V_{dc}}{z_o} = -mg + F_r. \tag{16}$$

Eliminating the term $C_0 q/z_0$ from the two force balance equations, we obtain the ratio of the radiation pressure force to the droplet weight as follows

$$\frac{F_r}{mg} = \left(1 - \frac{V_{dc}}{V_{dc,0}}\right). \tag{17}$$

Consequently, two voltage measurements are sufficient to determine the force-to-weight ratio. The weight, in turn, was established from knowledge of the droplet density and phase function measurements to determine the droplet radius. Thus, the radiation pressure is given by

$$p_r = \frac{F_r}{\pi a^2} = \frac{4}{3}\rho ag\left(1 - \frac{V_{dc}}{V_{dc,0}}\right). \tag{18}$$

Figure 7 shows the radiation pressure on a 1-octadecene droplet as a function of laser power reported by Allen and her coworkers.[27] The droplet slowly evaporated during the course of the experiment, and the droplet radii are indicated on the figure.

The radiation pressure on a levitated droplet was also studied by Roll et al.[28] using optical levitation rather than EDB levitation. A droplet of dibutyl phthalate (DBP) was suspended at the beam waste of a Gaussian laser beam, and the droplet size was determined by recording and analyzing MDRs. There was excellent agreement between theory and experiment, and resonance effects were clearly observed in the results.

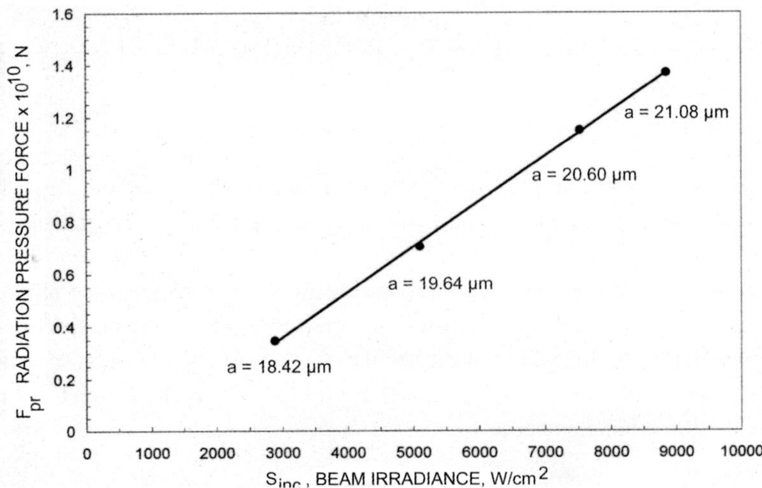

Fig. 7. The radiation pressure on a 1-octadecene droplet as a function of laser power obtained by Allen et al.[27]. (Reproduced with permission of Elsevier Publishing Company).

Mie theory includes the solution for the interior electric and magnetic fields for a sphere. These are relevant to particle heating, the photophoretic force, Raman scattering, and other optical phenomena. The electric vector for the interior has the components

$$E_r = \frac{\cos\phi}{k^2 r^2} E_{inc} \sum_n i^{n-1}(2n+1)d_n \pi_n (\cos\theta)krj_n(kr)\sin\theta , \tag{19}$$

$$E_\theta = \frac{\cos\phi}{kr} E_{inc} \sum_n i^{n-1}\frac{(2n+1)}{n(n+1)}\left[\begin{array}{l}d_n\tau_n(\cos\theta)\left[krj_n(kr)\right]' + \\ ic_n\pi_n(\cos\theta)krj_n(kr)\end{array}\right], \tag{20}$$

and

$$E_\phi = -\frac{\sin\phi}{kr} E_{inc} \sum_n i^{n-1}\frac{(2n+1)}{n(n+1)}\left[\begin{array}{l}d_n\pi_n(\cos\theta)\left[krj_n(kr)\right]' + \\ ic_n\tau_n(\cos\theta)krj_n(kr)\end{array}\right], \tag{21}$$

in which the coefficients c_n and d_n are given by

$$c_n = \frac{j_n(mx)\left[xh_n^{(1)}(x)\right]' - h_n^{(1)}(x)\left[xj_n(x)\right]'}{j_n(mx)\left[xh_n^{(1)}(x)\right]' - h_n^{(1)}(x)\left[mxj_n(mx)\right]'} , \tag{22}$$

and

$$d_n = \frac{mj_n(x)\left[xh_n^{(1)}(x)\right]' - mh_n^{(1)}(x)\left[xj_n(x)\right]'}{m^2 j_n(mx)\left[xh_n^{(1)}(x)\right]' - h_n^{(1)}(x)\left[mxj_n(mx)\right]'} . \tag{23}$$

The denominators of c_n and b_n are the same, and those of d_n and a_n are identical. Consequently, MDRs also affect the interior electromagnetic field, particle heating, and the Raman effect.

In studies of droplet evaporation by measuring optical resonances Allen et al.[27] observed that the evaporation rate of 1-octadecene droplets increased as the laser power increased. For diffusion-controlled, quasi-steady, time-independent droplet and gas temperatures Eq. (9) can be integrated to give

$$a^2 = a_0^2 - \frac{2D_{ij}p_i^0(T_s)M_i}{\rho RT}(t - t_0) , \tag{24}$$

where a_0 is the droplet radius at time t_0. Thus, a plot of a^2 versus time should yield a straight line with slope $S_{ij} = S_{ij} = -2D_{ij}p_i^0(T_s)M_i / \rho RT$. Figure 8 shows the effect of the laser power on the evaporation of 1-octadecene droplets levitated in an EDB using a laser wavelength of 488 nm. The rate of evaporation, indicated by the slopes of the lines, increased significantly as the laser power was increased. Although 1-octadecene weakly absorbs at wavelengths in the visible region of the spectrum, the increase in evaporation rate can be attributed to electromagnetic heating of the droplet.

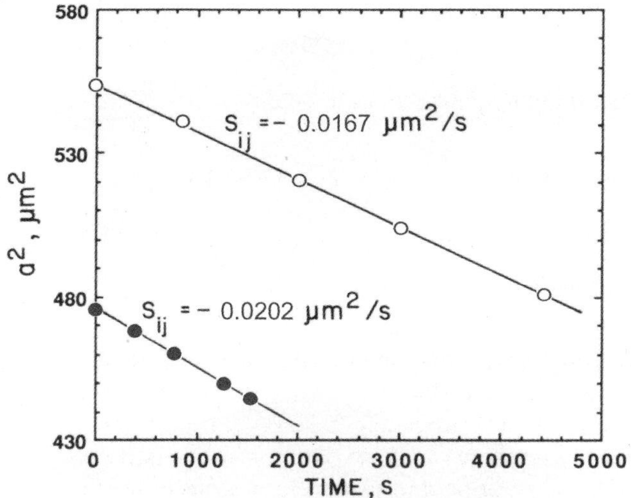

Fig. 8.   Evaporation data for 1-octadecene using a laser irradiance of ~1600 W/cm² (open circles) and 8900 W/cm² (filled circles) from Allen et al.[27]. (Reproduced with permission of Elsevier publishing company).

The heat source function for electromagnetic heating of a sphere is given by

$$Q(r,\theta,\phi) = \frac{1}{2}\sigma\, \mathbf{E}_i \cdot \mathbf{E}_i{}^*, \tag{25}$$

in which σ is the electrical conductivity of the sphere defined by

$$\sigma = \frac{4\pi}{\lambda_{inc}\mu c}\operatorname{Re}\{N_1)\}\operatorname{Im}\{N_1\} . \tag{26}$$

Here μ is the magnetic permeability of the dielectric medium, and c is the velocity of light.

The heat source functions, $Q(r,\theta,0°)$ and $Q(r,\theta,90°)$, for a 1-octadecene droplet corresponding to sizes in the experiments of Allen et al. are presented in Figs. 9a and 9b. For both plots $x = 265.3$ ($a = 20.58$ μm) and the complex refractive index is $N_1 = 1.4521+i1.3\times10^{-7}$. The imaginary component of the refractive index was determined by comparing computed evaporation rates with the measured rates. The computations involved solution of the steady state thermal energy equation subject to appropriate boundary conditions and heat sources such as those shown in Fig. 9. The thermal energy equation that was solved is

$$\kappa\left[\frac{1}{r^2}\frac{\partial}{\partial r}\left(r^2\frac{\partial T}{\partial r}\right) + \frac{1}{r^2\sin\theta}\frac{\partial}{\partial\theta}\left(\sin\theta\frac{\partial T}{\partial\theta}\right) + \frac{1}{r^2\sin^2\theta}\frac{\partial^2 T}{\partial\phi^2}\right] = -Q(r,\theta,\phi), \qquad (27)$$

in which κ is the thermal conductivity of the droplet.

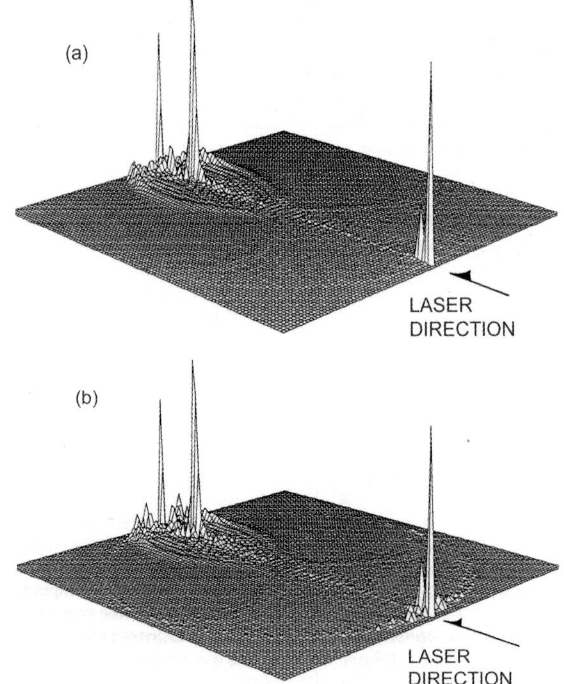

Fig. 9.   Heat source functions computed by Allen et al.[27] (a) $Q(r,\theta,0°)$ and (b) $Q(r,\theta,90°)$ for a 1-octadecene droplet with size parameter $x = 265.3$ and $N_1 = 1.4521+i1.3\times10^{-7}$. (Reproduced with permission of Elsevier Publishing Company).

The heat source functions show narrow peaks just inside the illuminated side of the sphere and farther into the sphere where internal focusing has led to spikes in the heat source. From the solution for the surface temperature of the droplet the vapor pressure was calculated, and the evaporation rate was computed. Table 1 compares the measured and calculated radius changes, da^2/dt, for a 1-octadecene droplet evaporating in nitrogen at a bulk temperature of 297.2 K. Also presented in the table are the results computed assuming uniform heat generation in the droplet. The assumption of uniform Q slightly over predicts the evaporation rate in this case.

Table 1. Evaporation results for a 1-octadecene droplet.[27]

Radius μm	Beam Irradiance W/cm^2	Measured da^2/dt $\mu m^2/s$	Calculated da^2/dt $\mu m^2/s$	da^2/dt for uniform Q $\mu m^2/s$
23.55	~1670	-0.0167	-0.0167	–
18.42	2880	-0.0174	-0.0175	–0.0178
19.64	5100	-0.0182	-0.0183	–0.0189
20.60	7530	-0.0206	-0.0202	–0.0216
21.08	8850	-0.0202	-0.0202	–0.0215

4. Inelastic scattering

Since Thurn and Kiefer[29,30] reported observations of structural resonances (MDRs) in Raman spectra of optically levitated droplets there has developed an extensive literature on the subject of microparticle Raman and fluorescence spectroscopies. Schweiger[31] reviewed the earlier literature (up to 1989) and the principles of Raman scattering. It is sufficient here to provide some of the major results of theory that are needed to interpret experimental results. A conventional pictorial representation of transitions leading to Raman scattering is shown in Fig. 10.

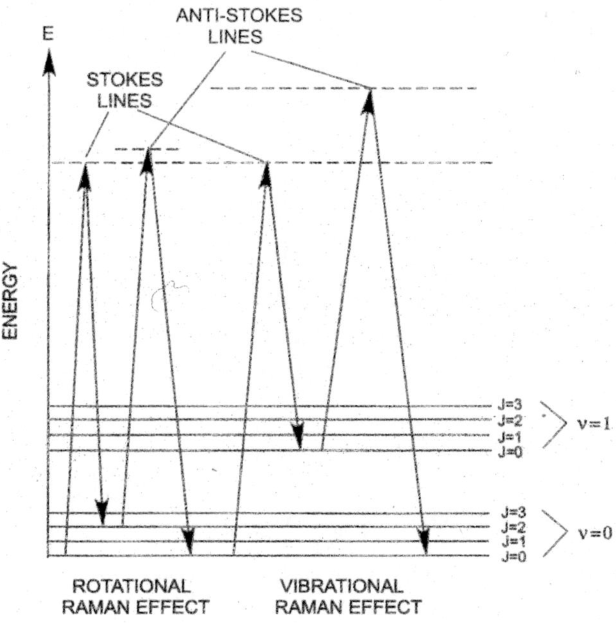

Fig. 10. Transition paths in Raman scattering.

The classical treatment of the Raman effect is attractive in its relative simplicity and correctly accounts for Stokes and anti-Stokes lines, but a quantum mechanical treatment is needed to characterize the intensities of Raman peaks. In the classical treatment a dipole moment is considered to be induced in a molecule by the electric field of the illumination, and the resulting dipole radiation produces the Raman effect. The dipole moment, **p**, is related to the electric vector, **E**, by (see Wilson et al.[32])

$$\mathbf{p} = \alpha\mathbf{E}, \tag{28}$$

in which α is the polarizability tensor.

The electric field associated with the incident wave may be written as

$$\mathbf{E} = E_{inc}\,\mathbf{e}\cos 2\pi\nu_{inc}t\,, \tag{29}$$

in which **e** is a unit vector in the direction of the electric field, E_{inc} is the amplitude of the incident wave, and ν_{inc} is its frequency. The classical treatment leads to terms in the dipole moment involving ν_{inc} and shifted frequencies $(\nu_{inc} - \nu_i)$ and $(\nu_{inc} + \nu_i)$. The first term represents Rayleigh scattering and indicates that the scattered light has the same frequency as the incident wave. The terms that include frequencies $(\nu_{inc} - \nu_i)$ and $(\nu_{inc} + \nu_i)$ represent Stokes scattering and anti-Stokes scattering, respectively.

For a linear oscillator the total power radiated by the *ith* dipole is given by (see Jackson[33])

$$P = \frac{16\pi^4\nu_i^4}{3c^3}\left|\mathbf{p}\right|^2. \tag{30}$$

Consequently, the power of the emitted radiation is proportional to E_{inc}. As the internal electric fields of spheres are affected by MDRs, it can be expected that Raman scattering from spheres will exhibit MDR effects as shown by Thurn and Kiefer.

This classical treatment leads to the conclusion that Stokes and anti-Stokes lines should be given by the ratio[32]

$$\frac{I_{Stokes}}{I_{anti-Stokes}} = \frac{\left(\nu_{inc} - \nu_i\right)^4}{\left(\nu_{inc} + \nu_i\right)^4}. \tag{31}$$

This is not observed experimentally because the population of molecules in the ground vibrational state is much larger than the population in an excited state at low temperatures. The quantum mechanical treatment of the Raman effect yields, for $\exp(h\nu_i/k_BT) \gg 1$, the ratio[3]

$$\frac{I_{Stokes}}{I_{anti-Stokes}} = \frac{\left(\nu_{inc} - \nu_i\right)^4}{\left(\nu_{inc} + \nu_i\right)^4}\exp(h\nu_i / k_BT)\,, \tag{32}$$

where h is Planck's constant, and k_B is the Boltzmann constant.

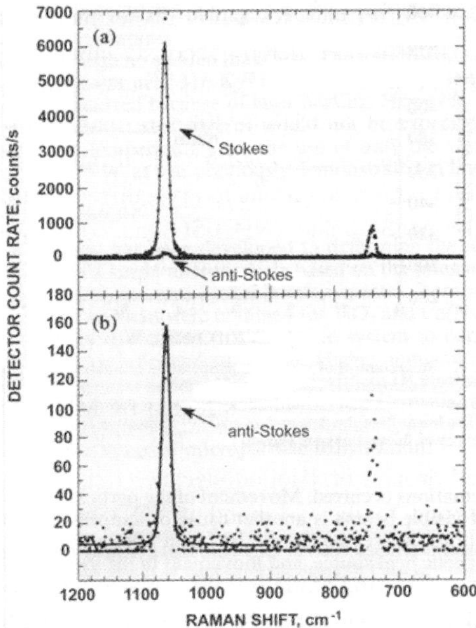

Fig. 11.   Stokes and anti-Stokes spectra obtained by Rassat and Davis[34] for a levitated 52-μm Ca(NO$_3$)$_2$ particle heated with an IR laser (λ = 10.6 μm) and illuminated with a 0.5-W laser beam (λ = 488 nm). The anti-Stokes signals in (a) are expanded in (b). (Reproduced with permission of the Society for Applied Spectroscopy).

5.   Microparticle Raman spectroscopy

Numerous investigators have reported Raman measurements of small levitated particles, mostly Stokes scattering. Figure 12 shows a representative Raman system coupled to a particle levitation system. The levitator can be an acoustic levitator, an optical levitator, or an electrodynamic balance. The sample can also be a droplet chain generated by a vibrating orifice as used for a variety of optical studies cited by Hill and Chang .[35]

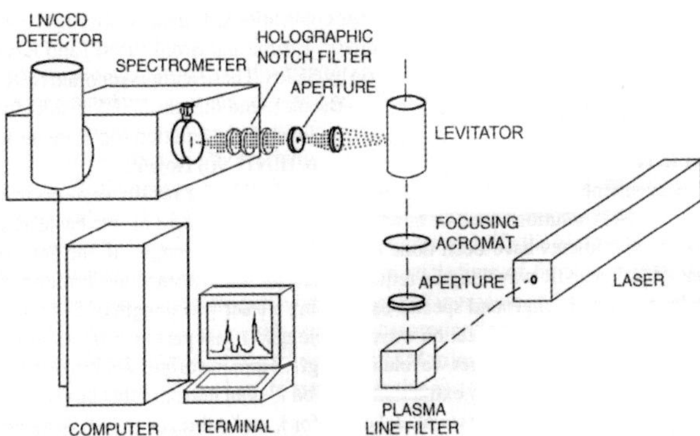

Fig. 12.   A Raman system coupled with a particle levitator from the author's laboratory.

A unique application of droplet levitation experiments is the study of highly supersaturated solutions that would heterogeneously nucleate when in contact with surfaces. Zhang and Chan[36,37] explored the molecular structural properties of water molecules in the hydrated layers of sodium, lithium, magnesium, and other salts using Raman measurements of electrodynamically levitated droplets. The mole ratios of water to salt varied from relatively dilute solutions to crystallized particles. Figure 13 presents some of their results for $MgSO_4$ and $(NH_4)_2SO_4$ aqueous droplets. The $(NH_4)_2SO_4$ spectra clearly show Raman bands at ~458 and ~618 cm^{-1} that correspond to the anti-symmetric stretching modes v_4 of SO_4^{2-}. The $MgSO_4$ spectra show a blue shift from 983 to 1007 cm^{-1} of the peak position of the symmetric stretching mode v_1 of SO_4^{2-} as the mole ratio decreases from 17.29 to 1.54.

Raman scattering from microdroplets is affected by MDRs as first demonstrated by Thurn and Kiefer.[29,30] This is illustrated in Fig. 14 which shows results for the evaporation of a levitated droplet of 1-iodododecane (IDD) obtained by Aardahl et al.[38] The Raman data correspond to the C-I bond in IDD having a wavenumber shift of 511 cm^{-1}, and the two lower traces show the measured and computed input resonances in the elastic scattering for a scattering angle $\theta = 90°$. Narrow peaks in the elastic scattering intensity and even some of the broader bands show up in the inelastic scattering pattern.

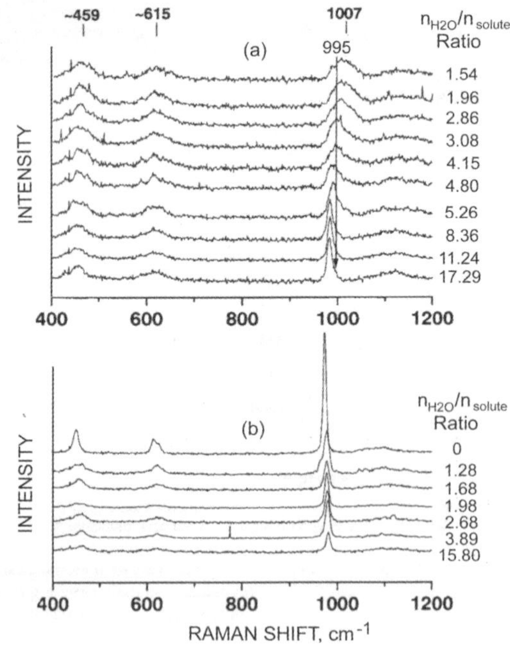

Fig. 13.　Raman spectra for droplets of (a) $MgSO_4$ and (b) $(NH_4)_2SO_4$ for various water-solute ratios from Zhang and Chan.[36] (Reproduced with permission of the American Chemical Society).

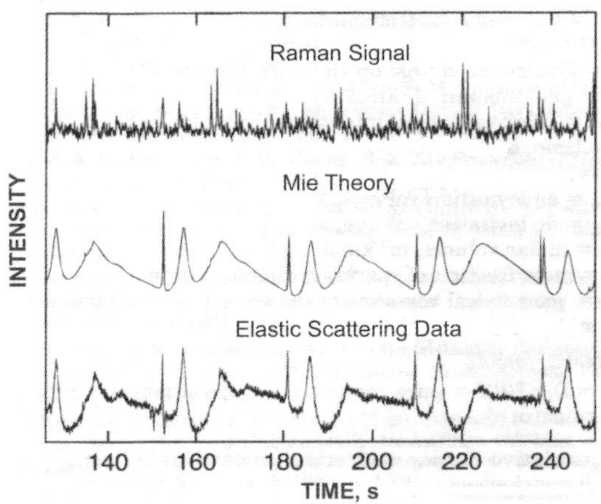

Fig. 14.   The effect of MDRs on Raman scattering reported by Aardahl et al.[38] during the evaporation of a droplet of 1-iodododecane. (Reproduced with permission of the American Chemical Society).

Microparticle Raman spectroscopy has been used to study chemical reactions of levitated particles. The first such studies were reported by Buehler and Davis[39,40] who employed Raman spectroscopy to follow the chemical reaction of bromine with an EDB-levitated droplet of 1-octadecene. Rassat and Davis[41] and Aardahl et al.[42,43] explored reactions between SO_2 and levitated sorbent particles using Raman measurements to follow the chemical reactions, and Li et al.[44] followed the formation of levitated TiO_2-coated microspheres produced by alkoxide chemical reaction by means of Raman measurements. Aardahl et al.[45] also applied Raman spectroscopy to follow the chemical reaction between EDB-levitated particles of Na_2CO_3 and $(NH_4)_2SO_4$ in a humid atmosphere after they collided.

Kiefer and his coworkers[46,47] investigated chemical reactions in optically levitated microdroplets using Raman measurements. These include acid-base reactions[46] (NH_3 reacting with a capric acid/heptanol droplet) and radical polymerization and copolymerization reactions.[47] In addition to Raman measurements they[47] recorded MDRs and radiation pressure. Figure 15 shows a sequence of Raman data for the copolymerization of styrene with unsaturated polyester resin (UP-resin). The peaks at 1596 and 1628 cm^{-1} correspond to aromatic C=C and olefinic C=C vibrations of styrene, respectively, and the peak at 1655 cm^{-1} is attributed to the C=C double bond of UP-resin. The peak at 1628 cm^{-1} is seen to disappear as polymerization proceeds. The peaks labeled with asterisks are due to MDRs. The MDR effects shown in Figs. 14 and 15 indicate the importance of taking them into account in the interpretation of Raman data for microspheres because the peaks associated with MDRs do not correspond to additional vibrational bonds in the Raman spectrum.

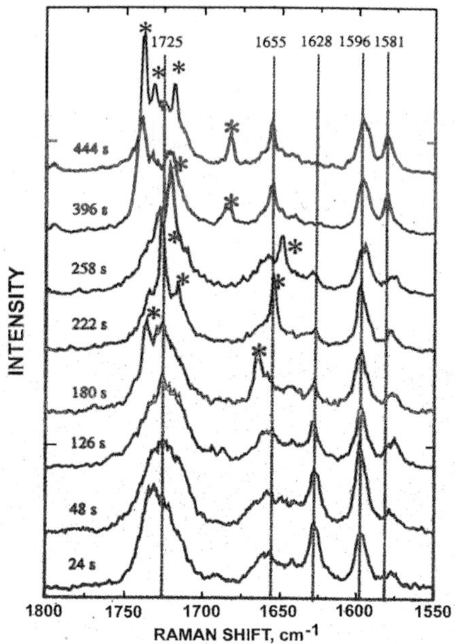

Fig. 15.   A sequence of Raman spectra obtained by Musick et al.[47] during the copolymerization of an optically levitated microdroplet consisting of styrene and UP-resin. (Reproduced with permission of the Society for Applied Spectroscopy).

In work closely related to Raman measurements Bhanti and Ray[48] applied fluorescence scattering to measure instantaneous rates of photochemical reactions in an EDB-levitated droplet.   The technique was used to explore the *trans*-to-*cis* photoisomerization of thioindigo dye in a solution of the dye in silicone oil.   The apparatus used is similar to that shown in Fig. 12.   An additional study of photochemical reaction by means of fluorescence measurements of a levitated particle was carried out by Ward et al.[49] who used the fluorophore auramine-O (dimethylaminodiphenylamine hydrochloride) to track the polymerization of acrylamide monomer. Auramine-O does not fluoresce in low-viscosity media, but exhibits increasing fluorescence as the viscosity increases.   By recording the fluorescence intensity at a wavelength of 515 nm as polymerization proceeded they followed the process. They also recorded phase functions to determine the microsphere size as a function of time. Chang and his associates[50-53] developed instrumentation and techniques to detect airborne biological particles and organic carbon by means of measuring fluorescence spectra of single particles in a train or sequence of particles.

Hydrated lime, $Ca(OH)_2$, is used to remove SO_2 from stack gases in coal-fired power plants. The hydrated lime also reacts with CO_2, so Chen et al.[54] examined the chemical reaction between CO_2 and EDB-levitated particles of $Ca(OH)_2$. Although the uptake of CO_2 by hydrated lime involves a sequence of reactions, the overall reaction is

$$Ca(OH)_2(s) + CO_2(g) \rightarrow CaCO_3(s) + H_2O(ads).$$

(33)

At low relative humidity values (RH < 70%) no reaction was observed, but at high RH the reaction proceeded with approximately 60% conversion of the $Ca(OH)_2$ indicated by simultaneous gravimetric analysis based on levitation voltage measurements [see Eq. (10)].

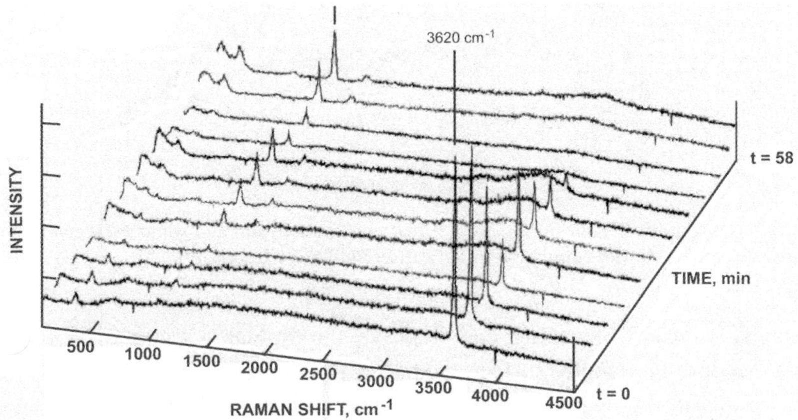

Fig. 16.   A sequence of Raman spectra for the reaction between a hydrated lime particle and CO_2 obtained for a relative humidity of 88% from Chen et al.[54]   (Aerosol Science & Technology: "Carbon dioxide uptake by hydrated lime aerosol particles," Aerosol Sci. Technol. 38, 588-597. Copyright 2004. Mount Laurel, NJ. Reprinted with permission.)

Figure 16 presents a sequence of Raman spectra obtained with RH = 88%. The strong Raman peak at 3620 cm^{-1} was attributed to the O-H bond in $Ca(OH)_2$, and the peak at 1042 cm^{-1} was shown to be associated with $CaCO_3$. The figure shows the decrease in the 3620 cm^{-1} peak with time as the peak at 1042 cm^{-1} increased until no further change was observed.

6. Biological particle detection

As indicated by the studies of Chang and his coworkers[50-53] biological materials usually exhibit strong fluorescence.   Davis and his coworkers[55-57] explored the use of Raman spectroscopy to characterize pollen and bacteria. The fluorescence of all of the pollen studied by Laucks et al.,[55] including cottonwood, cypress, Kentucky blue grass, Johnson grass, paper mulberry, ragweed, and sweet vernal grass, masked the Raman peaks unless either oxygen bleaching or photochemical bleaching was carried out. Figure 17 shows the effect of oxygen bleaching on the spectrum of sweet vernal grass pollen. In the presence of air, peaks at 1550 and 2330 cm^{-1} are clearly seen, and there is a broad base signal due to fluorescence. The peaks at 1550 and 2330 cm^{-1} correspond to oxygen and nitrogen, respectively. When oxygen flowed through the balance chamber, displacing the air, the broad base was reduced, and the nitrogen peak did not appear. In both cases there is a strong band centered at 2920 cm^{-1} that is associated with C-H bonds, but weaker Raman peaks are masked.

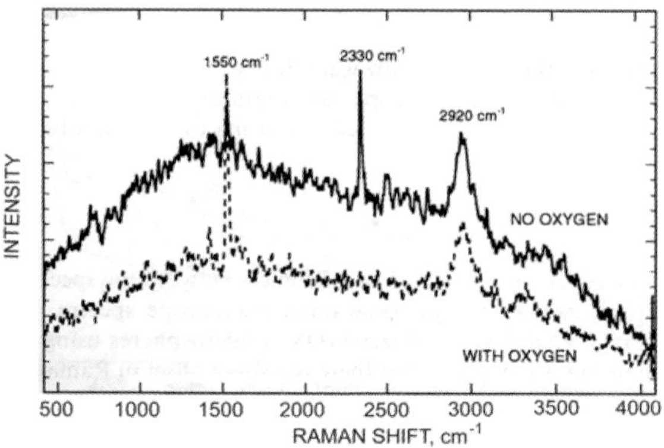

Fig. 17.   Raman spectra for EDB-levitated sweet vernal grass pollen with and without oxygen bleaching from Laucks et al.[55] (Reproduced with permission from Elsevier Publishing Company).

Although extensive bleaching can improve the Raman spectra of pollen, a more effective procedure is to apply surface-enhanced Raman spectroscopy (SERS). Since the discovery of SERS by Jeanmaire and Van Duyne[58] the phenomenon has been used extensively in biological applications. A remarkable increase in the intensity of the inelastically scattered light (by a factor of 10^6-10^{14}) can occur when a molecule is in contact with or very near the surface of a nanoscale metal such as silver or gold.

Sengupta et al.[56,57] applied SERS to the characterization of biological particles by replacing the levitation system shown in Fig. 12 with a cuvette containing an aqueous suspension of the biological material and nanocolloidal silver particles. The silver particles attached to the surfaces of the biological samples to produce significant enhancement of the Raman signal as well as suppression of the fluorescence. Figure 18 presents a sample of their data for cottonwood pollen and *E. coli* bacteria. There are significant differences between spectra for pollen and bacteria. But for pollen in the same family, such as sweet vernal grass and Kentucky bluegrass of the family Poaceae, and for bacteria such as *E. coli* and *Enterococcus sp.* the differences are subtle. In this case, quantitative differences can be determined using more sophisticated analysis of the data such as discriminant function analysis and hierarchical cluster analysis as used by Jarvis and Goodacre.[59]

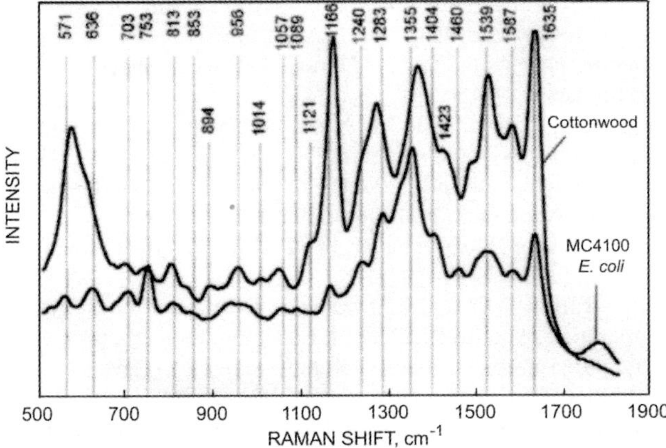

Fig. 18.   Raman spectra of cottonwood pollen and E. coli bacteria obtained by Sengupta et al.[56] using SERS. (Reproduced with permission of the Society for Applied Spectroscopy).

Acknowledgments

This contribution to the festschrift in honor of Professor Richard Chang was stimulated by discussions we had at the 7th International Congress on Optical Particle Characterization in Kyoto, Japan in July 2004. Professor Chang and I were privileged to give plenary lectures at that meeting, and we had ample opportunity to discuss our research. I was also honored by Professor Chang to serve as outside reader of the doctoral dissertation on stimulated Raman scattering in microdroplets of Dr. Ali Serpengüzel in 1992, and I thank Dr. Serpengüzel for inviting me to contribute to this festschrift.

References

1. W. Paul and M. Räther, "Der elektrische Massenfilter," Z. Phys. 140, 262-273 (1955).
2. R. F. Wuerker, H. Shelton and R. V. Langmuir, "Electrodynamic containment of charged particles," J. Appl. Phys. 30, 3422-349 (1959).
3. J. Davis and G. Schweiger, "The Airborne Microparticle," Springer Verlag, Heidelberg, 2002.
4. R. H. Frickel, R. E. Shaffer and J. B. Stamatoff, "Chambers for the electrodynamic containment of charged aerosol particles," Report No. ARCSL-TR-77041, Chemical Systems Laboratory, Aberdeen Proving Ground, MD (1978).
5. H. Dehmelt, "Radiofrequency spectroscopy of stored ions," I. Storage, Adv. At. Mol. Phys. 3, 53-72 (1967).
6. Zheng, "Thermophoretic force measurements of spherical and non-spherical particles," Ph.D. Dissertation, University of Washington (2000).
7. Lord Rayleigh, "On the scattering of light by small particles," Phil. Mag. 41, 447-454 (1871).
8. Mie, "Beiträge zur Optik trüber Medien, speziell kolloidaler Metallösungen," Ann. Phys. 25, 77-445 (1908).
9. C. van de Hulst, "Light Scattering by Small Particles," John Wiley & Sons, New York, 1957.
10. M. Kerker, "The Scattering of Light and Other Electromagnetic Radiation," Academic Press, London, 1969.

11. C. F. Bohren and D. R. Huffman, "Absorption and Scattering of Light by Small Particles," John Wiley & Sons, New York, 1983.

12. P. W. Barber and R. K. Chang (eds.), "Optical Effects Associated with Small Particles," World Scientific, Singapore, 1988.

13. K. Ray, A. Souryi, E. J. Davis and T. M. Allen, "Precision of light scattering techniques for measuring optical parameters of microspheres," Appl. Opt. 30, 3974-3983 (1991).

14. B. Richardson, R. L. Hightower and A. L. Pigg, "Optical measurement of the evaporation of sulfuric acid droplets," Appl. Opt. 25, 1226-1229 (1986).

15. C. Taflin and E. J. Davis, "Mass transfer from a sphere at intermediate Peclet numbers," Chem. Eng. Commun. 55, 199-210 (1987).

16. T. M. Allen, D. C. Taflin, D.C. and E. J. Davis, "Determination of activity coefficients via microdroplet evaporation experiments," Ind. Eng. Chem. Res. 29, 682-690 (1990).

17. Tu and A. K. Ray, "Measurement of activity coefficients from unsteady state evaporation and growth of microdroplets," Chem. Eng. Commun. 192, 474-498 (2005).

18. H. Tu and A. K. Ray, "Validation of activity coefficient models using resonances in light scattering from evaporating multicomponent droplets," Ind. Eng. Chem. Res. 47, 3963-3973 (2008).

19. D. C. Taflin, T. L. Ward and E. J. Davis, "Electrified droplet fission and the Rayleigh limit," Langmuir 5, 376-384 (1989).

20. Lord Rayleigh, "On the equilibrium of liquid conducting masses charged with electricity," Phil. Mag. 14, 184-186 (1882).

21. K.-Y. Li, H. Tu and A. K. Ray, "Charge limits on droplets during evaporation," Langmuir 21, 3786-3794 (2005).

22. L. Aden and M. Kerker, "Scattering of electromagnetic waves from two concentric spheres," J. Appl. Phys. 22, 1242-1246 (1951).

23. H. Tu and A. K. Ray, "Investigation of concentrically and eccentrically layered droplets by light scattering," Appl. Opt. 45, 7652-7656 (2006).

24. K. Ray and R. Nandakumar, "Simultaneous determination of size and wavelength-dependent refractive indices of thin-layered droplets from optical resonances," Appl. Opt. 34, 7759-7770 (1995).

25. Ashkin and J. M. Dziedzic, "Optical levitation of micron sized spheres," Appl. Phys. Lett. 19, 283-285 (1971).

26. Ashkin and J. M. Dziedzic, "Observations of resonance in the radiation pressure on dielectric spheres," Phys. Rev. Lett. 38, 1351-1354 (1977).

27. T. M. Allen, M. F. Buehler and E. J. Davis, "Radiometric effects on absorbing microspheres," J. Colloid Interface Sci. 142, 343-356 (1991).

28. G. Roll, T. Kaiser and G. Schweiger, "Optical trap sedimentation cell – A new technique for the sizing of microparticles," J. Aerosol Sci. 27, 105-117 (1996).

29. R. Thurn and W. Kiefer, "Raman-microsampling technique applying levitation by radiation pressure," Appl. Spectrosc. 30, 78-83 (1984).

30. R. Thurn and W. Kiefer, "Observations of structural resonances in the Raman spectra of optically levitated dielectric microspheres," Appl. Opt. 24, 1515-1519 (1984).

31. G. Schweiger, "Raman scattering on single aerosol particles and on flowing aerosols," J. Aerosol Sci. 21, 483-509 (1990).

32. B. Wilson, Jr., J. C. Decius and P. Cross, "Molecular Vibrations, The Theory of Infrared and Raman Vibrational Spectra," Dover Publications, New York, 1955.

33. D. Jackson, "Classical Electrodynamics," 2nd edition, Wiley, New York, 1975.

34. S. D. Rassat and E. J. Davis, "Temperature measurement of single levitated microparticles using Stokes/anti-Stokes Raman intensity ratios," Appl. Spectrosc. 48, 1498-1505 (1994).

35. S. C. Hill and R. K. Chang, "Nonlinear Optics in Droplets," in "Studies in Classical and Quantum Nonlinear Optics," O. Keller (ed.), Nova Science Publishers, Commack, New York, 1995.

36. Y.-H. Zhang and C. K. Chan, "Study of contact ion pairs of supersaturated magnesium sulfate solutions using Raman scattering of levitated single droplets," J. Phys. Chem. A 104, 9191-9196 (2000).

37. Y.-H. Zhang and C. K. Chan, "Observations of water monomers in supersaturated NaClO4, LiClO4, and $Mg(ClO_4)_2$ droplets using Raman spectroscopy," J. Phys. Chem. A 107, 5956-5962 (2003).

38. L. Aardahl, W. R. Foss and E. J. Davis, "Elastic and inelastic light scattering from distilling microdroplets for thermodynamics studies," Ind. Eng. Chem. Res. 35, 2834-2841 (1996).

39. J. Davis and M. F. Buehler, "Chemical reactions with single microparticles," Mat. Res. Soc. Bulletin 15, 26-33 (1990).

40. M. F. Buehler and E. J. Davis, "A study of gas/aerosol chemical reactions by microdroplet Raman spectroscopy: the bromine/1-octadecene reaction," Colloids Surfaces 79, 137-149 (1993).

41. S. D. Rassat and E. J. Davis, "Chemical reaction between sulfur dioxide and a calcium oxide aerosol particle," J. Aerosol Sci. 23, 765-780 (1992).

42. L. Aardahl and E. J. Davis, "Gas/aerosol chemical reactions in the NaOH-SO2-H2O system," Appl. Spectrosc. 50, 71-77 (1996).

43. L. Aardahl and E. J. Davis, "Raman spectroscopy studies of reactions between sulfur dioxide and microparticles of hydroxides," Mat. Res. Soc. Symp. Proc. 432, 47-53 (1997).

44. W. Li, S. D. Rassat, W. R. Foss and E. J. Davis, "Formation and properties of aerocolloidal TiO_2-coated microspheres produced by alkoxide droplet reaction," J Colloid Interface Sci 162, 267-278 (1994).

45. L. Aardahl, J. F. Widmann and E. J. Davis, "Raman analysis of chemical reactions resulting from the collision of micrometer-sized particles," Appl. Spectrosc. 52, 47-54 (1998).

46. M. Trunk, J. Popp, M. Lankers and W. Kiefer, "Microchemistry: time dependence of an acid-base reaction in a single optically levitated microdroplet," Chem. Phys. Lett. 264, 233-237 (1997).

47. Musick, J. Popp, M. Trunk and W. Kiefer, "Investigations of radical polymerization and copolymerization reactions in optically levitated microdroplets by simultaneous Raman spectroscopy, Mie scattering, and radiation pressure measurements," Appl. Spectrosc. 52, 692-701 (1998).

48. Bhanti and A. K. Ray, "In situ measurement of photochemical reactions in microdroplets," J. Aerosol Sci. 30 279-288 (1999).

49. T. L. Ward, S. H. Zhang, T. Allen and E. J. Davis, Photochemical Polymerization of Acrylamide Aerosol Particles, J. Colloid Interface Sci. 118, 343-355 (1987).

50. P. Nachman, G. Chen, R. G. Pinnick, S. C. Hill, R. K. Chang, M. W. Mayo and G. L. Fernandez, "Conditional-sampling spectrograph detection system for fluorescence measurements of individual airborne biological particles," Appl. Opt. 35, 1069-1076 (1996).

51. R. G. Pinnick, S. C. Hill, P. Nachman, G. Videen, G. Chen and R. K. Chang, "Aerosol fluorescence spectrum analyzer for rapid measurement of single micrometer-sized airborne biological particles," Aerosol Sci. Technol. 28, 95-104 (1998).

52. N. F. Fell Jr., R. G. Pinnick, S. C. Hill, G. Videen, S. Niles, R. K. Chang, S. Holler, Y. Pan, J. R. Bottiger and B. V. Bronk, "Concentration, size, and excitation power effects on fluorescence from microdroplets and microparticles containing tryptophan and bacteria," Proc. SPIE 3533, 52-63 (1998).

53. R. G. Pinnick, S. C. Hill, Y.-L. Pan and R. K. Chang, "Fluorescence spectra of atmospheric aerosol at Adelphi, Maryland, USA: measurement and classification of single particles containing organic carbon," Atmos. Environ. 38, 1657-1672 (2004).

54. Chen, M. L. Laucks and E. J. Davis, "Carbon dioxide uptake by hydrated lime aerosol particles," Aerosol Sci. Technol. 38, 588-597 (2004).

55. L. Laucks, G. Roll, G. Schweiger and E. J. Davis, "Physical and chemical (Raman) characterization of bioaerosols - pollen," J. Aerosol Sci. 31, 307-319 (2000).

56. Sengupta, M. L. Laucks and E. J. Davis, "Surface-enhanced Raman spectroscopy of bacteria and pollen," Appl. Spectrosc. 59, 1016-1023 (2005).

57. Sengupta, M. L. Laucks, N. Dildine, E. Drapala and E. J. Davis, "Bioaerosol characterization by surface-enhanced Raman spectroscopy (SERS)," J. Aerosol Sci. 36, 651-664 (2005).

58. L. Jeanmaire and R. Van Duye, "Surface Raman spectro-electrochemistry. Part I. Heterocyclic, aromatic, and aliphatic amines adsorbed on the anodized silver electrode," J. Electroanal. Chem. Interfacial Electrochem. 84, 1-20 (1977).

59. R. M. Jarvis and R. Goodacre, "Discrimination of bacteria using surface-enhanced Raman spectroscopy," Anal. Chem. 76, 40-47 (2004).

CHAPTER 6

PHYSICAL CHEMISTRY AND BIOPHYSICS OF
SINGLE TRAPPED MICROPARTICLES

CLAUDIU DEM[1], MICHAEL SCHMITT[1], WOLFGANG KIEFER[2] and JÜRGEN POPP[1,3]

[1]Institute of Physical Chemistry, Friedrich-Schiller University Jena,
Helmholtzweg 4, D-07743 Jena, Germany
[2]Institute of Physical Chemistry, University Würzburg, Am Hubland,
D-97074 Würzburg, Germany
[3]Institute of Photonic Technology, Jena,
Albert Einstein Strasse 9, D-07745 Jena, Germany
juergen.popp@ipht-jena.de

Microparticles, particularly in the form of spheres and cylinders with radii larger than the wavelength of light, as well as coated gas bubbles, are at the center of various fields of study that include linear and nonlinear optics, combustion diagnostics, fuel dynamics, colloid chemistry, atmospheric science, telecommunications, and pulmonary medicine. The spectroscopy of single microparticles is feasible nowadays due to the development of various optical and electromagnetic trapping techniques. While data derived from elastic scattering, such as the angular distribution of the scattered radiation or the radiation pressure acting on spherical resonators, e.g., microdroplets, provides mainly information about the morphology of the particle, inelastic light scattering, e.g., Raman spectroscopy, yields additional information concerning the chemical composition of the material under investigation. Trapping techniques allow to obtain Raman spectra of single particles, whose sizes are of the order of or larger than the wavelength of the exciting light. However, in scattering systems with well-defined geometries, e.g., cylindrical, spherical, or spheroidal cavities, the use of Raman spectroscopy as a diagnostic probe becomes complicated due to morphology-dependent resonances (MDRs) of the cavity. Such cavity resonances may give rise to sharp peaks in a Raman spectrum that are not present in bulk Raman spectra. These peaks result from resonance-induced enhancements to the Raman scattering. The physical nature of these resonances can be described for dielectric particles by means of the well-known Lorenz-Mie theory. These MDRs can be used together with Raman data for a comprehensive study of the physical properties as well as the time dependence of chemical reactions. Here, we present a short review of our own work on combined inelastic/elastic (Raman/Mie) light scattering studies and their applications to several microchemical reactions as well as on elastic light scattering on a femtosecond timescale. A few representative examples have been chosen to demonstrate the power of such light scattering studies of microparticles trapped by optical or electrodynamical forces.

1. Introduction

In the last decade, the study of micrometer-sized solid and liquid particles experienced a tremendous development because of their increased significance in a wide range of scientific disciplines, e.g., atmospheric chemistry, pollution research, or pharmaceutical industry.[1,2] The investigation of these microparticles implies the determination of the microparticles' size, refractive index, and shape. These parameters characterize not only solids but also liquid micrometer-sized particles. However, for liquid particles, the temperature, the evaporation, and condensation behavior might also play an important role, influencing the reaction phenomena at the droplet interface. The interaction of a particle with its surrounding medium might lead to complex physical processes which have to be well-understood. Especially, in pharmaceutical industry and in particular in the area of "drug design", such an interaction can lead to unexpected changes of the physical and chemical parameters of a particle over time with unpredictable consequences.[3] Thus, in addition to the physical parameters it is also important to characterize the chemical structure of microparticles.

Recent developments in the field of laser technology, light detection, and single particle traps provide a valuable technology platform to study microparticles. In doing so, the combination of light scattering techniques and simplified particle systems, *e.g.*, single particles or droplet chains, allow for a non-invasive and non-destructive characterization of the physical and chemical properties of such particle systems.[4,5]

We consider a particle characterized by its diameter and refractive index suspended in a medium. The interaction of a coherent, monochromatic electromagnetic wave with this particle leads to light scattering due to the change of the refractive index at the particle-medium interface. Micrometer-sized particles characterized by high geometrical symmetry, *e.g.*, spheres and spheroids, act as an optical cavity. Depending on the illumination wavelength, particle diameter, and the refractive index, the light-particle interaction leads to well-defined resonance frequencies which are referred to in literature as morphology-dependent resonances or MDRs.[6] Such modes were observed particularly when the injected laser light travels inside the droplet (rim) due to an almost total reflection on the droplet/medium interface. As a consequence, for a certain ratio of the particle diameter with respect to the light wavelength, a standing wave is formed inside the droplet (the wavepacket undergoes constructive interference with itself). This kind of interaction was first mentioned by Lorenz and Mie, and is therefore known in literature as Lorenz-Mie theory.[4] The theoretical development of this kind of particle-light interaction provides the internal and external scattered fields obtained by solving Maxwell's equations. The Lorenz-Mie theory can be successfully applied to particles with a diameter comparable to the illumination wavelength. For particles with a diameter much smaller than the wavelength, the Rayleigh theory describes the scattering of the incoming light more appropriately, whereas large particles are better described by the use of geometrical optics. The optical phenomena associated with the scattering theories mentioned here, *i.e.*, Lorenz-Mie and Rayleigh, occur without frequency changes of the scattered photons.

Not all light-particle interaction processes are characterized by this frequency invariance. Raman and fluorescence processes lead to scattered and emitted photons at frequencies that differ from the frequency of the incoming wave.[7] In a Raman process, the interaction of the exciting photons with a molecule leads to inelastically scattered photons at a longer wavelength (lower energy) than the incoming photons, that are called

Stokes-Raman scattering, and to photons scattered at shorter wavelength (higher energy), referred to as anti-Stokes-Raman scattering signal. Both scattering processes provide the same spectroscopic information, but due to the vibrational energy level occupation, the Stokes-Raman signal is usually more intense than the anti-Stokes-Raman signal and therefore easier to detect.

In the next sections, we review some aspects of the experimental work on microparticles diagnostics performed in our laboratories with special emphasis on Raman and Lorenz-Mie scattering processes. A detailed description of the theoretical developments regarding the Lorenz-Mie and Raman scattering can be found elsewhere and will not be presented here (see Brandt in Ref. 7 and references cited therein).

First, we report on investigations of single micrometer-sized particles. The investigation of the phase function yields information about the physical parameters of the particle, *e.g.*, diameter and refractive index, while the Raman spectra yield information about the chemical composition of the suspended microparticle. This chemical information is used to study several microchemical reactions.

In the second part, we describe the use of MDRs in Raman and Lorenz-Mie scattering. This combination was successfully applied to investigate structural and temporal changes of microparticles. Using MDRs, one can determine both particle size and refractive index by connecting the experimental results with theoretical, calculated MDR spectra – such an approach has been applied to evaporation/condensation experiments. We also discuss here the properties of the elastically scattered light on a femtosecond timescale.

Finally, we describe our recent efforts in the areas of spray drying and polymorphism characterization. These attempts build a bridge between the pure fundamental scientific interest on microparticle investigation and some challenging requirements emerging from pharmaceutical industry.

2. Single particle experiments

The development of new applications in single-particle research increased tremendously with recent improvements in instrumentations capable of suspending microparticles and stabilizing them in a stationary position. Micrometer-sized particles can be levitated by radiation pressure, sound waves, and by electromagnetic and electrostatic fields. Each technique presents advantages and disadvantages that are briefly outlined in the following section.

2.1. *Optical levitation*

The development of applications utilizing optical levitation started with the pioneering work of Ashkin.[8] By illuminating a particle from below with a laser beam he observed that the radiation pressure could compensate the gravitational force, leading to particle levitation. Furthermore, Ashkin distinguished between strong and weak trapping for a highly and a barely divergent beam, respectively. Ashkin *et al.* also observed the formation of optical resonances, *i.e.*, MDRs, for an optically levitated droplet.[9] In a later work, the group of Ashkin modified their optical trap to a single-beam gradient trap to test the trapping of colloidal particles, viruses and bacteria.[10, 11]

The first Raman measurements on optically levitated droplets were performed by Thurn and Kiefer,[12,13] studying a water/glycerol mixture. Further investigations on glycerol droplets were performed by means of Mie and Raman scattering,[14] wherein the droplet Raman spectra showed a different signal than the bulk substance, *i.e.*, the OH vibration region (3100-3500 cm^{-1}) was superimposed by sharp peaks. These sharp peaks have been identified by the authors as output MDRs. In order to temporally resolve the spectral position of the observed MDRs and to increase the signal-to-noise ratio, short integration times were used. Because the Lorenz-Mie size parameter, which is the ratio of the droplet circumference to the vacuum wavelength of light, does not change for a particular resonance if the refractive index is constant, it was possible to investigate the evaporation behavior of a suspended droplet. The MDR modes show a clear dependence on the droplet radius: If the radius decreases, the resonances move to smaller relative Raman wavenumbers, *i.e.*, higher absolute wavenumbers or shorter wavelengths. The droplet evaporation was investigated by plotting the Raman shift as a function of time. Furthermore, for some values of the Lorenz-Mie size parameter, an intensity increase of the Raman spectrum could be observed. This can be explained by considering a resonant coupling of the laser with the input MDR frequency. Such a resonant input coupling increases the internal electric field intensity. A classification of the input and output frequencies, respectively, can be found in the work of Van de Hulst[15] and Barber and Chang,[16] who used a series of internal (c_n, d_n) and scattered (a_n, b_n) electromagnetic (EM) field coefficients to describe the light scattering. The Lorenz-Mie theory offers the complete solution of Maxwell's equations and can therefore be applied to simulate the experimental results. In doing so, it is possible to distinguish between the input and output resonances. The results can be further verified by simulating the measured Raman spectra.

The information about the evaporation and refractive index time-dependency was used to investigate the temperature behavior of an evaporating droplet. Popp *et al.* showed that the spectral position of the output resonances suddenly changes, when the droplet eigen-frequency coincides with the input laser frequency.[17] An explanation of this effect is based on the droplet absorption efficiency. The input resonances tend to increase the temperature of the droplet, leading to a decrease in the refractive index and with it in the Lorenz-Mie size parameter (which is a fixed number for each mode for a given refractive index), yielding to a change in the spectral position of the recorded MDRs (see Fig. 1). The changes within the Raman spectra can also be used in a backward approach to obtain information about the droplet temperature by assuming a steady state energy balance between the absorbed light and the thermal heat loss and a negligible evaporative cooling. The good agreement between the measurements and the predicted values supports the assumption that the observed sudden steps can be attributed to input-resonance-induced jumps in the droplet temperature.

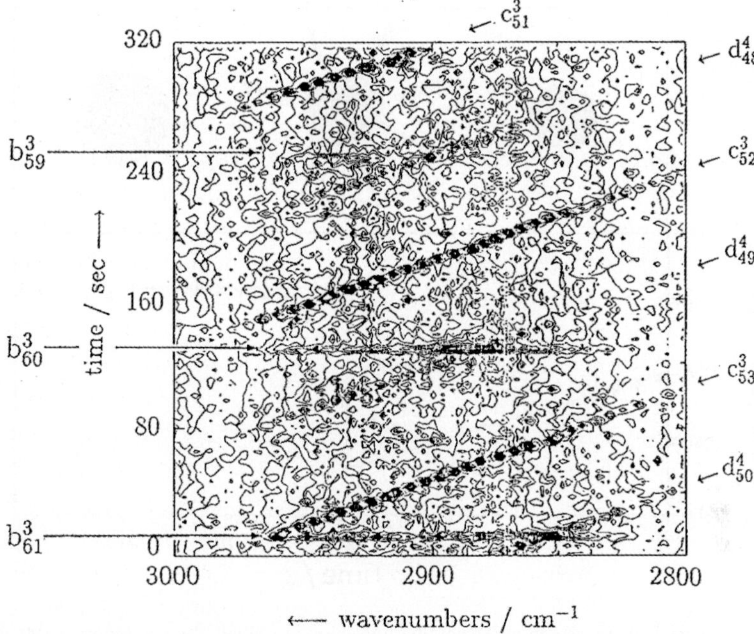

Fig. 1. Measured contour plot of the CH stretching region of an optically levitated glycerol droplet. The movement of the superimposed MDRs due to droplet evaporation is clearly visible. The MDRs are assigned by the multipole coefficients c_n and d_n of the internal electric field (Reprinted with permission from Popp *et al. J. Mol. Struct.* **1995**, *348*, 281. Copyright (2010) Elsevier).

Trunk *et al.* studied the growth of a salt-coated particle generated by a microchemical acid-base reaction between a single levitated capric acid/heptanol droplet (liquid phase) and the surrounding ammonia.[18] By means of Raman spectroscopy, it could be shown that a solid thin shell of the ammonium salt of capric acid was formed at the microparticle surface (see Fig. 2 upper panel). The observed Raman spectra were superimposed by sharp peaks, *i.e.*, MDRs.

The experimental setup allowed the recording of Raman spectra with an accumulation time of about one second. Such short accumulation times permit to locate exactly the positions of the MDRs from an evaporating droplet, and thus to monitor a chemical reaction. A significant change of the c_{82}^3 mode in the MDR environment was observed (see Fig. 2 lower panel). It was assumed that this change is due to the interaction of the microdroplet with the surrounding gaseous ammonia. This change, commonly referred to as sudden jump, indicates the formation of a layer of the ammonium salt of capric acid with a thickness of around 45 nm. The reaction time is around 10 s, after that, no further evaporation of the binary droplet was observed (see Fig. 2).

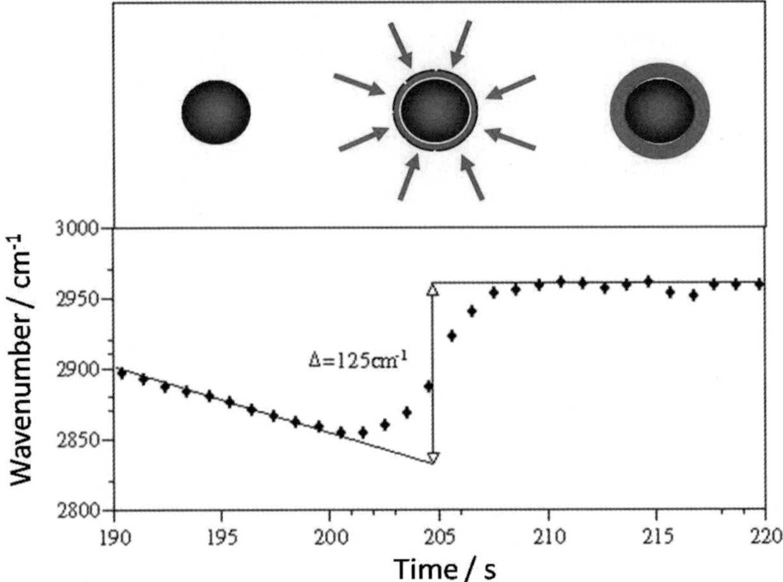

Fig. 2. Upper panel: schematic sketch of the acid-base reaction in a single droplet: between 190 and 200 s the microdroplet is evaporating: at 200 s ammonia (arrows) is added and reacts at the droplet surface; after 210 s the coated droplet is depicted. Lower panel: relative wavenumber position change of the c^3_{82} mode (Reprinted with permission from Trunk *et al. Chem. Phys. Lett.* **1997,** *264,* 233. Copyright (2010) Elsevier).

In an attempt to increase the complexity of the investigated system, a three-component mixture of water, glycerol and ammonium sulfate was also investigated by means of MDRs. These investigations were carried out by monitoring both the Raman spectra and the elastically scattered light.[19] By analyzing the temporal changes of the wavenumber position of the MDRs, changes of the droplet radius, and the refractive index were monitored. Refractive index changes are assumed to occur due to changes in the chemical composition during the evaporation process. Thus, it could be clearly shown that the liquid-to-solid transition in a droplet can be observed by means of MDRs.

MDRs have not only been observed for single, optically levitated droplets but also for an emulsified material, *e.g.,* toluene/polymer.[20] In contrast to the spectrum of the bulk substance, the spectrum of the emulsion droplet is characterized by sharp bands. The results revealed that the spectral position of the output MDRs move to smaller relative wavenumbers, as previously mentioned for optically levitated evaporating droplets. The MDR wavenumber position changes about 11 cm^{-1} within 2 minutes for the evaporating emulsion droplet under investigation. Furthermore, an enhancement of the fluorescence emission caused by the excitation of the MDRs was observed. This enhancement of light emission is strong enough to illuminate not only the edge region of the particle, but also to excite fluorescence in an adjacent particle. Changes of the particle shape were also observed by monitoring the spectral width of the MDR signals.

The theoretical aspects regarding the enhancement of light emission have also been considered.[21, 22] Barber and Chang showed that for resonant size parameters one term of the internal field increases, dominating all others, and resulting in the formation of a standing wave around the droplet circumference.[16] By means of geometrical optics, this phenomenon can be seen as the total internal reflection (TIR) of light at the droplet-medium interface. The lifetime of a resonant mode is defined by the quality factor which has theoretical values of up to 10^{20} while the experimental values are found in the range of 10^5–10^8. The TIR implies as a direct consequence the "storage" of light, which can propagate around the droplet circumference, providing a mechanism for optical feedback. The size and the ratio of the refractive indices of the liquid and the medium must follow certain rules in order to achieve a constructive interference between the waves. The droplet acts as a high-Q optical cavity at a specific wavelength, giving rise to the so-called input MDRs that are also observed in the fluorescence experiments. The MDRs lead to the amplification of the spontaneous scattering processes. Therefore, it should be possible to observe a similar mechanism for the inelastically scattered waves (*e.g.*, wavelength shifted by Raman frequency). Once coupled, the intensity will reach a threshold value after a few cycles, and exit the droplet. This modified signal, commonly referred to as stimulated Raman scattering (SRS), contains not only spectral information, but also information about the size of the investigated droplet.[23-25]

The MDRs were evidenced mainly in experiments using a droplet chain (see section III). A comprehensive review of the non-linear effects associated with MDR modes can be found in Fields *et al.*[23] and Reid *et al.*[26]

2.2. *Acoustic levitation*

Since optical levitation is limited to very small particles of up to 50 μm in diameter, alternative levitation methods are needed to balance heavy and larger particles. Such an alternative was presented in form of an acoustical trap by Brandt,[27] the principle being first mentioned by Fuks.[28] An oscillating fluid will put a suspended particle (within the fluid) in oscillation. The particle will be trapped in a stable equilibrium, if the time-averaged force generated due to the sound pressure variations (oscillations) is larger than the gravitational force.

The possibility of using a large frequency range opened the way for many applications. The theoretical calculations of Hertz[29] demonstrated the potential of acoustical systems for trapping sub-micrometer-sized particles. At present, acoustic levitation is applied in fluid dynamics, evaporation processes,[30] physical chemistry,[31] and analytical chemistry,[32] to mention just the important applications. Sprynchak *et al.*[33] observed a deformation of the acoustically levitated droplets, which was correlated with an increase/decrease in the intensity of the inelastically scattered signal depending on the deformation direction.

2.3. *Electrodynamic levitation*

In analogy to the acoustic field, electric fields are also oscillating. As outlined in the previous sections, optical and acoustical levitation can be successfully applied to study liquid droplets or microparticles, *i.e.*, systems that are characterised by a high symmetry. For systems, where the spherical symmetry is not preserved anymore, *i.e.*, during the evaporation/crystallisation process of a particle, these levitation methods reach their limits. Therefore, adequate methods are required to compensate the decreasing ability for stable trapping. Beside the electrostatic levitation,[2] the electrodynamic levitation method is best suited for studying systems of lower symmetry. In doing so, the electrodynamic balance is realized by a combination of the Millikan oil drop experiment and the electric mass filter of Paul and Raether.[34] The applied devices superimpose a continuous and an alternating potential for a stable trapping of charged microparticles.

In an electrodynamic balance, the DC field is used to equal the vertical forces by applying a potential difference between the endcap electrodes. Unfortunately, such a construction cannot avoid a radial drift of the suspended particles. Therefore, it is necessary to insert at least a third electrode between the endcap electrodes connected to an alternating potential source. In principle, the alternating potential exerts no time-average force on a particle suspended at the null point. However, as the particle tends to leave the null point, the force acting on the particle increases as the particle moves towards to the centre. Based on Paul's work,[34] Würker *et al.*[35] developed a hyperboloidal configuration, which is the most commonly applied trap for aerosol studies. However, there are also a lot of other configurations adapted for specific applications.[36-40] These electrodynamic balance configurations have been mainly used in combination with light scattering instruments to characterise the suspended particles.

The electrodynamic levitation method is the technique best suited for the investigation of solid particles with low smoothness. We have demonstrated that the combination of electrodynamic levitation and optical methods represents a powerful tool for the investigation of heterogeneous chemical reactions.[38, 39, 41, 42] *E.g.*, Fig. 3 displays Raman spectra monitoring the condensation of water vapour on a single magnesium chloride microparticle ($MgCl_2 \cdot 6H_2O$) levitated in an electrodynamic balance.[39] The solid-liquid transition can be observed by examining the loss of the OH vibrational modes. Furthermore, the solvation of SO_2 and NO_x was studied by means of inelastic light scattering. In the case of SO_2 solvation, the time-dependent Raman measurements showed that the reaction product depends on the surrounding gas concentration as well as on the water amount. The NO_x uptake leads to the formation of magnesium nitrate in the ionic form. The reaction can be followed by examining the total symmetric NO_3^- vibrational mode. Additionally, the study showed that a liquid droplet is capable of dissolving gases such as SO_2 and NO_x.[39]

Beside experiments where the task was to determine the physical and/or chemical properties of levitated particles, another research direction deals with the improvement of electrodynamic traps. Because the underlying spectroscopic methods are in general already well-suited for the investigation of aerosol particles, research is most commonly focussed on the "auxiliary" devices in order to improve signal quality or simplify the measurement procedure. For example, by using a common geometry – like the hyperbolic, ring and/or plate electrodes – more than one particle can be trapped at the same time. The electrical charge required in order to levitate the particles, leads to particle-particle interaction (Coulomb force) that might ruin any investigation. As pointed out by Holler *et al.*,[43] such common geometrical configurations present the disadvantage of coupled axial and radial electric field components.

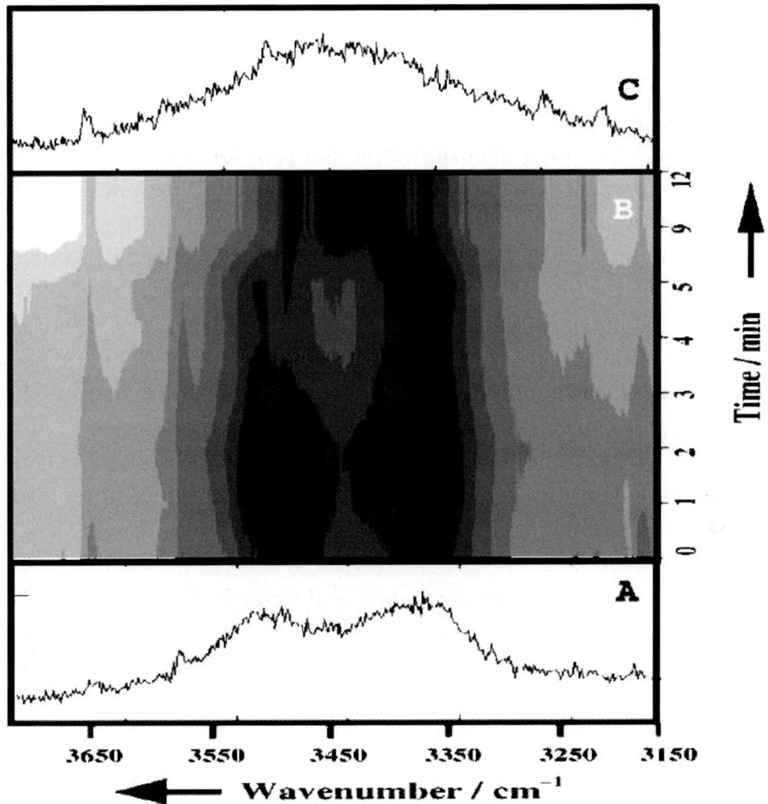

Fig. 3. Panel A: The Raman spectrum of an electrodynamically levitated $MgCl_2 \cdot 6H_2O$ particle. Panel B: A two-dimensional plot of the time-dependent Raman spectra in the OH wavenumber region. Panel C: Raman spectrum of the microdroplet at the end of measurement (Reprinted with permission from Musick *et al. PCCP* **1999**, *1*, 5497. Copyright (2010) Royal Society of Chemistry).

To overcome this disadvantage, instead of using circular electrodes a linear trap was designed and developed.[44] The middle ring (AC) electrode in a hyperbolic trap was replaced by a system consisting of a cage of four or six linear electrodes. In order to compensate the weight of the suspended particle, two DC tube electrodes were inserted concentrically to the AC electrode cage. In doing so, the axial and radial electrical field components could be decoupled. The advantages of this approach are significant:

Firstly, the manipulation of a particle cloud could be extremely improved. By means of a self-developed software program and specially adapted optics, it was possible to select a single particle from a particle cloud and to store it in the electrodynamic balance. Secondly, the linearity of the system allows transforming a particle cloud (a two- or three-dimensional system) into a linear particle array.

These properties are especially important for bio-aerosol investigations, since they lead to a much better signal-to-noise ratio due to a better confinement of the stored particle.

3. Droplet chain experiments

3.1. *Droplet chain experiments on the nanosecond time scale*

Focusing a laser beam on a droplet might lead to a focusing of the beam inside the drop forming a "hot-spot." The high intensity in the hot spot leads to a temperature increase and the irradiated substance can ionize. In order to "freeze" a certain physical state of a droplet the single particle approach can be replaced by a droplet chain.

Most commonly, such droplet chains are generated by a vibrating orifice generator (VOAG), a design proposed by Berglund and Liu.[45] The principle is simple: By means of a vibrating perforated plate, oscillations caused by a piezoelectric crystal are induced into a liquid jet. These oscillations break up the jet in a controlled manner resulting in the generation of liquid droplets of identical diameters. Based on the theoretical work performed by Fillmore *et al.*,[46] different groups improved the performance characteristics of such an aerosol generator. Trunk *et al.* presented a design for liquid feeding lowering the costs.[47] DePonte *et al.* presented a novel design of a droplet beam source based on the radial compression of a free liquid jet through a concentrically co-flowing gas.[48]

Weierstall *et al.* examined conventional Rayleigh sources generating protein droplets with a diameter of around 4 µm passing undamaged through this source.[49]

Applications in pharmacy and biotechnology do not always require the generation of droplet streams with a high generation rate. Ulmke *et al.* developed a new type of droplet generator, the droplet-on-demand generator (DODG).[50] These kinds of generators were first used in applications including ink-jet printing.[51]

Usually, the size of the generated droplets can be influenced by changing the orifice diameter, liquid feed pressure and the frequency of the orifice vibration. The generator also allows for the production of solid particles. One just has to dissolve the starting material, generate the droplets and bring them into a drying column to let the solvent evaporate. Both liquid and solid particles, are able to acquire some electrical charge, mainly by induction charging, and therefore are suited to stably trap them within an electrodynamic trap. This combination of electrodynamical balance and VOAG has been widely used to study the physical and chemical properties of different aerosols.[2] In the following, we will focus on some of our own work on droplet chains.

As shown above, the illumination of an optically levitated particle with a coherent light wave leads to the formation of MDRs. These MDRs are determined by the particle size and the refractive indices of particle and medium. We observed a fluorescence increase[20] and concluded that the optical feedback from MDRs can lead to lower detection limits for spontaneous processes. However, not only the spontaneous (*i.e.*, linear) processes can be enhanced *via* the optical feedback provided by the MDRs. Nonlinear optical processes can also be excited/pumped by the high intensity "hot spots" in microdroplets. The radiation generated in a "hot spot" can couple into a MDR, travel around the sphere rim, and reinteract with the "hot spot" region. A stimulated Raman signal (SRS) is generated, if the gain provided by the "hot spot" is greater than the round-trip loss. The SRS can be further improved by exciting a single normal mode, a scheme referred to as external seeding: A seed laser is injected into the droplet, *a priori* to the pump laser, at the frequency shift of the minority species.[52, 24, 25] There are three possible different pump geometries, which are shown in Fig. 4:

A   Nonresonant pumping – the frequency of the input laser does not coincide with a MDR frequency, giving rise to conventional stimulated Raman scattering (SRS),
B   Resonant pumping – the frequency of the input laser coincides with a MDR frequency, giving rise to stimulated Stokes-Raman scattering (SSRS) and stimulated anti-Stokes-Raman scattering (SARS)
C   Internal pumping – by an on resonant internally generated signal.

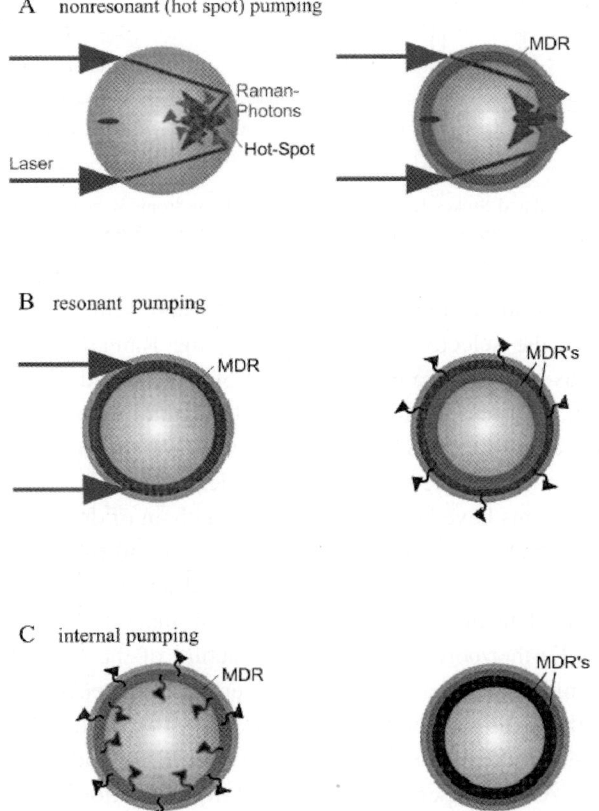

Fig. 4.   Possible pumping geometries for the SRS (Reprinted with permission from Victor *et al. PCCP* **1999**, *1*, 5491. Copyright (2010) Royal Society of Chemistry).

External seeding of stimulated Stokes-Raman scattering (SSRS) has been applied to reduce the detection limit of a minority species in multicomponent microdroplets.[53,54,24,25] Here, the "pumping linearity" was used to investigate the system; *i.e.*, the probed SSRS signal of the minority species (which appears at v_{min}) pumps the SSRS of the majority species (at v_{maj}). Therefore, the Raman signal of the minority species will appear as a linear combination of both component vibrations at $\nu_{min+maj}$. Such a combination signal can easily be distinguished from the elastically scattered light of the seed laser, as shown in Fig. 5.[53]

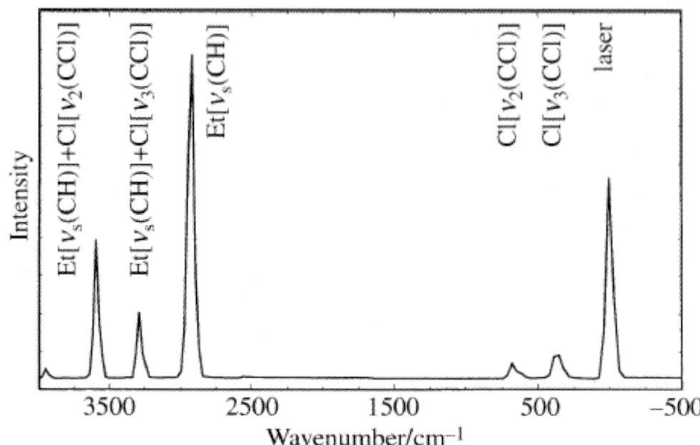

Fig. 5. "Pump only" stimulated Stokes Raman spectrum of 25 μm droplets consisting of 15% chloroform and 85% ethanol (Reprinted with permission from Popp *et al. J. Mol. Struct.* **1999**, *480-481*, 323. Copyright (2010) Elsevier).

By using the external seeding techniques (SSRS and stimulated anti-Stokes-Raman scattering - SARS), the selective enhancement of the Raman signal arising from the minority species was probed for a binary mixture of toluene and ethanol.[53, 54, 24, 25] Due to the external seeding, it is possible to achieve a high population level for SSRS generation, which serves to increase the SSRS and SARS signals. Enhancing the SARS signals leads to new possibilities to investigate fluorescent particles.

Similar measurements have been performed for a chain of droplets of binary mixtures of 3-chlorophenol (which can be seen as a model compound for pesticides) in ethanol.[55] The exponential dependence between the SSRS signal and the concentration leads to an improvement of the detection limit by an order of magnitude in comparison to the conventional SRS. Furthermore, the chemical structure of the studied samples could be investigated by scanning the tunable seed laser over different spectral regions.

In summary, the external seeding of SRS provides a way of studying size and composition changes within binary and ternary organic and inorganic droplets of 10-50 μm in radius. All the experiments presented in this section were carried out using a pulsed nanosecond laser.

3.2. *Investigations of the formation of MDRs on a femtosecond time scale*

The behaviour of ultrashort laser pulses coupled into the resonant modes of spherical cavities was explored both theoretically and experimentally.[56-58] The investigation of microparticles by means of femtosecond pulses, whose spatial length is smaller than the diameter of the particle, offers the advantage of distinguishing between the temporal behaviour of reflection, diffraction, refraction, and coupling into MDRs.

The calculations have been performed considering a simplified system formed by a spherical microparticle that interacts only with a single incoming fs laser pulse.[56] Lorenz-Mie theory was used to calculate the frequency spectrum of the elastically scattered light. Furthermore, the algorithm was improved by including different optical effects, *e.g.* MDRs, reflection, refraction, and diffraction.

The experimental setup is presented in Fig. 6. A single glass sphere suspended in an electrodynamic trap was illuminated by a pulsed Ti:Sapphire laser system operating at 800 nm (pulse length of approximately 100 fs, pulse energy 10 nJ, repetition rate 78 MHz). The scattered light was collected at 90° and directed to the spectrometer through an optical fiber.

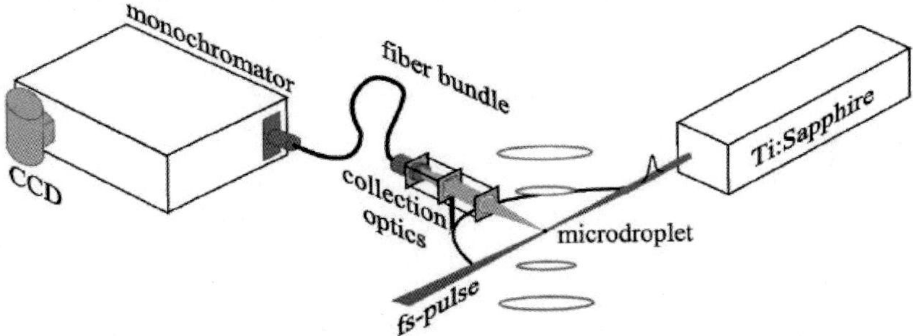

Fig. 6. Experimental arrangement to study the coupling of ultrashort laser pulses into resonant modes of spherical cavities. The fs laser pulses are focused onto a glass sphere which was stably levitated by an electrodynamic trap (Reprinted with permission from Sbanski *et al. J. Opt. Soc. Am. A* **2000**, *17A*, 313. Copyright (2010) Optical Society of America).

The scattered intensity was recorded over an angle range from 85° to 95°. As shown in Fig. 7, a good agreement between the experimental and theoretical results was found. Furthermore, the experimentally recorded peaks were assigned to the theoretically calculated MDRs. Due to the time scale of the experiment, the effect of accumulating intensity in MDRs from a sequence of laser pulses can be neglected – the repetition rate of 78 MHz corresponds to a pulse separation of around 12 ns which was larger than the longest observed resonance lifetimes.

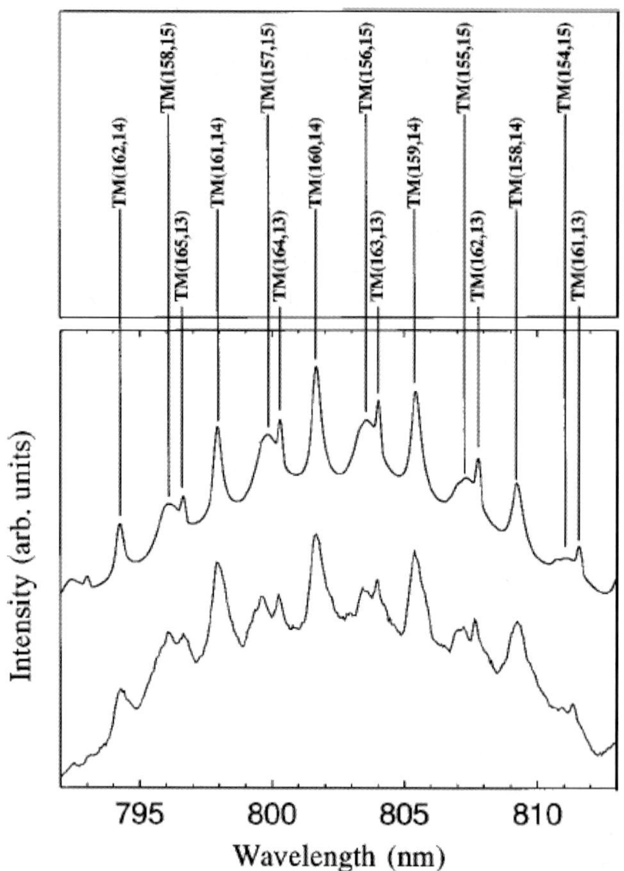

Fig. 7. Scattered radiation at 90° vs. wavelength for the experimental spectrum (lower curve) and the calculated spectrum (upper curve). (Reprinted with permission from Sbanski *et al. J. Opt. Soc. Am. A* **2000**, *17A*, 313. Copyright (2010) Optical Society of America).

Another study dealt with the behaviour of the scattered light in the time domain. The theoretically calculated time dependency of the scattered intensity at an angle of 90° is presented in Fig. 8. It can be seen that the elastically scattered light is composed of a series of pulses with different intensities: An initial pulse followed by a high-intensity pulse and a series of smaller peaks characterised by a decreasing intensity (see inset of Fig. 8 showing an expanded y-axis to emphasize the temporal behaviour). Based on geometrical optics, the first peak was identified as reflected light – this signal arrives first at the detector which implies the shortest optical path length. The intense peak was attributed to light refraction. Furthermore, the calculated time dependent spectra present additional peaks with distinct delay times in the picosecond range. Comparative studies for 0°, 180° and 90° scattering suggest that the light has been trapped in the particle and circumnavigates inside the microsphere – clockwise and counterclockwise. To clarify this aspect, further investigations have been performed for scattering angles of 45°, 90° and

135°, as depicted in Fig. 9. The observed transients can only be explained if pulses propagating on MDRs are considered.

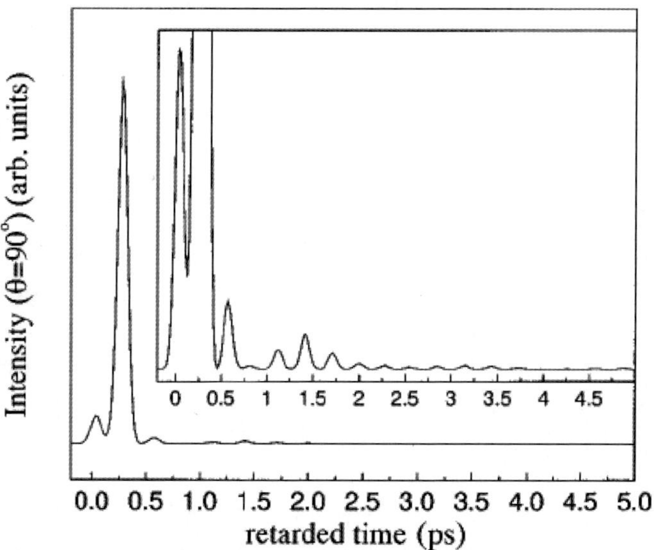

Fig. 8.   Time variation of the scattering intensity for an exact angle of 90° (calculated). The inset shows the same graph but with an expanded intensity- (y-) axis to highlight the temporal behavior. (Reprinted with permission from Sbanski *et al. J. Opt. Soc. Am. A* **2000**, *17A*, 313. Copyright (2010) Optical Society of America).

These theoretical results have been confirmed by experimental studies.[28] Here, the microparticle was mounted on a tip of a glass fiber and illuminated by a fs laser pulse. Fitting the experimental spectrum with the theoretical model, diameter and refractive index of the immobilized sphere could be determined with high accuracy. The excellent agreement between the experimental and theoretical results proved that the observed time dependence is due to the excitation of MDRs.[56] The storage time can be influenced either by changing the size or the refractive index of the microparticle.

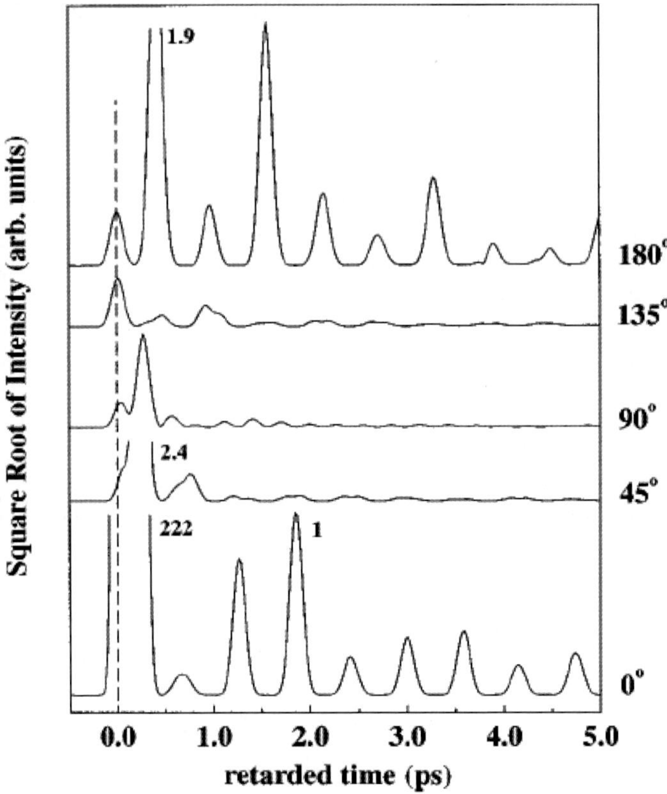

Fig. 9. Scattered light calculated as a function of the retarded time for different scattering angles. The bold numbers at the truncated peaks refer to their intensities. (Reprinted with permission from Sbanski *et al. J. Opt. Soc. Am. A* **2000**, *17A*, 313. Copyright (2010) Optical Society of America).

4. Technological developments for pharmaceutical industry

The experiments presented in sections II and III investigated the elementary processes associated with single particles or chains of droplets, respectively. We could show that scattering methods (Lorenz-Mie and Raman scattering) are perfectly suited to study the physical and chemical properties of the presented systems.

In the following, we will present some examples of how these single particle studies can be used for pharmaceutical applications.

4.1. *Sizing coated particles*

We were able to determine the core and shell radii of pharmaceutical relevant microcapsules containing gas bubbles by simulating the experimental phase functions by means of the Lorenz-Mie theory.[59] A dual illumination procedure by a laser and white light source allows for a systematic analysis of the generated phase functions (in the range from 0° to 170° with a resolution of 0.2°) using an autocorrelation function and fast

Fourier transformation (FFT). The comparison with the theoretically calculated gauge curves leads to the determination of the unknown radii. The accuracy in determining the unknown core and shell radii was approximately between ±2% and ±5%, respectively.

4.2. *Spray drying and polymorphism studies*

Spray-drying is a process that is especially suited for developing pharmaceuticals and excipients for pulmonary drug delivery.[3, 60] Spray-drying is a one step process involving the transformation of liquids, suspensions or emulsions into a fine powder. This spray drying process is achieved by atomisation of the feed solution followed by drying of the resulted droplets through contact with hot drying air or other gases leading to rapid drying times below <1 s. The properties of particles generated via such a spray-drying process are determined by the design and operation of the spray-dryer and also the chemical and physical properties of the feed.

The evaporation of the solvent from a spray involves both heat and mass transfer starting instantaneously after atomization as the droplet meets the hot drying gas. The rapid drying process predominantly leads to amorphous materials due to the rapid solidification. Since different physical forms of substance (material) exhibit different physicochemical properties, a careful detection and control of the amorphous amount is one of the main issues in the production process of pharmaceuticals for pulmonary drug delivery. Especially the structure of the generated particles plays an essential role concerning the bioavailability of the administrated drugs.[3] Therefore, understanding the spray drying process represents one of the most important issues to generate solid particles, *i.e.*, solid drug formulations of known crystalline structure.

Due to the complex behaviour of the drying process and the non-uniform conditions within a spray drier, it is not manageable to reduce the evaporation rate to allow for a proper crystallization. Thus, in order to avoid the complex problems taking place within a conventional spray drier, a simplified approach to investigate the droplet evaporation and crystallisation dynamics of D-mannitol droplets has been used. D-mannitol is an excipient for dry powder formulations.[37]

In a first step, single particles were investigated under controlled conditions, *i.e.*, for well-defined temperature and relative humidity, while in a second step chains of droplets have been studied. Single levitated particles were probed by means of Mie scattering (using a 60° scattering geometry) in order to investigate the time variation of the droplet diameter. Furthermore, the chemical composition of the resulting solid particles was surveyed by means of Raman spectroscopy. Single particle Raman spectra of D-mannitol for its three polymorphic forms labelled according to the nomenclature introduced by Burger *et al.*[61] are shown in Fig. 10.

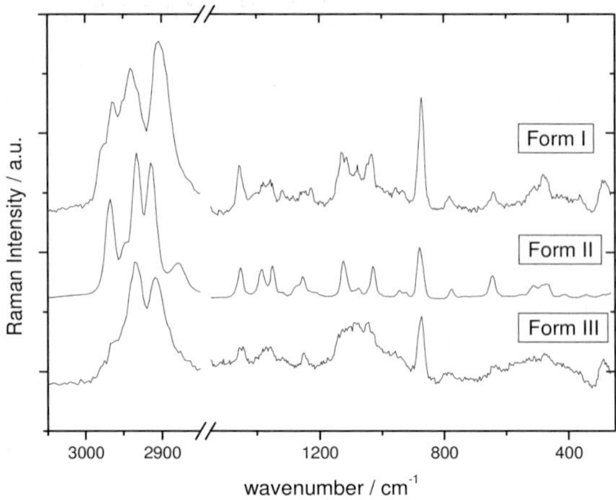

Fig. 10. Single particle Raman spectra of D-mannitol for its three polymorphic forms (Reprinted with permission from Dem *et al. Proc. Resp. Drug Del. X* **2006**, 257. Copyright (2010) RDD Online).

By identifying the polymorphic form present for a special temperature and relative humidity, a "phase diagram" of D-mannitol can be constructed by introducing boundary lines (see Fig. 11).

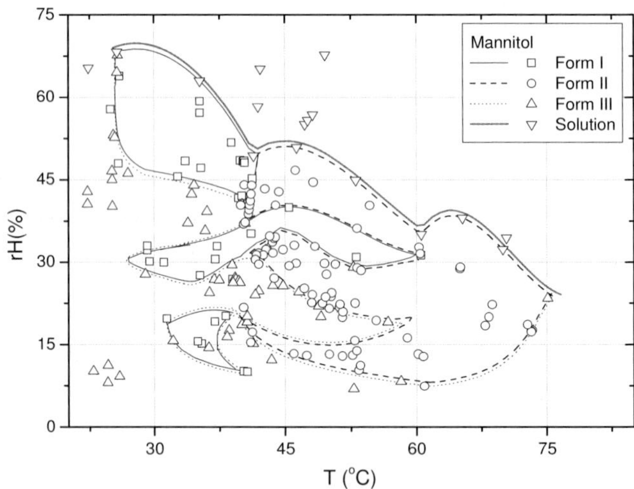

Fig. 11. Experimentally derived "phase map" of D-mannitol constructed by introducing boundary lines. Forms I – III refer to the 3 polymorphic forms of D-mannitol (see Fig. 10). (Reprinted with permission from Dem *et al. Proc. Resp. Drug Del. X* **2006**, 257. Copyright (2010) RDD Online).

However, this single particle approach is still not representative for the spray drying process. In an attempt to better understand the complexity of a real spray dryer, we studied in a second step the temporal evolution of a chain of droplets for well-defined temperature and relative humidity conditions. These droplets were also investigated by means of Lorenz-Mie and Raman scattering. The high droplet output of a vibrating orifice generator (in the kHz range) leads to droplet-droplet interactions affecting the evaporation process. To minimise such interactions, a droplet-on-demand generator (Microdrop GmbH) was used to produce droplets with a diameter of about 90 μm at low frequency in comparison to a normal VOAG. This dispenser head allowed a variable droplet rate which is of importance for the droplet-droplet distance.

The chains were injected into an aluminium tube with an internal diameter of 50 mm, containing a laminar air flow of defined temperature, relative humidity, and velocity. Different temperature and relative humidity gradients were established along the channel, exhibiting an almost linear dependence on the system length. The optical ports allowed the investigation of the free falling droplets either by analysing the phase function[37] of the elastically scattered light or by means of online or offline Raman spectroscopy. Sets of particles were generated and investigated under different environmental conditions.

Combining the derived polymorphic map and also representing the parameter sets (temperature and relative humidity) for the on-/offline Raman measurements, we were able to correlate the single particle and droplet chain experiments.

Furthermore, in order to prove our approach, we generated solid D-mannitol particles using a nebulizer. The obtained results matched the derived "polymorphic map." The initial droplet diameter was about 25 μm and the resulting solid particles had a diameter of around 3 μm, thus being suitable for pulmonary inhalation. These experiments open new possibilities in the controlled design of pharmaceutical excipients, the so-called "drug engineering."

5. Conclusions

We reviewed and discussed recent experimental and theoretical developments of elastically and inelastically scattering processes applied on simplified particle systems, *e.g.*, single particles, or droplet chains. The scattering techniques Lorenz-Mie and Raman are perfectly suited to precisely determine the particle's size and refractive index as well as the chemical composition, *e.g.*, polymorphism. Most of the experimental results could be verified by theoretical simulations. By applying different analysis algorithms, we were able to simulate the experimental scattering data acquired for different illumination times ranging from nanoseconds to femtoseconds.

Finally, we presented two examples demonstrating the great potential of light scattering experiments on small particles for industrial applications. Especially, the polymorphism study indicates a way to model/optimize the droplet kinetics in complex environments.

Acknowledgments

Financial support from the Deutsche Forschungsgemeinschaft and by the Fonds der Chemischen Industrie is gratefully acknowledged.

This paper is dedicated to Prof. Dr. Richard K. Chang for his vast contributions to the field of optics and photonics in microparticles.

References

1. Lafere, J. 1999. Aerosol Technology: Properties, Behaviour and Measurement of Airborne Particles by William C. Hinds.
2. Schweiger, G., and Davis, E. J. 2002. The Airborne Microparticle: Its Physics, Chemistry, Optics, and Transport Phenomena.
3. Hickey, A. J. 2004. Pharmaceutical Inhalation Aerosol Technology, Second Edition, Revised and Expanded. Informa Healthcare, London.
4. Mie, G. 1908. Contributions to the Optics of Turbid Media, Especially Colloidal Metal Solutions. Annalen der Physik 25, pp. 377-445.
5. Raman, C. V., and Krishnan, K. S. 1928. A new type of secondary radiation. Nature 121, pp. 501-502.
6. Ashkin, A., and Dziedzic, J. M. 1981. Observation of optical resonances of dielectric spheres by light scattering. Applied Optics 20, pp. 1803-1814.
7. Popp, J., and Kiefer, W. 2000. Fundamentals of Raman Spectroscopy, 13104. Encyclopedia of Analytical Chemistry. Wiley, Chichester.
8. Ashkin, A. 1970. Acceleration and trapping of particles by radiation pressure. Physical Review Letters 24, pp. 156-159.
9. Ashkin, A., and Dziedzic, J. M. 1977. Observation of resonances in the radiation pressure on dielectric spheres. Physical Review Letters 38, pp. 1351-1354.
10. Ashkin, A., and Dziedzic, J. M. 1987. Optical trapping and manipulation of viruses and bacteria. Science (New York, N.Y.) 235, pp. 1517-1520.
11. Ashkin, A., Dziedzic, J. M., and Yamane, T. 1987. Optical trapping and manipulation of single cells using infrared laser beams. Nature 330, pp. 769-771.
12. Thurn, R., and Kiefer, W. 1984. Raman-microsampling technique applying optical levitation by radiation pressure. Applied Spectroscopy 38, pp. 78-83.
13. Thurn, R., and Kiefer, W. 1985. Structural resonances observed in the Raman spectra of optically levitated liquid droplets. Applied Optics 24, pp. 1515-1519.
14. Popp, J., Lankers, M., Schaschek, K., Kiefer, W., and Hodges, J. T. 1995. Observation of sudden temperature jumps in optically levitated microdroplets due to morphology-dependent input resonances. Applied Optics 34, pp. 2380-2386.
15. Van de Hulst, H. C. 1982. Light Scattering by Small Particles.
16. Barber, P. W., and Chang, R. K. 1988. Optical effects associated with small particles. World Scientific, Singapore.
17. Popp, J., Trunk, M., Hartmann, I., Lankers, M., and Kiefer, W. 1995. Characterization of the interaction between morphology dependent in- and output resonances in laser trapped microparticles. Journal of Molecular Structure 348, pp. 281-284.
18. Trunk, M., Popp, J., Lankers, M., and Kiefer, W. 1997. Microchemistry: Time dependence of an acid-base reaction in a single optically levitated microdroplet. Chemical Physics Letters 264, pp. 233-237.
19. Trunk, M., Popp, J., and Kiefer, W. 1998. Investigations of the composition changes of an evaporating, single binary-mixture microdroplet by inelastic and elastic light scattering. Chemical Physics Letters 284, pp. 377-381.
20. Popp, J., Lankers, M., Trunk, M., Hartmann, I., Urlaub, E., and Kiefer, W. 1998. High-precision determination of size, refractive index, and dispersion of single microparticles from morphology-dependent resonances in optical processes. Applied Spectroscopy 52, pp. 284-291.
21. Schaschek, K., Popp, J., and Kiefer, W. 1993a. Morphology dependent resonances in Raman spectra of optically levitated microparticles: determination of radius and evaporation rate of single glycerol/water droplets by means of internal mode assignment. Berichte der Bunsen-Gesellschaft 97, pp. 1007-1011.

22. Schaschek, K., Popp, J., and Kiefer, W. 1993b. Observation of morphology-dependent input and output resonances in time-dependent Raman spectra of optically levitated microdroplets. Journal of Raman Spectroscopy 24, pp. 69-75.

23. Fields, M. H., Popp, J., and Chang, R. K. 2000. Progress in Optics. Elsevier.

24. Roman, V. E., Popp, J., Fields, M. H., and Kiefer, W. 1999a. Minority species detection in aerosols by stimulated anti-Stokes-Raman scattering and external seeding. Applied Optics 38, pp. 1418-1422.

25. Roman, V. E., Popp, J., Fields, M. H., and Kiefer, W. 1999b. Species identification of multicomponent microdroplets by seeding stimulated Raman scattering. Journal of the Optical Society of America B 16, pp. 370-375.

26. Reid, J. P., and Mitchem, L. 2006. Laser probing of single-aerosol droplet dynamics. Annual Review of Physical Chemistry 57, pp. 245-271.

27. Brandt, E. H. 2001. Acoustic physics: suspended by sound. Nature 413, pp. 474-475.

28. Fuks, N. A. 1960. Evaporation and Droplet Growth in Gaseous Media.

29. Hertz, H. M. 1995. Standing-wave acoustic trap for nonintrusive positioning of microparticles. Journal of Applied Physics 78, pp. 4845-4849.

30. Yarin, A. L., Brenn, G., Kastner, O., Rensink, D., and Tropea, C. 1999. Evaporation of acoustically levitated droplets. Journal of Fluid Mechanics 399, pp. 151-204.

31. Tuckermann, R., Neidhart, B., Lierke, E. G., and Bauerecker, S. 2002. Trapping of heavy gases in stationary ultrasonic fields. Chemical Physics Letters 363, pp. 349-354.

32. Eberhardt, R., and Neidhart, B. 1999. Acoustic levitation device for sample pretreatment in microanalysis and trace analysis. Fresenius' Journal of Analytical Chemistry 365, pp. 475-479.

33. Sprynchak, V., Esen, C., and Schweiger, G. 2003. Enhancement of Raman scattering by deformation of microparticles. Optics Letters 28, pp. 221-223.

34. Paul, W., and Raether, M. 1955. Das elektrische Massenfilter Zeitschrift für Physik 140, pp. 262-273.

35. Wuerker, R. F., Shelton, H., and Langmuir, R. V. 1959. Electrodynamic Containment of Charged Particles. Journal of Applied Physics 30, pp. 342-349.

36. Arnold, S., and Folan, L. M. 1987. Spherical void electrodynamic levitator. Review of Scientific Instruments 58, pp. 1732-1735.

37. Dem, C., Egen, M., Krüger, M., and Popp, J. 2006. 257. Respiratory Drug Delivery X, Boca Raton, FL, USA.

38. Musick, J., Kiefer, W., and Popp, J. 2000. Chemical reactions of single levitated inorganic salt particles with ammonia gas. Applied Spectroscopy 54, pp. 1136-1141.

39. Musick, J., and Popp, J. 1999. Investigations of chemical reactions between single levitated magnesium chloride microdroplets with SO_2 and NO_x by means of Raman spectroscopy and elastic light scattering. Physical Chemistry Chemical Physics 1, pp. 5497-5502.

40. Widmann, J. F., Aardahl, C. L., and Davis, E. J. 1998. Microparticle Raman spectroscopy. Trends in Analytical Chemistry 17, pp. 339-345.

41. Musick, J., Popp, J., and Kiefer, W. 1999. Raman spectroscopic and elastic light scattering investigations of chemical reactions in single electrodynamically levitated microparticles. Journal of Molecular Structure 480-481, pp. 317-321.

42. Musick, J., Popp, J., and Kiefer, W. 2000. Observation of a phase transition in an electrodynamically levitated NH4NO3 microparticle by Mie and Raman scattering. Journal of Raman Spectroscopy 31, pp. 217-219.

43. Holler, S., Arnold, S., Wotherspoon, N., and Korn, A. 1995. Phased injection of microparticles in a Paul trap near atmospheric pressure. Review of Scientific Instruments 66, pp. 4389-4390.

44. Dem, C. 2003. Design and construction of a device for light scattering studies on airborne particles Institute of Physical Chemistry. Julius-Maximilians-Universität, Würzburg.

45. Berglund, R. N., and Liu, B. Y. H. 1973. Generation of monodisperse aerosol standards. Environmental Science and Technology 7, pp. 147-153.

46. Fillmore, G. L., Buehner, W. L., and West, D. L. 1977. Drop Charging and Deflection in an Electrostatic Ink Jet Printer. IBM Journal of Research and Development 21, pp. 37-47.

47. Trunk, M., Lankers, M., Popp, J., and Kiefer, W. 1994. Simple and inexpensive design for a uniform-size droplet generator. Applied Spectroscopy 48, pp. 1291-1293.

48. DePonte, D. P., Weierstall, U., Schmidt, K., Warner, J., Starodub, D., Spence, J. C. H., and Doak, R. B. 2008. Gas dynamic virtual nozzle for generation of microscopic droplet streams. Journal of Physics D: Applied Physics 41, pp. 195505/195501-195505/195507.

49. Weierstall, U., Doak, R. B., Spence, J. C. H., Starodub, D., Shapiro, D., Kennedy, P., Warner, J., Hembree, G. G., Fromme, P., and Chapman, H. N. 2008. Droplet streams for serial crystallography of proteins. Experiments in Fluids 44, pp. 675-689.

50. Ulmke, H., Wriedt, T., Lohner, H., and Bauckhage, K. 1999. Precision Engineering - Nanotechnology, 290-293. 1st International Euspen Conference. Shaker Verlag, Aachen.

51. Döring, M. 1982. Ink-jet printing. Philips Tech. Rev. 40, pp. 192-198.

52. Kwok, A. S., and Chang, R. K. 1992. Fluorescence seeding of weaker-gain Raman modes in microdroplets: enhancement of stimulated Raman scattering. Optics Letters 17, pp. 1262-1264.

53. Popp, J., and Roman, V. 1999. Species detection in single microparticles using nonlinear Raman scattering. Journal of Molecular Structure 480-481, pp. 323-327.

54. Roman, V. E., and Popp, J. 1999. In situ microparticle diagnostics by stimulated Raman scattering. Physical Chemistry Chemical Physics 1, pp. 5491-5495.

55. Schluecker, S., Roman, V., Kiefer, W., and Popp, J. 2001. Detection of pesticide model compounds in ethanolic and aqueous microdroplets by nonlinear Raman spectroscopy. Analytical Chemistry 73, pp. 3146-3152.

56. Sbanski, O., Roman, V. E., Kiefer, W., and Popp, J. 2000a. Elastic light scattering from single microparticles on a femtosecond time scale. Journal of the Optical Society of America A 17, pp. 313-319.

57. Sbanski, O., Roman, V. E., Kiefer, W., and Popp, J. 2000b. Morphology-dependent resonances in a dielectric microsphere and femtosecond laser pulses. Journal of the Chinese Chemical Society (Taipei) 47, pp. 863-864.

58. Siebert, T., Sbanski, O., Schmitt, M., Engel, V., Kiefer, W., and Popp, J. 2003. The mechanism of light storage in spherical microcavities explored on a femtosecond time scale. Optics Communications 216, pp. 321-327.

59. Sbanski, O., Kiefer, W., Popp, J., Lankers, M., and Rossling, G. 2000. Sizing of polymer-coated spherical air bubbles. Applied Spectroscopy 54, pp. 1075-1083.

60. Masters, K. 1972. Spray Drying: an Introduction to Principles, Operational Practice, and Applications

61. Burger, A., Henck, J.-O., Hetz, S., Rollinger, J. M., Weissnicht, A. A., and Stottner, H. 2000. Energy/temperature diagram and compression behavior of the polymorphs of D-mannitol. Journal of Pharmaceutical Sciences 89, pp. 457-468.

CHAPTER 7

CAVITY-ENHANCED EMISSION IN FLUORESCENT MICROSPHERES: REVISITING RKC'S FIRST MDR EXPERIMENT

ALFRED S. KWOK

Department of Physics and Astronomy
Pomona College, Claremont, CA 91711 USA
ASK14747@pomona.edu

We revisit Prof. Chang's early experiments on morphology-dependent resonances (MDRs) in this chapter. We observed MDRs in the lasing spectra of dye-coated polystyrene microspheres and the fluorescence spectra in quantum dot-coated microspheres. An Ar^+ laser is used to excite lasing/fluorescence from the dye molecules and the quantum dots while a near-IR laser tweezer beam is used to trap individual microspheres and control their positions with respect to the Ar^+ beam. Photobleaching prevented us from observing the transition from spontaneous to stimulated emission in quantum dot-coated microspheres.

1. Introduction

Structural resonances in the fluorescence, or spontaneous emission, spectra of 10-µm fluorescent polystyrene spheres were first observed in Prof. Chang's lab in 1980.[1] Fluorescence is enhanced at wavelengths corresponding to the normal modes, or the "whispering gallery modes" of the microsphere. A few years later, lasing from 60-µm ethanol droplets containing the laser dye Rhodamine 6G was observed,[2] and the term "morphology-dependent resonances (MDRs)" was coined to describe the resonant modes at which lasing in these microdroplets occurred. Prof. Chang's lab then moved on to study various nonlinear optical processes in microdroplets in the late 1980's.[3] In addition, Prof. Chang's early MDR work inspired the InGaAs microdisk laser, with a lasing threshold of < 100 µW pump power.[4] Others in this volume have also written about the applications of MDRs in laser diagnostics of aerosols.[5]

Besides a discontinuity in the dependence of the output power on the input pump power, the lasing threshold, or the transition from spontaneous to stimulated emission, is characterized by a change in the emission linewidth. Although it has been shown that the emission peak from a single-mode microdisk laser sharpens above the lasing threshold,[6] the emission lineshape from fluorescent microspheres have not been studied. At what input pump power do MDRs start emerging from the broad background fluorescence continuum? How do the linewidths of the MDRs change as the emission from the microsphere crosses the lasing threshold? Do the linewidths of the fluorescence MDRs remain constant below the lasing threshold and decreases monotonically above the lasing threshold? We describe our attempt to answer these questions by studying the spectra of cavity-enhanced emission from dye-coated and quantum-dot coated microspheres in this chapter.

The fluorescence spectra in Prof. Chang's first fluorescence-MDR paper[1] was recorded when a microsphere drifted into the focal point of the pump/excitation laser beam. We use a laser tweezer to stabilize the position of an individual microsphere in our experiments so that we can increase the integration time. Commercially available fluorescent polystyrene microspheres are usually embedded with the popular biological label FITCI and are very susceptible to photobleaching. We therefore coat our microspheres with Alexa Fluor 488 (A488) or CdSe nanocrystals. A488 is a dye which resembles Rhodamine 6G in both its structure and its fluorescence spectrum, which extends from 500 nm to > 600 nm with a peak at $\approx$ 525 nm, but is much more photostable. A488 functionalized with a succinimidyl ester moiety (Molecular Probes) is added to a suspension of $\approx$10 µm diameter amine(NH2)-functionalized polystyrene microspheres (Interfacial Dynamics Corp), and covalent amide(C-N) bonds are formed between A488 and the amines, which mainly reside on the surface of the microspheres. The A488-coated microsphere suspension is then centrifuged. The supernatant containing the unbonded A488 is aspirated, and the coated microspheres are re-suspended in water. We repeat this centrifugation – aspiration – re-suspension procedure five times, which results in a colloid of A488-coated polystyrene microspheres in water with no background fluorescence. Mercaptoethylamine (MEA) is added at a concentration of 0.2 M to the colloidal suspension as an anti-photobleaching agent. CdSe nanocrystals, or quantum dots, have been replacing organic fluorophores as biological labels in the past ten years because of their high brightness and superior photostability, and the emission spectrum from a quantum dot can be easily changed by changing its size. The biotin-streptavidin linkage, which is frequently used in Molecular Biology, is used to coat polystyrene microspheres with CdSe quantum dots (qdots). Qdots functionalized with streptavidins (Quantum Dot Corp.) are added to a suspension of $\approx$10 µm diameter biotin-functionalized polystyrene microspheres (Spherotech Inc.), and the above centrifugation–aspiration–re-suspension process is used to produce a colloid of "background free" qdot-coated polystyrene spheres. Although the biotin-streptavidin linkage is non-covalent, it is considered to be as strong as a covalent bond. Since polystyrene is a long chain polymer, a polystyrene sphere is really a ball of "folded polymer;" the surface of each of our polystyrene sphere is considered a three-dimensional "fluffy" surface.[7]

2. Experimental setup

A schematic of our experimental setup is shown in Fig. 1. A near infrared diode laser beam at 832 nm (Melles Griot 56ICS115) is used for trapping microspheres, and an Ar^+ laser at 488 nm is used to excite fluorescence/lasing from individual microspheres. A dichroic mirror external to the microscope is used to combine both laser beams, which then enters an inverted microscope (Leica DMIRB) with a trinocular head. A Plan

FLOUTAR 100× 1.3 N.A. oil immersion objective focuses the two laser beams into a sample chamber. We place gimbal-mounted mirrors at the respective eyepoints of the two laser beam-paths so that we can independently translate the laser beams in the sample chamber by steering the mirrors without losing any laser power at the sample. The near IR diode laser beam overfills the back aperture of the microscope objective to ensure the formation of a strong trap.

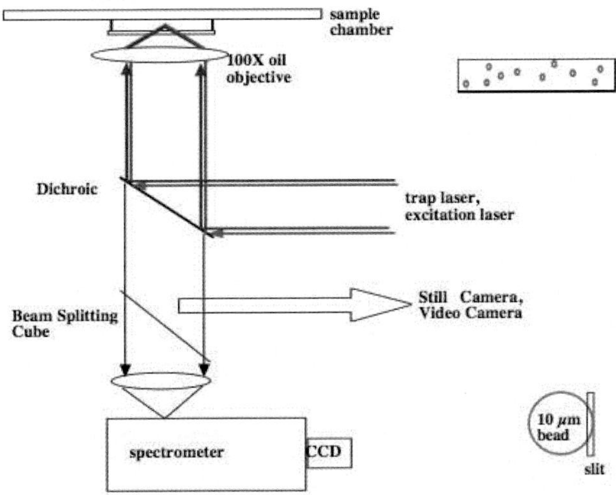

Fig. 1. Schematic of the experimental setup (from Fig. 1 in Ref. 10, permitted by OSA).

The scattered light from a trapped microsphere is collected by the microscope objective, and is then imaged onto a digital still camera (Nikon Coolpix 995), a video camera (GBC) mounted on the trinocular head, or an imaging spectrograph (Acton SP-300) mounted on the camera port of the microscope. A small fraction of the scattered laser diode light is transmitted by a dichroic mirror in the beam path and allows us to monitor the position of our laser trap with the video camera. All the back-scattered light is imaged onto the entrance slit of the spectrograph when the inelastic emission spectra from a microsphere is recorded. A bandpass filter (Chroma Technology E720SP) is used to filter out the elastically scattered light from both lasers, and a liquid-nitrogen cooled CCD camera (Roper Scientific Spec-10:100) is mounted at the exit plane of the spectrograph to record the inelastic spectra. The magnified image of the microsphere at the entrance slit of the spectrograph is 3 mm; however, we have to limit the entrance slit of the spectrograph to 30 μm to maximize the spectral resolution of our system. We therefore use the laser tweezer to position the microsphere such that we are only collecting scattered radiation from an edge of the microsphere.

3. Lasing in dye-coated microspheres

Fig. 2 shows the inelastic emission spectra from four A488-coated microspheres at four different excitation laser intensities (I_{ex}'s). Each spectrum was collected when the excitation Ar$^+$ laser was positioned at the center of the microsphere and was integrated for a period between one and three minutes, during which the microsphere exhibited only Brownian motion.

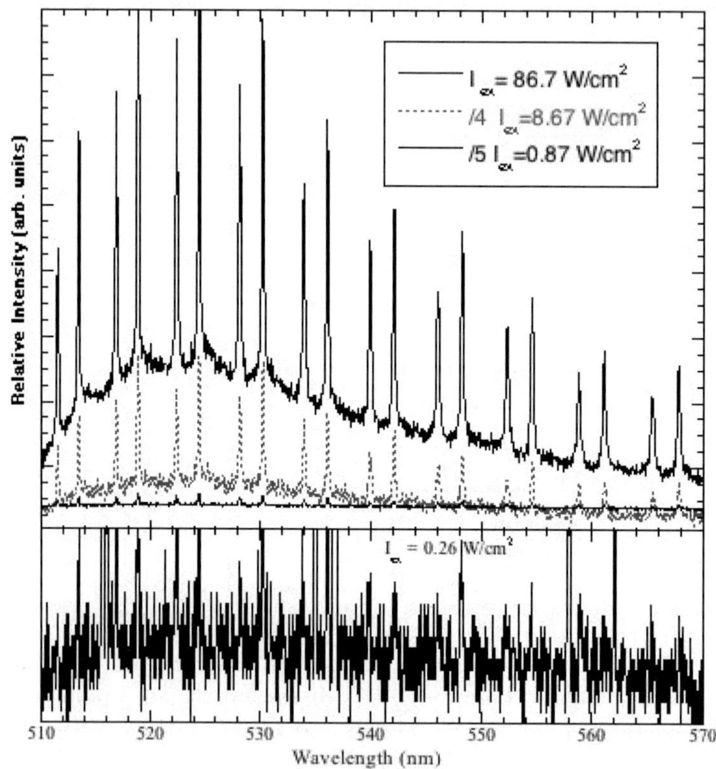

Fig. 2. Normalized inelastic emission spectra from A488-coated polystyrene microspheres.

At 0.87 W/cm$^2 \leq I_{ex} \leq$ 86.7 W/cm^2, the inelastic emission spectra consist of two sets of peaks above a broader background that are evenly spaced. The broader background corresponds to A488 fluorescence emitted into free space modes. The narrow peaks correspond to A488 emission coupled to the lowest radial order MDRs (TE and TM) of the microsphere. These MDRs are localized to a radial region that is closest to the microsphere's surface and thus couple efficiently to emission from A488, which resides mainly on the surface of the microsphere. Similar MDR-enhanced emission can be observed when the excitation laser was positioned at the edge of the microsphere that was furthest away from the edge of the microsphere where the spectra were collected. However, when $I_{ex} >$ 250 W/cm^2, the A488-coated microsphere photobleached despite the addition of the anti-bleach agent MEA.

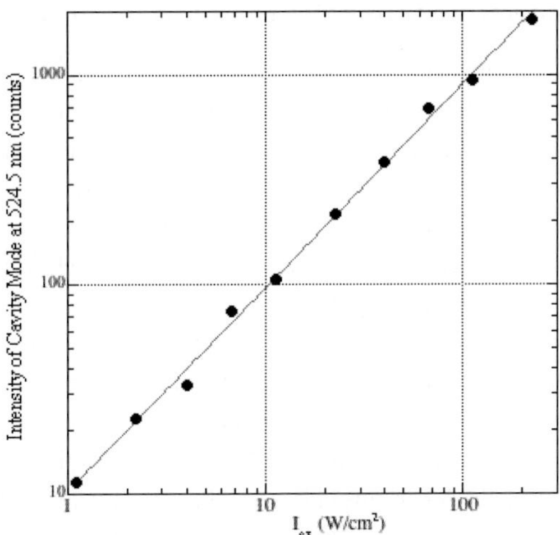

Fig. 3.   Intensity of MDR at 524.5nm vs. I_{ex}.

It is very likely that the MDR-enhanced inelastic emission is due to lasing from A488. Fig. 3 shows that the intensity of the MDR at 524.5 nm is linearly proportional to the excitation laser intensity, which is consistent with the property of a laser cavity mode above the lasing threshold.   The lack of a tunable laser source prevents us from measuring the cold-cavity Q of our A488-coated microsphere. Using a Lorenz-Mie resonance algorithm,[8] we calculated that the cold-cavity Q of the lowest radial order MDRs in the 600 – 500 nm region is between 2000 and 300 for a perfectly spherical 10-μm microsphere with the refractive index of polystyrene (1.59) that is suspended in water. The extended nature of the non-uniform surface of our A488-coated microsphere further suggests that the cold-cavity Q of an A488-coated microsphere is likely to be less than what we calculated above. However, all the MDRs in Fig. 2 have a FWHM of 0.2 nm to 0.3 nm, which is less than a typical cold-cavity linewidth of $\Delta\lambda = \lambda/Q = 524.5$ nm/1000 = 0.5 nm. It is well known that line-narrowing of stimulated emission can occur above the lasing threshold.[9] Therefore, the MDR-enhanced inelastic emission from A488 in Fig. 2 has to be due to lasing from A488. In addition, the image of the microsphere shows two bright spots (see Fig. 4b in Ref. 10), which have typically been associated with laser spots arising from two counter-propagating laser beams at the rim of a spherical microcavity.[11]

We reduced the excitation laser intensity to observe the broadening how the line-narrowed lasing MDRs broaden below the lasing threshold. Unfortunately, we cannot observe any MDRs at lower I_{ex}'s. The bottom panel of Fig. 2 shows an inelastic spectrum from an A488-coated microsphere at $I_{ex} = 0.29$ W/cm^2 with an integration time of three minutes.   The single-pixel or two-pixel wide irregularly-spaced spikes correspond to cosmic rays incident on the CCD camera. When we increase the integration time, a substantial number of cosmic ray peaks are collected in each spectrum, making it impossible to resolve any MDRs in the spectrum.

4. MDR-enhanced fluorescence from Qdot-coated microsphere

Since we were unable to observe MDRs below the lasing threshold in dye-coated microspheres, we continued our study of MDR lineshapes in qdot-coated microspheres. The high quantum efficiency of quantum dots should allow strong fluorescence MDR to be readily observed in quantum dot-coated microspheres. We should then be able to study line narrowing of fluorescence/lasing MDRs as the pump power crosses the lasing threshold.   Although we readily observed fluorescence MDRs from qdot-coated microspheres (Fig. 4), we were unable to observe lasing from the qdot-coated microspheres. The volume fraction of semiconductor material in a sample has to be above a certain threshold for stimulated emission to occur in qdot-containing samples,[12] and the volume fraction of quantum dots in our qdot-coated microspheres is too low for stimulated emission to occur in the quantum dots at the excitation intensities used. We tried to increase the excitation laser intensity to observe stimulated emission from the qdot-coated microspheres but the quantum dots photobleached, despite the manufacturer's claim that they do not.   The qdot-coated microsphere glows very intensely for a fleeting moment and then decays into a non-luminescent dark state upon excitation by an intense excitation laser beam.

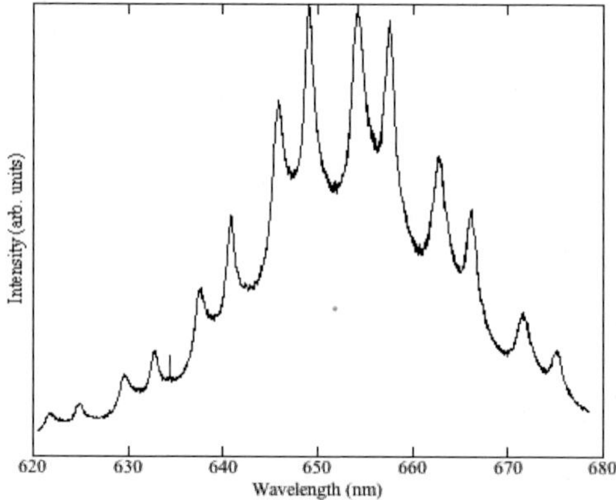

Fig. 4.   Fluorescence MDRs from a qdot-coated microsphere.

Fig. 4 shows a typical inelastic scattering spectrum from a ≈ 10 μm qdot-coated microsphere at 0.1 W/cm^2 < I_{ex} < 100 W/cm^2. The spectrum contains two sets of peaks, and the spacing between the peaks within each set is $\approx$ 10 nm, which is commensurate with the spacing between the MDRs of a 10 μm polystyrene microsphere suspended in water in this wavelength region. However, the integrated intensity of the background emission is much stronger than the integrated intensity of the MDRs, indicating that most of the inelastic scattering couple to free space modes of the microsphere. Therefore, these MDRs are likely to be just spontaneous emission coupled into MDRs of the microsphere.

Acknowledgments

The author would like to acknowledge the contribution of the Pomona College student research assistants in his lab in 2003 – 2005: Perry Schiro and Robert Gerrity collected the data in Figs. 2 and 4 respectively, while Elizabeth Smith's senior thesis research (results not shown) contributed to further understanding of MDR-enhanced emission in qdot-coated microsphere. In addition, he would like to thank Dr. Steven Hill for providing the Lorenz-Mie resonance code described in Ref. 8 and for many helpful discussions.

References

1. R.E. Benner, P.W. Barber, J.F. Owens and R.K. Chang, "Observations of structure resonances in the fluorescence spectra from microspheres," Phys. Rev. Lett. **44** 475 (1980).
2. H.-M. Tzeng, K.F. Wall, M.B. Long and R.K. Chang, "Lasing emission from individual droplets at wavelengths corresponding to morphology-dependent resonances," Opt. Lett. **9** 499 (1984).
3. R.K. Chang, "Some of Bloembergen's nonlinear optical effects revisited in single micrometer-sized droplets," in Resonances, World Scientific (1990).
4. S.L. McCall, A.F.J. Levi, R.E. Slusher, and R.A. Logan, "Whispering-gallery mode microdisk lasers," App. Phys. Lett. **60** 289 (1992).
5. for example, see Chapters 5, 6 and 11
6. R.E. Slusher, A.F.J. Levi, U. Mohideen, S.L. McCall, S.J. Pearton and R.A. Logan, "Threshold characteristics of semiconductor microdisk lasers," App. Phys. Lett. **63**, 1310 (1993).
7. Private communications, Larry Yamaoka, Invitrogen Inc.
8. S.C. Hill, C.K. Rushforth, R.E. Benner and P.R. Conwell, "Sizing dielectric spheres and cylinders by aligning measured and computer resonance locations: algorithm for multiple orders," App. Opt. **24**, 2380-2390 (1985).
9. A.L. Schawlow and C.H. Townes, "Infrared and optical masers," Phy. Rev. **112**, 1940-1949 (1958).
10. P.G. Schiro and A. S. Kwok, "Cavity-enhanced emission from a dye-coated microsphere," Opt. Express **12** 2857 (2004).
11. A. Serpenguzel, S. Kucuksenel and R. K. Chang, "Microdroplet identification and size measurement in sprays with lasing images," Opt. Express **10**, 1118-1132 (2002).
12. Klimov et al., Science **290** 314 (2000).

CHAPTER 8

THEORY OF RAMAN AMPLIFICATION IN MICROSPHERES

MIKHAIL V. JOURAVLEV

Center for Superfunctional Materials,
Department of Chemistry, Pohang University of Science
and Technology, Pohang 790-784, Korea
jouravl@rambler.ru

GERSHON KURIZKI

Department of Chemical Physics, Weizmann Institute of Science, Rehovot 76100 Israel
gershon.kurizki@weizmann.ac.il

We investigate the scattering threshold and cavity-enhanced gain in nonlinear spheres with second- or third- order permittivity. Pairs of pump-driven and signal modes are considered, satisfying morphology-dependent resonance conditions. The thresholds and gain coefficients of amplified and stimulated Raman scattering, parametric downconversion and analogous parametric processes in microspheres are derived and evaluated under typical conditions. Applications may include the measurement of chemical impurity concentrations or the creation of low-threshold optical parametric amplifiers using microspheres.

1. Introduction

Richard Chang has pioneered experimental research of Raman scattering by a transparent dielectric sphere, a subject of both fundamental and applied interest in nonlinear optics.[1,2,3,8,11,14,16,20,21] On the fundamental side, it is an example of amplification or harmonic generation that are determined by the spatial extent, boundary conditions and symmetry of the medium. On the applied side, spectral analysis of stimulated Raman scattering (SRS) by microspheres is a powerful means of gaining information about their size, chemical composition and impurities concentration and may be used for aerosol particle identification. Yet quantitative evaluation of the spectra of such inelastic scattering is challenging, because the radiation field within the microparticle depends in a complicated way on its size, refractive index and wavelength, at wavelengths near the morphology-dependent resonances (MDRs), also known as Mie scattering resonances.[2,3] MDRs have been observed by Richard Chang in the Raman spectra of spherical aerosol particles and droplets.[21,20]

Stimulated Raman scattering (SRS) under input resonance conditions were first explained by introducing the spatial overlap of interacting mode functions in the equations of stimulated scattering in Ref.[14] It is well established that spherical liquid droplets undergoing laser irradiation exhibit threshold behavior for stimulated and spontaneous Raman scattering that requires spatial overlap between the interacting partial modes.[5,16]

Here we develop the theory of microcavity-enhanced Raman gain, and calculate the thresholds of spontaneous and stimulated (coherent) Raman amplification for nonlinear dielectric spheres with large MDR orders. To this end, previously suggested approaches[15,30] are further developed to render the explicit dependence of the gain and threshold on experimental parameters. The present calculation involves the eigenfrequencies and electromagnetic modes of a sphere and their spatial overlap. The general formulae obtained here for the threshold and gain are valid in a resonator of any shape. However, for each shape it is necessary to know the appropriate solution of the problem of diffraction by the dielectric resonator and to obtain its eigenfunctions from the boundary conditions. Analytical solutions for spherical eigenmodes,[3,22,26] are used here to obtain closed-form expressions for the gain spectrum and the threshold conditions.

We are concerned with dielectric spherical media with quadratic, $\chi^{(2)}$, or cubic, $\chi^{(3)}$, nonlinear susceptibilities. A simplified model of two interacting modes satisfying both the input and output resonance conditions may be used to evaluate the threshold intensity and gain for parametric amplification, or for Raman conversion of a pump into a Stokes field. In general, an infinite number of modes will be coupled in such a spherical resonator. However, we may ignore all but the two nondegenerate modes with the highest Q-factors and largest intermode coupling coefficients, obtainable from either experimental or numerical data.[4,7,23,24]

The most crucial variable describing the interaction of light with dielectric microsphere is the size parameter $\rho = ka$, where a is the radius of sphere and k is wave number in the dielectric medium. The interest lies principally with input resonance conditions such that the laser pump (or idler) and the Stokes (or signal) modes both correspond to high-Q modes of the cavity.[20,23] When the size parameter satisfies the equation of $\rho \simeq n$, n being a large integer, the scattering exhibits a complicated angular pattern due to interference among many partial waves and has sharp peaks, corresponding to n-th order MDRs.[24] Such resonances can be extremely narrow, with Q-factors as high as 10^8 for perfectly round, homogeneous and transparent silica spheres[12] and spherical aerosol droplets.[18] The highest-Q MDRs, whose electromagnetic energies concentrated in a narrow surface layer,[6] governs the nonlinear properties. Special attention in our calculations is paid to the dependence of threshold intensity and enhanced Raman gain upon the MDR orders. Enhancement of the Raman scattering is explained here by the Raman power integrated over the MDR volume, as determined by the characteristics of the mode-overlap integral.

This chapter is organized as follows: in Sec. II we derive the basic expressions for threshold intensities and cavity enhanced gain for Raman amplification under input and output resonance conditions, as well as the functional dependence of Raman gain on the active molecules concentration. In Sec. III we describe the implementation of these expressions in typical cases and the pertinent numerical calculations. In Sec. IV we discuss the results of the calculations leading to a decrease of the threshold pump.

We then survey some possible applications of these results and summarize our conclusions in Sec. V.

2. Basic expressions for threshold and gain

The interactions between modes in a nonlinear cavity are described by the field-interaction Hamiltonian density in the form of [25,31]:

$$H_{int} = -\frac{2}{3} \sum_{ijk} \chi^{(2)}_{ijk} \vec{E}_i \vec{E}_j \vec{E}_k \tag{1}$$

or

$$H_{int} = - \sum_{ijkl} \chi^{(3)}_{ijkl} \vec{E}_i \vec{E}_j \vec{E}_k \vec{E}_l \tag{2}$$

where: i, j, k or i, j, k, l indicate three or four interacting partial waves. Because of the energy conservation condition, the fields $\vec{E}_i$, $\vec{E}_j$, $\vec{E}_k$, $\vec{E}_l$ in (1) and (2) are centered at cavity eigenfrequencies satisfying $\omega_i + \omega_j = \omega_k$ or $\omega_i + \omega_j = \omega_k + \omega_l$ respectively. The coefficients $\chi^{(2)}_{ijk}$ and $\chi^{(3)}_{ijkl}$ are the corresponding effective nonlinear susceptibilities. The electromagnetic fields inside the cavity can be quantized as:

$$\vec{E}(\vec{r},t) = - \sum_j i(\frac{\hbar\omega_j}{2\varepsilon_j})^{1/2}(a_j^\dagger(t) - a_j(t))\vec{E}_j(\vec{r}) \tag{3}$$

where $a_j^\dagger$ and a_j are the creation and annihilation operators for the j-th mode, ε_j is the dielectric permittivity. The TE and TM field eigenfunctions in (3) are the solutions of the Helmholtz equation:

$$\vec{\nabla} \times \vec{\nabla} \times \vec{E}_j(\vec{r}) + k_j^2 \vec{E}_j(\vec{r}) = 0 \tag{4}$$

and

$$\nabla^2 \vec{E}_j(\vec{r}) + k_j^2 \vec{E}_j(\vec{r}) = 0, \tag{5}$$

respectively, obeying the cavity boundary conditions and satisfying the normalization condition within the cavity volume V:

$$\int_V \vec{E}_i(\vec{r})\vec{E}_j(\vec{r})dV = \delta_{ij}. \tag{6}$$

We shall quantize the Raman-amplified modes or those parametrically generated or amplified at frequencies ω_i, ω_j upon replacing $\vec{E}_i(\vec{r},t)$, $\vec{E}_j(\vec{r},t)$ by their corresponding quantum mechanical operators. By contrast, we shall treat the field mode excited by a laser pump classically, and assume it remains undepleted by the nonlinear interaction. This quantum approach is widely employed for parametric nonlinear processes in cavity electrodynamics.[25]

After some standard manipulations, the Hamiltonians (1) and (2) for nonlinear interactions in the cavity may be cast in the energy-conserved form[8,28,29,30]: responsible for Raman scattering or coupling between the pump and the Stokes waves,

$$H_{Raman} = \hbar \sum_{i,j} \left\{ S_{ij}^{(n)} a_i a_j^\dagger + hc \right\}. \tag{7}$$

Here hc denotes Hermitian conjugation: $a \leftrightarrow a^\dagger$. In the Raman Hamiltonian, $S_{ij}^{(n)}$ are the n-th order integral coefficients ($n = 2$) or ($n = 3$) of nonlinear coupling of the form [31,5,25]:

$$S_{ij}^{(2)} = \frac{b_k(\omega_i\omega_j)^{1/2}}{6m^2} \chi_{ijk}^{(2)} \int_V \vec{E}_i(\vec{r})\vec{E}_j(\vec{r})\vec{E}_k(\vec{r})dV \tag{8}$$

for $\chi^{(2)}$ processes, or

$$S_{ij}^{(3)} = \frac{b_k^2(\omega_i\omega_j)^{1/2}}{2m^2} \chi_{ijkk}^{(3)} \int_V \vec{E}_i(\vec{r})\vec{E}_j(\vec{r})\vec{E}_k(\vec{r})\vec{E}_k(\vec{r})dV \tag{9}$$

for $\chi^{(3)}$ processes. Here m is the refractive index, b_k is the partial wave amplitude of the ω_k mode within the cavity (Mie-scattering solution), for either TE or TM modes. The Raman Hamiltonian terms (7) have the form $a_i^\dagger a_j$ and $a_i a_j^\dagger$, corresponding to the transfer of energy by single-photon exchange between the pump and Stokes, respectively. The properties of integral coefficients $S_{ij}^{(2)}$ and $S_{ij}^{(3)}$ are defined by selection rules derived from the theory of angular momentum (Sec.III).[27]

In the Raman Hamiltonian (7), upon substituting the photonic state $|\Psi\rangle = \sum_{n_i,n_j} C_{n_i,n_j}|n_i, n_j\rangle$, with n_i,n_j photon numbers in the i, j-cavity modes, into the Schrödinger equation in the interaction picture and applying the Wigner-Weisskopf method[13], we obtain the integrodifferential equation for the parametric interaction of modes p and s with eigenfrequencies ω_p, ω_s, corresponding to MDR distance $\omega_{ps} = |\omega_p - \omega_s|$[30]:

$$\dot{C}_s(t) = -n_p(n_s + 1)V\gamma_s \int_0^\infty d\omega (S_{ps}^{(n)})^2 \rho(\omega)L(\omega - \omega_s, \Gamma)G(t,\omega) \tag{10}$$

Here n_p, n_s are the respective photon numbers, γ_s is the ω_s-mode linewidth, $\rho = m^3\omega^2/(\pi^2 c^3)$ denotes the free-space density of states, $L(\omega - \omega_s, \Gamma)$ is a Lorentzian centered at ω_s, Γ being the homogeneous linewidth of the Raman process. The last factor in the integral of (10) is

$$G(t,\omega) = \int_0^t C_s(t') \exp\left[(i(\omega - \omega_{ps}) - \gamma_s)(t - t')\right] dt'. \tag{11}$$

Analogous equations are obtainable for $\dot{C}_p$, upon replacing the indices $s \leftrightarrow p$. We shall assume in what follows that the pump field is intense and undepleted, as opposed to the Stokes field. The factor n_p in (11) leads to the expected linear dependence of the transition rate on laser intensity, while the factor $n_s + 1$ leads to stimulated scattering through the contribution of n_s and to spontaneous scattering through the contribution of unity. This dependence is reminiscent of the stimulated and spontaneous contributions to the single-photon emission rate of an atom, as described by the Einstein A and B coefficients.[9] Equations (10),(11) yield multi-exponential damping at times long enough for the oscillation in Eq.(11) to subside. The resulting solution for $C_s(t)$ will be written under the following conditions: 1) $\Gamma >> \gamma_{s,p}$, which is appropriate for high-Q modes of the resonator; 2) we single out two interacting modes within the homogeneous Γ-width, one being the pump mode and the other the resonant signal (Stokes) mode, namely, both the input and the output resonance conditions hold.[21,23] The input resonance condition is satisfied for a broadband input pump, which spans several high-Q MDRs, whereas the output resonance condition is always satisfied, since the bandwidth of Raman scattering spans at least several high-Q MDRs. The high-Q modes modulate the free-space mode density $\rho(\omega)$ by sharp Lorentzian peaks. Near ω_s, the relevant Lorentzian has width γ_s. We then obtain from (10),(11)[30]:

$$C_s(t) = (p_{1,s} - p_{2,s})^{-1}(p_{1,s} \exp(-p_{1,s}t) + p_{2,s} \exp(-p_{2,s}t)) \tag{12}$$

where $p_{1,s}$ and $p_{2,s}$ are the roots of the secular equation:

$$p_s(p_s + \gamma_s - i\xi) + \beta_s = 0 \tag{13}$$

with $\xi = \omega_p - \omega_{ps} - \omega_s$ and

$$\beta_s = (S_{ps}^{(n)})^2 n_p(n_s + 1)\gamma_s V \rho_s \omega_s \Gamma^{-1}. \tag{14}$$

The Raman transition rate will be evaluated for the two-photon resonant case, $\xi = 0$. In the underdamped limit $\beta_s >> (\gamma_s/2)^2$, we obtain $p_{1,2} = \gamma_s/2 \pm i\beta_s^{1/2}$. In this limit there is an oscillatory modulation of the Raman decay rate at the frequency $\beta_s^{1/2}$. We shall be concerned with the overdamped limit $(\gamma_s/2)^2 >> \beta_s$, yielding $|p_{1,s}| \approx \beta_s/\gamma_s$. The condition for Raman amplification inside a cavity is that the corresponding Raman rate $|p_s|$ be greater than the transverse relaxation rate of the effective two-mode model, given here by the homogeneous Raman linewidth Γ. The threshold of Raman amplification is then determined by

$$\Gamma = |p_s| = \beta_s/\gamma_s. \tag{15}$$

This threshold condition implies, using Eq.(14), that the threshold numbers of photons in the p and s modes are:

$$n_p(n_s + 1)|_{th} = \frac{3\Gamma^2}{\rho_s V \omega_s (S_{ps}^{(n)})^2}. \tag{16}$$

The corresponding occupation photon number n_p in the p mode inside the resonator in the spontaneous regime of amplification, when $n_s \ll 1$, has the form:

$$n_p^{spo}\big|_{th} = \frac{3\omega_p \Gamma^2}{\rho_s V \omega_s (S_{ps}^{(n)})^2}.$$ (17)

In order to obtain the Raman gain we use the rate equation for the occupation number of photons in the p and s modes[5]:

$$\frac{dn_s}{dt} = D n_p (n_s + 1).$$ (18)

Here D is the rate constant, determined as follows in the spontaneous regime, $n_s \ll 1$: one photon from the p-mode is lost for each Stokes photon that is created. Hence, the cavity enhanced decay rate β_s/γ_s in (15), (16) corresponds to the spontaneous rate:

$$D_{spo} = \frac{\rho_s V \omega_s (S_{ps}^{(n)})^2}{\Gamma}.$$ (19)

The Stokes intensity in the spontaneous regime is proportional to the effective amplification length l_{eff} in the spherical resonator, traversed at the velocity c/m during the photon lifetime in the mode, yielding[32]:

$$n_s\big|_{spo} = \frac{m}{c} D_{spo} n_p l_{eff}.$$ (20)

The Raman gain per unit length in the spontaneous amplification regime is correspondingly:

$$G^{spo} = \frac{D_{spo} m n_p}{c}.$$ (21)

The total gain scales with l_{eff}, which can be expressed via the Q-factor of the p-mode:

$$l_{eff} = 2cQ_p/m\omega_p,$$ (22)

where

$$\frac{1}{Q_p} = \left(\frac{1}{Q_{scat}} + \frac{1}{Q_{abs}} \right)_p,$$ (23)

$(Q_{scat})_p$ and $(Q_{abs})_p$ being the Q-factors of scattering loss and absorption loss in the p-mode. We can now derive the threshold incident intensity P_{th}^{sp} of the pump for spontaneous Raman amplification using the basic relation for any Q-factor[1], namely, that a Q-factor is the ratio of the field energy inside the mode to the incident power, multiplied by the leakage rate. This yields, at threshold:

$$Q_p = \frac{\pi \hbar \omega_p^2 n_p^{spo}\big|_{th}}{P_{th}^{spo} \sigma_{ext} Q_p},$$ (24)

where $n_p^{spo}|_{th}$ is given by (17) and σ_{eff} is the extinction cross-section of the p-mode. The threshold incident intensity of the pump for spontaneous Raman amplification is then obtained from Eqs.(17)-(20), yielding

$$P_{th}^{spo} = \frac{3\pi\hbar\omega_p^2\Gamma^2}{\sigma_{ext}Q_p^2\rho_s V\omega_s(S_{ps}^{(n)})^2},$$
(25)

σ_{ext} being the extinction cross section of the pump mode. The threshold intensity of the incident laser pump is thus inversely proportional to the coupling of partial modes squared, $(S_{ps}^{(n)})^2$, and to the Q-factor of the pump mode squared.

It is advantageous to relate the gain coefficient for spontaneous Raman amplification to the extinction (scattering) cross section σ_{ext}. Using formulae (21)-(24) and (25), we obtain for the gain normalized to the threshold intensity pump:

$$G_{norm}^{spo} = \frac{G^{spo}}{P_p|_{th}^{spo}} = \frac{2mQ_p(\rho_s V\omega_s)(S_{ps}^{(n)})^2\sigma_{ext}}{3c\hbar\omega_p^2\Gamma}.$$
(26)

The useful new results (25),(26) allow us to relate the threshold and gain relevant experimental parameters.

In order to obtain the threshold intensity of stimulated Raman amplification in the limit $n_s \gg 1$, we employ the Manley-Rowe (energy-conservation) condition, relating the numbers of Stokes and pump photons[5,25]:

$$n_s|_{st} = \left(\frac{\omega_p Q_s}{\omega_s Q_p}\right) n_p$$
(27)

The threshold intensity of stimulated amplification, derived analogously to (25) for $n_p,\, n_s \gg 1$, is then

$$P_{th}^{st} = \frac{3^{1/2}\pi\hbar\omega_p^2\Gamma}{\sigma_{ext}Q_p^2(\omega_p\rho_s V)^{1/2}(S_{ps}^{(n)})^2}\left(\frac{Q_p}{Q_s}\right)^{1/2}.$$
(28)

The threshold p-mode gain per s-mode photon can be introduced by means of (17),(18) and yield the rate constant:

$$D_{st} = \frac{Q_s\rho_s V\omega_p(S_{ps}^{(n)})^2}{Q_p\Gamma}.$$
(29)

The stimulated Raman gain normalized to the threshold intensity (28) is found to be:

$$G_{norm}^{st} = \frac{G^{st}}{P_{th}^{st}} = \frac{mQ_s\rho_s V(S_{ps}^{(n)})^2\sigma_{ext}}{c\Gamma\hbar\omega_p}$$
(30)

The threshold intensity $P_{th}^{spo,st}$ of p-mode excitation as well as the Raman scattering cross section σ_R are obtained from experimental data[5,25]. Thus, using (25) or (28), we have an effective tool for measuring the concentration of Raman active molecules N in the form:

$$N = \frac{G_{norm}^{spo,st}P_{th}^{spo,st}}{\sigma_R}$$
(31)

In Sec.III and IV we explicitly evaluate the factors $G_{norm}^{spo,st}$ in nonlinear spheres.

3. Integral coefficients and interacting modes

We now pursue the calculation of the coupling between interacting electromagnetic modes and the selection rules governing this coupling. Integral coefficients of coupling between modes in a microsphere in Hamiltonian (7), may be written as [31]:

$$S_{ij}^{(2)} = \frac{b_k(\omega_i\omega_j)^{1/2}}{6m^2}C_{ijk}^{(2)} \tag{32}$$

and

$$S_{ij}^{(3)} = \frac{b_k b_q(\omega_i\omega_j)^{1/2}}{2m^2}C_{ijkq}^{(3)} \tag{33}$$

for second and third order nonlinearity respectively, where b_k are given by Ref.[26], and $C_{ijk}^{(2)}$ and $C_{ijkq}^{(3)}$ are the volume integrals:

$$C_{ijk}^{(2)} = \chi_{ijk}^{(2)} \int_V \vec{E}_i(\vec{r})\vec{E}_j(\vec{r})\vec{E}_k(\vec{r})dV \tag{34}$$

with $\chi_{ijk}^2 = \chi^{(2)}(\omega_i + \omega_j = \omega_k)$ or permutations thereof[31].

$$C_{ijkq}^{(3)} = \chi_{ijkq}^{(3)} \int_V \vec{E}_i(\vec{r})\vec{E}_j(\vec{r})\vec{E}_k(\vec{r})\vec{E}_q(\vec{r})dV \tag{35}$$

and with $\chi_{ijkq}^{(3)} = \chi^{(3)}(\omega_i, -\omega_i, \omega_k, -\omega_k)$ or permutations thereof. Under input-output resonance conditions, only $\omega_i = \omega_s$ and $\omega_k = \omega_p$ differ from each other, while $\omega_j = \pm\omega_s$, $\omega_q = \pm\omega_p$. These integrals are separable into the radial and angular parts. The scalar form is:

$$\int_V E_{ai}E_{bj}E_{ck}dV = \int_0^{r_0} R_{ai}(k_ir)R_{bj}(k_jr)R_{ck}(k_kr)r^2 dr$$

$$\int_0^{\pi} \Theta_{ai}(\theta)\Theta_{bj}(\theta)\Theta_{ck}(\theta)sin(\theta)d\theta$$

$$\int_0^{2\pi} \exp i(m_{ai} + m_{bj} + m_{ck})d\varphi \tag{36}$$

for $C_{ijk}^{(2)}$ and

$$\int_V E_{ai}E_{bj}E_{ck}E_{dq}dV = \int_0^{r_0} R_{ai}(k_ir)R_{bj}(k_jr)R_{ck}(k_kr)R_{dq}(k_qr)r^2 dr$$

$$\int_0^{\pi} \Theta_{ai}(\theta)\Theta_{bj}(\theta)\Theta_{ck}(\theta)\Theta_{dq}(\theta)sin(\theta)d\theta$$

$$\int_0^{2\pi} \exp i(m_{ai} + m_{bj} + m_{ck} + m_{dq})d\varphi \tag{37}$$

for $C_{ijkq}^{(3)}$. Here the indices i,j,k,q correspond to the radial (r) mode orders, whereas a,b,c,d, label the polar θ index and the azimuthal φ index and of either the TE or TM modes. The well-known angular parts of the eigenfunctions of TE and TM modes Θ_{ai}, Θ_{bj}, Θ_{ck}, Θ_{dq} and the corresponding resonant size-parameters are given in Ref.[26] We have developed numerical algorithms that evaluate the radial part of the integral coefficients in Eq. (8), (9) by Simpson methods of integration, the angular part involving Clebsch-Gordan integrals and the associated Legendre polynomials by the Gauss-Hermite and Tschebishev methods. The Riccati-Bessel, spherical Bessel and Neumann functions were calculated by well-known recurrence methods. With such algorithms we were able to perform the calculation up to size parameter 170 and mode number $n = 170$. The most appreciable couplings involve two modes of the form: $TE_n^m - TM_n^m$, $TE_n^m - TE_n^m$ and $TM_n^m - TM_n^m$, traveling inside the microsphere near the surface.

The integral coefficients for such pairs of modes differ from zero only in the case of phase factor $\exp(\pm \sum_i m_{ai}\varphi) \neq 0$. This yields the following selection rule for a sphere with second order nonlinearity in the form:

$$|m_{ai} \pm m_{bj} \pm m_{ck}| = 0 \tag{38}$$

For two-mode $\chi^{(2)}$-nonlinear interaction this implies $m_p = 2m_s$. For a sphere with third order nonlinearity:

$$|m_{ai} \pm m_{bj} \pm m_{ck} \pm m_{dl}| = 0 \tag{39}$$

which implies for two-mode interaction: $m_p = m_s$. The addition rules of orbital momenta $\vec{\ell}_{ai} \pm \vec{\ell}_{bj} \pm \vec{\ell}_{ck}$ are more complicated.[27] These rules, allowing us to select interacting modes, are alternative to the phase-matching or selection rules for the same processes in bulk or in planar cavities.[5]

4. Results

In Fig.1 and Fig.2 the spontaneous Raman threshold intensity (Eq.(25)) and gain (Eq.(26)) are presented for pairs of modes obeying the selection rules of Sec. III. Here we concentrate on the threshold intensity of spontaneous Raman scattering and the threshold of stimulated Brillouin scattering (SBS) for modes whose amplitudes and resonance half-widths are given in Ref. 10. The threshold pump intensity for spontaneous Raman scattering (Eq. (25)) is found to be 0.6 MW/cm^2, well below the threshold of stimulated Brillouin scattering in a glass sphere, which is of order 9.5 MW/cm^2. The threshold intensity calculated by us for mode-locked $Ti : Al_2O_3$ laser pump operating at 840 nm for particle sizes between 1.5 and 5 μm ranges from 0.3 to 2.6 MW/cm^2. This finding is in agreement with the experiments of high Q-factor dielectric resonators such as spherical aerosol droplets, where the output signal from the particle was 10^3 times larger than the signal from the bulk[17], as explained by the gain presented below. The different thresholds and interacting modes involved allow us to separate these nonlinear processes from each other.

In Fig. 3 and Fig. 4, the threshold intensities and gain for stimulated Raman scattering are presented in dependence on the MDR order. The threshold intensity

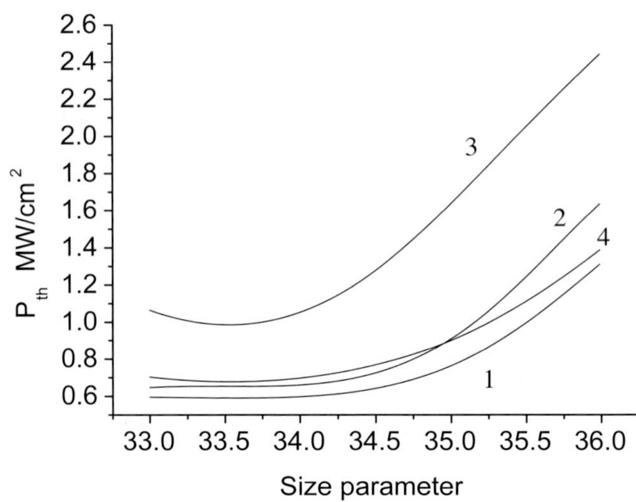

Fig. 1. The threshold intensity for spontaneous Raman scattering from fused silica microsphere irradiated by laser pump of wavelength 840 nm. 1. $TE^0_{42} - TE^1_{45}$, 2. $TE^0_{42} - TE^1_{44}$, 3. $TE^0_{42} - TE^1_{43}$, 4. $TE^0_{42} - TE^1_{42}$.

for stimulated Raman amplification in $Er : Yb$-doped phosphate glass microspheres with diameter $57\mu m$ irradiated by laser pump 1.06 μm is found to be very small, of order 10 W/cm^2, or even less. The gain of stimulated amplification ((Eq.(30)) can be dramatically enhanced relative to its counterpart in bulk by $3.5 \cdot 10^4$ times.

5. Discussion

We have presented the unified theory of spontaneous and stimulated Raman scattering in microspherical nonlinear cavities, pioneered by Richard Chang. We have analyzed the decrease of the threshold intensity and increase of the gain under conditions of input and output resonances, satisfied by nonlinearly coupled modes near morphology dependent resonances (MDRs). Our approach is based on the expansion of the internal field of the coupled modes in the basis of eigenfunctions of the stationary diffraction problem, yielding their internal partial wave amplitudes. The coupling of these partial waves has been accounted for by the integral coefficients, associated with second or third order nonlinearity. It has been shown that in the amplification of Stokes modes, the threshold of excitation and gain can reveal the concentration of the active molecules.

Various applications of the foregoing results are foreseen:

1) Measurements of the stimulated threshold may be used to estimate the concentration of Raman active molecules or nanoparticles (inclusions) embedded in microspherical resonators or aerosol droplets.

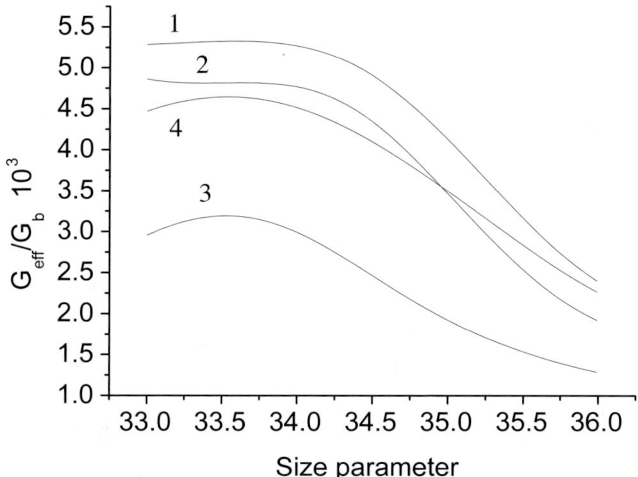

Fig. 2.   The cavity enhanced gain normalized to its value in bulk for the same process as Fig1. 1. $TE_{42}^0 - TE_{45}^1$, 2. $TE_{42}^0 - TE_{44}^1$, 3. $TE_{42}^0 - TE_{43}^1$, 4. $TE_{42}^0 - TE_{42}^1$.

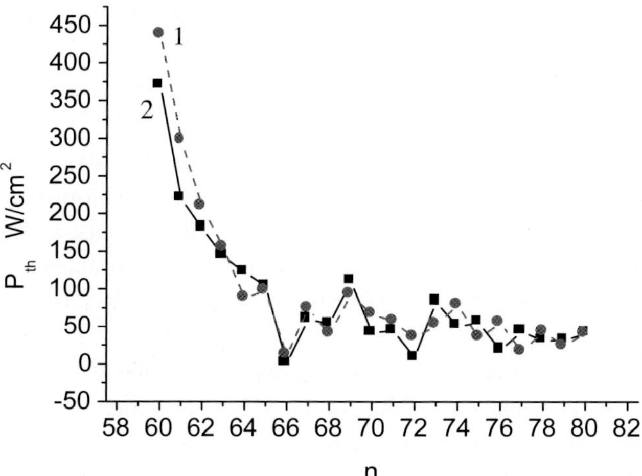

Fig. 3.   The threshold of stimulated Raman amplification for Er:Yb-doped phosphate glass microsphere (diameter of $57\mu m$) for different MDR orders n: 1. $TE_{2n}^0 - TE_{2n}^1$ modes, 2. $TE_{2n}^0 - TE_{2n+1}^1$ modes. The laser pump has the wavelength $1.064\mu m$.

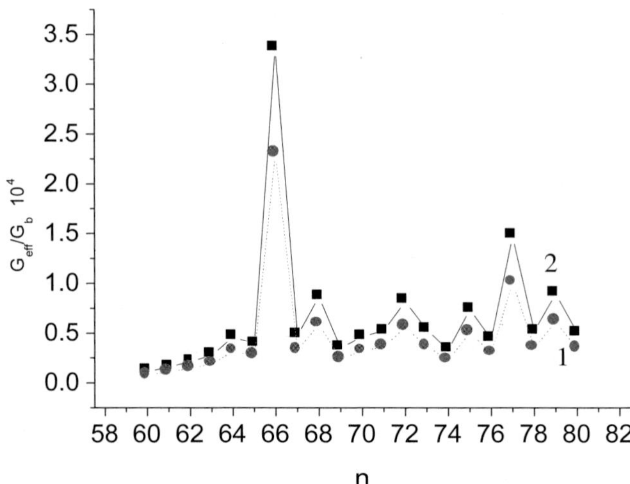

Fig. 4.   The stimulated cavity-enhanced gain normalized to its bulk value for the same process and modes as in Fig.3.

2) Alternatively, they could be used for experimental estimation of Q-factors, temperature and $\chi^{(2)}$ and $\chi^{(3)}$ nonlinear susceptibility.

3) The enormous enhancement of gain, by 3 or 4 orders of magnitude, suggests applications in the context of nonlinear optical microscopy, based on Raman amplification.

4) The ultralow thresholds of stimulated Raman amplification in the two-mode regime, typically less than 10 W/cm^2, may lead to the development of optical microamplifiers based on solid spheres and on aerosol droplets in a wide range of wavelengths. Glass spheres with Raman active inclusions may act as high-gain/low-threshold Raman amplifiers with THz Stokes shift, pumped by sunlight or laser radiation.

5) Raman amplification may be used for atmospheric aerosols identification by providing information on surface concentration of any chemical substances, from the threshold intensity of the nonlinear scattering processes.

6) The surface mode pairs with strong nonlinear coupling have very low thresholds and large gain. As was shown in Ref. 19 there is line broadening of the modes with size parameters $33 \leq \rho \leq 36$ if the microsphere is doped with latex nanoparticles. The concentration of any absorbing or scattering nanoparticles could be estimated by formula (31). The threshold for Raman amplification will increase due to losses incurred by absorbing nanoparticles. However, if the nanoparticles are non-absorbing, the effect of line-broadening may keep the Q-factors intact and cause

the splitting of resonant modes. In this case, the integral coefficients of interaction between "split" Stokes and pump (or signal and idler) modes may actually increase! This intriguing possibility calls for further investigation, whose practical aim may be the creation of broad-band microsphere amplifiers.

Acknowledgments

We acknowledge the support of the EU (through the MIDAS), ISF, and Minerva.

References

1. Affolter, P. and Eliasson, B. (1973). *IEEE Trans. Microwave Theory Tech.* **MTT-21**, p. 573.
2. Barber, P.W. and Chang, R.K.*Optical Effects Associated with Small Particles* (World Scientific, Singapore, 1988).
3. Barber, P.W. and Hill, S.C. *Light Scattering by Particles: Computational Methods* (World Scientific, Singapore, 1990).
4. Biswas, A., Latifi, H., Armstrong, R. and Pinnik, R. (1989). *Phys. Rev. B* **40**, p. 7413.
5. Boyd, R.W. *Nonlinear Optics* (Academic Press, Boston, 1984).
6. Braginsky, V., Gorodetsky, M. and Ilchenko, V. (1989). *Phys. Lett. A* **137**, p. 393.
7. Braunstein, D., Khazanov, A., Koganov, G. and R.Shuker (1996). *Phys. Rev. A* **53**, p. 3565.
8. Campillo, A., Eversole, J. and H.-B.Lin (1991). *Phys. Rev. Lett.* **67**, p. 437.
9. Ching, S., Lai, H. and Young, K. (1987). *J.Opt.Soc.Am.B* **4**, p. 1995.
10. Chitanvis, S. M. and Cantrell, C. D. (1989). *J.Opt.Soc.Am.B* **6**, p. 1326.
11. Eversole, J., Lin, H. and Campillo, A. (1995). *J.Opt.Soc.Am.B* **12**, p. 287.
12. Gorodetsky, M., Pryamikov, A. and Ilchenko., V. (2000). *J.Opt.Soc.Am.B* **17**, p. 1051.
13. Heitler, W. *Quantum Theory of Radiation* (Oxford University Press, 1949).
14. Kurizki, G. and Nitzan, A. (1988). *Phys. Rev. A* **38**, p. 267.
15. Lai, H., Leung, P. and Young, K. (1988). *Phys. Rev. A* **37**, p. 1597.
16. Lin, H.-B. and Campillo, A. (1994). *Phys. Rev. Lett.* **73**, p. 2440.
17. Lin, H.-B. and Campillo, A. (1997). *Opt. Comm.* **133**, p. 287.
18. Lin, H. B., Huston, A. L., Eversole, J. D. and Campillo, A. J. (1990). *J.Opt.Soc.Am.B* **7**, p. 2079.
19. Pellegrin, S., Kozhekin, A., Sarfati, A., Akulin, V. M. and G.Kurizki (2001). *Phys. Rev. A* **63**, p. 033814.
20. Quian, S. and Chang, R. (1989). *Phys. Rev. Lett.* **56**, p. 926.
21. Quian, S.-X., Snow, J. and Chang, R. (1985). *Opt. Lett.* **10**, p. 499.
22. Roll, G., Kaiser, T. and Schweiger, G. (1999). *J.Opt.Soc.Am.A* **16**, p. 882.
23. Schweiger, G. (1990). *J.Raman.Spectrosc.* **21**, p. 165.
24. Schweiger, G. (1991). *J.Opt.Soc.Am.B* **8**, p. 1770.
25. Shen, Y.T. *The Principles of Nonlinear Optics* (Wiley, New York, 1984).
26. Stratton, J.A.*Electromagnetic Theory* (McGraw-Hill, New-York, 1941).
27. Varshalovich, D., Moskalev, A. and Khersonsky, V.*Quantum Theory of Angular Momentum* (World Scientific, Singapore, 1988).
28. Wu, Y. (1996). *Phys. Rev. A* **54**, p. 1586.
29. Wu, Y. and Yang, X. (2003). *Opt. Lett.* **28**, p. 1793.
30. Wu, Y., Yang, X. and Leung, P. (1999). *Opt. Lett.* **24**, p. 345.

31. Yariv, A. *Quantum Electronics* (Wiley, New York, 1989).
32. Zhang, J.-Z., Leach, D. and Chang, R. (1988). *Opt. Lett.* **13**, p. 270.

PART III

LINEAR & NONLINEAR SPECTROSCOPY OF BIOAEROSOLS

CHAPTER 9

FLUORESCENCE-BASED CLASSIFICATION WITH SELECTIVE COLLECTION AND IDENTIFICATION OF INDIVIDUAL AIRBORNE BIOAEROSOL PARTICLES

HERMES C. HUANG,[1] YONG-LE PAN,[2] STEVEN C. HILL,[2] and RONALD G. PINNICK[2]

[1]*Department of Applied Physics and Center for Laser Diagnostics, Yale University,
New Haven, Connecticut 06520, USA*
[2]*U.S. Army Research Laboratory,
2800 Powder Mill Road, Adelphi, MD 20783-1197, USA*
yongle.pan@arl.army.mil

The development of techniques for bioaerosol detection and characterization has flourished during the last decade. A brief summary of the advancements of Prof. Richard K. Chang and his research group at Yale University, together with collaborators at the US Army Research Laboratory (ARL), is given here. We focus on the development of the Single-Particle Fluorescence Spectrometer (SPFS). The SPFS is capable of real-time, in-*situ* monitoring, classification, sorting, and collection of bioaerosols. The SPFS rapidly samples single aerosol particles having sizes in the 1- to 10-μm-diameter range, illuminates them one-by-one with a pulsed UV laser, disperses the fluorescence generated, measures the fluorescence spectra and elastic scattering, and uses these measurements to classify the particles. Bioaerosol particles can be sorted one-by-one, according to their fluorescence spectra, using an aerodynamic puffer, and collected to yield an enriched aerosol sample defined by the particles' fluorescence "signature." A system to further identify specific particle types in the enriched sample has also been investigated. The intent is to provide a system capable of monitoring harmful aerosols in the highly variable and complex atmospheric environment. Here we describe our investigations of ultraviolet-laser-induced fluorescence (UV-LIF) of aerosols, the evolution of the SPFS technology, and the application of the SPFS to characterization of atmospheric aerosols at Adelphi, MD, New Haven, CT, and Las Cruces, NM, USA.

1. Introduction

An aerosol is defined as a suspension of particles in a gas. The particles may be solid, liquid or a mixture of both. "Aerosol" is often used to refer to the particles in the suspension, e.g., "aerosol collector" or "aerosol detector," because it becomes tedious to repeatedly say, "aerosol particles." Typically, ambient aerosol particles have sizes ranging from a few nm to a few hundred micrometers, and can be composed of a wide variety of materials.

Aerosol particles that are composed of a significant fraction of biological materials are termed bioaerosols. Airborne bacteria, viruses, toxins and allergens, fungal spores, are important bioaerosols of interest for detection and characterization.[1-5] Bioaerosols have a tremendous economic impact in the transmission of diseases of humans[2-4] (e.g., flu, severe acute respiratory syndrome (SARS)), other animals, agricultural crops,[5] and other plants, and in causing asthma and other allergies. There is a long history of the intentional dissemination of harmful bioaerosols such as Bacillus anthracis (anthrax), and of investigations of other bioagents for potential release, e.g., Yersinia pestis (plague), Coxiella burnetii (Q fever), Variola virus (smallpox), and biological toxins such as Botulinum. In 1979, an accidental anthrax release from a biowarfare facility in Sverdlovsk, in the former Union of Soviet Socialist Republics (USSR), resulted in 66 fatalities and livestock deaths up to 50 km from the source.[3] In 2001, anthrax contaminated mail in the USA caused 4 deaths and sickened 22 people.[4]

There is a need for improved methods and instrumentation for rapidly characterizing harmful bioaerosols. Such instruments could be used in studying and monitoring disease dissemination. In some cases an early detection and warning system for harmful bioaerosols could trigger responses such as avoiding locations having airborne allergens, to emergency responses which may include evacuation, quarantine, and decontamination (SARS, smallpox, tuberculosis, anthrax). The challenge for detection schemes is finding and characterizing the minute concentration of harmful particles entrained in a dominant concentration of innocuous background aerosol, while maintaining low-false-alarm rates so that alarms are credible. Industries, universities, and government institutions have expended considerable effort in research for real-time in-situ monitoring, classifying, sorting, and identification of bioaerosol threats.

A variety of approaches have been tried in attempts to realize effective early detection for harmful bioaerosols. Optical techniques include measurements of: fluorescence (total and spectrally dispersed),[6-36] Raman spectroscopy,[37-42] and infrared (IR) absorption spectroscopy[43] to obtain information about a particle's chemical properties and composition; laser-induced breakdown spectroscopy (LIBS),[44-45] and mass spectrometry[46-47] to obtain information on a particle's elemental and molecular composition; and elastic scattering intensity, angular distribution, and polarization[48-51] to estimate an aerosol particle's properties such as size, shape, and symmetry features, and even to attempt to obtain refractive index. In general, researchers obtain aerosol information on single particles using these methods, or a combination of complementary methods. Among these techniques, elastic scattering has the largest cross section, and fluorescence the next largest. However, fluorescence is far more informative than elastic scattering for detecting bioaersols. A number of prototype systems based on fluorescence or fluorescence plus elastic scattering have been built.[10,16,18,20,21,27-29,34,35] Commercial aerosol fluorescence systems are or have been sold by companies such as TSI, ICx, and Northrop-Grumman.

Our Single-Particle Fluorescence Spectrometer (SPFS) technology has unique characteristics: it classifies aerosol particles based on their dispersed fluorescence spectrum (280-600 nm), not simply two or three bands, using a 32-anode photomultiplier tube (PMT) detector.[21,32,34] The SPFS can sample and classify aerosol particles at a high rate (maximum frame rate 90,000 spectra/sec.), and can collect an enriched sample of targeted particles with enrichment factor $> 10^3$.

In this chapter, the development of the SPFS system and its application to atmospheric aerosol measurement is described. We begin with an overview of the basics of how fluorescence can be used for discriminating bioaerosols from non-bioaerosols. We then describe the evolution of the SPFS and present a summary of our cluster analysis of fluorescence spectra from ambient aerosols. We finish with an overview of our methods for aerodynamic collection of targeted particles onto a substrate and into solution.

2. Fluorescence of molecules in biological materials

Biological molecules are typically very large, and consequently the vibronic states and the electronic manifolds are dense and result in broad, diffuse fluorescence emission bands. Furthermore, many similar bio-molecules are ubiquitous among living organisms. Consequently, the fluorescence signal can only be used as a classifier, and not as a species-level identifier. For example, among the 20 amino acids, only 3 amino acids (tryptophan, phenylalanine, and tyrosine) give rise to laser-induced fluorescence (LIF). The molecules responsible for most of the fluorescence in most biological cells[13] are listed as follows:

(i) The amino acids tryptophan, tyrosine, and phenylalanine. Each of these constitutes 1% to 5% of the dry weight of some typical bacteria. The excitation and emission maxima of these fluorophors in solution are at approximately 255 nm and 282 nm for phenylalanine; 275 nm and 303 nm for tyrosine, and 280 nm and 348 nm for tryptophan. In proteins that contain tryptophan and one of these other fluorescent amino acids, it is common for the energy absorbed by phenylalanine and tyrosine to be transferred to tryptophan, and to appear around 350 nm. In a protein, different tyrosine or tryptophan molecules can exhibit different fluorescence depending upon their proximity to other amino acids in the same or other protein molecules. That proximity is determined by the three-dimensional structure of the protein and can vary with pH, ion concentrations, temperature, and growth conditions. Because of this, the amount of variation in fluorescence of mixtures can be much larger than would be expected by looking only at the individual fluorophors.

(ii) The reduced nicotinamide adenine dinucleotides, NADH and NADPH, which have excitation and emission maxima around 340 nm and 450 nm, respectively.

(iii) The flavin compounds such as riboflavin, the flavin adenine dinucleotides (FAD and FADH), flavin mononucleotides (FMN), and a variety of flavoproteins. In aqueous solutions flavins have excitation and emission maxima around 450 and 520 nm. There are also a large number of other fluorescent molecules in different biological materials: chlorophylls in plants and many bacteria, cellulose, lignins and lignans in many plants, ferulic acids, and coumarins. (Some discussion of this is in Ref. 32).

In order to find the most suitable excitation wavelengths that can excite a relatively strong fluorescence for the particular interesting bioaerosol particles, excitation/emission (EEM) spectra of some biological materials have been reported, e.g., Ref. 36.

3. Single-particle fluorescence spectrometer: laboratory systems

The first demonstration that the intrinsic laser-induced fluorescence (LIF) of single, micron-sized, biological particles, flowing in air, could be measured was reported by Pinnick et al., at the Scientific Conference on Obscuration and Aerosol Research, in June 1994 at the Edgewood Area, Aberdeen Proving Ground, MD. The work was published the next year.[6]

At the time of that meeting, Richard Chang, Ron Pinnick, Steve Hill and Mike Mayo met there to finalize their decision to form a team to work on the bioaerosol detection problem, emphasizing LIF. Richard argued strongly for measuring spectra, not simply fluorescence in one or two bands. Mike Mayo from the US Airforce (USAF) at Edgewood Chemical Biological Center (ECBC) led in funding both Yale and Army Research Laboratory (ARL) to work on LIF aerosol detectors for harmful bioaerosols, and the Yale/ARL/USAF team collaboration began.

The team's first prototype SPFS was constructed at White Sands Missile Range, NM. The elastic scattering and fluorescence emission from single particles were measured as they flowed through an optical cell placed within the cavity of an argon-ion laser.[7] The fluorescence was imaged onto the slit of a spectrometer, and the resulting spectra measured with a charge-coupled device (CCD) camera. Using this prototype, discrimination of polystyrene microspheres of different size and doped with different color dyes was demonstrated. For biological aerosols, the 488-nm laser beam predominantly excited fluorescence from flavins and possibly from chlorophyll. Fluorescence spectra were also measured from aerosolized bacteria, but signals were integrated over long times to allow for summation of signals from many aerosol particles. Although single-particle spectra could not be demonstrated for bioaerosols, this effort suggested that a fluorescence spectrometer could be developed to discriminate, single micron-sized bioaerosol particles flowing through a sampling instrument.

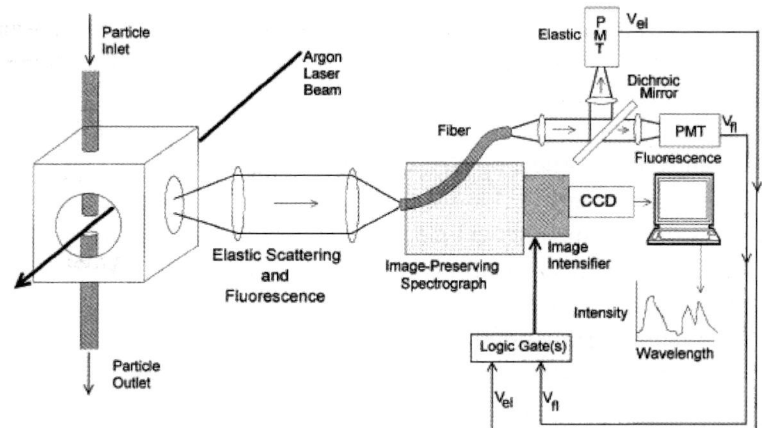

Fig. 1.   Schematic of conditional-sampling spectrograph system. As a particle traverses the chamber, the combined elastic and fluorescence signal from the fiber is used to trigger the image intensifier in front of the CCD camera, which records the particle fluorescence spectrum. (From Fig.1 in Ref. 8, permitted by OSA).

Early major improvements to the prototype were the addition of a gateable image intensifier and a trigger system.[8] An optical fiber collected light as the particle entered the very top edge of the laser beam, and brought the light to a dichroic mirror and two PMT detectors. Using the dichroic mirror, one photomultiplier measured scattering intensity, and the second measured total fluorescence intensity (see Fig. 1). The CCD camera was activated only when thresholds were exceeded on both channels. Using this conditional-sampling scheme, every frame which the CCD camera recorded contained the fluorescence spectrum from a particle. Additionally, the thresholds could be set so that nonfluorescent interferent particles could be ignored. A test case using a mixture of kaolin and Bacillus subtilis (*B. subtilis)* was used to demonstrate how one could ignore the signals from the kaolin, which would not meet threshold requirements on the total fluorescence PMT, and only record spectra from the *B. subtilis*.

In the next improvement of the particle fluorescence spectrometer,[9,11] ultraviolet (UV) light was used to excite the fluorescence in order to observe the spectral contributions from tryptophan, tyrosine, and other molecules that are excited only in the UV. Tryptophan and tyrosine occur in all cells, with each typically contributing at least 1% of the dry weight of bacteria and most other cells. They also occur in most proteins, which typically do not include other fluorophores unless they are not washed well. The argon-ion laser was then only used for triggering the UV laser and intensifier. When the scattering and fluorescence was detected from a particle of interest, a frequency-quadrupled Nd:YAG UV laser (wavelength 266 nm) was triggered to fire, the image intensifier was gated on, and the CCD camera recorded the spectrum.

In the particle fluorescence spectrometers described above, only a single trigger beam was used, and so the particle position was somewhat poorly defined and this increased the laser power required and decreased the particle count rate usable for good spectra. Besides, the fluorescence was collected using lenses which had significant chromatic aberrations in such a wide spectral range (300-600 nm). These aberrations reduced the quality of the spectra significantly, because only a small fraction of the emission spectral range could be focused well at the input slit of the spectrometer. The resulting spectra were distorted.

To circumvent these problems we upgraded the particle fluorescence spectrometer design by employing a reflecting Schwarzschild objective (which has no chromatic aberration, and a larger numerical aperture) to capture the fluorescence, and by requiring particles to pass through the intersection of two continuous-wave (CW) crossed red laser diodes, configured to define a small trigger volume within the optical cell.[12,13] Because the air flow from the nozzle assembly is laminar near the nozzle and provides for somewhat well-defined particle trajectories, the crossed trigger beams precisely define the position of targeted particles (to within a few hundreds of μm), and enable precise timing of the pulsed UV laser (see Fig. 2). Consequently, each targeted particle is illuminated by more uniform laser fluence. Furthermore, the precise timing and laminar flow enables aerodynamic deflection of pre-selected particles downstream from the interrogation region (discussed later). The crossed-beam system reduced the power and space requirements for the triggering system, as the diode lasers were much smaller and used less power than the argon laser. However, there are two main disadvantages. One was that the diode laser scattering could only be used as a trigger and so the trigger lasers did not generate usable LIF, as the argon laser did. LIF excited by 488-nm light is useful information, even if only in one channel. Another disadvantage was that the alignment of the system was far more difficult than with the previous systems because: (a) the Schwarzschild reflecting objective has a high numerical aperture (to collect sufficient fluorescence from even 1-micrometer bacterial cells), and consequently has a very small focal volume, and (b) the diode lasers had to be focused to a relatively small region.

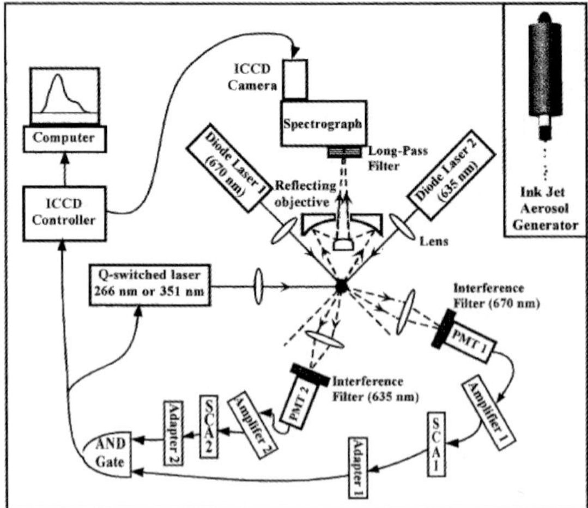

Fig. 2.   Schematic of UV SPFS. Two crossed diode lasers define a detection volume. When a particle crosses through the detection volume, simultaneous scattering is seen by the two photomultipliers, which then triggers the UV laser and the intensified CCD (ICCD) camera in order to record a UV fluorescence spectrum for the particle (From Fig.2 in Ref. 13, permitted by Wiley).

However, alignment difficulties were overcome, and using this new triggering system, both the third and fourth harmonic of the Nd:YAG laser (355 nm and 266 nm) were used to measure fluorescence from single aerosol particles.[12,13] We found that some types of bioaerosols were easier to distinguish from each other using 266 nm as an excitation wavelength, while in other cases 355 nm was better.

A significant issue remained. The system was limited in particle count rate by the readout time of the CCD camera. When Hamamatsu developed their 32-anode PMT, Richard Chang was especially interested in finding a way to use them to increase acquisition speed. Richard and Pan suggested to VTech Engineering that they build a high-speed data acquisition system according to our specifications, to be used for high-speed single-particle fluorescence spectral measurements. A 32-anode photomultiplier system was developed jointly with VTech Engineering.[21] The initial setup yielded a frame rate of 1.4 kHz, with very minimal noise and crosstalk characteristics. The technology has continued to improve, and the version now used by Chang's associates has a maximum acquisition rate of 90 kHz. A separate company, Vertilon, was formed by VTech to market the acquisition board, and their most recent high-speed acquisition board is capable of a 390 kHz trigger rate. These high acquisition rates now make it feasible to acquire spectra for aerosol particles with high density, so long as the probability of two particles being in the sample volume at the same time is small.

The high-speed acquisition board manufactured by VTech also had the advantage of an on-board hardware processing unit. This unit allows for rapid comparison of the spectra against set criteria, and for a trigger signal to be sent from the board if the criteria are met. The capability of on-board processing enabled the triggering of a second-stage sorter-puffer subsystem based on UV-fluorescence data. The aerodynamic puffer[30] can deflect aerosol particles with pre-selected fluorescence characteristics (described in section 5).

The fast electronics detection technology for ultra-weak spectra that we proposed and improved has been used by other research groups (Naval Research Laboratories (NRL), Los Alamos National Laboratory, and Sandia National Laboratory), and in products for bioaerosol detection (Wyatt Technol., MSP Corp.). It has spread to other fields such as confocal microscopy, flow cytometry, laser ranging and detection (LIDAR), multi-angle light scattering, and positron emission tomography (PET) (http://www.vertilon.com).

4. Single-particle fluorescence spectrometer: sampling systems

Prior to 2002, the LIF-aerosol measurements of the Yale/ARL group were of particles forced through the sampling region using slightly pressured air (a few millibars). A sampling system requires an airtight box. The first sampling system of our group was built as collaboration between the Yale/ARL team and Richard's former student Justin Hartings, who was then working at Ft. Detrick, MD. The system was assembled and tested at Yale.[27] To investigate this sampling particle fluorescence spectrometer's capability to measure particles emanating from non-humans (e.g., allergens), Richard brought in his dog to the laboratory. The instrument could detect the increase in fluorescent particles when Richard's dog walked around wagging his tail; thus this new sampler was termed the "Dog Sniffer." It also sniffed cigarettes very well.[52]

The SPFS technology continued to improve. The original box (18" on a side) was replaced by a smaller cell about 2" on a side (but with more optics outside). The ICCD detector was replaced by the 32-anode PMT detector with VTech Electronics, which can sample and classify aerosol particles at a high rate (maximum frame rate 90,000 spectra/sec.).[34]

The most recent improvement to the system employs two excitation lasers with the capability to measure the corresponding (dual) fluorescence spectra with the same detection system. As shown previously,[13] some particles are easier to distinguish using 266-nm excitation, while others are easier to distinguish using 355-nm excitation.

In this advancement, we designed an arrangement whereby both 263-nm and 351-nm lasers were triggered sequentially to excite fluorescence from a single aerosol particle, at two slightly-separated positions along its trajectory through the optical cell.[33] In this setup, the light scattered from the particle is imaged (with 16x magnification) onto the slit of the spectrometer, which has a split filter placed in front. The split filter prohibits the elastically-scattered light from reaching the detector (at each wavelength) but passes the fluorescence (see Fig. 3). The 32-anode PMT senses the fluorescence signal from the first laser pulse, and is then read and reset (within 12 microseconds) to sense subsequent fluorescence signal from the second laser pulse. Thus, for every particle sampled by the SPFS, two spectra can be recorded, one for each excitation wavelength. Using the two spectra, we are able to better distinguish between different types of bioaerosols, both from reference samples and from the ambient air.

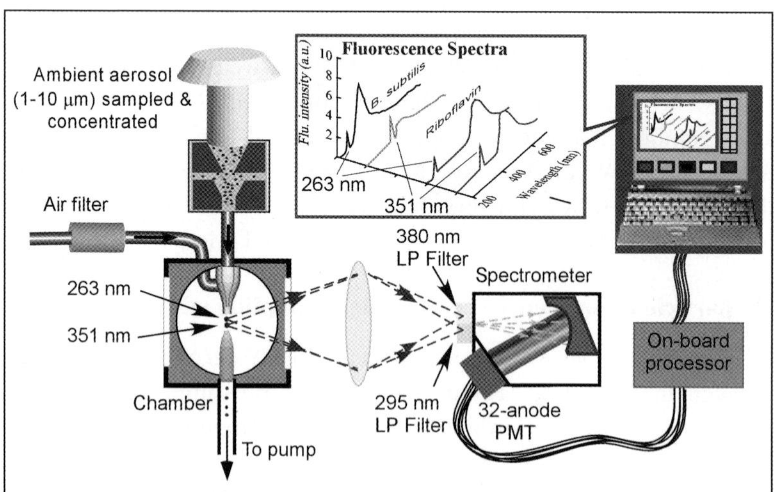

Fig. 3.   Schematic of dual-wavelength UV particle spectrometer. Two crossed CW laser diodes (not shown) define a trigger volume, and particles passing through it are interrogated sequentially by two different-wavelength probe lasers. The resulting fluorescence spectra for each particle are recorded with a 32-anode photomultiplier (From Fig.1 in Ref. 33, permitted by OSA).

5. Investigations of fluorescence properties of bioaerosols

In trying to improve the SPFS for discriminating between various types of bioaerosols, organic carbon aerosols, and other particles, we have studied the fluorescence properties of these aerosol particles.

The angular distribution of fluorescence from molecules in aerosols is larger in the backward direction (e.g., toward the laser source) and differs greatly from that of elastic scattering, which peaks in the forward direction.[19,23,24] Multiphoton-excited fluorescence tends to be emitted even more strongly in the backward direction. Plasma emission, which limits the maximum laser fluence that can be used to illuminate the bioaerosols, also is emitted most strongly in the backward direction.[24,26] Although we have not used laser-induced plasma emission as a diagnostic for bioaerosols, other researchers have.[44,45] The dependence of fluorescence from aerosols is not linear with particle size (diameter, cross section, or volume), or illumination intensity, or concentration of fluorophors.[22,23]

At the incident laser intensities employed with the SPFS, the emission can be nonlinear because of photochemical changes. Tryptophan is especially well known for its photosensitivity. We noticed a most interesting effect with riboflavin, where the fluorescence spectrum changed with time during a single laser pulse, when the illuminating laser intensity was above a certain level. In studying this odd phenomenon in more detail we found that the riboflavin was converted, during the laser pulse, into lumiflavin and lumichrome.[23]

6. Measurement and analysis of atmospheric aerosol

We investigated the single-particle fluorescence spectra of atmospheric aerosols, both as an approach for fundamental investigations of the physics and chemistry of atmospheric particles, and to determine the atmospheric background, against which LIF bioagent detectors must operate. We recorded ambient aerosol data in Adelphi, MD, New Haven, CT, and Las Cruces, NM.[29,32] All several million spectra were measured in a variety of atmospheric conditions. In an effort to categorize the fluorescence spectra from ambient aerosols, an unstructured hierarchical cluster analysis[53] was performed on the data. To perform the analysis, each spectrum is treated as a large multidimensional vector, with only the anodes which correspond with the fluorescence signal being used (roughly from 300 nm to 600 nm in wavelength) in the analysis. The spectra were unit normalized, so that the sum of the intensities in the fluorescent range equaled 1. This allowed the use of the dot product as a comparison of how similar two spectra were to each other, i.e., identical spectra would have a dot product of 1, and as the similarity decreased the dot product would also decrease. The hierarchical clustering was then performed by combining spectra with the largest dot products. The dot products of every pair of spectra are compared, and the pair with the highest dot product is combined into a new cluster, which is then assigned the averaged and renormalized spectra of the two spectra being combined. The process is then repeated, until the largest dot product found is below a set threshold.

We found that even though we took atmospheric aerosol spectra from three different locations, the final cluster templates were similar (see Fig. 4); the main difference between the different sites was the percentage of aerosol particles which fall within each cluster. This result suggests that the cluster templates may be sufficiently robust to be applied to different geographic locations having different regional climate. We speculate that the similarity in cluster templates may be in part due to long-range atmospheric transport homogenizing ambient aerosols, man-made aerosol sources such as combustion being similar in all locations, and dominant fluorophors in biological aerosols being the same, even if the actual species are different.

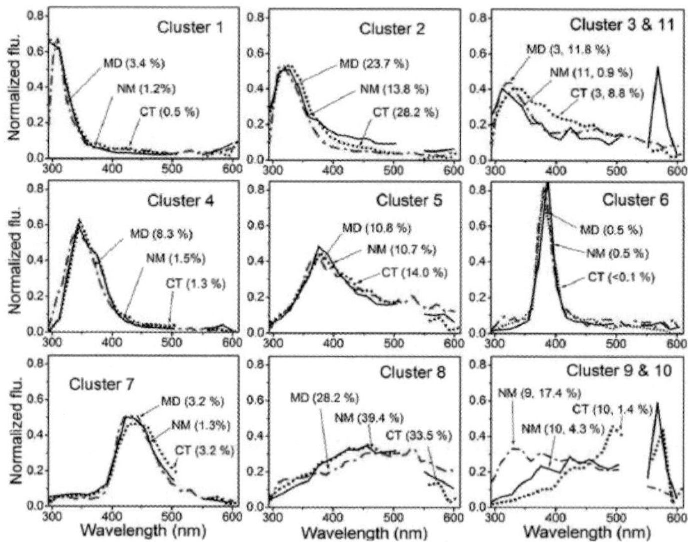

Fig. 4.   Clusters from ambient aerosol spectra measured at Adelphi, MD, Las Cruces, NM, and New Haven, CT. The number in parentheses is the percentage of all fluorescent particles from the site which is grouped into that cluster (From Fig. 4 in Ref. 32, permitted by the American Geophysical Union).

The computation of these cluster templates is a computationally intensive process, because the dot product of every possible pair of spectra must be calculated. The number of spectra which can be clustered using a non-optimized FORTRAN code is limited by computer memory on our personal computer (PC) to a few tens of thousand (when to save computational time the dot products for all pairs are stored to reduce computational time). We investigated other cluster analysis methods, and were able to write an algorithm which is able to handle about an order of magnitude more spectra, and currently is limited by memory constraints. This new method is based around the k-means algorithm.[54] The k-means algorithm is an optimization algorithm, where one starts with a set of clusters, and continues to move data from one cluster to another to maximize the similarity between the data and the cluster centers. Because this method depends on some initial clustering as a starting point and to determine how many clusters are solved for, we used a quick hierarchical clustering to seed the k-means optimization algorithm. In our quick hierarchical clustering, a random spectrum is chosen, and all spectra which have a dot product higher than a set threshold with that chosen spectrum is combined into that cluster. The process continues until all spectra have been put into a cluster. The k-means algorithm is then used to optimize the clusters. Because the initial clustering tends to create many more clusters than are necessary, the final step is to go back through all of the clusters and combine any clusters whose means have a dot product above a set threshold. This process proved to be quite good at finding clusters, while being able to handle a much larger set of spectra using less computational time.

7. Aerodynamic deflection of particles

To improve discrimination among bioaerosols and other particles, we invented an aerodynamic puffer to deflect, from the aerosol stream, specific particles after their fluorescence has been measured.[31] The puffer can deflect single particles from an aerosol stream flowing through the spectrometer, cued by a specific fluorescence signature, as determined by the on-board processing unit of the VTech control board. The deflected particles are impacted onto a substrate (or into a microfluidic well), and can subsequently be analyzed using other forms of spectroscopy, microscopy, or biochemical assay.

The original puffer was an electromagnetically actuated pulse valve, which controlled the release of compressed air at 18 psi. Upon opening and closing quickly (~20 µs), a short burst (60 µs) of gas was released which was used to deflect aerosol particles. This first puffer had a maximum rate of approximately 20 Hz, and with it, we were able to demonstrate enrichment factors of over 10^3.[30]

Because of the high speed at which the air is emitted from the puffer nozzle, the air in the flowing puff is turbulent. This causes some spread in the trajectories of the deflected particles. In order to localize the area where the puffed particles would impact on a substrate, so they would be easy to find for subsequent analyses, a refocusing nozzle was also designed.[34] We have demonstrated that with the combination of puffer and refocusing nozzles, deflected micron-sized particles can be impacted onto a spot of approximately 1-mm diameter (see Fig. 5).

In the latest advance of the puffer technology, we adapted a piezoelectric valve to replace the bulky and slow electromagnetic valve. This piezoelectric valve has the advantage of being smaller, requiring less power, and more rapid operation (maximum rate of 1000 Hz).

As an example of how the sorter-puffer technology might be used, in the Defense Advanced Research Projects Agency (DARPA) funded Spectral Sensing of Biological Aerosols (SSBA) project, we teamed up with ITT and ChemImage in trials conducted at Johns Hopkins University (JHU) Applied Physics Laboratory (APL). ChemImage performed Raman spectroscopy on the test aerosols in an attempt to identify threat aerosols mixed with various interferents. We were able to sort and collect bio-enriched aerosol samples on a substrate. These enriched samples greatly helped ChemImage to measure Raman spectra, analyze the data, and then identify these stimulant bioaerosol particles that were hidden in a background of complicated interferents.

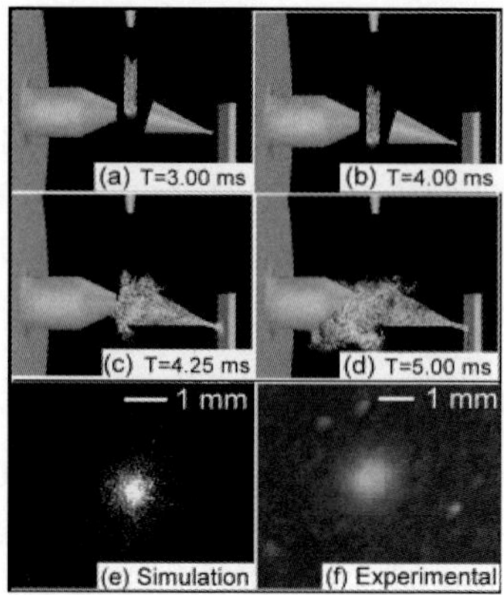

Fig. 5.   Computational fluid dynamics (CFD) simulation and experimental results of the puffer with refocusing nozzle (From Fig. 4 in Ref. 55, permitted by AAAR).

8.   Aerodynamic deflection of particles into a liquid solution for bioanalysis

One of the most reliable diagnostics for the identification of biological cells is biochemical assays. While biochemical assays alone are not ideal as a warning system because of the use of consumable reagents and relatively low throughput compared with an all-optical system, as a second-stage identifier it has the potential to significantly reduce false alarms. If the fluorescence spectra are first used to enrich the particles that will be analyzed biochemically, the number of particles analyzed can be greatly reduced, and reagent usage can be decreased. Because biochemical assays are done in an aqueous solution, the first requirement for an automated second-stage biochemical assay is for specific aerosols to be deposited into an aqueous solution.

We used a microfluidic cell as the platform for the second-stage biochemical assay. Because of the small volume of fluids in the microfluidic cell, reagent use is very small, and assays are performed on a particle-by-particle basis. On such a platform, single-particle biochemical assays for identification of specific species could be accomplished. For the purposes of testing this approach, we used a commercially available microfluidic chip, the Agilent cell fluorescence Labchip.[56] This microfluidic chip is made from etched glass and is embedded in a plastic holder. It is designed to perform six flow cytometry experiments in parallel. We assembled a microscopy setup suitable for measuring fluorescence from single bacteria. Using this microscope, syringe pumps, photodetectors and electronics, we were able to operate the chip and perform antibody binding assays. The channel cross-section in the Labchip is 75 μm × 25 μm. The chip has a focusing flow which concentrated the sample to about half of the channel width. This still meant that in order to see the ~2-μm-long bacterial cells and their fluorescence above the ambient background, we had to use a higher magnification and only view a small portion of the channel. The specific system we chose was *E. coli* bacteria and fluorescein-labeled antibodies targeting that strain of *E. coli*. The 496-nm beam from an argon-ion laser was used to illuminate the viewing region in the microfluidic cell. As cells flowed through the sample volume, the *E. coli* cells lit up because of the fluorophors on the antibodies bound to the exterior of the cells. Other cell which bound the antibodies only weakly remained dark.

The first method we tried to collect the airborne particles into solution was to use existing puffer technology to push aerosol particles into the input well of the microfluidic cell. Shooting a droplet of liquid to try to catch the aerosol particle by impaction proved to be too difficult, but simply using a puff of air to deflect particles towards the microfluidic cell did work. In the comparative test which we ran, we had the microfluidic cell set up with fluorescein-labeled *E. coli* antibody solution. When *E. coli* cells were directly pipetted into the input well, the number of fluorescent counts was approximately 8% of the amount expected, based on the concentration of the *E. coli* cells. A primary reason for the low percentage of cells seen is the small field of view of the microscopy setup relative to the channel size. *E. coli* cells were then aerosolized via the inkjet aerosol generator, and light scattering from a CW laser diode was used to trigger the puffer. Using the puffer to try to collect the *E. coli* cells into solution, we found that approximately 2% of the scattering triggers resulted in a fluorescence signal being seen in the Labchip.

A separate method we tried for collecting particles into solution was electrospray. An electrospray[57] is a monodisperse spray of highly charged droplets which can be generated by applying a high electric field to an ionic fluid from a small capillary tube. The resulting spray can be used to charge an aerosol flowing through the spray and after the aerosol is charged, the electric field will quickly deflect it. In our setup, the ground electrode of the electrospray was embedded within the input well of the Labchip, so that the spray was pulled into the input well by the electric field which swept charged particles into the well. In tests with dye droplets (so that the droplets could be imaged and distinguished from the cloud of droplets from the electrospray) we showed that this method was effective at deflecting the droplets, and that electrospray could charge the droplets well.

Acknowledgments

We acknowledge funding from the US Army Research Laboratory (ARL), the Joint Services Technology Office for Chemical and Biological Defense (JSTO-CBD), the US Air Force Research Laboratory (AFRL), the Edgewood Chemical Biological Center (ECBC) , the US Defense Threat Reduction Agency (DTRA), the US Defense Advanced Research Projects Agency (DARPA), and the US Department of Energy (DOE).

References

1. C. S. Cox, and C. M. Wathes eds. *Bioaersols Handbook* (Lewis, Boca Raton, 1995).
2. H. A. Burge eds. *Bioaerosols* (CRC, Boca Raton, 1995).
3. M. Meselson, J. Guillemin, M. Hugh-Jones et al, *Science* **266**, 1202 (1994).
4. C. P. Weis, A. J. Intrepido, A. K. Miller et al, *Journal of the American Medical Association,* **11(288),** 2853.
5. D. E. Aylor, *Ecology,* **84**, 1989 (2003).
6. R. G. Pinnick, S. C. Hill, P. Nachman et al, *Aerosol Science and Technology,* **23**, 653 (1995).

7. S. C. Hill, R. G. Pinnick, P. Nachman, G. Chen, R. K. Chang et al., *Applied Optics*, **34**, 7149 (1995).
8. P. Nachman, G. Chen, R. G. Pinnick, S. C. Hill, R. K. Chang et al., *Applied Optics*, **35**, 1069 (1996).
9. G. Chen, P. Nachman, R. G. Pinnick, S. C. Hill, R. K. Chang, *Optics Letters*, **21**, 1307 (1996).
10. P. P. Hairston, J. Ho, F. R. Quant, *Journal of Aerosol Science*, **28**, 471 (1997).
11. R. G. Pinnick, S. C. Hill, P. Nachman, G. Videen, G. Chen et al., *Aerosol Science and Technology*, **28**, 95 (1998).
12. Y. L. Pan, S. Holler, R. K. Chang, S. C. Hill, R. G Pinnick et al, *Optics Letters*, **24**, 116 (1999).
13. S. C. Hill, R. G. Pinnick, S. Niles, Y. L. Pan, S. Holler et al, *Field Analytical Chemistry and Technolog*, **3**, 221 (1999).
14. M. Seaver, J. D. Eversole, J. J. Hardgrove, W. K. Cary, D. C. Roselle, *Aerosol Science and Technolog*, **30**, 174 (1999).
15. J. D. Eversole, J. J. Hardgrove, W. K. Cary, D. P. Choulas, M. Seaver, *Field Analytical Chemistry and Technology*, **3**, 249 (1999).
16. F. L. Reyes, T. M. Jeys, N. R. Newbury, C. A. Primmerman, G. S. Rowe et al, *Field Analytical Chemistry and Technology*, **3**, 240 (1999)..
17. Y. S. Cheng, E. B. Barr, B. J. Fan, P. J. Hargis, D. J. Rader et al, *Aerosol Science and Technology*, **30**, 186 (1999).
18. P. H. Kaye, J. E. Barton, E. Hirst, J. M. Clark, *Applied Optics*, **39**, 3738 (2000).
19. S. C. Hill, V. Boutou, J. Yu, S. Ramstein, J. P. Wolf, Y. L. Pan et al, *Physical Review Letters*, **85**, 54 (2000).
20. J. D. Eversole, W. K. Cary, C. S. Scotto, R. Pierson, M. Spence et al, *Field Analytical Chemistry and Technology*, **5**, 205 (2001).
21. Y. L. Pan, P. Cobler, S. Rhodes, A. Potter, T. Chou et al, *Review of Scientific Instruments*, **72**, 1831 (2001).
22. S. C. Hill, R. P. Pinnick, S. Niles et al, *Applied Optics*, **40**, 3005 (2001).
23. Y. L. Pan, R. G. Pinnick, S. C. Hill, S. Niles et al, *Applied Physics B*, **72**, 449 (2001).
24. V. Boutou, C. Favre, S. C. Hill, Y. L. Pan, *Applied Physics B*, **75**, 145 (2002).
25. Y. L. Pan, S. C. Hill, J. P. Wolf, S. Holler et al, *Applied Optics*, **41**, 2994 (2002).
26. C. Favre, V. Boutou, S. C. Hill, W. Zimmer et al, *Physical Review Letters*, **89**, 035002-1 (2002).
27. Y. L. Pan, J. Hartings, R. G. Pinnick, S. C. Hill et al, *Aerosol Science and Technology*, **37**, 628 (2003).
28. V. Sivaprakasam, A. L. Huston, C Scotto, J. D. Eversole, *Optics Express*, **12**, 4457 (2004).
29. R. G. Pinnick, S. C. Hill, Y. L. Pan, R. K. Chang, *Atmospheric Environment*, **38**, 1657 (2004).
30. Y. L. Pan, V. Boutou, J. R. Bottiger et al, *Aerosol Science and Technology*, **38**, 598 (2004).
31. Y. L. Pan, J. D. Eversole, P. H. Kaye et al, in *Optics of Biological Particles*, eds. A. Hoekstra, V. Maltsev, and G. Videen (Springer, December 2006), p. 63-164.
32. Y. L. Pan, R. G. Pinnick, S. C. Hill et al, *Journal of Geophysical Research - Atmospheres*, **112**, D24S19 (2007).
33. H. C. Huang, Y. L. Pan, S. C. Hill, R. G. Pinnick, R. K. Chang, *Optics Express*, **16**, 16523 (2008).
34. Y. L. Pan, R. G. Pinnick, S. C. Hill, R. K. Chang, *Environmental Science and Technology*, **43**, 429 (2009).
35. V. Sivaprakasam, T. Pletcher, J. E. Tucker et al, *Applied. Optics*, **48**, B126 (2009).
36. S. C. Hill, M. W. Mayo, R. K. Chang, *US Army Research Laboratory Technical Report* (ARL-TR-4722, 2008), p. 1-49.
37. W. H. Nelson, R. Dasari, M. Feld, J. F. Sperry, *Applied Spectroscopy*, **58**, 1408 (2004).
38. T. D. Alexander, P. M. Pellegrino, J. B. Gillespie, *Applied Spectroscopy*, **57**, 1340 (2003).
39. C. Xie, Y. Q. Li, *J. Applied Physics*, **93**, 2982 (2003).
40. A. P. Esposito, C. E. Talley, T. Huser et al, *Applied Spectroscopy*, **57**, 868 (2003).
41. K. S. Kalasinsky, T. Hadfield, A. A. Shea et al, *Analytical Chemistry*, **77**, 4390 (2007).

42. R. M. Jarvis, A. Brooker, R. Goodacre, *Faraday Discussions*, **132**, 281 (2006).

43. A. Ben-David, H. Ren, *Applied Optics*, **42**, 4887 (2003).

44. B. Hettinger, V. Hohreiter, M. Swingle, D. W. Hahn, *Applied Spectroscopy*, **60**, 237 (2006).

45. J. D. Hybl, G. A. Lithgow, S. G. Buckley, *Applied Spectroscopy*, **57**, 1207 (2003).

46. H. J. Tobias, M. P. Schafer, M. Pitesky et al, *Applied and Environmental Microbiology*, **71**. 6086 (2005).

47. S. C. Russell, G. Czerwieniec, C. Lebrilla, *Analytical Chemistry*, **77**, 4734 (2005).

48. P. H. Kaye, N. A. Eyles, I. K. Ludlow, J. M. Clark, *Atmospheric Environment Part A-General Topics*, **25**, 645 (1991).

49. S. Holler, Y. L. Pan, R. K. Chang et al, *Optics Letters*, **23**, 1489 (1998).

50. Y. L. Pan, K. B Aptowicz, R. K Chang, M. Hart, J. D. Eversole, *Optics Letters*, **28**, 589 (2003).

51. K. B. Aptowicz, R. G. Pinnick, S. C. Hill, Y. L. Pan, R. K. Chang, *Journal of Geophysical Research - Atmospheres*, **111**, D12212 (2006).

52. At about 9:30 PM we realized we did not have cigarettes to test the system, and so Richard took one of us to buy some. Richard found it humorous that when asked what type of cigarette we wanted to buy, our response was, "The cheapest ones."

53. D. M. Murphy, A. M. Middlebrook, M. Warshawsky, *Aerosol Science and Technology*, **37**, 382 (2003).

54. J. A. Hartigan, *Clustering Algorithms* (John Wiley and Sons, New York, 1975).

55. M. Frain, D. P. Schmidt, Y. L. Pan, R. K. Chang, *Aerosol Science and Technology*, **40**, 218 (2006).

56. Z. Palkova, L. Vachova, M. Valer, T. Preckel, *Cytometry*, **59A**, 246 (2004).

57. L. DeJuan, J. F. DelaMora, *Journal of Colloid and Interface Science*, **186**, 280 (1997).

CHAPTER 10

DISCERNING SINGLE PARTICLE MORPHOLOGY FROM TWO-DIMENSIONAL LIGHT SCATTERING PATTERNS

STEPHEN HOLLER [1] and KEVIN B. APTOWICZ[2]

[1]Thermo Fisher Scientific, Redwood City, CA 94065, USA
[2]Department of Physics, West Chester University,
West Chester, Pennsylvania 19383 USA
stephen.holler@thermofisher.com
kaptowicz@wcupa.edu

Elastic scattering of light from micron-sized aerosol particles provides a wealth of information about the scatterer including shape, size, internal structure, surface roughness, composition, and orientation. The ability to understand and interpret properly the light scattering properties of irregular and nonspherical particles (including aggregates) is important to fields as diverse as global warming and food processing. Incident radiation is elastically scattered over 4π sr, however due to the complexity of the angular scattering and limitations on computing power, measurements and numerical comparisons are traditionally performed at a fixed azimuthal angle, while varying the polar angle ($0°$ – $180°$). This limited angular coverage excludes much of the information contained in the elastic scattering signal. The emergence of inexpensive detector arrays permits the measurement of two-dimensional angular optical scattering (TAOS) patterns with wider angular coverage (over 2π sr). In addition, increases in computational power make it feasible to model complex particle morphologies with numerical simulations. Over the past two decades, researchers at Yale University, University of Hertfordshire, the US Army Research Laboratory (ARL), and Porton Down have developed techniques for maximizing the collection of light scattered from nonspherical particles as well as data analysis techniques capable of extracting single particle morphology from the information-rich TAOS patterns. This chapter will present their scientific achievements, discuss current research efforts, and outline the major questions remaining in the field.

1. Introduction

The elastic scattering of light by objects is perhaps one of the most important physical phenomena there is. This particular interaction of light with matter provides for one of the most basic and fundamentally important human senses: vision. Without the ubiquitous scattering of light by material objects, we would be unable to visualize the world around us and make the requisite observations critical to understanding important physical phenomena.

Elastic scattering of light by micron-sized aerosol particles in the atmosphere is responsible for spectacular optical phenomena: from blue skies and spectacular sunsets to rainbows and halos, light scattering from a myriad of particle shapes produce impressive optical displays.[1] However, light scattering from airborne particulate matter can also have serious consequences, particularly on the radiation balance of the planet.[2] Understanding light scattering interactions and the elucidation of aerosol morphology may one day lead to improved climate models and the mitigation of anthropogenic causes of global climate change.[3] Furthermore, the use of light scattering from micron-sized particles may also serve a role in identifying airborne pathogenic matter (particularly in the case of biological weapons), aid in the identification of biological matter in extraterrestrial environments, and provide assurance as to the quality of urban air and water supplies.[4,5]

Since Gustav Mie first quantitatively described light scattering by spheres in 1908,[6] the vast majority of work in the field has been limited to measurements and calculations, which are one-dimensional. By one-dimensional, we mean that in the traditional light scattering geometry (Fig. 1(A)) data is obtained with a fixed azimuthal angle (typically $\phi = 90°$), while varying the polar angle ($0° \leq \theta \leq 180°$). While this may be adequate for single spherical particles or angle-averaged measurements of clouds of particles, such measurements exclude the richness of information available to the observer in the case of a single nonspherical particle.

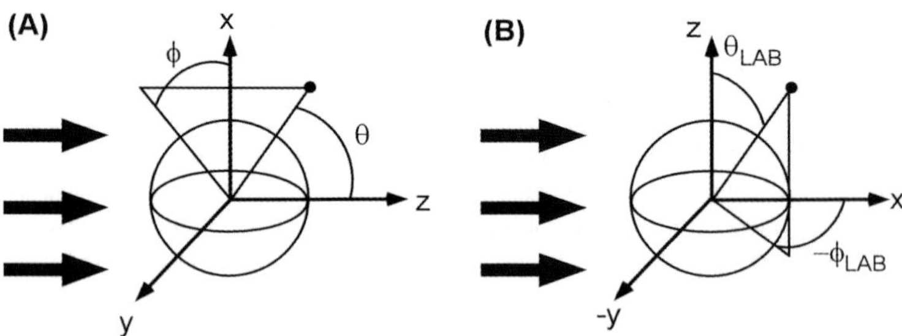

Fig. 1. (A) Traditional geometry used in light scattering calculations. (B) Coordinate frame for some experimental measurements in which particles fall in the −z direction. Unless denoted by the subscript lab, all angles refer to the traditional coordinate frame.

The limitations arising from one-dimensional light scattering measurements may be likened to observing the world through a narrow slit; only of small portion of the environment may be observed at any given instant. By widening the slit to a window, vastly more information can be observed at a single glance, providing a detailed snapshot of the environment at the time of the measurement.

Two-dimensional Angular Optical Scattering (TAOS) measurements seek to take advantage of advances in array detectors and computing power to observe large swaths of the angular scattering patterns associated with single, primarily nonspherical, aerosol particles in order to compare with theoretical computations, and employ sophisticated algorithms to elucidate classifying characteristics.[7] In what follows, we overview the TAOS technique and describe various approaches that have been utilized to make measurements of angular scattering patterns from a variety of aerosol particles. We discuss the intricate patterns observed from a myriad of particle shapes and sizes, approaches that enhance the amount of information obtained, computational techniques that demonstrate good qualitative agreement with the observed data, and analysis methods employed for discerning distinguishing characteristics for identifying and classifying unknown particles.

2. Measurement techniques

The acquisition of TAOS data requires high-numerical-aperture (NA) optics and large detector arrays to collect the scattered radiation spanning a large angular range. The emergence of inexpensive charge coupled device (CCD) cameras and the availability of precision optical components have made the implementation of TAOS measurements accessible.

In an ideal world, scattered light would be collected over 4π sr, however such a large angular coverage is not possible. Therefore, practical considerations come into play when determining the angular range to detect. Drive at night with a dusty windshield and the anisotropy of elastically scattered light becomes uncomfortably obvious as light from opposing headlights scatters towards you eyes. The high degree of forward scattering from small particles and the practical restrictions on the experimental configuration significantly impact the necessary considerations for the TAOS collection geometry. Measuring scattering in the forward direction provides gains in the scattered intensity, but is dominated by diffractive effects yielding little information other than aerosol size and shape. The backward scattered light exhibits greater dependency on the composition of the aerosol (*e.g.*, refractive index, absorption), but is a much weaker signal. In addition, the choice of the detector array and laser source(s) needs to be taken into account when deciding the collection angle. In the following sections, we highlight several collection geometries that have been used to measure TAOS patterns from single aerosol particles and briefly discuss their strengths and weaknesses.

2.1. *Abbe sine condition*

One convenient approach to collecting the elastically scattered light from an aerosol particle is to use a lens in such a way that the Abbe sine condition is satisfied. When this occurs, a small region of the object plane near the optic axis of the system is imaged sharply despite the angular divergence of rays entering the imaging system.[8]
In the case of TAOS measurements, where the image plane (*i.e.*, the CCD array) is placed far from the lens, the Abbe sine condition takes a simple form:

$$f \sin \gamma = h,\tag{1}$$

where f is the focal length of the lens, γ is the angle of the scattered ray makes with the lens' optic axis, and h is the displacement from the optic axis of the ray emerging from the lens. This expression is arrived at by applying Fermat's principle to the optical system and evaluating the line integrals for the optical path length. The consideration that the image plane is far from the lens reduces the general expression to the convenient form expressed by Eq. 1.

With respect to TAOS measurements, light scattered from a particle positioned at the focal point (f) of the lens enters the lens at an angle (γ) and emerges with an off-axis distance h proportional to the sine of the scattered angle. The choice of sine over tangent is subtle, but is related to the shape of the principal plane associated with the focal length of the lens. By considering a spherical principal plane of radius f, it becomes clear that the height of the emergent ray is proportional to the sine of the incident angle.

The advantage of the Abbe sine condition is that it provides a convenient means of transforming the scattered angle into a linear displacement across the face of the array detector. However, the physical dimension of the collection optics provides a practical limitation on the degree of angular coverage. Additionally, this mechanical obstruction precludes the measurement of TAOS patterns close to the forward and backward directions, $\theta = 0°$ and $180°$ respectively in the traditional scattering geometry (Fig. 1(A)). For these reasons, other collection geometries are considered.

2.2. *Off-axis parabolic mirrors*

An alternative collection geometry that allows for light scattering measurements to be made close (*i.e.,* within $5°$) to the forward and backward direction employs an off-axis parabolic mirror. These optical elements are mirrors having a curvature associated with a section of a paraboloid of revolution. Collimated light parallel to the rotation axis that impinges on the mirror surface will focus to a point corresponding to the vertex of the paraboloid. Depending on the segment of the paraboloid that is chosen, the distance from the mirror edge and the paraboloid vertex varies and, correspondingly, so does the off-axis angle.

In TAOS measurements, the particles of interest are illuminated at the focal point of the parabolic mirror so that the collected scattered light emerges from the mirror collimated and is directed towards the CCD camera.[9] The principal advantage of the off-axis mirror is that it can be positioned in such a way that it does not obstruct the illuminating beam, but allows for the collection of scattered signal in the near-forward or near-backward direction. Another advantage of this configuration is that it utilizes a reflective optical component thus eliminating chromatic aberrations common in refractive optical components. Unfortunately, TAOS configurations incorporating off-axis parabolic mirrors still lack wide angular coverage, thereby limiting the solid angle of collection. To achieve greater solid angle coverage, another approach is required.

2.3. *Ellipsoidal collection optics*

An elegant approach that provides wide solid angle coverage and the potential for simultaneous TAOS measurements in the forward and backward directions incorporates an ellipsoidal reflector.[10,11,12] In this geometry, the scattered radiation from a particle illuminated at the focal point within the ellipsoidal reflector is focused to the secondary focal point. Here a spatial filter can be placed to eliminate any stray light and provide a clean TAOS image. By placing a detector a distance z behind the spatial filter, we may invoke the Fraunhofer approximation and observe the far-field scattering pattern. The Fraunhofer approximation is given by[13]

$$z > \frac{kD^2}{2}, \tag{2}$$

where k is the wave vector of the incident beam and D is the diameter of the scatterer.

The angular coverage afforded by the ellipsoidal reflector is significantly greater than that achievable using either the Abbe sine condition or the parabolic reflector, even for the fastest optics available. Furthermore, illumination of the particle transverse to the symmetry axis of the reflector allows large portions of the forward and backward scattering signal to be obtained. One disadvantage of the ellipsoidal reflector is the short depth-of-field, which can lead to aberrations in the TAOS image for particles illuminated slightly (i.e., tens of microns) off the focal point. However, judicious use of cross-beam triggering can minimize the aberrations by recording only TAOS images associated with those particles which are well within the focal volume.

3. TAOS measurements

Two-dimensional Angular Optical Scattering (TAOS) measurements have been made from a myriad of particle morphologies in order to attempt to discern characteristic features that may be employed in discriminating and classifying unknown scatterers. In this section, we survey some of the TAOS measurements that have been made using the light collection geometries discussed in §2. Of particular interest are nonspherical and absorbing particles, including aggregates and spores, since these particle types are prominent in environmental contaminants, climate forcing, and biological weapons.

3.1. *Spheres and spheroids*

The physics governing scattering from spherical particles is well understood and serves as a useful tool for validating the collection geometries.[14,15] Quantitative comparisons among experiment and Lorenz-Mie theory are easily made using widely available computational tools.

TAOS patterns obtained at a wavelength of 3.41 μm from spherical droplets of H_2O, D_2O, and their 50/50 mixtures are shown in Fig. 2(a)-(c). At this wavelength H_2O is more absorbing than D_2O. Below each of the images are slices of the experimental data where the subtle effect of changing the complex refractive index is evident.[16] In addition, the numerical simulation based on Lorenz-Mie theory is shown and serves to validate the TAOS approach to studying light scattering from individual aerosol particles.[17]

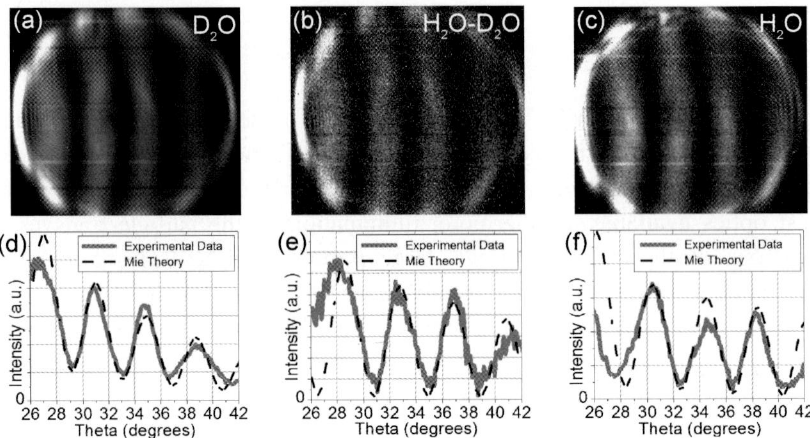

Fig. 2.   (a)-(c) Measured TAOS patterns from liquid droplets comprised of D_2O, H_2O, and a 50/50 mixture of each. (d)-(f) Slices through the respective images along with Lorenz-Mie calculations showing good agreement.

Light scattering from smooth spherical particles produce a well-known and easily characterized ring pattern in both the forward and backward direction. However, this is not necessarily the case for smooth particles having a spheroidal geometry. Marston and coworkers have demonstrated this effect in a series of light scattering experiments on acoustically levitated oblate spheroids.[18] A striking feature observed in light scattering from spheroidal particles is akin to the rainbow, and known as a hyperbolic umbilic diffraction catastrophe.[19]

Fig. 3 shows a fluorescent image of the tilted prolate spheroid (aspect ratio D/H = 1.2) and a portion of the backward scattering signal observed from a prolate spheroid tilted 23° towards the incident beam.[7] The observed pattern resembles the hyperbolic umbilic diffraction catastrophe. This particular catastrophe has been well-studied in both optical and acoustic scattering phenomena.[20] As a generalization of the rainbow, it may be considered geometrically as the caustic formed by the merger of four rays: two rainbow rays (*i.e.*, in the plane of the major axis) and two skew rays (*i.e.*, out of plane of the major axis).

In addition to the observed caustics, this TAOS image shows a fracturing of the elongated rings associated with a prolate spheroid illuminated orthogonal to its major axis. The break-up of the rings is the result of path length differences as scattered rays traverse the interior of the tilted particle and subsequently interfere in the far-field. T-matrix computations for a similar (*i.e.*, same aspect ratio) scatterer qualitatively show the same features, with one characteristic feature being the angular location of the cusp point. For oblate spheroids the cusp point has been shown to be a function of the aspect ratio;[21] the same is true for prolate spheroids. Here the local curvature induced by the particle's aspect ratio affects the emerging wavefront leading to the observed caustic. Thus the location of the cusp point in the TAOS pattern may serve as an indicator of the aspect ratio of the spheroidal particle.

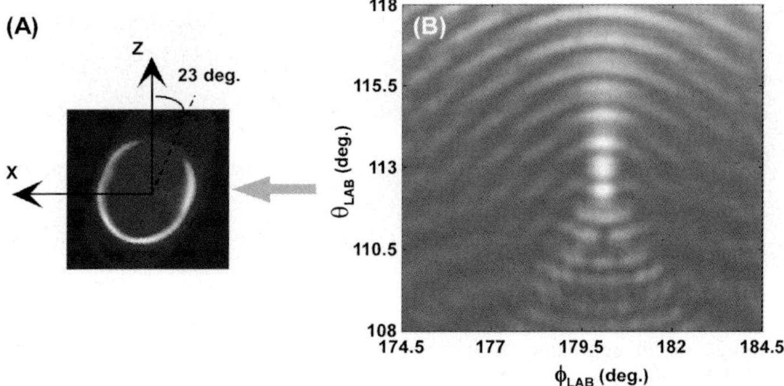

Fig. 3. (A) Fluorescent image of the tilted spheroid. (B) Backward scattering TAOS image from a tilted prolate spheroid. The resulting patterns shows a distinctive inverted 'Y' along with a breaking up of the rings despite the smooth spheroidal geometry.

3.2. *Aggregates and spores*

The vast majority of airborne particulate matter is neither spherical nor smooth. Complex morphology including surface roughness and aggregation are common among aerosol particles. For example, environmentally important aerosols such as black carbon soot particles can form fractal aggregates, while frozen water takes on a variety of shapes such as hexagonal columns, platelets, and rosettes. The airborne transmission of disease such as influenza involves the inhalation of pathogenic biological particles. In the case of biological weapons, airborne pathogens, such as Bacillus anthracis, the causative agent of Anthrax, are sporulated; approximately cylindrical with a coarse surface.

An example of TAOS data from spores of *Bacillus subtilis* (BG), a simulant for *Bacillus anthracis*, is shown in Figure 4. The SEM image Fig. 4(A) shows a close packing of the spores, cylinders approximately 1-μm long. The TAOS images seen in Fig. 4(B) & (C) correspond to the forward and backward scattering, respectively, in the laboratory coordinate frame (Fig. 1(B)), and were obtained by illuminating the aggregate with 532-nm laser light. The surface roughness associated with such particles is significant compared to the incident wavelength and results in a speckle pattern. The forward scattering data exhibits a trace of ring-like pattern out to ϕ_{Lab} ~23°, but shows significant break-up at larger scattering angles. This is not surprising since the general cluster geometry has a spherical shape and forward scattering is dominated by diffractive effects. The islands of intensity seen in these images are representative of TAOS observations from aggregates, and may be used to elucidate information about particle morphology.

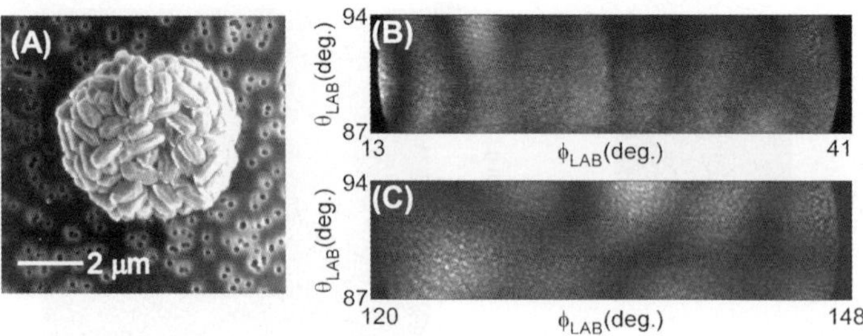

Fig. 4. (A) SEM image of an aggregate of *Bacillus subtilis* spores. Representative TAOS images, in the laboratory coordinate frame (Fig. 1(B)), from such aggregates in the forward (B) and backward (C) directions.

3.3. *Multicolor TAOS*

The major factor that influences the observed TAOS patterns is the shape of the aerosol being illuminated. However, dispersion within the material that comprises the aerosol particle has an influence on the scattered radiation. Absorption of the incident light by the scatterer results in a diminished contrast between peaks and valleys in the measured TAOS pattern, as has been shown in Fig. 2. Multiple measurements taken in different spectral regions serve to elucidate the inherent absorption of the aerosol and thus provide a key indicator of the constituent properties of the particle.[16] Such measurements are particularly important when measuring scattering from biological particles, or environmentally relevant black or brown carbon particles.[2,22] In addition, by varying the wavelength of light illuminating the scatterer one probes different length scales of the scatterer's morphology, which can also lead to drastic changes in the TAOS pattern.

The emergence of powerful semiconductor lasers and the availability of inexpensive color CCD cameras have enabled multi-color TAOS data to be recorded simultaneously with a single device. This eliminates the technical challenges associated with simultaneous acquisition of data by multiple camera systems.

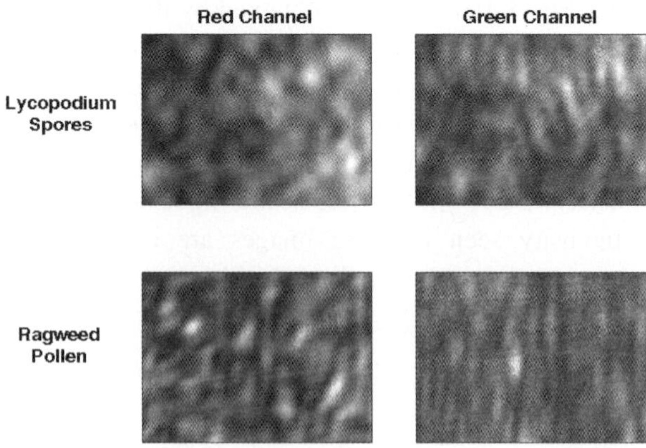

Fig. 5. TAOS patterns in the near-forward direction ($13° < \phi_{Lab} < 28°$ & $80° < \theta_{Lab} < 100°$) acquired simultaneously with red and green incident light from Lycopodium spores (top) and Ragweed Pollen (bottom) particles.

Fig. 5 shows TAOS data from Lycopodium spores and Ragweed pollen particles obtained in the near-forward direction ($13° < \phi_{Lab} < 28°$ & $80° < \theta_{Lab} < 100°$) with a single color CCD camera while being simultaneously illuminated by two diode laser sources (532 nm and 660 nm). Both particle types have a general spherical shape, but exhibit significant surface roughness and corrugation that lead to the speckle pattern seen in the images. The density of the speckle pattern is greater in the green channel (RIGHT), which is consistent with a larger optical size parameter. Despite the apparent randomness, classification based on a multivariate analysis of properties inherent to the TAOS image is possible and is discussed in §4.4.3.

4. Data analysis

Several approaches have been implemented in the analysis of TAOS data in order to elucidate information regarding the scatterer morphology. These approaches have employed image processing routines and statistical algorithms to characterize and classify aerosol particles based on the observed TAOS patterns. In this section, we highlight different analysis techniques and illustrate the results of each approach.

4.1. *Watershed routine*

A quick comparison of TAOS data from different particle sizes illustrates the variability in the peaks and valleys in the scattered intensity that ripple across the observed solid angle. As expected, the density correlates with particle size so that a first, albeit rough, measure of particle size may be obtained by quantifying the density of the peaks and valleys in the recorded TAOS image.

This quantification may be performed through the application of a watershed routine commonly used in image processing. The routine involves pre-processing of the image through normalization and the application of a blur filter to remove any high-frequency artifacts associated with the experimental measurement (*e.g.*, etalon effects, diffraction from dust on the CCD elements). The normalized and suitably smoothed image is then analyzed by counting the number of valleys (*i.e.*, depressions in the signal intensity) that meet or exceed a pre-defined threshold level. The number of peaks is obtained by inversion of the image and application of the same watershed counting routine. In both cases, watershed regions that touch the boundary of the image are excluded.

4.1.1. *Watershed Results*

The watershed routine has been applied to TAOS patterns obtained from light scattering from aggregates of 1.33-μm-diameter polystyrene latex (PSL) spheres. The aggregates were generated with an ink-jet aerosol generator[23] that produces a 50-μm-diameter water droplet containing a prescribed concentration of PSL spheres. Upon evaporation of the water, the PSL spheres form an aggregate whose diameter depends on the initial concentration of the droplet generating solution. TAOS data was recorded from aggregates ranging in diameter from 5 μm to 19 μm.

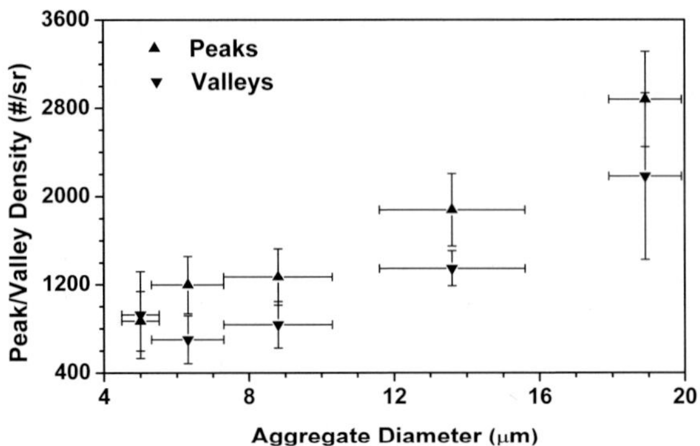

Fig. 6.   Results of the watershed routine analysis of series of TAOS images obtained from different diameter aggregates comprised of 1.33-μm diameter polystyrene latex spheres.

The results of the watershed routine for the PSL aggregates studied are shown in Fig. 6. The points represent the mean value while the error bars denote the spread in the measurements from the dataset. TAOS data were obtained from forward scattering measurements in the angular range of $14° \leq \phi_{LAB} \leq 41°$ and $87° \leq \theta_{LAB} \leq 94°$. The counts obtained from the watershed routine were normalized by the observed solid angle to provide a density of peaks or valleys. A clear trend is seen in Fig. 6 as the density of both peaks and valleys increases with increasing aggregate diameter.

4.2. *Symmetry analysis*

While the watershed routine provides an indication of the particle size based on the density of the speckle pattern in the TAOS image, it does not provide much information regarding the shape of the scatterer. A useful measure of particle shape may be determined by examining the symmetry of the observed scattering pattern. This requires a measurement technique like the ellipsoidal collection geometry discussed in §2.3 that encompasses a large azimuthal range (Fig. 1(A)).

Several groups have employed symmetry analyses to their light scattering measurements to discern information about particle size and shape.[24,25,26] Although each group has a different method for analyzing the scattering pattern, each approach considers the rotational symmetry of the pattern. In one scheme, a series of three detectors are positioned at the same polar scattering angle (θ) but at different azimuthal angles ($\phi = 0°$, 120°, and 240°).[25] To quantify the asymmetry of the scattered light, they introduced an *asymmetry factor* (A_f), based upon the relative magnitudes of the three detector signals. The general form for A_f is defined to be:

$$A_f = \frac{F}{<D>} \sqrt{\sum_{i=1}^{N}(<D> - D_i)^2} \,, \tag{3}$$

where D_i is the signal recorded by the ith detector, $<D>$ is the mean detector signal, N is the number of detectors, and F is a scaling factor (that assures the maximum value of A_f is 100) defined as

$$F = 100 \cdot \left[N(N-1) \right]^{-1/2} . \tag{4}$$

Another measure of particle symmetry is known as the sphericity index (SPX). This parameter may be expressed in terms of the asymmetry factor as:

$$SPX = 1 - \frac{A_f}{F(N-1)} , \tag{5}$$

and has been employed in an eight-detector scheme to measure the properties of ambient aerosols, and found that <10% of particles in the 0.2 – 0.8-μm range were nonspherical.[27,28]

A third gauge of the scattering pattern symmetry is the Degree of Symmetry (D_{sym}) defined as

$$D_{sym} = 1 - \sum_{pixel\,subset} \left| \frac{I^N(\theta,\phi) - I^N(\theta,\phi+\pi)}{2} \right| , \tag{6}$$

where $I^N(\theta,\phi)$ is the single pixel intensity value at angle pair (θ, ϕ).[26] This metric differs from A_f and SPX in that it was developed to analyze high-resolution patterns generated with linearly, rather than circularly, polarized light. Generally speaking, if a particle possesses mirror symmetry about the polarization plane of the incident beam, the corresponding TAOS pattern will also exhibit mirror symmetry resulting in $D_{sym} = 1$, otherwise D_{sym} is less than one.

4.2.1. *Degree of symmetry results*

The discrimination and classification of unknown aerosol particles, particularly those of biological origin that pose a pathogenic threat is important for obvious reasons. Fortunately, those biological aerosols that pose the greatest threat are nonspherical, allowing them, based on the D_{sym} parameter, at least to be distinguished from spherical particulate matter. A survey of over 6,500 particles (including ambient aerosols, well-characterized spheres, and threat particles) was conducted to demonstrate the role that D_{sym} can play in real-world discrimination of aerosol particles.[26]

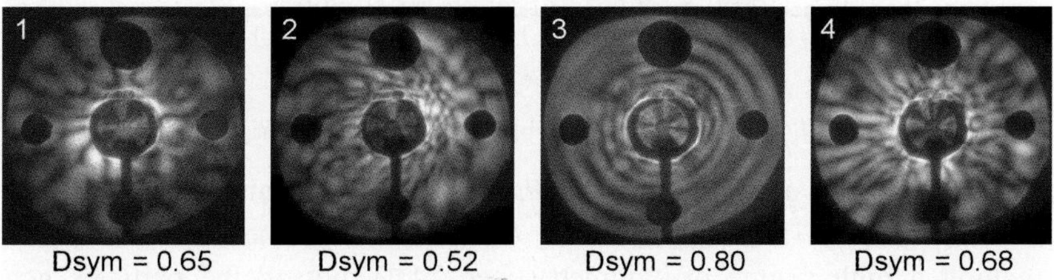

Fig. 7.   TAOS images and computed D_{sym} values for different ambient aerosol particles.

Fig. 7 shows examples of TAOS data obtained with the ellipsoidal collection geometry from various ambient aerosol particles. The five dark spots are artifacts corresponding to the apertures in the ellipsoidal mirror. The variation of the extent of speckle seen in the images points to the degree of complexity found in naturally occurring aerosol particles. Also displayed in Fig. 7 is the computed value of D_{sym} for each of these images. Pattern 3, which appears to be generated from a sphere-like scatterer, has the highest D_{sym} value.

The degree of symmetry was computed for the broad dataset studied. Fig. 8 summarizes the results. As expected, the Dioctyl phthalate (DOP) droplets and PSL spheres exhibit a high D_{sym} value, approaching 0.9. Instrumental artifacts, as well as experimental error (*e.g.*, small mirror deformations, slight misalignments), serve to reduce the observed value of D_{sym}, for spherical particles, from the ideal value of 1.0.

The degree of symmetry of ambient aerosols has a wide variability, with the bulk of the particles having D_{sym} in the range of 0.6 to 0.7. This is not surprising considering that pollen, dust, and other ambient particles are generally nonspherical having hard edges, pits, and corrugated surfaces. However, close examination of the ambient particle data indicates a bimodal distribution in the D_{sym} value, with 16% of the particles exhibiting sphere-like scattering properties, while the remaining data are attributed to nonspherical characteristics of the measured particles.

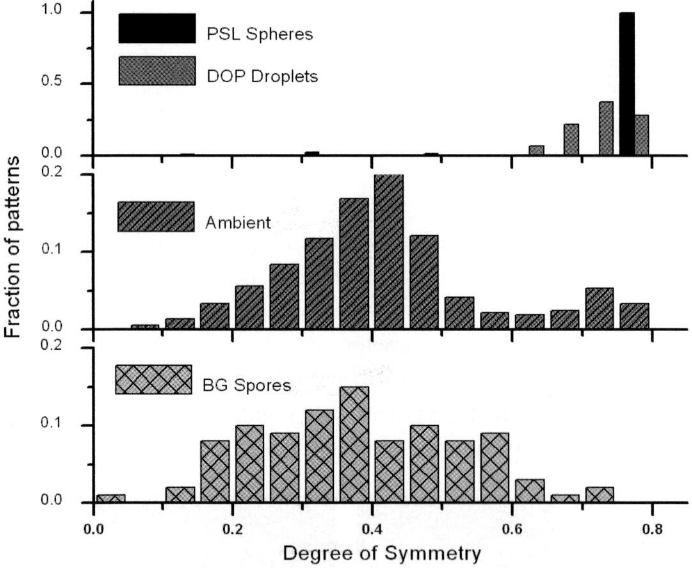

Fig. 8.   Histograms of the Degree of Symmetry for different particles.

The analysis of powder-dispersed *Bacillus subtilis* spores also showed a significant degree of variability ($0.15 \leq D_{sym} \leq 0.83$). Although TAOS measurements from single spores exhibit a high degree of symmetry (see §4.4), not all the particles generated contained single spores, but may have been comprised of several spores forming aggregates that lack the symmetry associated with single spores. Despite this, coupled with other techniques (*e.g.*, fluorescence), the D_{sym} parameter can provide a useful indicator for the presence of pathogenic bio-aerosols entrained in an ambient airstream.

4.3. *T-matrix for aggregates and spores*

The T-matrix (transition matrix) technique is one of the most widely used methods for rigorously calculating light scattering from nonspherical particles. Since its implementation,[29,30] it has been used to calculate light scattering for a variety of nonspherical particles.[31] Its point-group symmetries have been studied,[32] and it has even been cast as a T-matrix in spheroidal coordinates.[33] Mishchenko *et al.*[34] have presented an excellent review of T-matrix computations for nonspherical particles.

The problem of light scattering by an aggregate of spheres has generated great interest in recent years,[35,36,37,38] while increased computational power combined with innovative experiments[7,39,40,41] has made it easier to compare theoretical and experimental results. One approach employed to make such comparisons involves the formulation of a global T-matrix.

Consider N isotropic, homogeneous spheres of radius a and refractive index n_1 arranged in a compact cluster. The origin of each sphere O_i is defined in the principal coordinate system by a position vector r_i. The relative position vector between any two spheres i and j is r_{ij}. Invoking the superposition principle, the total external field may be considered as the sum of the incident and scattered fields.

Because of the close proximity of the spheres, multiple scattering plays an important role, and the effective incident field for the ith scatterer is comprised not only of the source field, but also the scattered fields from all other particles in the cluster

$$\vec{E}^{\,i}_{\,eff} = \vec{E}_{\,inc} + \sum_{\substack{j=1 \\ j \neq i}}^{N} \vec{E}^{\,j}_{\,sca} \, . \tag{7}$$

The T-matrix formalism relates the scattered field of the ith sphere to the effective incident field. However, in Eqn. 7 the incident field must be expanded in the principal coordinate system, while the scattered field from each sphere is expressed in the local coordinate system.

In order to deal with the issue of multiple scattering, both fields must be expressed in terms of the ith scatterer coordinate system. By applying the translation-addition theorem for vector spherical harmonics,[42,43] the problem can be reduced to a group of N coupled linear equations whose unknown variables are the expansion coefficients of the scattered field of each individual sphere.

The N-scatterer T-matrix of the ith scatterer ($T^{i(N)}$) contains all the relevant information regarding multiple scattering effects among the particles that comprise the aggregate. The incident field is then linked to the scattered field of the ith scatterer through the T-matrix:

$$f^{\,i(N)} = T^{\,i(N)} a^i = T^{\,i(N)} \beta(i,0) a^0 \, , \tag{8}$$

where $\beta(i,0)$ is a translation matrix[42,43] for the incident field, and $f^{\,i(N)}$ and a^i are defined in the coordinate system of the ith scatterer.

Translating everything back into the principal coordinate system allows for the generation of the global T-matrix ($T^{T(N)}$), which relates the incident field coefficients to the coefficients associated with the total scattered field:

$$f^{T(N)} = T^{T(N)} a^0. \tag{9}$$

Here the global T-matrix is related to the ith scatterer T-matrix through the translation matrix:

$$T^{T(N)} = \sum_{i=1}^{N} \beta(0,i) T^{i(N)} \beta(i,0). \tag{10}$$

The global T-matrix is independent of the incident field, depending solely on the morphology and configuration of the constituent particles of the aggregate.

4.3.1. *Computational comparisons – aggregates*

Quantitative comparisons among experimental TAOS data from single aggregate particles and corresponding calculations are difficult due to the degrees of freedom associated with real-world clusters. Two key variables that prohibit direct comparisons are the packing structure of the aggregate and the orientation relative to the illuminating beam. Despite the complexity of the problem, it is possible to make qualitative comparisons among experimental and theoretical TAOS patterns.

Forward- and backward-scattering TAOS images were recorded from aggregates comprised of 2.29-μm-diameter polystyrene latex spheres so that qualitative comparisons could be made using the T-matrix approach.[44] Fig. 9(a) shows a scanning electron microscope (SEM) image of a representative aggregate. Statistical analyses of over 200 aggregate particles indicated a median diameter of 6.7 μm while the most probable number of PSL comprising the cluster was determined to be 18. Fig. 9(b) shows a model aggregate comprised of 13 spheres in a close-packed arrangement having the same optical characteristics (*i.e.*, optical size and refractive index) as the experimental aggregate. T-matrix calculations were performed using the model aggregate oriented at different tilt angles (0°, 5°, and 45°) relative to the incident beam.

Fig. 9.   (a) SEM image of an aggregate comprised of 2.29-μm-diameter PSL; (b) Model aggregate comprised of 13 spheres having the same optical characteristics as the PSL spheres used in TAOS experiments.

Comparisons among the experimental data and the theoretical calculations were performed by scanning a solid angle that corresponded with experimental observation in the computational space to find a pattern that resembled an observed TAOS image. Fig. 10 shows examples of representative forward and backward TAOS images along with equivalent portions of the T-matrix computation that compare well, qualitatively, with the observations. For the numerical data, the angle through which the model cluster was tilted is indicated above the corresponding TAOS pattern calculation.

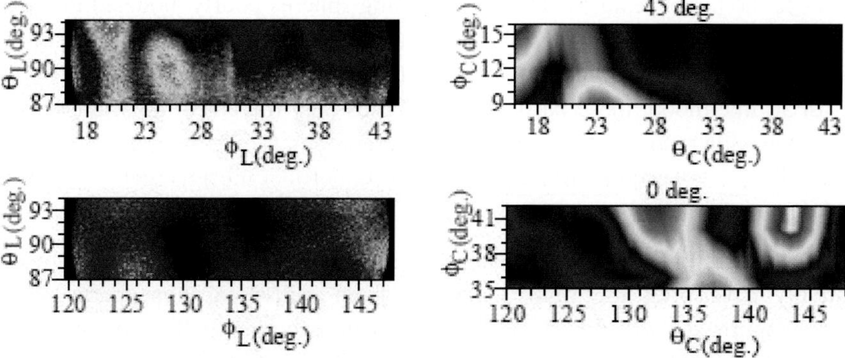

Fig. 10.   Experimental (left) and computational (right) TAOS patterns from aggregates of PSL spheres. The T-matrix calculations demonstrate good qualitative agreement with observations.

4.3.2. *Computational comparisons – spores*

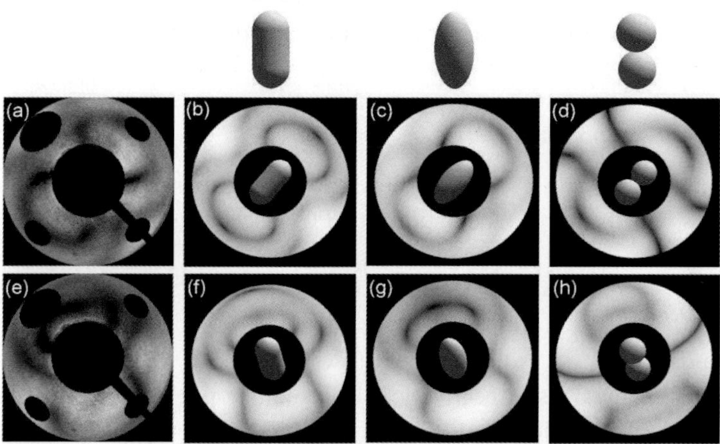

Fig. 11.   (a & e) Experimental TAOS data using the ellipsoidal reflector geometry with the patterns rotated so incident laser beam is vertically polarized. Scattering based on numerical simulation from (b & f) cylinder with spherical endcaps, (c & g) spheroid, and (d & h) two touching spheres. The scattered intensity is plotted on a logarithmic scale with maximum intensity coded white. The overlaid images of the simulation particles show their orientation.

The T-matrix method has also been applied to modeling scattering from single BG spores. In particular, recent research focused on shapes that accurately reproduce the scattering pattern from a BG spore. The shapes that were analyzed were a cylinder with spherical endcaps, an ellipsoid, and two touching spheres as shown in Fig. 11.[45] The black spot in the middle and the four dark regions near the edge are artifacts associated with the ellipsoidal collection geometry, which detects scattered light from $77° \leq \theta \leq 130°$ and $0° \leq \phi \leq 360°$ with the center corresponding to $\theta = 180°$. The cylinder with spherical endcaps most accurately captured the scattering behavior of a single BG spore. The scattering patterns from the spheroid lacked some of the finer features but were able to reproduce the larger features. The two touching spheres poorly modeled the scattering behavior of a BG spore. Note that the patterns are rotated indicating that the spores did not have their long axis aligned to the air-stream which carried it through the sample region. In all likelihood, these particles tumbled as they traversed the scattering volume leading to the wide range of observed D_{sym} values.

4.4. *Multivariate statistics*

Although the computational approaches discussed in §4.3 provide a means for qualitatively describing the light scattering from individual nonspherical particles, particularly aggregates, the difficulty associated with experimentally knowing the precise orientation and arrangement of constituent particles makes quantitative comparisons nearly impossible. Orientational averaging can only provide general information about dispersed clouds of nonspherical particles. Despite the fact that the apparent randomness of TAOS patterns precludes the use of quantitative modeling techniques, multivariate statistical algorithms have demonstrated the ability to extract useful information from TAOS data from individual particles, and subsequently classify based on this information.[46,47,48]

4.4.1. *Principal component analysis*

Principal Component Analysis (PCA) allows a raw data matrix X to be represented by fewer variables, *i.e.*, the principal components (PCs), according to

$$X = \sum_{a=1}^{A} t_a \otimes p_a = E = T \cdot P + E . \tag{11}$$

Here the summation limit A is the number of PCs retained, T ($M{\times}A$) and P ($A{\times}N$) are the scores and loadings matrices respectively; t_a is a column of T; the vector p_a (*i.e.*, the PC of index a) is a row of P; $\otimes$ denotes the Kronecker product; $E = [e_{ij}]$ is the error matrix of differences between the original data X and the modeled data ($T{\cdot}P$), where modeling implies approximation.

PCA effectively serves to reduce the dimensionality of a data set. The PCs are, in fact, the eigenvectors of the covariance matrix of the data, thus making each PC uncorrelated with any of the others. The corresponding eigenvalues provide a measure of the information content described by the associated PC. Hence, when the first few PCs already represent a large proportion of the total information in the raw data, the remaining PCs may be considered to represent primarily noise in the data and therefore can be ignored.

The graphical representation of the scores matrix (T) has been shown to aid with the preliminary investigation of relationships among groups of objects. Namely, data points with similar features have close coordinates.

4.4.2. *Disciminant function analysis*

Discriminant Function Analysis (DFA)[49,50,51] is another multivariate technique that is employed to determine a direction in which to project the data (*e.g.*, PC scores) that maximizes separation between classes. Applied to the Scores matrix obtained from Principal Component Analysis, the discriminant function $D_\ell(i)$ is the affine map

$$D_\ell(i) = \alpha_{\ell,1} t_{i,1} + \alpha_{\ell,2} t_{i,2} + \ldots + \alpha_{\ell,A} t_{i,A} + \alpha_{\ell,0}, \tag{12}$$

where i is the TAOS pattern index, the α's are the unknown coefficients, the PC scores (t_i) are regarded as independent variables, and ℓ is the function index. The function index ranges from 1 to the lesser of $g - 1$ (the number of dependent groups $-$ 1) or k (the number of independent variables). It is important to note that the number of discriminant functions is less than the number of classes for discrimination. In general, one fewer function is needed. The exception is when there are fewer variables than classes.[52]

The unknown coefficients (α) are determined by solving the general eigenvalue problem:

$$(S - W) \cdot V = \lambda \cdot W \cdot V, \tag{13}$$

where S is the Total Sums of Squares and Cross-products matrix, W is the Within-groups Sums of Squares and Cross-products matrix, V is the unscaled matrix of discriminant function coefficients and λ is the diagonal matrix of eigenvalues. The solution to Eq. 13 is such that the distance between classes is maximized.

Though mathematically distinct, each discriminant function provides a dimension that differentiates a case into categories of the dependent variables based on its values on the independent variables. The first function is the most powerful differentiating dimension. However, subsequent functions may also represent additional significant dimensions of differentiation.

Analogous to principal component analysis, the relative percentage of a discriminant function equals a function's eigenvalue divided by the sum of all eigenvalues of all discriminant functions in the model. Thus it is the percent of discriminating power for the model associated with a given discriminant function. Relative percentage is used to tell how many functions are important. One may find that only the first two or so eigenvalues are of importance.

4.4.3. *Multivariate statistical results*

The application of multivariate statistical algorithms to key features extracted from the TAOS images lead to transformations that aid in classifying unknown aerosol particles based on known data sets. Previously, TAOS data from aggregates and other nonspherical particles has been analyzed by applying multivariate algorithms to the power spectrum of the image.[48] Recent work on multicolor TAOS data (Fig. 5) has taken a different approach by seeking out other features in the data that may enhance discrimination capabilities.

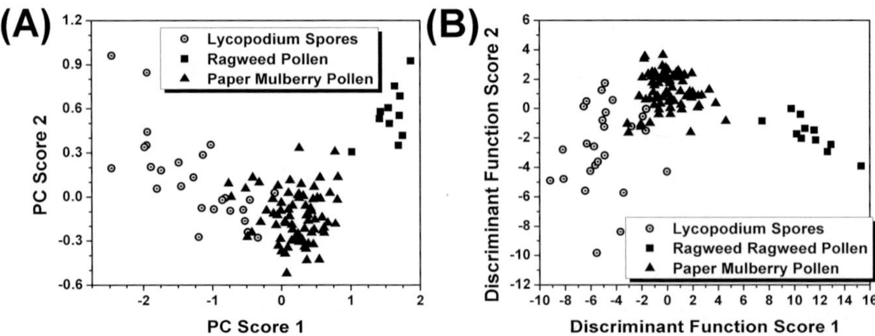

Fig. 12.   (A) Plot of the first two principal component scores obtained from analysis of multicolor TAOS data from particles comprised of Lycopodium Spores, Ragweed Pollen, and Paper Mulberry Pollen. (B) Plot of discriminant function scores obtained from subsequent analysis of PCA scores for the same particle data.

Among the properties analyzed for incorporation into the multivariate algorithms were the mean intensity and entropy values for each of the illuminating color channels. The mean intensity provides an indication of the scattering cross-section by quantifying the amount of light incident on the CCD detector in a given solid angle, while the entropy is a statistical measure of randomness that is often used to characterize the texture of an image. Preliminary results of the multivariate analysis of TAOS data obtained from aggregates comprised of either Lycopodium spores, Ragweed pollen, or Paper Mulberry pollen is presented in Fig. 12. The first two principal component scores are shown in Fig. 12(A), while the two discriminant function scores obtained by subsequent analysis of the PC scores are plotted in Fig. 12(B). Separation of the classes, although not perfect, is apparent in both of these plots. Also, consistent with previous discussions, the discriminant function scores show a transformation of the PC scores data into a tighter arrangement of the classes. The ability to distinguish among the particles based on properties inherent to the scattering provides a potential means for identifying unknown particles when compared against a known database has significant implications for security applications.

5. Summary

The complex data derived from two-dimensional light scattering images of aerosol particles provides important insights in the morphology of the aerosol. Whether the particle is a smooth spheroid or a corrugated aggregate, TAOS data provides a wealth of information that may be analyzed in various ways to discern characteristics. The experimental techniques, particles studied, and data analysis tools presented herein merely scratch the surface of what is possible in the characterization and classification of particles. However, we have shown that TAOS data provides access to useful information otherwise lost in traditional one-dimensional scattering measurements. The widespread availability of inexpensive lasers sources and array detectors has opened a window through which the larger scattering picture may be seen.

Acknowledgments

We are grateful to Professor Richard Kounai Chang for providing us the opportunity to work in his lab and learn a great deal from him, both scientifically and otherwise. Our time with Professor Chang has provided us with the audacity to tackle complex problems for which simple solutions do not exist, the wisdom to know when this occurs, and the ingenuity to extract meaningful information anyway; these skills will forever serve us well.

One of us (SH) wishes to acknowledge support from the US Army Corps of Engineers (USACE) Engineer Research and Development Center (ERDC) through the Small Business Innovation Research (SBIR) program (contract # W912HZ-06-C-0001) for recent work on multicolor TAOS instrument development.

References

1. M. Minnaert, The Nature of Light and Colour in the Open Air (Dover, New York, 1954)
2. M.Z. Jacobson, Nature 409, 695 (2001).
3. P. Yang, Q. Feng, G. Hong, et al., Aerosol Sci. 38, 995 (2007)
4. W.S. Bickel, J.F. Davidson, D.R. Huffman, et al., Proc. Nat. Acad. Sci. USA 73, 486 (1976)
5. Sindoni, R. Saija, M.A. Iati, et al., Opt. Exp. 14, 6942 (2006)
6. G. Mie, Ann. Phys. 25, 377 (1908)
7. S. Holler, Y.L. Pan, R.K. Chang, et al., Opt. Lett. 23, 1489 (1998)
8. M. Born and E. Wolf, Principles of Optics, 6e (Oxford, Pergamon Press, 1980)
9. S. Holler, M. Surbek, Y.L. Pan, et al., Opt. Lett. 24, 1185 (1999)
10. P.H. Kaye, E. Hirst, J. M. Clark et al., J Aerosol Sci 23, 597 (1992)
11. Y.L. Pan, K.B. Aptowicz, and R.K. Chang, Opt. Lett. 28, 589 (2003)
12. G.E. Fernandes, Y.L. Pan, R.K. Chang, et al., Opt. Lett. 31, 3034 (2006)
13. J.W. Goodman, Introduction to Fourier Optics, 2e, (McGraw Hill, New York, 1996)
14. H.C. van de Hulst, Light Scattering by Small Particles (Dover, New York, 1981)
15. C.F. Bohren and D.R. Huffman, Absorption and Scattering of Light by Small Particles (John Wiley & Sons, New York, 1983)
16. K.B. Aptowicz, Y.L. Pan, R.K. Chang, et al., Opt. Lett. 29, 1965 (2004)
17. P.W. Barber and S.C. Hill, Light Scattering by Particles: Computational Methods (World Scientific, Singapore, 1990)
18. P.L. Martson and E.H. Trinh, Nature 312, 529 (1984)
19. J.F. Nye, Nature 312, 531 (1984)
20. M.V. Berry and C. Upstill, in Progress in Optics XVIII (E. Wolf, ed. North-Holland, Amsterdam, 1980), pp. 257-346
21. P.L. Marston, Opt. Lett. 10, 588 (1985)
22. D.T.L. Alexander, P.A. Crozier, and J.R. Anderson, Science 321, 833 (2008)
23. J.R. Bottiger, P.J. Deluca, E.W. Stuebing, et al., J. Aerosol Sci. 29, s.1, s965 (1998)
24. F.T. Gucker, J. Tuma, H.M. Lin, et al., Aerosol Sci. 4, 389 (1973)
25. P.H. Kaye, J.E. Barton, E. Hirst, et al., Appl. Opt. 39, 3738 (2000)
26. K.B. Aptowicz, R.G. Pinnick, S.C. Hill, et al., J. Geophys. Res. 111, D12212 (2006)
27. W.D. Dick, P.J. Ziemann, P.F. Huan, et al., Meas. Sci. Technol. 9, 183 (1998)
28. B.A. Sachweh, W.D. Dick, and P.H. McMurry, Aerosol Sci. Technol. 23, 373 (1995)
29. P. C. Waterman, Proc. IEEE 53, 805 (1965).
30. P. C. Waterman, Phys. Rev. D 3, 825 (1971).
31. M.I. Mishchenko, L.D. Travis, and A.A. Lacis, Scattering, Absorption, and Emission of Light by Small Particles (Cambridge University Press, Cambridge, 2002)
32. F.M. Schulz, K. Stamnes, and J.J. Stamnes, J. Opt. Soc. Am. A 16, 853 (1999).

33. F.M. Schulz, K. Stamnes, and J.J. Stamnes, Appl. Opt. 37, 7875 (1998)
34. M.I. Mishechenko, L.D. Travis, and D.W. Mackowski, J. Quant. Spectrosc. Radiat. Transfer 55, 535 (1996).
35. Y.-L. Xu, Appl. Opt. 36, 9496 (1997)
36. G. Videen, W. Sun, and Q. Fu, Opt. Commun. 156, 5 (1998)
37. K. A. Fuller and G. W. Kattawar, Opt. Lett. 13, 1063 (1988).
38. D. W. Mackowski and M. I. Mishchenko, J. Opt. Soc. Am. A 13, 2266 (1996).
39. Y.-L. Xu and B.Å.S. Gustafson, Appl. Opt. 36, 8026 (1997)
40. J. R. Bottiger, E. S. Fry, and R. C. Thompson, in Light Scattering by Irregularly Shaped Particles, D. W. Schuerman ed. (Plenum Press, New York, 1979).
41. R.H. Zerull, B.Å.S. Gustafson, K. Schulz, et al., Appl. Opt. 32, 4088 (1993).
42. S. Stein, Quart. Appl. Math. 19, 15 (1961).
43. R. Cruzan, Quart. Appl. Math. 20, 33 (1962).
44. S. Holler, J.-C. Auger, B. Stout, et al., Appl. Opt. 39, 6873 (2000)
45. J.-C. Auger, K.B. Aptowicz, R.G. Pinnick, et al., Opt. Lett. 32, 3358 (2007)
46. G.F. Crosta, M.C. Camatini, S. Zomer, et al., Opt. Exp. 8, 302 (2001)
47. G.F. Crosta, S. Zomer, Y.L. Pan, et al., Opt. Eng. 42, 2689 (2003)
48. S. Holler, S. Zomer, G.F. Crosta, et al., Appl. Opt. 43, 6198 (2004)
49. J. Mike, Classification Algorithms, (Collins, London, 1985)
50. W.R. Klecka, Discriminant Analysis: Quantitative Applications in the Social Sciences Series, No. 19 (Sage Publications, Thousand Oaks, CA, 1980)
51. H. Shen, J.F. Carter, R.G. Brereton, et al., The Analyst 128, 287 (2003)
52. S.K. Kachigan, Multivariate Statistical Analysis: A Conceptual Introduction, 2e (Radius Press, New York, 1991)

CHAPTER 11

FEMTOSECOND SPECTROSCOPY
FOR BIOSENSING

JEAN-PIERRE WOLF

GAP-Biophotonics, University of Geneva, 20,
Rue Ecole de Medecine, CH 1211 Geneve 4, Switzerland
Jean-Pierre.Wolf@unige.ch

Femtosecond spectroscopy opens new perspectives in bioaerosol sensing. On one side, the induced nonlinear light emission from the particles is remarkably peaked in the backward direction, which is favorable for remote detection, and on the other, quantum control schemes allow discrimination from major interference such as soot and other organic non-biological particles.

1. Introduction

Rapid detection and identification of pathogens in air and water are important safety issues. A major challenge for bioaerosol detectors is to provide a fast (typ. minutes) and selective response for discriminating pathogenic from non-pathogenic particles and minimize false alarm rates. Biochemical identification procedures such as polymerase chain reaction (PCR) are highly selective but slow while commonly used optical techniques provide information in "real time" but are genetically unspecific. Several optical systems, based on fluorescence and/or elastic scattering have been developed for distinguishing bio- from non-bio-aerosols (see chapters 5, 9, 10). A major drawback of linear spectroscopy is, however, frequent false alarms triggered by other organic particles, such as diesel soot and other organic compounds.[1,2] Fig. 1 shows, for example, the fluorescence spectra of diesel fuel, soot particles, the amino acid tryptophan (Trp) and *Bacillus subtilis (B. subtilis)*.

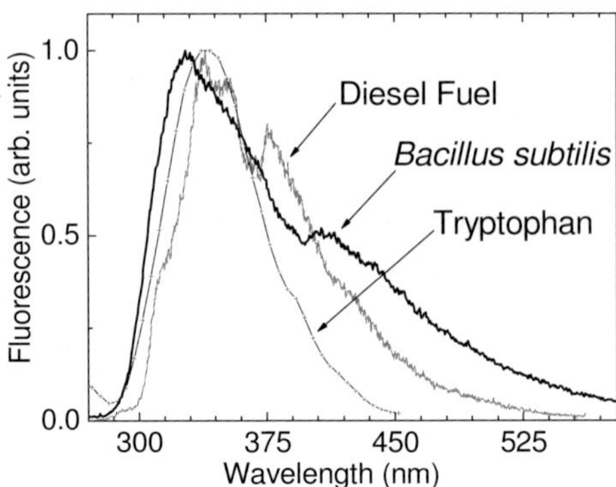

Fig. 1.   Comparison of the fluorescence spectra of Tryptophan, *B. Subtilis* and Diesel fuel.

As shown in the figure, these spectra all significantly overlap so that major interference occurs. The main contribution in the Ultraviolet-Visible (UV-Vis) fluorescence (around 340 nm) of biological particles is due to tryptophan. The longer wavelength tail of the fluorescence is attributed to the emission of nicotinamide adenine dinucleotide (NAD; around 450 nm) and flavins, i.e., riboflavin, flavin mononucleotide (FMN), and flavin adenine dinucleotide (FAD) around 560 nm.[1]

Due to the interference with polycyclic aromatic hydrocarbons (PAHs) containing soot and diesel, the identification of bioaerosols in a background of traffic related particles (typical of urban conditions) is therefore extremely difficult. It is even more elusive to expect discrimination of different types of bacteria.[3] An interesting approach was recently reported,[4] which combines optical and biochemical analyses. A fluorescence/scattering device was used to sort "on-line" bioaerosols from other particles, which could then be subsequently chemically analyzed in situ.

Optical techniques are also attractive as they can provide information remotely. The lidar (light detection and ranging) technique[5-7] allows for mapping aerosols in three-dimension (3-D) over several kilometers, similar to optical radar.

Lidars are able to detect the release and spread of potentially harmful plumes (such as pathogen releases or legionella from cooling towers) at large distance and thus allow taking measures in time for protecting populations or identifying unknown sources. So far, lidar detection of bioaerosols has been demonstrated either using elastic scattering[7] or UV-laser induced fluorescence (UV-LIF).[7,8] However, the distinction between bio- and non-bio- aerosols was either impossible (elastic scattering only) or unsatisfactory for LIF-LIDARs (interference with pollens and traffic related soot and PAHs).

A possible way of overcoming these difficulties is to excite the fluorescence with ultrashort laser pulses in order to access specific molecular dynamics features. Recent experiments using coherent control and multiphoton ultrafast spectroscopy have shown the ability of discriminating between molecular species that have similar one-photon absorption and emission spectra.[9,10] Two-photon excited fluorescence (2PEF) and pulse shaping techniques should allow for selective enhancement of the fluorescence of one molecule versus another that has almost identical spectra.[11] This approach, called "Optimal dynamic discrimination (ODD)" provides the basis for generating optimal signals for identifying similar agents.

In this chapter, we show that femtosecond pump-probe spectroscopy allows for distinguishing biological microparticles from PAH-containing ones. More precisely, we could distinguish amino acids (Trp) and flavins (riboflavin RbF, FMN and FAD) from PAHs (naphthalene) and diesel fuel in the liquid phase using "pump-probe depletion" (PPD) technique. We also show that the non-linear properties of aerosols and droplets generate a strong backward enhancement of the emitted light, which is most favorable for remote sensing applications. Finally, laser induced breakdown spectroscopy (LIBS) is also envisaged as an alternative to multi-photon excited fluorescence (MPEF).

2. Femtosecond spectroscopy of microparticles

The most prominent feature of nonlinear processes in aerosol particles using femtosecond lasers is a strong localization of the emitting molecules within the particle, and subsequent backward enhancement of the emitted light.[12-15] This unexpected behavior is very attractive for remote detection schemes, such as lidar applications.[16]

Localization is achieved by non-linear processes, which typically involve the n-th power of the internal intensity $I^n(r)$ (r for position inside the particle, Fig. 2). The backward enhancement can be explained by the reciprocity (or "time reversal") principle: Re-emission from regions with high $I^n(r)$ tends to return toward the illuminating source by essentially retracing the direction of the incident beam that gave rise to the high-intensity points.

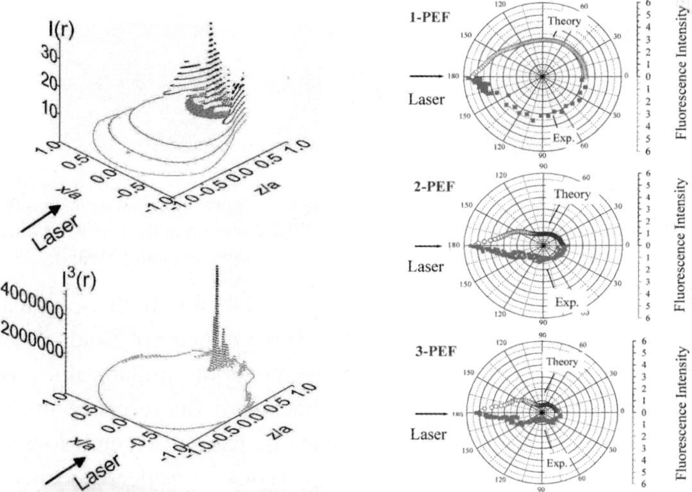

Fig. 2.   Backward-enhanced MPEF from spherical microparticles.[12] Left column: Molecular excitation within droplets, proportional to $I^n(r)$ with n =1 and 3. Right column: Angular distribution of MPEF emission for 1, 2 and 3 photon excitation (reprinted from ref. 12, copyright 2000 by the American Physical Society).

More precisely, we investigated, both theoretically and experimentally, incoherent multiphoton processes involving n = 1 to 5 photons.[14] For n = 1, 2, 3, MPEF occurs in bioaerosols because of natural fluorophors such as amino acids (tryptophan, tyrosin), NAD, and flavins. The strongly anisotropic MPEF emission was demonstrated on individual microdroplets containing tryptophan, riboflavin, or other synthetic fluorophors.[12-15] Fig. 2 (right column) shows the angular distribution of the MPEF emission and the comparison between experimental and theoretical (Lorenz-Mie calculations) results for the one- (400 nm) (upper), two- (800 nm) (center) and three-photon (1.2 μm) (lower) excitation process. They show that fluorescence emission is at maximum in the direction toward the exciting source. The directionality is dependent on the increase of n, because the excitation process involves the nth power of the intensity $I^n(r)$. The ratio $R_f = P(180°)/P(90°)$ increases from 1.8 to 9 when n changes from 1 to 3 (P is the emitted light power). For 3PEF, fluorescence from aerosol microparticles is therefore mainly backwards emitted, which is ideal for lidar experiments, as demonstrated in the field using the TERAMOBILE system.[16,17]

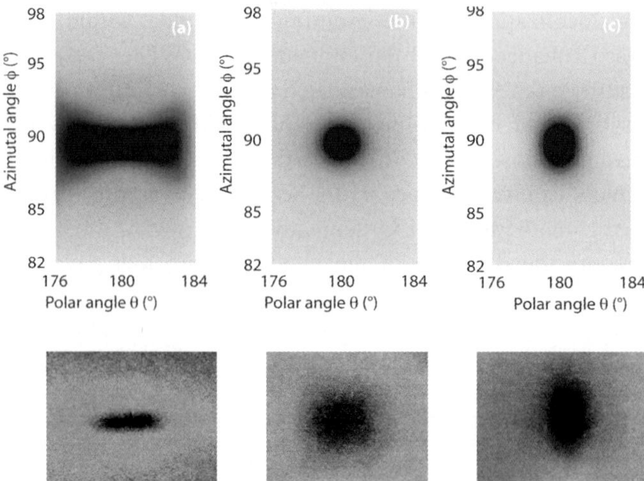

Fig. 3.   Backward-enhanced 2PEF in spheroidal microdroplets.[18] Upper: simulations using ray-tracing Lower: Experimental measurements using a CCD camera capturing 2PEF emission in the backward direction. (a) oblate (b) spherical (c) prolate (From ref. 18, with kind permission from Springer Science Media).

This backward enhancement has been observed for both spherical and non-spherical[15,18] microparticles, such as 2PEF from dye-doped spheroids.[18] Within aspect ratios ranging from 0.8 to 1.2, backward enhancements of similar values as for spheres are obtained, although the round shape of the backward fluorescence image (Fig. 3) is somewhat affected by the shape of the particle. In particular, multifoci are produced within the spheroid. Due to the large aspect ratios considered, exact Lorenz-Mie calculations could not be achieved. Novel ray tracing approaches were therefore developed in order to determine both the intensity distribution (and its square $I^2(r)$) and the reemission efficiencies. Although limited to rather large droplets (with a radius of

50 µm in our case), the ray tracing approach satisfactorily reproduced the experimental data, as shown in Fig. 3.

Similar experiments were conducted on clusters of spheres and solid tryptophan microparticles. Backward enhancement was observed in any of the studied cases, demonstrating that the effect understood as an illustration of the time-reversal principle is generic.[15]

For n = 5 photons, laser induced breakdown (LIB) in water microdroplets occurs, initiated by multiphoton ionization (MPI). The ionization potential of water molecules is 6.5 eV, so that 5 photons are required at a laser wavelength of 800 nm to initiate the process of plasma formation. Both localization and backward enhancement strongly increased with the order n of the multiphoton process, exceeding $R_f = 35$ for n = 5.[13] As for MPEF, LIBS has the potential of providing information about the aerosols composition, as was demonstrated for bacteria (see section 4).

3. Pump-probe measurement of particle size: ballistic trajectories in microdroplets

The very small spatial extension of femtosecond pulses (15 fs corresponds to 3.4 µm in water, i.e. the circumference of a 0.6-µm radius droplet) can be used for measuring the size of microparticles.[19, 20]

An experimental scheme for this approach consists of creating an optical correlator between two wavepackets, centered at wavelengths $\lambda_1 = 1200$ nm, and $\lambda_2 = 600$ nm, which circulate on ballistic orbits (see Fig. 4). 2PEF is then recorded as a function of the time delay between the two wavepackets in order to quantify the path length traveled within the particle. With this method, the radii of droplets up to 670 µm could be precisely measured.[19]

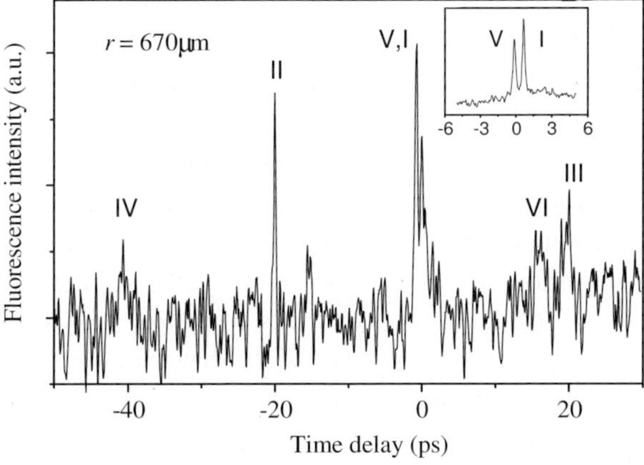

Fig. 4. Measurement of the ballistic trajectories of femtosecond pulses within microdroplets using 2PEF pump-probe technique. Each peak of the series I, II, III, IV corresponds to a roundtrip of the travelling pulse. V and VI correspond to other types of trajectories than Morphology Dependent Resonances (MDRs), such as rainbow[19] (reprinted from ref. 19, copyright 2001 by the American Physical Society).

The time delay between the two pulses gives access to ballistic path lengths, so that lengths different from a multiple of the orbital roundtrip indicates the contribution of other trajectories than MDRs. In order to better understand this behavior, the impact parameter (*a* in Fig. 5) of the laser onto the droplet was modified, so that evanescent coupling in surface modes could be distinguished from refractive penetration onto inner ballistic trajectories.

Detailed analysis of the results showed that for large impact parameters, only surface modes contributed, while for smaller ones, light bullets traveled on rainbow trajectories.

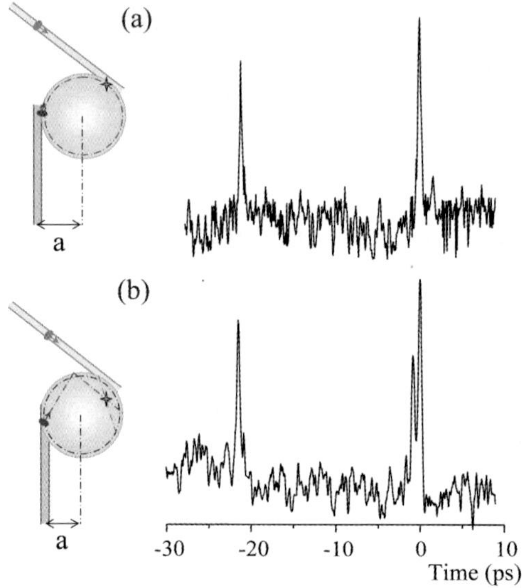

Fig. 5. Control of the ballistic trajectories used by the travelling pulse by modifying the impact parameter:[19] (a) evanescent coupling (b) evanescent + refractive coupling (reprinted from ref. 19, copyright 2001 by the American Physical Society).

The sizing application of this method was applied for a wide range of sizes, but could not be performed experimentally for droplets with radii smaller than 100 μm, due to experimental limitations.

In order to address lidar applications of the pump-probe approach for measuring remotely both size and composition of atmospheric aerosols, time resolved Lorenz-Mie calculations have been carried out (with 50 femtosecond pulses).[20] In a pump-probe 2PEF lidar experiment, the composition would be addressed by the excitation/fluorescence signatures and the size by the time delay between the two exciting pulses. The high peak contrast obtained by the numerical simulations shows that measurements should be feasible for microparticles even of sub-micron size.

4. Femtosecond LIBS in bioaerosols

As mentioned earlier, thanks to the high intensity/energy ratio provided by femtosecond laser pulses, aerosols are not significantly deformed during the induced light emission.

Therefore, even higher order nonlinear processes such as laser induced breakdown and plasma emission are enhanced in the backward direction.[13,14] The resulting "nanoplasma" is, for instance, highly localized at the focal line within the droplet and the associated plasma lines can be efficiently recorded in a lidar arrangement.

Although nanosecond-LIBS (nano-LIBS) has already been applied to the study of bacteria,[21, 22] femtosecond lasers open new perspectives in this respect. The plasma temperature is indeed, much lower in the case of femtosecond excitation, which strongly reduces the blackbody background and interfering lines from excited N_2 and O_2 from the air. This allows performing time gated detection with very short delays, and thus observing much richer and cleaner spectra from the biological sample. This crucial advantage is shown in Fig. 6, where the K line emitted by a sample of Escherichia coli (E. coli) is clearly detected in femtosecond-LIBS and almost unobservable under nanosecond -laser excitation.[23]

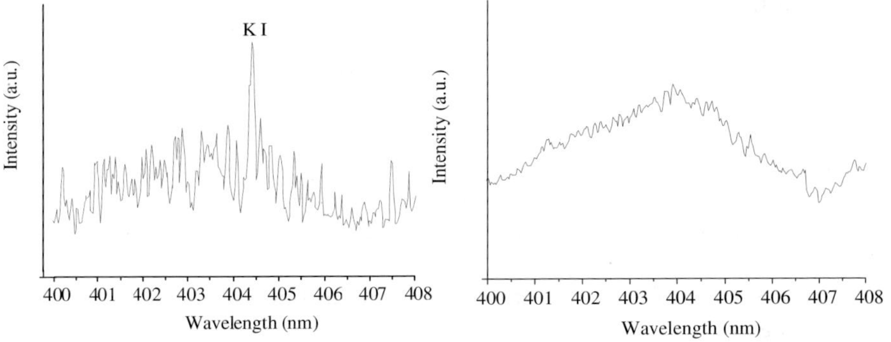

Fig. 6.   Comparison of fs-LIBS (left) and ns-LIBS results on the K line of Escherichia coli[23] (reprinted with permission from ref 23, copyright 2006, American Institute of Physics).

The low thermal background in femtosecond-LIBS allowed the recording of 20-50 lines for each bacterial sample considered in the study (Acinetobacter, E. coli, Erwinia, Shewanella, and B. subtilis). A systematic sorting with sophisticated algorithms is in progress in order to evaluate whether the spectra are sufficiently different for univoquely identifying each species.[24] The results are promising, as can be seen in Fig. 7, where significant difference between E. coli and B. subtilis are observed for the Li line intensity (also observed for the Ca line). This difference can be understood by the typical difference of the cell wall structure between a Gram+ and a Gram- bacterium. The ratio of these lines as compared to Na for example constitutes an "all-optical Gram test."

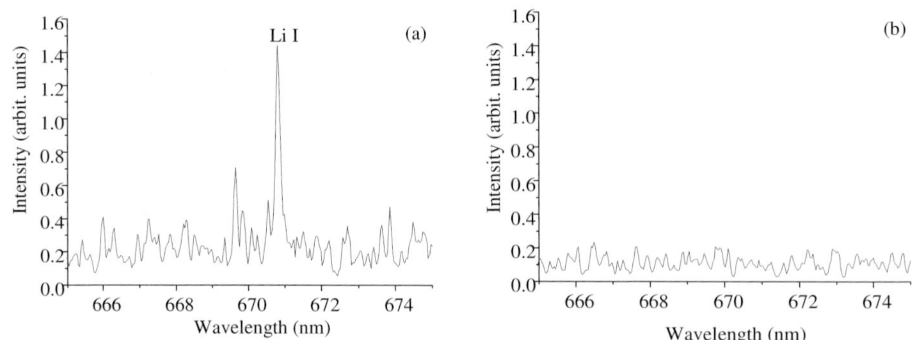

Fig. 7.   The "all-optical Gram test": B. subtilis vs. E. coli (reprinted with permission from ref 23, copyright 2006, American Institute of Physics).

Low-temperature plasma is not the only advantage of fs-LIBS: the ablation process itself seems different. fs-LIBS acts more as a direct bond-braking and evaporation process than as a thermal evaporation. This particular ablation process could be put into evidence since not only atomic and ionic lines were observed but also molecular signatures such as CN or C2.[25] It was shown in particular that these molecular species are directly ablated from the sample, and not created by recombination of C atoms or ions with nitrogen from the air (which occurs for ns excitation). Obtaining molecular signatures in addition to trace elements is a significant improvement of the method. The presence of CN molecules is, for instance, a good indicator for a biological material.

5.   Pump-probe femtosecond spectroscopy for identifying bacteria and biomolecules

As already mentioned, a major drawback inherent in LIF instruments is the lack of selectivity because UV-Vis fluorescence is incapable of discriminating different molecules with similar absorption and fluorescence signatures. While mineral and carbon black particles do not fluoresce significantly, aromatics and polycyclic aromatic hydrocarbons (PAHs) from organic particles and diesel soot strongly interfere with biological fluorophors such as amino acids. The similarity between the spectral signatures of PAH and biological molecules under UV-Vis excitation is due to similar π-electrons from carbonic rings. Therefore, PAHs (such as naphthalene) exhibit absorption and emission bands similar to those of amino acids like tyrosine or tryptophan (Trp). Some spectral shifts are present because of differences in specific bonds and the number of aromatic rings, but the broad featureless nature of the bands renders them almost indistinguishable.

In order to discriminate these fluorescing molecules, we applied a novel femtosecond pump-probe depletion (PPD) concept (Fig. 7), which is based on the time-resolved observation of the competition between excited-state absorption (ESA) into higher-lying excited states and fluorescence into the ground state. This approach makes use of two physical processes beyond that available in the usual linear fluorescence spectroscopy:

(1) the dynamics in the intermediate pumped state (S_1) and (2) the coupling efficiency to higher-lying excited states (S_n).

As sketched in Fig. 8, a first femtosecond pump pulse (at 270 nm for Trp and PAHs, 405 nm for flavins), resonant with the first absorption band of the fluorophores, coherently excites them from the ground state S_0 to a set of vibronic levels $S_1\{v'\}$. The vibronic excitation relaxes by internal energy redistribution to lower $\{v\}$ modes. Fluorescence relaxation to the ground state occurs within a lifetime of several nanoseconds. Meanwhile, a second 810-nm femtosecond "re-pump" pulse is used to transfer part of the $S_1\{v\}$ population to higher-lying electronic states S_n.

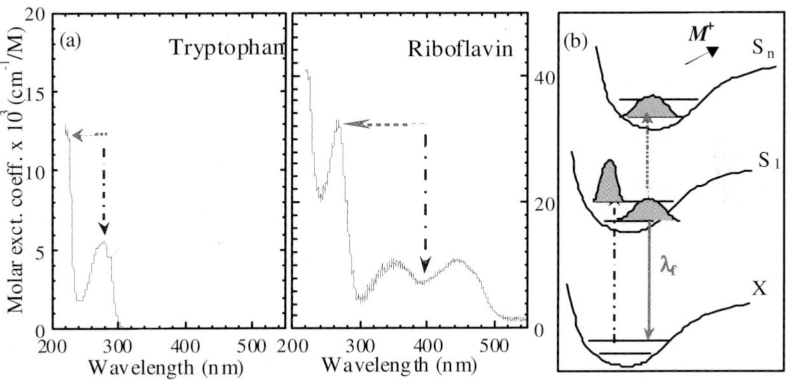

Fig. 8.   Absorption spectra of tryptophan (a) and riboflavin (b). (c) PPD scheme in Trp, flavins and polycyclic aromatics. The pump pulse brings the molecules in their first excited state S_1. The S_1 population (and therefore the fluorescence) is depleted by the second pump pulse.[26] (Reprinted with permission of Elsevier from ref 26).

The depletion of the S_1 population under investigation depends on both the molecular dynamics in this intermediate state and the transition probability to S_n. The relaxation from the intermediate excited state may be associated with different processes, including charge transfer, conformational relaxation[27,28] and intersystem crossing with repulsive $\pi\sigma^*$ states.[29]

S_n states are both autoionizing and relaxing radiationless into S_0. By varying the temporal delay Δt between the UV-Vis and the infrared (IR) pulses, the dynamics of the internal energy redistribution within the intermediate excited potential hypersurface S_1 is explored. The S_1 population and the fluorescence signal are therefore depleted as a function of Δt. As different species have distinct S_1 hypersurfaces, discriminating signals can be obtained.[30,31]

Fig. 9 (left) shows the PPD dynamics of S_1 in Trp as compared to diesel fuel and naphthalene in cyclohexane, one of the most abundant fluorescing PAHs in diesel fuel. While depletion reaches as much as 50 % in Trp, diesel fuel and naphthalene appear almost unaffected (within a few percent), at least on these timescales.[26, 30, 31]

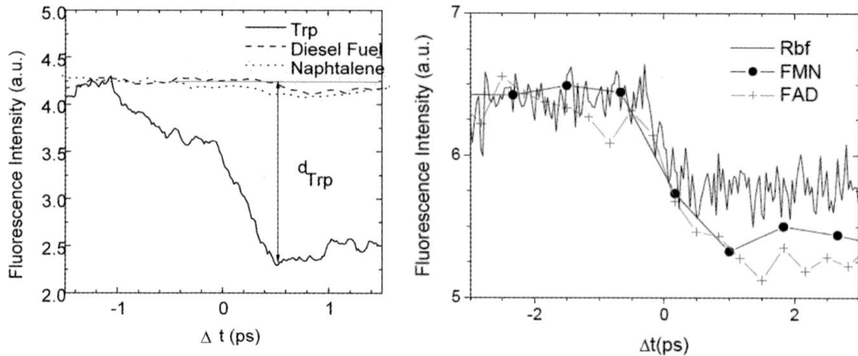

Fig. 9.   Left: PPD experiment on Trp and PAHs, demonstrating discrimination capability between the amino acid and other aromatic molecules. Right: Similar results obtained in flavins[31]   (from ref 31, reprinted with permission of the Royal Society of Chemistry).

The depletion factor δ is defined as $\delta = (P_{undepleted} - P_{depleted})/P_{undepleted}$. (where P is the fluorescence power). This remarkable difference allows for efficient discrimination between Trp and organic species, although they exhibit very similar linear excitation/fluorescence spectra.

Two reasons might be invoked to understand this difference: (1) the intermediate-state dynamics are predominantly influenced by the NH- and CO- groups of the amino acid backbone and (2) the ionization efficiency is lower for the PAHs. Further electronic structure calculations are required to better understand the process, especially on the higher-lying S_n potential surfaces.

Fluorescence depletion has been obtained as well for Riboflavine (Rbf), FMN, and FAD (Fig. 9, right). However, the depletion in this case is only about 15% (with a maximum intensity of 5×10^{11} W/cm^2 at 810 nm).

In order to more closely approach the application of detecting and discriminating bioagents from organic particles, we applied PPD spectroscopy to live bacteria ($\lambda_1 = 270$ nm and $\lambda_2 = 810$ nm), such as E. coli, Enterococcus and B. subtilis. Artefacts due to preparation methods have been discarded by using a variety of samples, i.e., lyophilized cells and spores, suspended either in pure or in biologically buffered water, i.e., typically 10^7-10^9 bacteria per cc. The bacteria containing solutions replaced the Trp or flavin containing solutions of the formerly described experiment. The observed pump-probe depletion results are remarkably robust (Fig. 10), with similar depletion values for all the considered bacteria (results for Enterococcus, not shown in the figure, are identical), although the Trp microenvironment within the bacteria proteins is very different from water.

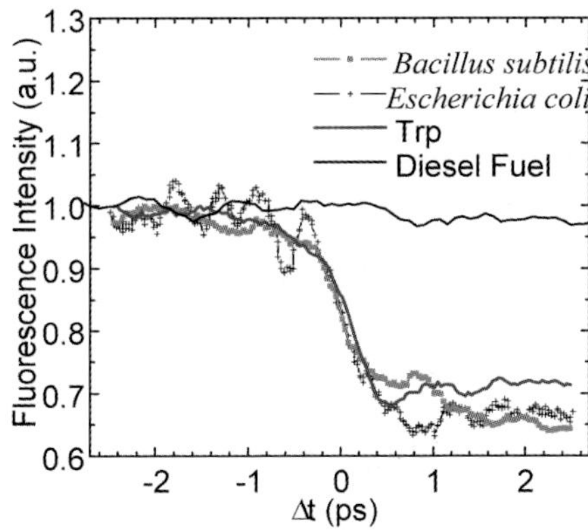

Fig. 10.   Discrimination between bacteria and diesel fuel using PPD ultrafast spectroscopy[26]   (Reprinted with permission from Elsevier from ref 26).

On the other hand the very similar depletion behaviour for all bacteria and Trp also shows the limitations of PPD spectroscopy in the present configuration. Biomolecules can be distinguished from other aromatics, but PPD is unable to discriminate two different bacteria in solution. We are currently exploring new experimental ODD approaches to reach this goal.[11]

Primarily, MPEF-LIDAR is advantageous as compared to linear LIF-LIDAR for the following reasons: (1) MPEF is enhanced in the backward direction and (2) the transmission of the atmosphere is much higher for longer wavelengths. For example, if we consider the detection of Trp with 3-PEF, the transmission of the atmosphere is typically 0.6 km^{-1} at 270 nm, whereas it is 3×10^{-3} km^{-1} at 810 nm (for a clear atmosphere, depending on the background ozone concentration). This compensates the lower 3-PEF cross-section compared to the 1-PEF cross-section at distances larger than a couple of kilometers. The most attractive feature of MPEF is, however, the possibility of using pump-probe techniques in order to discriminate bioaerosols from background interferents such as traffic related soot or PAHs.

In order to get closer to the real application, we then performed ultrafast PPD spectroscopy in bioaerosols and in particular water microdroplets that contain Trp and/or FMN. The droplets radius was about 20 μm, which is larger than the size of single bacteria (1 μm) or even bacteria clusters (typically 10 μm), but still constitutes an acceptable model.

The most impressive result of these experiments is the very high PPD efficiency as compared to depletion ratios in liquids. The depletion factor δ reaches 80 % for Trp droplets and 60 % for FMN droplets (to be compared to 50% and 15% in liquids, respectively). Some tentative explanations could be invoked for this unexpectedly high efficiency, but the definitive reason is not clear yet: (1) The spatial overlap between pump and probe pulses might be enhanced by the shape of the droplet, (2) The spherical shape induces hot spots inside the droplet where intensities are up to 100 times higher than the incident one,[13,14] but the total hot spot volume is rather small (3) There might be some surface effects (orientation of molecules on the surface) that would enhance the two-photon absorption.

We finally applied this technique to 20-µm-radius water droplets containing typically 100 live bacteria (*E. coli*). As shown in Fig. 11, the depletion factor δ is again greatly enhanced as compared to bacteria in bulk water: 60 % depletion in the microdroplet and 20 % in solution ($\lambda_1 = 270$ nm, $\lambda_2 = 810$ nm).

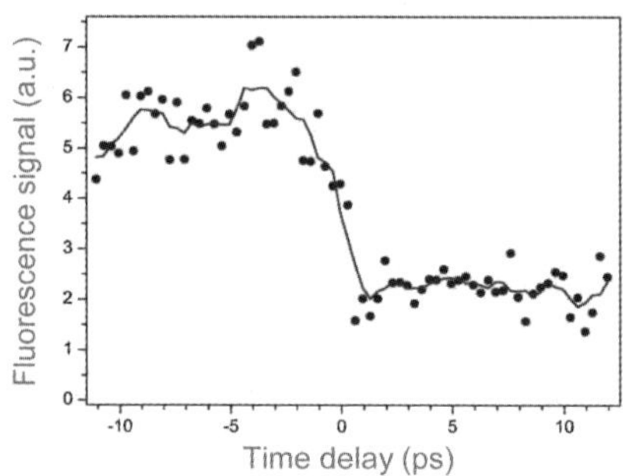

Fig. 11.　PPD spectroscopy in 20-µm radius droplets containing about 100 E. coli bacteria (from ref 31, reprinted with permission of the Royal Society of Chemistry).

This experiment is also interesting for field applications, as bacteria and viruses are efficiently transmitted by droplets of saliva while coughing, breathing, and speaking.

6.　Conclusion

Femtosecond spectroscopy offers numerous advantages as compared to linear spectroscopy. We showed that particles emission are backward enhanced, that simultaneous sizing can be performed thanks to the very small spatial extension of the laser pulses, and molecular dynamics can be resolved in order to discriminate molecules with nearly identical spectra. For instance, we can imagine to use the discrimination capability of PPD as a basis for a novel selective bioaerosol detection technique that avoids interference from traffic related background organic particles in air.

PPD spectroscopy is an attractive technique for discriminating bioaerosols from other organics. However, PPD spectroscopy is unable to discriminate one type of biological aerosol from another (Fig. 10). A possible reason is that the averaged dynamics in the excited states of the biofluorophors (embedded in complex proteins) are quite similar in all living organisms. A natural approach is therefore to extend the PPD technique to coherent control, where the amplitude and phase of every spectral component of the exciting pulses are shaped in order to best-fit the potential hypersurfaces. The method is then extremely sensitive to the details of the potential hypersurfaces, which provides unprecedented selectivity.

For this approach, a large number of parameters (corresponding to the amplitude and phase of each spectral component within the exciting laser pulse(s)) have to be controlled. In 1992, the concept of "optimal control" was introduced,[32,33] in which a feedback loop optimizes the laser pulse characteristics to reach most efficiently the desired target. Excellent results have been obtained using coherent control schemes in atomic and molecular systems. Our current and future work will be dedicated to these novel opportunities: applying coherent control schemes for efficiently discriminating pathogens from the 10^6-10^8 non-pathogen bacteria usually present in a cubic meter of air.

Acknowledgments

The author gratefully acknowledges the members of the groups of R. K. Chang at Yale University (with whom we performed much of this work), S. C. Hill and R. Pinnick at Army Research Laboratories, H. Rabitz at Princeton University and L. Woeste at Free University Berlin. We also wish to thank F. Courvoisier, L. Guyon, V. Boutou, M. Baudelet and J. Yu at University of Lyon 1 and L. Bonacina, A. Rondi, J. Extermann and M. Moret at University of Geneva. We also gratefully acknowledge the support of the Swiss National Foundation for Research (Grant numbers 200021-111688 and 200020-124689), as well as the European Commission Secretariat for Education and Research for their support within the framework of the COST P21 ("Physics of Droplets") and COST MP063 ("MicroCARS") programs.

References

1. S.C Hill, R.P. Pinnick, S. Niles *et al*, *Field Anal Chem. Tech.* **5**, 221 (1999)
2. R.G. Pinnick, S.C. Hill, Y.L. Pan *et al*, *Atmosph. Environ.* **38**, 1657 (2004)
3. Y.S. Cheng, E.B. Barr, B.J. Fan *et al*, *Aero.Sci.Tech.* **31**, 409 (1999)
4. Y. Pan , V. Boutou, J.R. Bottiger *et al*, *Aerosol Sci. Techn.* **38**, 598 (2004)
5. J. Kasparian, E. Fréjafon, P. Rambaldi, J.Yu, P. Ritter, P. Viscardi, J.P. Wolf, *Atmospheric Environment* **32(17)**, 2957 (1998)
6. M. Beniston, M. Beniston-Rebetez, H.J. Kölsch, P.Rairoux, J.P. Wolf, L. Wöste, *J. Geophys. Res.*, **95** (D7), 13 (1990)
7. C. Weitkamp, *LIDAR*, Springer-Verlag New York 2005
8. F.Immler, D. Engelbart, and O. Schrems, *Atmos.Chem.Phys. Disc.* **4**, 5831 (2004)
9. T. Brixner, P. Niklaus, and G. Gerber, *Nature* **414**, 57 (2001)
10. T. Brixner, B. Kiefer, and G. Gerber, *J. Chem. Phys.* **118**(8) (2003)
11. B. Li, H. Rabitz and J.P. Wolf, *J. Chem. Phys.* **122**, 154103-1-8 (2005)
12. S. C. Hill, V. Boutou, J. Yu, S. Ramstein, J.P. Wolf, Y. Pan, S. Holler, R. K. Chang, *Phys.Rev.Lett.* **85(1)**, 54 (2000)
13. C. Favre, V. Boutou, S. C. Hill, W. Zimmer, M. Krenz, H. Lambrecht, J. Yu, R. K. Chang, L. Woeste, J.P. Wolf, *Phys.Rev.Lett.* **89(3)**, 035002 (2002)
14. V. Boutou, C. Favre, S. C. Hill, Y. Pan, R. K. Chang, J.P. Wolf, *App.Phys.B* **75**, 145 (2002)
15. Y. Pan, S. C. Hill, J.P. Wolf, S. Holler, R.K. Chang, J.R. Bottiger, *Applied Optics* **41(15)**, 2994 (2002)
16. G. Méjean, J. Kasparian, J. Yu, S. Frey, E. Salmon, J.P. Wolf, *Applied Physics B* **78(5)**, 535 (2004)
17. J. Kasparian, M. Rodriguez, G. Méjean, J.Yu, E. Salmon, H. Wille, R. Bourayou, S. Frey, Y.B. Andre, A. Mysyrowicz, R. Sauerbrey, J.P. Wolf, L. Woeste, *Science* **301(5629)**, 61 (2003)

18. J. Kasparian, V. Boutou, J.P. Wolf, Y. Pan, R.K. Chang, *Appl. Phys. B* **91**, 167 (2008)

19. J.P. Wolf, Y. Pan, S. Holler, G. M. Turner, M. C. Beard, R.K. Chang, C. Schmuttenmaer, *Phys.Rev. A* **64**, 023808-1-5 (2001)

20. L. Méès, J.P. Wolf, G.Gouesbet, G. Gréhan, *Optics Comm.* **208,** 371 (2002)

21. S. Morel, N. Leon, P. Adam, and J. Amouroux, *Appl. Opt.* **42**, 6184 (2003).

22. P. B. Dixon and D.W. Hahn, *Anal Chem.* **77**, 631 (2005)

23. M.Baudelet, L. Guyon, J. Yu, J.-P. Wolf, Amodeo, E. Fréjafon, P. Laloi, *J. App. Physics* **99,** 084701 (2006)

24. M. Baudelet, M. Bossu, J. Jovelet, J. Yu, J.-P. Wolf, T. Amodeo, E. Fréjafon, P. Laloi, *App. Physics Lett.* **89**, 163903 (2006)

25. M. Baudelet, L. Guyon, J. Yu, J.-P. Wolf, T. Amodeo, E. Fréjafon, P. Laloi, *Appl. Phys. Lett.* **88**, 053901 (2006)

26. F. Courvoisier, L. Bonacina, J. Extermann, M. Roth, H. Rabitz, L. Guyon, C. Bonnet, B. Thuillier, V. Boutou, J.P. Wolf, *Faraday Discuss.***137**, 37 (2008)

27. J.T. Vivian and P.R. Callis, *Biophysical J.* **80**, 2093 (2001)

28. H.B. Steen, *J. Chem. Phys.* **61**(10), 3997 (1974)

29. Y. Iketaki, T. Watanabe, S-i. Ishiuchi *et al*, *Chem. Phys. Lett.* **372,** 773 (2003)

30. F. Courvoisier, V. Boutou, V. Wood *et al*, *App. Phys. Lett* **87(6),** 063901 (2005)

31. F. Courvoisier, V. Boutou, H. Rabitz *et al*, *J. Photochem. and Photobio. A* **180**, 300 (2006)

32. R. Judson, and H.Rabitz, *Phys.Rev.Lett.* **68**, 1500 (1992)

33. S.Warren, H. Rabitz and M. Daleh, *Science* **259,** 1581 (1993)

PART IV

OPTICAL MICROCAVITIES & NANOSTRUCTURES

CHAPTER 12

LASING IN RANDOM MEDIA

HUI CAO

Yale University, Department of Applied Physics, New Haven 06520 Connecticut, USA
hui.cao@yale.edu

In the past decade, an unconventional laser named "random laser" has drawn much attention. The laser cavity is formed not by mirrors, but by optical scattering in a disordered gain medium. This novel laser exhibits many features distinct from a conventional laser, which lead to unique applications.

1. Introduction

1.1. *"LASER" versus "LOSER"*

A photon, unlike an electron, can stimulate an excited atom to emit a second photon into the same electromagnetic mode. This stimulated emission process is the foundation for light amplification and oscillation (i.e. self-generation). Initially, LASER referred to Light Amplification by Stimulated Emission. Nowadays laser often means Light *Oscillation* by Stimulated Emission, which literally shall be called "LOSER" instead of "LASER". To distinguish the above two devices, the former is called laser amplifier, the latter laser oscillator.[1]

In a laser amplifier, input light is amplified when the net gain coefficient $g_{eff} = g_a - \alpha_r > 0$, where g_a and α_r represent gain and absorption coefficients respectively. In the absence of input light, photons spontaneously emitted by excited atoms are multiplied, giving amplified spontaneous emission (ASE).

Laser oscillation occurs when the photon generation rate exceeds photon loss rate in a system. If gain saturation were absent, the photon number in a laser oscillator would diverge in time. In other words, the rate equation for photon number would acquire an unstable solution above the oscillation threshold. In reality, gain saturation reduces the photon generation rate to the photon loss rate so that the number of photons in the oscillator remains a finite value.

1.2. *Random laser*

For a long time, optical scattering was considered to be detrimental to lasing because such scattering removes photons from the lasing modes of a conventional laser cavity. However, in a disordered medium with gain, light scattering plays a positive role in both laser amplification and laser oscillation. Multiple scattering increases the path length or dwell time of light in an active medium, thereby enhancing laser amplification. In addition, strong scattering increases the chance of light (of wavelength λ) returning to a coherence volume ($\sim \lambda^3$) it has visited previously, providing feedback for laser oscillation.

Since the pioneering work of Letokhov and coworkers[2], lasing in disordered media has been a subject of intense theoretical and experimental studies. It represents the process of light amplification by stimulated emission with feedback mediated by random fluctuation of dielectric constant in space. There are two kinds of feedback: one is *intensity* or *energy* feedback, the other is *field* or *amplitude* feedback.[3] The field feedback is phase sensitive (i.e. coherent), and therefore frequency dependent (i.e. resonant). It requires that light scattering be elastic and spatial distribution of dielectric constant be time-invariant. The intensity feedback is phase insensitive (i.e. incoherent) and frequency independent (i.e. non-resonant). It can happen, e.g., in the presence of inelastic scattering, mobile scatterers, dephasing, and nonlinearity. Based on the feedback mechanisms, random lasers are classified into two categories: (i) random lasers with incoherent and non-resonant feedback, (ii) random laser with coherent and resonant feedback.[4,5]

1.3. *Characteristic length scales for random laser*

Correlation radius R_c. In a disordered dielectric medium, the dielectric constant $\epsilon(\mathbf{r})$ fluctuates randomly in space. The spatial variation of $\epsilon(\mathbf{r})$ can be characterized statistically by the correlator $K(\Delta\mathbf{r}) \equiv \langle\epsilon(\mathbf{r})\epsilon(\mathbf{r}+\Delta\mathbf{r})\rangle$, where $\langle...\rangle$ represents ensemble average. When the random medium is isotropic, the width of $K(\Delta r)$ is called the correlation radius R_c. It reflects the length scale of spatial fluctuation of dielectric constant. If $R_c \gg \lambda$, light is deflected by long-range disorder. When R_c is comparable to or less than λ, light is scattered by short-range disorder.

Scattering mean free path l_s and transport mean free path l_t. The relevant length scales that describe light scattering process are the scattering mean free path l_s and the transport mean free path l_t. The scattering mean free path l_s is defined as the average distance that light travels between two consecutive scattering events. The transport mean free path l_t is defined as the average distance a wave travels before its direction of propagation is randomized. These two length scales are related:

$$l_t = \frac{l_s}{1 - \langle\cos\theta\rangle} .\tag{1}$$

$\langle\cos\theta\rangle$ is the average cosine of the scattering angle, which can be found from the differential scattering cross section. Rayleigh scattering is an example of $\langle\cos\theta\rangle = 0$ and $l_t = l_s$, while Mie scattering may have $\langle\cos\theta\rangle \approx 0.5$ and $l_t \approx 2l_s$.

Gain length l_g and amplification length l_{amp}. Light amplification by stimulated emission in a random medium is described by the gain length l_g and the amplification length l_{amp}. The gain length l_g is defined as the path length over which light intensity is amplified by a factor e. The amplification length l_{amp} is defined as the (rms) average distance between the beginning and ending point for paths of length l_g. In a homogeneous medium, light travels in a straight line, thus $l_{amp} = l_g$. In a diffusive sample, $l_{amp} = \sqrt{D\tau_{amp}}$, where D is the diffusion coefficient, $\tau_{amp} = l_g/v$, v is the speed of light. In a three-dimensional (3D) system, $D = v\,l_t/3$, thus

$$l_{amp} = \sqrt{\frac{l_t l_g}{3}} \ . \tag{2}$$

The gain length l_g is the analogue of the inelastic length l_i defined as the travel length over which light intensity is reduced to $1/e$ by absorption. Hence, the amplification length l_{amp} is analogous to the absorption length $l_{abs} = \sqrt{l_t\, l_i/3}$.

Dimensionality d and size L. Light transport in a random medium depends on its dimensionality d and size L. For a random medium of $d > 1$, L refers to its smallest dimension. The average trapping time of photons in a diffusive random medium $\tau_d = L^2/D$. In an active random medium, the gain volume may be smaller than the volume of entire random medium, e.g., when only part of the disordered medium is pumped. The gain volume is characterized by its dimension L_g, and $L_g \leq L$.

1.4. *Light localization*

There are three regimes for light transport in a 3D random medium: (i) ballistic regime, $L \leq l_s$; (ii) diffusive regime, $L \gg l_s \gg \lambda$; (iii) localization regime, $k\,l_s \simeq 1$ (k is the wave vector in the random medium).[6]

Light localization can also be understood in the mode picture. Quasimodes, also called quasi-states, are the eigenmodes of the Maxwell equations in a passive random medium. Due to the finite size of a dielectric medium and its open boundary, the frequency of a quasimode is a complex number: $\Omega = \omega_r + i\gamma$. γ is the decay rate of a quasimode as a result of its coupling to the environment. It also represents the linewidth of the quasimode in frequency. The Thouless number δ is defined as the ratio of average linewidth $\delta\nu = \langle\gamma\rangle$ to average frequency spacing $d\nu$ of adjacent quasimodes, $\delta \equiv \delta\nu/d\nu$. In the delocalization regime, the quasimodes are spatially extended over the entire random system. They have large decay rates and overlap in frequency, $\delta > 1$. In the localization regime, quasimodes are localized inside the system and have small decay rates. They do not overlap spectrally, thus $\delta < 1$. The localization threshold is set at $\delta = 1$, that is the Thouless criterion. Thus the localization transition corresponds to a transition from overlapping modes to non-overlapping modes.

The openness of a random laser, namely, its coupling to the environment, can be characterized by δ. Note that the value of δ is obtained in the absence of gain or absorption so that it describes solely light leakage from the random system.

Typical chaotic cavity lasers and photon localization lasers have non-overlapping modes, $\delta < 1$; whereas in the diffusive random lasers most quasimodes overlap, $\delta > 1$.

This article is a compendium of a wide range of experimental studies on light amplification and lasing in random media. Brief descriptions of some theoretical models and numerical calculations are given for qualitative explanations of experimental phenomena.

2. Random laser with incoherent feedback

2.1. *Laser with scattering reflector*

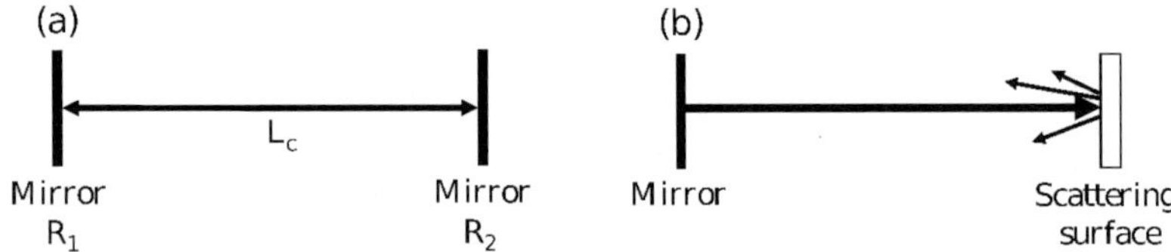

Fig. 1.   (a) Schematic of a Fabry-Perot cavity made of two parallel mirrors with reflectivity R_1 and R_2. The cavity length (distance between the two mirrors) is L_c. (b) One of the mirrors is replaced by a scattering surface which scatters light instead of reflecting it. (with copyright permission).

The two essential components of a laser are gain medium and cavity. The gain medium amplifies light through stimulated emission, and the cavity provides positive feedback. A simple laser cavity is a Fabry-Perot cavity shown in Fig. 1(a). The Fabry-Perot cavity is made of two mirrors in parallel. We assume the gain medium is uniformly distributed between the two mirrors. After traveling one round trip between the mirrors, light returns to its original position. The requirement of constructive inference determines the resonant frequencies, namely

$$2kL_c + \phi_1 + \phi_2 = 2\pi m \, , \tag{3}$$

where k is the wave vector, L_c is the cavity length, ϕ_1 and ϕ_2 represent the phases of the reflection coefficients of the two mirrors, m is an integer. Only light at the resonant frequencies experiences minimum loss and spends a long time in the cavity. The long dwell time in the cavity facilitates light amplification. When the optical gain balances the loss of a resonant mode, lasing oscillation occurs in this mode. The threshold condition is

$$R_1 \, R_2 \, e^{2gL_c} = 1 \, , r \tag{4}$$

where R_1 and R_2 represent the reflectivities of the two mirrors, g_{eff} is the net gain coefficient.

In 1966, Ambartsumyan *et al* realized a different type of laser cavity that provides non-resonant feedback. [2] They replaced one mirror of the Fabry-Perot cavity with a scattering surface, as shown in Fig. 1(b). Light in the cavity suffers multiple scattering, its direction is changed every time it is scattered. Thus light does not return to its original position after one round trip. The spatial resonances for the electromagnetic field are absent in such a cavity. The dwell time of light is not sensitive to its frequency. The feedback in such a laser is used merely to return part of the energy or photons to the gain medium, i.e., it is energy or intensity feedback.

The non-resonant feedback can also be interpreted in terms of "modes". When one end mirror of a Fabry-Perot cavity is replaced by a scattering surface, escape of emission from the cavity by scattering becomes the predominant loss mechanism for all modes. Instead of individual high-Q resonances there appear a large number of low-Q resonances which spectrally overlap and form a continuous spectrum. It corresponds to the occurrence of non-resonant feedback. The absence of resonant feedback means that the cavity spectrum tends to be continuous, i.e., it does not contain discrete components at selected resonant frequencies. The only resonant element left in this kind of laser is the amplification line of the active medium. With an increase of pumping intensity, the emission spectrum narrows continuously towards the center of the amplification line. However, the process of spectral narrowing is much slower than in ordinary lasers.[7] Since many modes in a laser cavity with non-resonant feedback interacts with the active medium as a whole, the statistical properties of laser emission is quite different from that of a ordinary laser. As shown by Ambartsumyan *et al*, the statistical properties of the emission of a laser with non-resonant feedback are very close to those of the emission from an extremely bright "black body" in a narrow range of the spectrum.[8] The emission of such a laser has no spatial coherence and is not stable in phase.

Because the only resonant element in a laser with non-resonant feedback is the amplification line of the gain medium, the mean frequency of laser emission does not depend on the dimensions of the laser but is determined only by the center frequency of the amplification line. If this frequency is sufficiently stable, the emission of this kind of laser has a stable mean frequency. Ambartsumyan *et al* proposed using the method of non-resonant feedback to produce an optical standard for frequency.[9] To realize it, they built continuous gas lasers with non-resonant feedback based on the scattering surface.[10]

2.2. *Photonic bomb*

In 1968, Letokhov took one step further and proposed self-generation of light in an active medium filled with scatterers.[11] When the photon mean free path is much smaller than the dimension of scattering medium, the motion of photons is diffusive. Letokhov solved the diffusion equation for the photon energy density $W(\vec{r}, t)$ in the presence of a uniform and linear gain.

$$\frac{\partial W(\vec{r}, t)}{\partial t} = D\nabla^2 W(\vec{r}, t) + \frac{v}{l_g} W(\vec{r}, t) \,, \tag{5}$$

where v is the transport velocity of light inside the scattering medium, l_g is the gain length, and D is the diffusion constant given by

$$D = \frac{v\,l_t}{3} .$$ (6)

The general solution to Eq.(5) can be written as

$$W(\vec{r}, t) = \sum_n a_n \Psi_n(\vec{r}) e^{-(DB_n^2 - v/l_g)t} ,$$ (7)

where $\Psi_n(\vec{r})$ and B_n are the eigenfunctions and eigenvalues of the following equation

$$\nabla^2 \Psi_n(\vec{r}) + B_n^2 \Psi_n(\vec{r}) = 0$$ (8)

with the boundary condition that $\Psi_n = 0$ at a distance z_e from the boundary. z_e is the extrapolation length. Usually z_e is much smaller than the dimension of scattering medium and it can be neglected. Hence, the boundary condition becomes that $\Psi_n = 0$ at the boundary of the random medium.

The solution of $W(\vec{r}, t)$ in Eq.(7) changes from exponential decay to exponential increase in time when crossing the threshold

$$DB_1^2 - \frac{v}{l_g} = 0 ,$$ (9)

where B_1 is the lowest eigenvalue of Eq. (8). If the scattering medium has the shape of a sphere of diameter L, $B_n = 2\pi n/L$, and the smallest eigenvalue $B_1 = 2\pi/L$. If the scattering medium is a cube whose side length is L, the smallest eigenvalue $B_1 = \sqrt{3}\pi/L$. Regardless the shape of the scattering medium, the lowest eigenvalue B_1 is on the order of $1/L$. Substituting $B_1 \approx 1/L$ into Eq.(9), the threshold condition predicts a critical volume

$$V_{cr} \approx L^3 \approx \left(\frac{l_t l_g}{3}\right)^{3/2} .$$ (10)

With the fixed gain length l_g and transport mean free path l_t, once the volume of the scattering medium V exceeds the critical volume V_{cr}, $W(\vec{r}, t)$ increases exponentially with t.

This can be understood intuitively in terms of two characteristic length scales. One is the generation length L_{gen}, which represents the average distance a photon travels before generating a second photon by stimulated emission. L_{gen} can be approximated by the gain length l_g. The other is the mean path length L_{pat} that a photon travels in the scattering medium before escaping through its boundary. $L_{pat} \sim vL^2/D$. When $V \geq V_{cr}$, $L_{pat} \geq L_{gen}$. This means on average every photon generates another photon before escaping the medium. This triggered a "chain reaction", i.e. one photon generates two photons, and two photons generate four photons, etc. Thus the photon number increases with time. This is the onset of photon self-generation. Because this process of photon generation is analogous to the multiplication of neutrons in an atomic bomb, this device is sometimes called a photonic bomb.

In reality the light intensity will not diverge (there is no "explosion") because gain depletion quickly sets in and l_g increases. Taking into account gain saturation, Letokhov calculated the emission linewidth and the generation dynamics. If the scattering centers are stationary, the limiting width of the generation spectrum is determined by the spontaneous emission. Otherwise, the Brownian motion of the scattering particles leads to a random variation (wandering) of the photon frequency as a result of the Doppler effect on the scattering particles. He also predicted damped oscillation (pulsation) in the transient process of generation.

2.3. *Powder laser*

In 1986 Markushev *et al* reported intense stimulated radiation from the $Na_5La_{1-x}Nd_x(MoO_4)_4$ powder under resonant pumping at low temperature (77 K).[12] When the pumping intensity exceeded a threshold, Nd^{3+} emission spectrum was narrowed to a single line, and the emission pulse duration was shortened by approximately four orders of magnitude. Later on, they reported similar phenomena in a wide range of Nd^{3+}-activated scattering materials, including La_2O_3, La_2O_2S, $Na_5La(MoO_4)_4$, La_3NbO_7, and $SrLa_2WO_7$.[13] The powder was pumped by 20 ns Q-switched laser pulse. When the pump energy reached a threshold (in the range 0.05 - 0.1 J/cm^2), a single emission pulse of 1-3 ns duration was observed. With a further increase of pump energy, the number of emission pulses increased to three or four. At a constant pumping intensity, the number of pulses, their duration, and the interval between them were governed by the properties of the materials. The emission spectrum above the threshold was related to the particle shape. In the powder of particles with various shape, there is only one narrow emission line at the center of a luminescence band, while in the powder of particles with one specific shape, the emission spectrum consists of several lines in the range of the luminescence band.[14] In all cases, the spectral width of the emission lines above threshold is on the order of 0.1 nm. The observed emission was very much like laser emission. Because the particle size ($\sim 10\mu m$) was much larger than the emission wavelength, Markushev *et al* speculated that individual particles served as effective resonators and lasing occurred in the modes formed by total internal reflection at the surface of a particle.[13] However, there might be some weak coupling between the neighboring particles. In the mixture of two powders with slightly shifted luminescence bands [e.g. $Na_5La_{1-z}Nd_z(MoO_4)_4$ powder with significantly different Nd^{3+} concentrations], the emission wavelength depended on the relative concentrations of components in the mixture and the excitation wavelength.[15] Briskina *et al* set up a model of coupled microcavities to interpret the experimental result.[16,17] They treated the powder as an aggregate of active optically coupled microcavities and calculated the modes formed by total internal reflection (in analog to the whispering-gallery modes). They found the quality factor of a coupled-particle cavity in the compact powder could be higher than that of a single-particle cavity due to the optical coupling. To confirm their model, they measured the spot of laser-like radiation from the powder of $Al_3Nd(BO_3)_4$ and NdP_5O_{14}.[18] The minimum spot size was 20-30 μm. When the particle size was 20 μm, the laser-like radiation was from a single particle or a few particles.

Later, the powder laser was realized with non-resonant pumping at room temperature.[19,20] The gain materials were extended from Nd^{3+}-doped powder to Ti:Sapphire powder[21], Pr^{3+}-doped powder [22], and pulverized LiF with color centers.[23] Despite the difference in material systems, the observed phenomena are similar: (i) drastic shortening of the emission pulse and spectral narrowing of the emission line above a pumping threshold; (ii) damped oscillation of the emission intensity under pulsed excitation; (iii) drifting of the stimulated emission frequency and hopping of emission line from one discrete frequency to another within the same series of pulses. Gouedard *et al* analyzed the spatial and temporal coherence of the powder laser.[19] From the contrast of the near-field speckle pattern, they concluded that the powder emission above the threshold is spatially incoherent. This result was explained by incoherent superposition of uncorrelated speckle patterns. Their time-resolved measurement showed the speckle pattern changed rapidly in time. The estimated coherence time ~ 10 ps indicated low temporal coherence of the powder emission. Noginov *et al* also performed quantitative measurement of the longitudinal and transverse coherence with interferometric techniques.[24] Using a Michelson (Twyman-Green) interferometer, they found the longitudinal coherence time of $Nd_{0.5}La_{0.5}Al_3(BO_3)_4$ powder (ceramics) emission was 56 ps at a pumping energy twice of the threshold. This value corresponded to 0.07 nm linewidth, in agreement with the result of direct spectroscopic linewidth measurement. They also examined the transversal spatial coherence using Young's double-slit interferometric scheme. The transversal coherence was not noticeable when the distance between two points on the emitting surface was approximately 85 μm or larger.

Despite the detailed experimental study of powder laser, the underlying mechanism was not fully understood. Gouedard *et al* conjectured that the grains of the powder emit collectively in a subnanosecond pulse with a kind of distributed feedback provided by multiple scattering.[19] Auzel and Goldner identified two processes of coherent light generation in powder: (i) amplification of spontaneous emission by stimulated emission, (ii) synchronized spontaneous emission namely superradiance and superfluorescence.[25−27] Noginov *et al* noticed the essential role played by photon diffusion in stimulated emission when comparing the powder laser with the single crystal laser.[20] The diffusive motion of light led to long path length of emission in the powder, resulting in a threshold. Wiersma *et al* proposed a model based on light diffusion with gain.[28] They considered an incident pump pulse and probe pulse onto a powder slab. The active material was approximated as four-level $(2,1,0',0)$ system with the radiative transition from level 1 to 0' and the pumping from level 0 to 2. Fast relaxation from level 2 to 1 and from level 0' to 0 made both level 2 and level 0' nearly unpopulated, thus the population of level 1 can be described by one rate equation. The whole system was described by three diffusion equations for the energy densities of the pump light $W_G(\vec{r}, t)$, the probe light $W_R(\vec{r}, t)$, the (amplified) spontaneous emission $W_A(\vec{r}, t)$, and one rate equation for the population density $N_1(\vec{r}, t)$ of level 1.

$$\frac{\partial W_G(\vec{r}, t)}{\partial t} = D\nabla^2 W_G(\vec{r}, t) - \sigma_{abs}v[N_t - N_1(\vec{r}, t)]W_G(\vec{r}, t) + \frac{1}{l_G}I_G(\vec{r}, t) , \quad (11)$$

$$\frac{\partial W_R(\vec{r}, t)}{\partial t} = D\nabla^2 W_R(\vec{r}, t) + \sigma_{em}v N_1(\vec{r}, t)W_R(\vec{r}, t) + \frac{1}{l_R}I_R(\vec{r}, t) , \quad (12)$$

$$\frac{\partial W_A(\vec{r}, t)}{\partial t} = D\nabla^2 W_A(\vec{r}, t) + \sigma_{em}v N_1(\vec{r}, t)W_A(\vec{r}, t) + \frac{1}{\tau_e}N_1(\vec{r}, t) , \quad (13)$$

$$\frac{\partial N_1(\vec{r}, t)}{\partial t} = \sigma_{abs}v[N_t - N_1(\vec{r}, t)]W_G(\vec{r}, t) \quad (14)$$
$$-\sigma_{em}v N_1(\vec{r}, t)[W_R(\vec{r}, t) + W_A(\vec{r}, t)] - \frac{1}{\tau_e}N_1(\vec{r}, t) ,$$

where σ_{abs} and σ_{em} are the absorption and emission cross sections, τ_e is the lifetime of level 1, $I_G(\vec{r}, t)$ and $I_R(\vec{r}, t)$ are the intensities of the incoming pump and probe pulses, l_G and l_R are the transport mean free paths at the pump and probe frequencies, N_t is the total concentration of four-level atoms. Wiersma *et al* numerically solved the above coupled nonlinear differential equations. Their simulation result reproduced the experimental observation of transient oscillation (spiking) of the emission intensity under pulsed excitation.

In the slab geometry, the critical volume predicted by Letokhov is reduced to the critical thickness $L_{cr} = \pi\sqrt{l_t l_g/3} = \pi l_{amp}$. For the fixed slab thickness L, there exists a critical amplification length $l_{cr} = L/\pi$. In the beginning of the pump pulse, the average amplification length l_{amp} decreases due to an increasing excitation level. Once l_{amp} crosses l_{cr}, the gain in the sample becomes larger than the loss through the boundaries and the system becomes unstable. This leads to a large increase of the amplified spontaneous emission (ASE) energy density. The characteristic time scale corresponding to the buildup of ASE is l_g/v. The large ASE energy density will deexcite the system again, which leads to an increase of l_{amp}. This deexcitation continues as long as the large ASE energy density is present. The characteristic time scale on which the ASE energy density diffuses out of the medium through the front or rear surface is given by L^2/D. On one hand, an overshoot of the excitation takes place because the deexcitation mechanism needs some time to set in. On the other hand, once the ASE has built up considerably, the ASE energy density can disappear only slowly due to the presence of multiple scattering, which leads to an undershoot below the threshold. These two processes result in transient oscillations of the outgoing ASE flux. The oscillations are damped because the increase of l_{amp} during the deexcitation is opposed by reexcitation owing to the presence of pump light. Therefore the system reaches the equilibrium situation $l_{amp} = l_{cr} = L/\pi$ after a few oscillations.

Both models based on light diffusion and intraparticle resonances reproduced the experimental phenomena. It is hard to tell whether the feedback in the powder laser is provided by multiple scattering or total internal reflection, because the gain medium and scattering elements are not separated in the powder. Lawandy *et al* separated the scattering and amplifying media in liquid solutions.[29] This separation allowed the scattering strength to be varied independently of the gain coefficient, and facilitated a systematic study of the scattering effect on feedback.

2.4. *Laser paint*

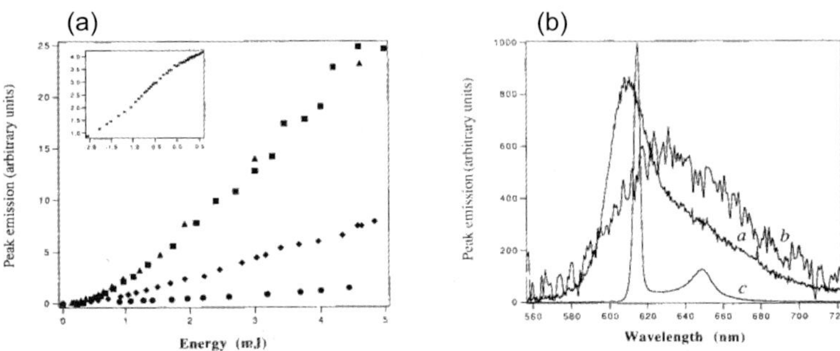

Fig. 2. (a) Emission intensity at the peak wavelength as a function of the pump pulse energy for four TiO_2 nanoparticle colloidal dye solutions. The TiO_2 particle densities are 1.4×10^9 cm^{-3} (circles), 7.2×10^9 cm^{-3} (diamonds), 2.8×10^{10} cm^{-3} (squares), and 8.6×10^{11} cm^{-3} (triangles). (b) Curve a is the emission spectrum of a 2.5×10^{-3}M solution of rhodamine 640 perchlorate in methanol pumped by 3-mJ (7-ns) pulses at 532nm. Curves b and c are emission spectra of the titanium dioxide (TiO_2) nanoparticle (2.8×10^{10} cm^{-3}) colloidal dye solution pumped by 2.2-μJ and 3-mJ (7-ns) pulses. The amplitude of spectrum b is scaled up by a factor of 10, whereas that of spectrum c is scaled down by a factor of 20. (with copyright permission).

In 1994, Lawandy *et al* observed laser-like emission from a methanol solution of Rhodamine 640 perchlorate dye and TiO_2 microparticles.[29] The dye molecules were optically excited by laser pulses and provide optical gain. The TiO_2 particles, with a mean diameter of 250 nm, were scattering centers. As shown in Fig. 2(a), the (input-output) plot of the peak emission intensity versus the pump energy exhibited a well-defined pumping threshold for the slope change. At the same threshold, the emission linewidth (FWHM: full width at half maximum) collapsed rapidly from 70 nm to 4 nm [Fig. 2(b)], and the duration of emission pulses was shortened dramatically from 4 ns to 100 ps. The threshold behavior suggested the existence of feedback. The relatively broad and featureless emission spectrum above the threshold indicated that the feedback was frequency-insensitive (non-resonant). In the solution, the feedback mechanism based on morphology-dependent resonance can be ruled out because the gain was outside the scatterer and individual scatterers were too small to serve as morphology-dependent resonators. It was found experimentally that the threshold was reduced by more than 2 orders of magnitude when the density of scattering particles was increased from 5×10^9 to 2.5×10^{12} cm^{-3} at the fixed dye concentration of 2.5×10^{-3} M.[30] The strong dependence of the threshold on the transport mean free path revealed that the feedback was related to scattering.[31–33]

However, light diffusion is negligible unless the smallest dimension of the scattering medium is much larger than the transport mean free path. Experimentally, when a spatially broad pump pulse is incident on a dye cell, a disk-shaped amplifying region is formed near the front window.[34] The thickness of the disk is determined by the penetration depth L_{pen} of the pump light. In Lawandy's experiment, L_{pen} is close to the transport mean free path. However, the actual sample thickness (i.e., the thickness of the entire suspension) is much larger than the transport mean free path. Hence, light transport in the suspension is diffusive. Nevertheless, the emitted photons could easily escape from the thin amplifying region. Part of them escapes through the front surface into the air, the rest went deeper into the unpumped region of the suspension. After multiple scattering (or random walk), some of these photons return to the active volume for more amplification. This return process provides energy feedback. When scattering is stronger, the return probability is higher, thus the feedback is stronger. However, incomplete feedback (less than 100% return probability) gives rise to loss. The lasing threshold is set by the condition that the photon loss rate is balanced by the photon generation rate in the amplifying region. On one hand, the total amount of gain or amplification is the product of the amplification per unit path length and the path length traveled through the amplifying volume. The frequency dependence of the amplification per unit path length gives the highest photon generation rate at the peak of gain spectrum. On the other hand, owing to the weak frequency dependence of the transport mean free path, the feedback is nearly frequency independent within the gain spectrum, as is the loss rate for photons. Therefore, with an increase of the pumping rate, the photon generation rate in the spectral region of maximum gain first reaches the photon loss rate, while outside this frequency region the photon generation rate is still below the loss rate. Then the photon density around the frequency of gain maximum builds up quickly. The sudden increase of photon density near the peak of gain spectrum results in the collapse of the emission linewidth.

A model based on the ring laser in the random phase limit was proposed by Balachandran and Lawandy to quantitatively explain the experimental data.[35] The amplifying volume is approximated as a sharply-bounded disk with homogeneous gain coefficient. In a Monte Carlo simulation of the random walk of photons, they calculated the return probabilities R_{t1} and R_{t2} of photons to the disk after being launched from the disk bottom either toward the disk interior or away from it, and the average total path length L_{pat}. The threshold gain g_{th} is determined by the steady-state condition

$$R_{t1}\, R_{t2}\, e^{g_{th}\, L_{pat}} = 1. \tag{15}$$

This condition is analogous to the threshold condition of a ring laser. Note that typical ring laser has a second condition on the round-trip phase shift: $kL_{pat} = 2\pi m$, which determines the lasing frequencies. In the scattering medium the phase condition can be ignored, because the diffusive feedback is non-resonant, i.e., it requires only light return to the gain volume instead of to its original position. Therefore this kind of laser is a random laser with non-resonant or incoherent feedback.

2.5. *Further developments*

The discovery of Lawandy *et al* triggered many experimental and theoretical studies that are briefly summarized here.

(1) *Lasing threshold.* The dependence of the lasing threshold on the dye concentration and the gain length was investigated.[32] Usually the threshold was reached at the point at which the pump transition was bleached. Such bleaching increased the penetration depth of the pump and consequently led to longer path lengths for the emitted light within the gain region which resulted in a reduced threshold.[36] The influence of the excitation spot diameter on the threshold was also examined.[37] In a suspension of TiO_2 scatterers in Sulforhodamine B dye, the threshold pump intensity increased by a factor of 70 when the excitation beam diameter got close to the mean free path. This is because a large pump beam spot produced large amplifying volume. The emitted light could travel a long path inside the active region and experienced more amplification before escaping. After the light went into the passive (unexcited) region, there was a large probability that it would return to the amplifying region because of the large pumped area. For a small excitation beam diameter, the emitted light would very likely leave the active volume after a short time, with a small chance of returning. This gave a larger photon loss rate and higher threshold. The amplification by stimulated emission was found to be strongest when the absorption length of the pump light and the transport mean free path had approximately the same magnitude.[38] A critical transport mean free path was identified for each beam diameter, below which the threshold was almost independent of the mean free path.[39] These results can be explained in terms of the spatial overlap of the gain volume and the diffusion volume. By solving the coupled rate and diffusion equations, Totsuka *et al* calculated the spatial distribution of the excited state population (gain volume) and the spatial spreading of the trajectories for the luminescence light (diffusion volume). When the gain volume was smaller than the diffusion volume of the luminescence light, the amplification was not efficient as the light propagated mostly through the gainless region. If the gain volume was larger than the diffusion volume of the luminescence light, the excitation pulse energy was not used efficiently for amplification. There existed an optimized condition under which the pulse energy was used most efficiently for stimulated emission.

(2) *Emission spectra.* The stimulated emission spectrum was shifted with respect to the luminescence spectrum. This spectral shift was explained by a simple ASE model accounting for absorption and emission at the transition between the ground and first singlet excited states of the dye.[40] Bichromatic emission was produced in a binary dye mixture in the presence of scatterers.[41] The dye molecules were of the donor-acceptor type, and the energy transfer between them gave double emission bands. The relative intensity of stimulated emission of the donor and the acceptor

depended on the scatterer density in addition to the pumping intensity and the concentration of the dyes. The narrow-linewidth bichromatic emission was also observed in the single dye solution with scatterers at large pumping intensity or high dye concentration.[42,43] John and Pang explained the bichromatic emission in terms of dye molecules' singlet and triplet transitions.[44] Using the physically reasonable estimates for the absorption and emission cross section for the single and triplet manifolds and the singlet-triplet intersystem crossing rate, they solved the nonlinear rate equations for the dye molecules. This led to a diffusion equation for the light intensity in the scattering medium with a nonlinear intensity-dependent gain coefficient. Their model could account for most experimental observations, e.g. the collapse of emission linewidth at a specific threshold pump intensity, the variation of the threshold intensity with the transport mean free path, the dependence of peak emission intensity on the transport mean free path, the dye concentration, and the pump intensity.

(3) *Dynamics.* One surprising result about the dynamics of stimulated emission from colloidal dye solutions is that the emission pulses can be much shorter than the pump pulses when the pumping rate is well above the threshold. For instance, 50 ps pulses of stimulated emission was obtained from the colloidal solution excited by 3 ns pulses.[30] The shortest emission pulses were ~ 20 ps long and produced by 10 ps pump pulses.[36] Berger *et al* modeled the dynamics of stimulated emission from random media using a Monte Carlo simulation of the random walk of pump and emitted photons.[45] They tracked the temporal and spectral evolution of emission by following the migration of photons and molecular excitation as determined purely by local probabilities. Their simulation results revealed a sharp transition to ultrafast, narrow line-width emission for a 10 ps incident pump pulse and a rapid approach to steady state for longer pump pulses. Using a different approach van Soest *et al* also studied the dynamics of stimulated emission.[46] They numerically solved the coupled diffusion equations for the pump light and the emitted light and the rate equation for the excited population. Their simulation result illustrated that the slow response of the population, compared to the light transport, started a relaxation oscillation at the threshold crossing.

(4) β *factor.* Compared with the traditional laser theory, the spontaneous emission coupling factor β was introduced for random laser.[47]. In a conventional laser, β is defined as the ratio of the rate of spontaneous emission into the lasing modes to the total rate of spontaneous emission. Its value is determined by the overlap in the wave-vector space between the spontaneous emission and laser field. In conventional macroscopic lasers, the spontaneous emission is isotropic while the cavity modes occupy small solid angles. The directional mismatch contributes to small β value (less than 10^{-5}). In the scattering medium, the diffusive feedback is non-directional, thus the spatial distinction between lasing and nonlasing modes vanishes, and the only criterion is the spectral overlap of the spontaneous emission spectrum with the lasing spectrum. This gives large β value (~ 0.1).

(5) *Control and optimization.* Liquid solutions are awkward to handle, e.g. the sedimentation of scattering particles in the solvent causes instability. Thus liquid solvents were replaced by polymers as host materials.[43] The polymer sheets containing laser dyes and TiO_2 nanoparticle scatterers were made with the cell-casting technique. The lasing phenomenon in solid dye solutions was similar to that in liquid dye solutions, despite that the different embedding environments affected the fluorescence characteristics of the dye.[48] Such kind of random laser is called a painted-on laser or photonic paint as the polymer film can be deposited on any substrate.[49,50] Many techniques developed for traditional lasers were exploited to optimize and control random lasers. For example, external-feedback was introduced to control the lasing threshold. De Oliverira *et al* placed a mirror close to the high-gain scattering medium, and measured the spectral line shapes of the emitted light as a function of the distance between them.[51] The main effect of the feedback from the mirror was to increase the lifetime of the photons inside the pump region, resulting in a reduction of the threshold pump energy. The injection-locking technique was also utilized to control the emission wavelength. Introducing a seed into the optically pumped scattering gain medium resulted in an intense isotropic emission whose wavelength and linewidth were locked to the seeding beam properties.[52] Moreover, multiple narrow-linewidth emission was obtained by pumping one laser paint with the output from another laser paint.[53] Lately, temperature tuning was employed to turn random lasers on and off. A liquid crystal was infiltrated into macro porous glass, and the diffusive feedback was controlled through a change of the refractive index of the liquid crystal with temperature.[54] In a different approach, a mixture with a lower critical temperature, which could be reversibly transformed between a transparent state and a highly scattering colloid with small temperature change, was used to tune the lasing threshold with temperature.[55] Finally, an external electric field was used to switch random lasing in dye-doped polymer dispersed liquid crystals from a three-dimensional random walk to a quasi-two-dimensional type of transport.([56] The laser emission was anisotropic and the polarization was enhanced due to strong scattering anisotropies.

(6) *Solid-state random lasers.* Recently there has been much progress in the development of solid-state random lasers.[57−61] For example, a low-threshold GaAs powder laser was realized.[62,63] Both the lasing threshold and the slope efficiency were significantly improved when the pulverized gain medium was mixed with a powder of optically inert material (without any absorption or emission). The employment of fiber-coupling scheme significantly improved the performance of random laser.[64] When the tip of a fiber is relocated from the surface to the depth of the powder volume, the lasing threshold is reduced twofold and the slope efficiency is increased fivefold. High absorption efficiency (85%) and high conversion efficiency of population inversion to stimulated emission (90%) make the fiber-coupled random laser a promising laser source.

3. Random laser with coherent feedback

3.1. *"Classical" vs. "quantum" random lasers*

Random lasers with coherent feedback can operate either in the classical optics regime or wave optics regime.[65] Let R_c be the correlation length of spatial fluctuation of the refractive index in a random medium. In the classical regime $R_c \gg \lambda$, whereas in the quantum/wave regime $R_c \leq \lambda$. They have analogues in chaotic cavity lasers whose cavity shape is irregular. In the classical regime, the spatial variation of the chaotic cavity shape is much larger than the wavelength of radiation, whereas in the quantum/wave regime the spatial variation is comparable to or smaller than the wavelength.

When the dielectric constant in a random medium varies over length scales much larger than λ, geometrical optics can be applied to describe light propagation in terms of ray trajectories. The majority of the ray trajectories are chaotic and open, yet unstable periodic orbits exist when the sample size is large enough. In an open system, unstable periodic orbits might trap light for longer time than chaotic trajectories. Thus lower optical gain is needed to realize lasing oscillation in certain "scar" modes that concentrate on some unstable periodic orbits.

When the dielectric constant fluctuates over length scales comparable to or even smaller than λ, ray optics no longer holds. The ray optics is replaced by wave optics that not only describes light scattering by short-range disorder, but also takes into account interference of scattered waves. The interference effect is crucial to light localization in a random medium, which is analogous to the (quantum) Anderson localization of electrons in a short-range potential. Even if the scattering mean free path is longer than the wavelength, light may still be trapped partially in a random medium via the process of multiple scattering and wave interference. Incomplete confinement can be compensated for by photon amplification when optical gain is introduced to a random medium, leading to lasing oscillation.

3.2. *Classical random laser with coherent feedback*

The classical type of random lasers with coherent feedback was first demonstrated by Vardeny and coworkers in weakly-disordered media such as π-conjugated polymer films[66,67], organic dye-doped gel films[68], synthetic opals infiltrated with π-conjugated polymers and dyes.[68–70] The long-range fluctuations of refractive indices in their polymer films are most likely caused by inhomogeneity of the film thickness. Since light is confined within a film due to waveguiding, larger thickness leads to higher effective index of refraction.

The samples were excited by short laser pulses, and emitted broad-band luminescence at low pumping. With increasing pump intensity the photoluminescence band narrows drastically. As the excitation intensity increases even further, the emission spectrum transforms into a fine structure that consists of a number of sharp peaks [Fig. 3(a)]. The spectral width of these peaks is less than 1nm. When the pump

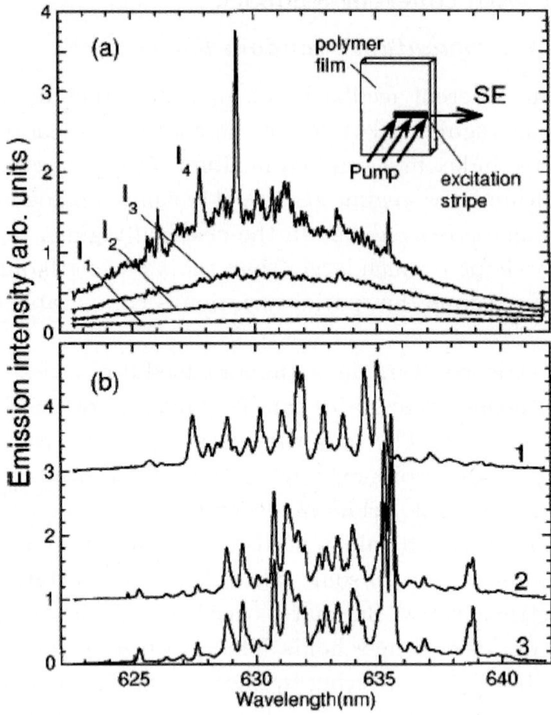

Fig. 3. Stimulated emission (SE) spectra of a DOO-PPV film obtained using a stripelike excitation area with length $L = 51$ mm and width $a = 530$ mm: (a) SE at different excitation intensities I: $I_1 = I_A = 1$ MW/cm2, $I_2 = 1.25 I_A$, $I_3 = 1.4 I_A$, $I_4 = 1.6 I_A$; the inset schematically shows the excitation geometry. (b) SE spectra (offset for clarity) measured sequentially at $I \simeq 2 I_B$ from the same DOO-PPV film: line 2 was obtained from a different excited area vertically shifted by 0.3 mm from that of line 1, line 3 was obtained after 3 min delay from the same area as line 2.

light excites a different sample or a different part of the same sample, the narrow peaks change completely. However, when the same part of the sample is excited repeatedly by different pump pulses, the spectral peaks are reproducible [Fig. 3(b)].

Polson *et al* suggested that the lasing modes in the polymer films are formed by total internal reflection at the boundaries of high refractive index regions.[65] The long-range fluctuation of the refractive index is similar to a ring resonator, in the sense that it gives rise to a number of resonant modes having close frequencies and quality factors. These modes are revealed in the emission spectrum, and their frequency spacings are determined by the cavity lengths. Hence, the size, L_r, of the underlying resonators can be estimated from the power Fourier transform (PFT) of lasing spectra. They indeed found $L_r \gg \lambda$ in their samples.

Alongside with the long-range fluctuation of refractive index, short-range disorder is also present in the polymer films. It is conjectured that the ability of most random resonators to trap light by total internal reflection is suppressed by the short-range disorder. The dramatic consequence of this suppression is the resonators that "survive" the short-range disorder are sparse, and consequently almost identical. Experimentally, despite the PFT of individual random lasing spectra exhibits position-specific multi-peak structures, averaging the PFTs over the sample positions does not smear these features, on the contrary, yields a series of distinct transform peaks. Moreover, the shape of the averaged PFT is universal, i.e., increasing the disorder and correspondingly reducing l_t do not change this shape: the average of the PFT spectra at different l_t scales with l_t to a universal curve.[65]

To understand the classical type of random laser with coherent feedback, Apalkov, Raikh and Shapiro conducted a theoretical analysis.[71] They believe the lasing modes are the almost localized states in the passive medium. Such states are formed due to some rare disorder configurations that can trap light for a long time in a sub-mean-free-path region in space. In the case of a continuous random potential, the almost localized states are confined to small rings of a sub mean-free-path size. The almost localized states are non-universal, i.e., their character and formation probability depend not only on the average strength of disorder, but also on the microscopic details of disorder.[72] Apalkov, Raikh and Shapiro calculated the areal density of the almost localized states in a film with fluctuating refractive index.[73] The rings formed by disorder can be viewed as waveguides that support the whispering-gallery type modes. Because of the azimuthal symmetry, these modes are characterized by the angular momentum, m. The areal density of ring resonators with quality factor Q can be expressed as $N_m(k\,l_t, Q) = N_0 \exp[-S_m(k\,l_t, Q)]$ for $k\,l_t > 1$. In the case of smooth disorder $kR_c \gg 1$, $S_m \sim l_t(\ln Q)^{4/3}/(kR_c^2 m^{1/3})$. In the opposite limit of short-range disorder, $S_m \sim k\,l_t \ln Q$. Therefore, when on average the light propagation is diffusive, the likelihood for finding an almost localized state increases sharply with the disorder correlation radius R_c for a given $k\,l_t$. Note that this conclusion applies only to a continuous (Gaussian) random potential. In the presence of a discrete lattice (the Anderson model), a new type of almost localized state is formed, whose formation probability is reduced by correlation in disorder.[74,75]

Apalkov and Raikh also investigated the fluctuation of the random lasing threshold.[76] The distribution of the threshold gain over the ensemble of statistically-independent finite-size samples is found to be universal. This universality stems from two results, (i) the lasing threshold in a given sample is determined by the highest-quality mode of all the random resonators present in the sample, (ii) the areal density of the random resonators decays sharply with the quality factor of the mode that they trap. In a 2D sample of area S, the distribution function of the threshold excitation intensity I_{th} is:

$$F_S(I_{th}) = \frac{\beta_S}{I_{th}} \left(\frac{I_{th}}{I_S} \right)^{-\beta_S} \exp\left[\left(\frac{I_{th}}{I_S} \right)^{-\beta_S} \right]. \tag{16}$$

The typical value I_S is related to the sample area S:

$$I_S \propto \exp\left\{-\left[\frac{\ln(S/S_0)}{G}\right]^{1/\alpha}\right\}. \tag{17}$$

S_0 is the typical area of a random resonator. The parameters α and G are determined by the intrinsic properties of the disordered medium and are independent of S. These two parameters play different roles: α is determined exclusively by the shape of the disorder correlator, G is a measure of the disorder strength. $\beta_S \propto [\ln(S/S_0)]^{(\alpha-1)/\alpha}$. The parameter β_S decreases with $k\,l_t$ as a power law, and the exponent depends on the microscopic properties of the disorder. For a weakly scattering medium, $\beta_S \gg 1$, and $F_S(I_{th})$ is close to a Gaussian distribution, $F_S(I_{th}) \propto \exp[-(\beta_S^2/2)(I_{th}/I_S - 1)^2\}$. When β_S is small, the distribution $F_S(I_{th})$ is broad and strongly asymmetric. It has a long tail towards the high thresholds, and falls off abruptly towards low thresholds.

3.3. *Quantum random laser with coherent feedback*

In the classical type of random laser with coherent feedback, formation of "closed periodic orbits" with small leakage results in light confinement. The interference effect plays a second role as it only determines the resonant frequencies in the periodic orbits. However, in the quantum/wave type of random lasers with coherent feedback, the random media have discrete scatterers and strong short-range disorder, thus the interference of scattered waves is essential to light trapping in a random medium. The active random media used for the quantum/wave type of random lasers can be divided into two categories: (i) aggregation of active scatterers such as ZnO powder; (ii) passive scatterers in continuous gain media, e.g. TiO_2 particles in dye solution. Both have their advantages and disadvantages. In (ii), gain media and scattering centers are separated, that allows independent variation of the amounts of scattering and gain. However, the scattering strength in (i) is usually higher than that in (ii), owing to larger contrast of refractive index and higher density of scatterers. In the following subsections, lasing in both types of random media are discussed with some examples.

3.3.1. *Lasing oscillation in semiconductor nanostructures*

Figure 4 shows the scanning electron microscope (SEM) images of some semiconductor nanostructures that are used for random laser experiments. The ZnO nanorods in Fig. 4(a) are grown on a sapphire substrate by metalorganic chemical vapor deposition (MOCVD).[77] The rods are uniform in diameter and height, but randomly located on the substrate. The average rod diameter is about 50nm. The ZnO nanorod array is a two-dimensional (2D) scattering system, as light is scattered by the nanorods in the plane perpendicular to the rods. When the layer of ZnO nanorods has a larger refractive index than the substrate, index guiding leads to

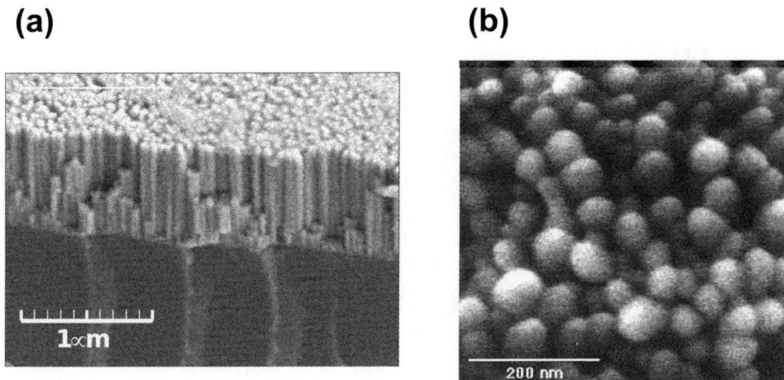

Fig. 4. SEM images of (a) ZnO nanorods on a sapphire substrate, (b) closely-packed ZnO nanoparticles.

light confinement in the third direction parallel to the rods. Figure 4(b) shows the ZnO nanoparticles synthesized in a wet chemical reaction.[78] The particles are polydisperse with an average size $\sim$ 100nm. They are random close packed with a filling fraction $\sim$ 50%. Since light is scattered by ZnO nanoparticles in all directions, the ZnO powder represents a 3D scattering system. The above two examples of random media have discrete scatterers of subwavelength size. The short-range disorder results in strong light scattering. In the ZnO powder, the transport mean free path $l_t \sim \lambda$. To introduce optical gain, the ZnO samples are optically pumped by the frequency-tripled output ($\lambda = 355$nm) of a mode-locked Nd:YAG laser (10Hz repetition rate, 20ps pulse width). The ZnO nanorods and nanoparticles become active scatterers.

Lasing with coherent feedback has been observed in both ZnO nanorods and nanoparticles. Since their lasing behavior are similar, only the measurement results of ZnO powder are presented next. Figure 5 shows the measured spectra and spatial distribution of emission in a ZnO powder film at two pumping intensities.[79] At low pumping level, the spectrum consists of a single broad spontaneous emission band. Its full width at half maximum (FWHM) is about 12 nm [Fig. 5(a)]. In Fig. 5(b), the spatial distribution of the spontaneous emission intensity is smooth across the excitation area. Due to pump intensity variation across the excitation spot, the spontaneous emission in the center of the excitation spot is stronger. When the pump intensity exceeds a threshold, discrete narrow peaks emerge in the emission spectrum [Fig. 5(c)]. The FWHM of these peaks is about 0.2 nm. Simultaneously, tiny bright spots appear in the image of the emitted light across the film [Fig. 5(d)]. The size of the bright spots is between 0.3 and 0.7 μm. When the pump intensity is increased further, additional sharp peaks emerge in the emission spectrum, and more bright spots appear in the image of the emitted light distribution. Above the threshold, the total emission intensity increases much more rapidly with the excitation intensity.

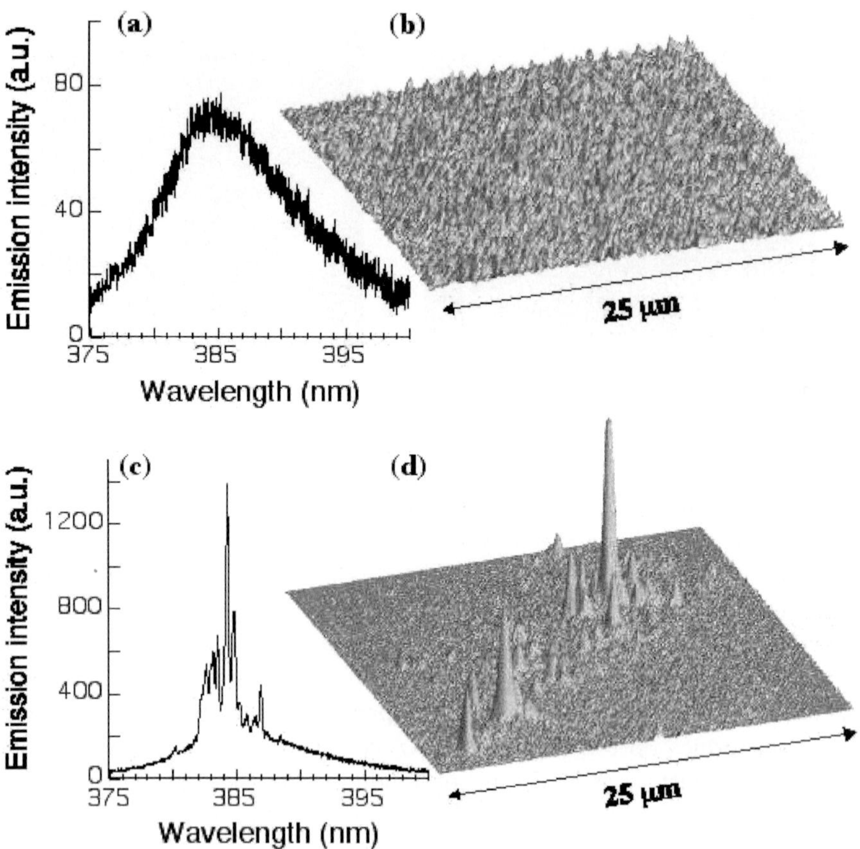

Fig. 5.　(a) and (c) are the measured spectra of emission from the ZnO powder. (b) and (d) are the measured spatial distribution of emission intensity on the sample surface. The incident pump pulse energy is 5.2 nJ for (a) and (b), and 12.5nJ for (c) and (d).

The frequencies of the sharp peaks depend on the sample position. As the excitation spot is moved across the sample, the frequencies of these peaks change completely. However, at a fixed sample position, the peak frequencies remain the same, while the peak heights fluctuate from shot to shot. These phenomena suggest that the discrete spectral peaks result from spatial resonances for light in the ZnO powder, and such resonances depend on the local configurations of nanoparticles. To find out the spatial size of each lasing mode, spectrally-resolved speckle analysis was employed to map the spatial profile of individual lasing modes at the sample surface.[80] The far-field speckle pattern of one lasing mode is recorded, then

Fourier-transformed to generate the spatial field correlation function in the near-field zone. Once above the lasing threshold, spatial coherence is established across the entire lasing mode. Hence, the spatial extend of the field correlation function directly reflects the mode size. The experimental data reveal that individual lasing modes have dimensions of a few micron. This can be understood as follows. Due to local variation in particle density and spatial configuration, there exist small regions of stronger scattering. Light can be trapped in these regions through the process of multiple scattering and wave interference. For a particular configuration of particles, only light at certain frequencies can be confined, because the interference effect is frequency sensitive. In a different part of the sample, the particle configuration is different, thus light is confined at different frequencies. However, the confinement is incomplete as light can escape through the sample surface. When the photon generation rate reaches the photon escape rate, lasing oscillation occurs at the local resonant frequencies, resulting in discrete lasing peaks in the emission spectrum.

The dependence of random lasing on the pump area A_p was investigated by Cao et al.[79] The lasing threshold decreases with increasing A_p, as it is more likely to find a stronger trapping site for light within a larger gain volume. At a fixed pump intensity, more lasing peaks appear when A_p increases. It is simply because there are more trapping sites for light. Eventually at very large pump area, the lasing peaks are so close to each other in frequency that they can no longer be resolved. However, when A_p is reduced to below a critical value, lasing oscillation stops. The critical pump area decreases with increasing pumping intensity.

The temporal evolution of emission was measured by a streak camera.[81] Below the lasing threshold, the decay time of the emission is 167 ps. When the pump intensity exceeds the threshold, the emission pulse is shortened dramatically. The initial decay of emission intensity is very fast, the decay time is 27 ps. After about 50 ps, the decay is slower. The later decay time is 167 ps, which equals the decay time below the threshold. The initial fast decay is caused by stimulated emission, and the later slow decay results from spontaneous emission and nonradiative recombination. As the pump intensity increases further, the initial stimulated emission becomes much stronger than the later spontaneous emission. The dynamics of individual lasing modes was also measured by combining a spectrometer with a streak camera. The time traces of individual lasing modes reveal that lasing in different modes is not synchronized. Just above the lasing threshold, relaxation oscillation is observed for some of the lasing modes. Since the pump pulse is shorter than both radiative and nonradiative recombination times of ZnO particles, lasing is in the transient regime. Recently, lasing in ZnO powder has been realized with 10ns pump pulses.[82] Since the pumping time is much longer than all the characteristic time scales in the system, the lasing oscillation could be regarded as quasi-continuous.

The quantum statistical property of laser emission from the ZnO powder was also probed in a photon counting experiment.[83] The photon number distribution for coherent light in a single electromagnetic mode satisfies the Poisson distribution, whereas the photon number distribution for chaotic light in a single-mode meets the

Bose-Einstein distribution.[84] However, the photon number distribution for chaotic light in multiple modes approaches Poisson distribution as the number of modes increases to infinity. Hence, it would be difficult to distinguish coherent light from chaotic light in a measurement of photon statistics of multimode. Experimentally the number of photons in a single electromagnetic mode is counted. The counting time is shorter than the inverse of frequency bandwidth of a lasing mode, and the collection angle in the far-field zone is less than one angular speckle. The photon number distribution in a single mode changes continuously from a Bose-Einstein distribution near the threshold to a Poisson distribution well above the threshold. The second-order correlation coefficient G_2 decreases gradually from 2 to 1. It is well known that for a single-mode chaotic light $G_2 = 2$, while for a single-mode coherent light $G_2 = 1$.[84] Hence, the coherent light is indeed generated in the highly-disordered ZnO powder.

Note that the photon statistics of a random laser with resonant feedback is very different from that of a random laser with nonresonant feedback. For a random laser with nonresonant feedback, lasing occurs simultaneously in a large number of modes which are spatially and/or spectrally overlapped. Mode coupling prevents quenching of photon number fluctuation in a single mode[8,10,85], resulting in excess photon noise.[86−90] Above the lasing threshold, the fluctuation of laser emission intensity is quenched by gain saturation.[91,92] The total number of photons in all lasing modes exhibits a fluctuation much smaller than that of the black-body radiation in the same number of modes. However, the number of photons in a single lasing mode is subject to large fluctuations as a result of mode coupling. Well above the lasing threshold, the amount of photon number fluctuation in each lasing mode is increased above the Poissonian value by an amount that depends on the number of lasing modes. In contrast, a random laser with resonant feedback has lasing modes well separated in frequency. Due to little mode coupling, the photon number fluctuation in each lasing mode can be quenched efficiently by gain saturation at the mode frequency.

3.3.2. *Random microlaser*

The small size of lasing modes in closely-packed ZnO nanoparticles indicates strong optical scattering not only supplies coherent feedback for lasing, but also leads to spatial confinement of laser light in micron-size volume.[93] This makes it possible to realize a new type of microlaser are made of disordered media.[94]

Figure 6(a) is a SEM image of a microcluster of ZnO nanoparticles. The size of the cluster is about 1.7 μm. It contains roughly 20000 nanoparticles. A single cluster is optically pumped by the third harmonic of a pulsed Nd:YAG laser. At low pumping level, the emission spectrum consists of a single broad spontaneous emission speak. Its FWHM is 12 nm. The spatial distribution of the spontaneous emission intensity is uniform across the cluster. When the pump intensity exceeds a threshold, a sharp peak emerges in the emission spectrum [Fig. 6(b)]. Its FWHM is

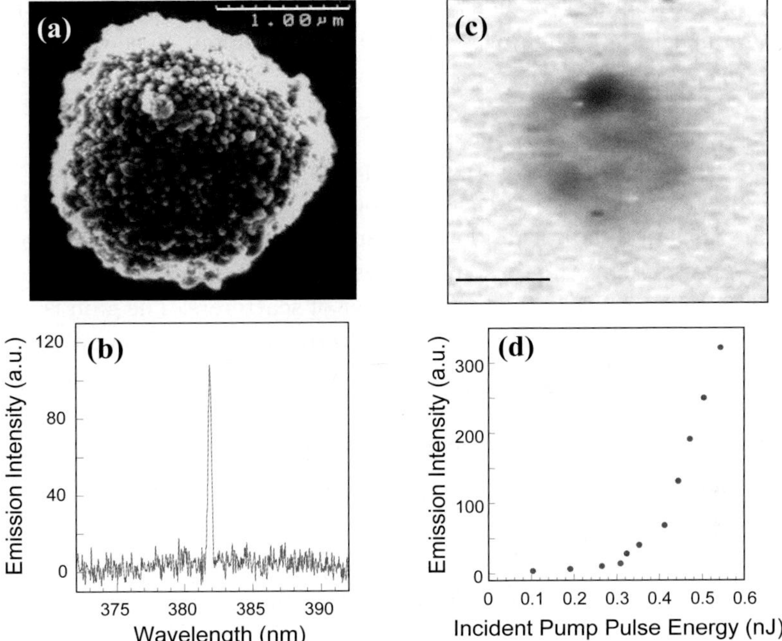

Fig. 6. (a) Scanning electron micrograph of a microcluster of ZnO nanocrystallites. (b) The spectrum of emission from the cluster at the incident pump pulse energy of 0.35 nJ. (c) Optical image of the emitted light distribution across the cluster. The incident pump pulse energy is 0.35 nJ. The scale bar is 1 μm.(d) Spectrally-integrated emission intensity as a function of the incident pump pulse energy.

0.2 nm. Simultaneously, a few bright spots appear in the image of the emitted light distribution in the cluster [Fig. 6(c)]. When the pump intensity increases further, a second sharp peak emerge in the emission spectrum. Correspondingly, additional bright spots appear in the image of the emitted light distribution. Note that the frequencies of the sharp peaks and the positions of the bright spots do not change from pump pulse to pulse (from shot to shot).

The total emission intensity is plotted against the pump intensity in Fig. 6(d). The curve exhibits a distinct change in slope at the threshold where sharp spectral peaks and bright spots appear. Well above the threshold, the total emission intensity increases almost linearly with the pump intensity. These data reveals lasing oscillation in the micron-size cluster.

Since the cluster is very small, optical reflection from the boundary of the cluster may have some contribution to light confinement in the cluster.[95,96] However, the laser cavity is not formed by total internal reflection at the boundary. Otherwise the spatial pattern of laser light would be a bright ring near the edge of the cluster.[97] The 3D optical confinement in a micron-size cluster is realized through multiple scattering and interference. Since interference effect is wavelength sensitive, only light at certain wavelengths can be confined in a cluster. In another cluster of different particle configuration, light at different wavelengths is confined. Therefore, the lasing frequencies are fingerprints of individual random clusters.

3.3.3. *Collective modes of resonant scatterers*

A simplified way of simulating the closely-packed ZnO nanoparticles or nanorods is to approximate individual particles or rods as dipolar oscillators.[98] Burin *et al* calculate the quasimodes in a random ensemble of point dipoles. The k-th dipolar oscillator is represented by its resonant frequency ω_k and transition dipole moments $\mathbf{d}_k$, where $k = 1, 2, ...N$. N is the total number of scatterers. The gain is introduced into each scatterer by adding an imaginary term $i\tilde{g}$ to its resonant frequency. A quasimode of this system represents a collective excitation of the coupled dipoles, thus it is also called a collective mode. The equation of motion for the k-th oscilla-tor's polarization component $\mathbf{p}_k = p_k \mathbf{d}_k / d_k$ can be written as:

$$-\Omega^2 p_k = -(\omega_k - i\tilde{g})^2 p_k + 2\omega_k d_k (\mathbf{d}_k \cdot \mathbf{E}_k). \tag{18}$$

$\mathbf{E}_k$ is the local electric field,

$$\mathbf{E}_k = \sum_{j \neq k} \mathbf{E}_{kj} + i\frac{2}{3}q^3 \mathbf{p}_k, \tag{19}$$

where $q = \Omega/c$, and $\mathbf{E}_{kj}$ represents the electric field generated by the j-th dipole at the location of k-th dipole. The solution to the Maxwell equations for electric field of a single dipole gives:

$$\mathbf{E}_{kj} = e^{iqR_{kj}} \frac{\mathbf{p}_j - 3\mathbf{n}_{kj}(\mathbf{n}_{kj} \cdot \mathbf{p}_j)}{R_{kj}^3}(1 - iqR_{kj}) + q^2 e^{iqR_{kj}} \frac{\mathbf{p}_j - \mathbf{n}_{kj}(\mathbf{n}_{kj} \cdot \mathbf{p}_j)}{R_{kj}}, \tag{20}$$

where $\mathbf{n}_{kj} = \mathbf{R}_{kj}/R_{kj}$, $\mathbf{R}_{kj}$ is the vector from the j-th dipole to the k-th dipole.

There are N solutions to the above equations for N coupled dipolar oscillators. Hence, there are N collective modes, each is characterized by a complex frequency Ω_α ($\alpha = 1, 2, ...N$). The imaginary part of Ω_α, γ_α, represents the decay rate of a collective mode caused by light leakage out of the system. In the absence of gain $\tilde{g} = 0$, the system is lossy and all the decay rates are positive. An increase of gain leads to a decrease of γ_α. At some finite value of gain $\tilde{g}_{th}$, the decay rate for some collective mode vanishes. This corresponds to the onset of lasing instability. The collective mode with the smallest decay rate in the passive system ($\tilde{g} = 0$) turns out to be the first lasing mode.

Burin *et al* numerically calculated the threshold gain $\tilde{g}_{th}$ in 2D random arrays of dipolar oscillators with N up to 1000. All the dipoles are assumed to have the same resonant frequency: $\omega_k = \omega_0$. They are positioned randomly within a circle of radius R_0. The average interdipole distance, normalized by the resonant wavelength $\lambda_0 = 2\pi c/\omega_0$, is described by $\eta = 2\pi R_0/\sqrt{N}$. The ensemble-averaged lasing threshold decreases with increasing N as $\langle g_{th} \rangle \propto N^{-\beta}$, where the exponent β is a function of η. It is equivalent to $1/A^\beta$ dependence on the sample area A for a fixed particle density with $\beta = 0.52$ for $\eta = 0.3$, $\beta = 0.51$ for $\eta = 1$, and $\beta = 0.335$ for $\eta = 3$.

The size of the lasing modes is usually smaller than that of the entire system. Hence, the decrease of $\tilde{g}_{th}$ with the system size results from the increase in the probability of finding optimum particle configurations for minimum γ_α rather than the formation of larger modes. Next dispersion (random variation) is gradually introduced to the resonant frequencies ω_k of dipolar oscillators. As the dispersion increases, the efficiency of forming collective modes with small γ_α decreases. The threshold gain $\tilde{g}_{th}$ becomes nearly size independent when ω_k deviates from each other by more than the near-neighbor dipolar coupling constant.

Therefore, collective excitations can be formed most efficiently when all the scatterers have identical resonant frequencies.[98] This condition was realized experimentally with ZnO particles of uniform shape and size.[99] Seelig *et al* developed a two-stage chemical reaction process to synthesize monodisperse ZnO nanospheres.[78] The mean diameter of ZnO spheres is varied from 85 nm to 617 nm. The dispersion of the sphere diameter is $5 - 8\%$. The ZnO spheres are closely packed with the volume fraction $\sim 58\%$. In the lasing experiment, the threshold pump intensity I_{th} decreases drastically with increasing sphere diameter from 85nm to 137nm. This rapid drop is replaced by a slow decrease as d_s increases from 137nm to 355nm. When d_s increases further to 617nm, I_{th} increases slightly. The variation of I_{th} with d_s follows roughly the trend of scattering cross section σ_{sc} of ZnO nanospheres. The range of d_s covers the first few Mie resonances at the ZnO emission wavelength. σ_{sc} exhibits a drastic increase with d_s before reaching the first Mie resonance at $d_s \simeq 200$nm. The value of σ_{sc} reaches the maximum at $d_s \sim 370$nm. Then it starts decreasing with a further increase of d_s to 617nm. At the Mie resonances, photon dwell time within individual scatterers is drastically increased. This leads to a significant enhancement of light amplification because optical gain exists inside the particles. Of course in such densely packed system, scattering particles cannot be considered independent, the resonances of individual scatterers are significantly modified by the interactions among them. Strong coupling of resonant scatterers lead to formation of collective modes with small as well as large decay rates.[100,101] The former serve as the lasing modes. The experimental data of I_{th} can be explained qualitatively in terms of σ_{sc}. The larger the scattering cross section of individual spheres, the stronger the coupling among them, the higher the chance of forming collective modes with smaller decay rates. Hence, the lasing threshold is lower.

3.3.4. *Time-dependent theory of random laser*

Simulation of random lasers above the threshold requires a time-dependent model that takes into account the gain saturation effect. Jiang and Soukoulis have developed a time-dependent theory for random laser that couples the Maxwell equations with the rate equations of electronic population.[102] The gain medium has four electronic levels. Electrons are pumped from level 0 to level 3, then relax quickly (with time constant τ_{32}) to level 2. Level 2 and level 1 are the upper and lower levels of the lasing transition at frequency ω_a. After radiative decay (with time constant τ_{21})

from level 2 to 1, electrons relax rapidly (with time constant τ_{10}) from level 1 back to level 0. If the dephasing time is much shorter than all other time scales in the system, the populations in four levels (N_3, N_2, N_1, N_0) satisfy the following rate equations:

$$\frac{dN_3(\mathbf{r},t)}{dt} = P_r(t)\, N_0(\mathbf{r},t) - \frac{N_3(\mathbf{r},t)}{\tau_{32}} \, ,$$

$$\frac{dN_2(\mathbf{r},t)}{dt} = \frac{N_3(\mathbf{r},t)}{\tau_{32}} + \frac{\mathbf{E}(\mathbf{r},t)}{\hbar\omega_a} \cdot \frac{d\mathbf{P}(\mathbf{r},t)}{dt} - \frac{N_2(\mathbf{r},t)}{\tau_{21}} \, ,$$

$$\frac{dN_1(\mathbf{r},t)}{dt} = \frac{N_2(\mathbf{r},t)}{\tau_{21}} - \frac{\mathbf{E}(\mathbf{r},t)}{\hbar\omega_a} \cdot \frac{d\mathbf{P}(\mathbf{r},t)}{dt} - \frac{N_1(\mathbf{r},t)}{\tau_{10}} \, ,$$

$$\frac{dN_0(\mathbf{r},t)}{dt} = \frac{N_1(\mathbf{r},t)}{\tau_{10}} - P_r(t)\, N_0(\mathbf{r},t) \, . \tag{21}$$

$P_r(t)$ represents the external pumping rate. $\mathbf{P}(\mathbf{r},t)$ is the polarization density that obeys the equation:

$$\frac{d^2\mathbf{P}(\mathbf{r},t)}{dt^2} + \Delta\omega_a \frac{d\mathbf{P}(\mathbf{r},t)}{dt} + \omega_a^2 \mathbf{P}(\mathbf{r},t) = \frac{\Gamma_r}{\Gamma_c}\frac{e^2}{m}\left[N_1(\mathbf{r},t) - N_2(\mathbf{r},t)\right]\mathbf{E}(\mathbf{r},t) \tag{22}$$

ω_a and $\Delta\omega_a$ represent the center frequency and linewidth of the atomic transition from level 2 to level 1. $\Gamma_r = 1/\tau_{21}$, $\Gamma_c = e^2\omega_a^2/6\pi\varepsilon_0 mc^3$, where e and m are electron charge and mass. $\mathbf{P}(\mathbf{r},t)$ introduces gain into Maxwell's equations:

$$\nabla \times \mathbf{E}(\mathbf{r},t) = -\frac{\partial \mathbf{B}(\mathbf{r},t)}{\partial t} \, , \tag{23}$$

$$\nabla \times \mathbf{H}(\mathbf{r},t) = \varepsilon(\mathbf{r})\frac{\partial \mathbf{E}(\mathbf{r},t)}{\partial t} + \frac{\partial \mathbf{P}(\mathbf{r},t)}{\partial t} \, , \tag{24}$$

where $\mathbf{B}(\mathbf{r},t) = \mu\,\mathbf{H}(\mathbf{r},t)$. The disorder is described by the spatial fluctuation of the dielectric constant $\varepsilon(\mathbf{r})$. Eqs. 22, 22, 24 are solved with the finite-difference time-domain (FDTD) method[103] to obtain the electromagnetic field distribution in the random medium. Fourier transforming $\mathbf{E}(\mathbf{r},t)$ gives the local emission spectrum. To simulate an open system, the random medium has a finite size and it is surrounded by air. The air is then surrounded by strongly-absorbing layers, e.g. the uniaxial perfectly matched layers, that absorb all the light escaping through the boundary of the random medium. Within a semiclassical framework, spontaneous emission can be included in the Maxwell equations as a noise current.

Jiang and Soukoulis simulated the lasing phenomenon in a 1D random system with a time-dependent theory.[102] A critical pumping rate exists for the appearance of lasing peaks in the spectrum. The number of lasing modes increases with the pumping rate and the length of the system. When the pumping rate increases even further, the number of lasing modes does not increase any more, but saturates to a constant value, which is proportional to the system size for a given randomness. This saturation is caused by spatial repulsion of lasing modes that results from gain competition and spatial localization of the lasing modes. This prediction was later confirmed experimentally.[104,105] The time-dependent theory is especially suitable for the simulation of laser dynamics. Soukoulis *et al* simulated the dynamic response and relaxation oscillation in random lasers.[81] The simulation results reproduced experimental observations and provides an understanding of the dynamic response of random lasers.

Vanneste and Sebbah calculated the spatial profile of lasing modes in 2D random media with the same method.[106,107] They compared the passive modes of a 2D localized system with the lasing modes when gain is activated, and found they are identical. When the external pump is focused, the lasing modes change with the location of the pump, in agreement with experimental observations.[79,108] Therefore, local pumping of the system allows selective excitation of individual localized modes. Jiang and Soukoulis also showed that a knowledge of the density of states and the eigenstates of a random system without gain, in conjuction with the frequency profile of the gain, can accurately predict the mode that lases first when optical gain is added.[109]

The advantage of the time-dependent theory is that it can simulate lasing in a real random structure. The numerical simulation gives the lasing spectra, the spatial distribution of lasing modes, and the dynamic response that can be compared directly with the experimental measurements. The problem is that simulation of large samples requires too much computing power and the running time is too long. So far the numerical simulations have been carried out only in 1D and 2D systems, even though the method can be applied to 3D systems. Furthermore, the simulation must be done for thousands of samples with different configurations before any statistical conclusion can be drawn.[110]

3.3.5. *Lasing modes in diffusive samples*

The above experimental and theoretical studies are focused on random systems near or in the localization regime where $l_t \sim \lambda$. The coherent feedback for lasing, supplied by multiple scattering, is expected to be strong. However, lasing with coherent feedback is also observed in diffusive samples with $l_t \gg \lambda$, e.g., in polymers doped with dye molecules and microparticles.[104] In such samples, the microparticles are the scattering centers, and the excited dye molecules provide optical gain. By varying the particle density, the scattering strength changes from $l_t \sim \lambda$ to $l_t \gg \lambda$. As l_t increases, the lasing threshold I_{th} rises quickly. The strong dependence of the lasing threshold on the transport mean free path confirms the essential contribution of scattering to lasing oscillation. With a decrease in optical scattering, the feedback provided by scattering becomes weaker, and the lasing threshold is increased. Curvefitting of experimental data gives $I_{th} \propto l_t^{0.52}$. At a fixed pump intensity, the number of lasing modes increases with decreasing l_t. When l_t approaches λ, the lasing threshold pump intensity drops rapidly, and the number of lasing modes increases dramatically. This result agrees with John and Pang's prediction of a dramatic threshold reduction in the regime $l_t \to \lambda$ of incipient photon localization[44].

In the diffusion regime $l_t \gg \lambda$, the coherent feedback supplied by scattering is weak, and light leakage from the sample is large. Nevertheless, lasing oscillation can still occur as long as the optical gain is high enough to compensate for the leakage loss. One important question is whether the lasing modes are still the quasimodes of the passive system as in the localization regime. In the absence of gain, the quasimodes of a diffusive system strongly overlap in space and frequency. The spectral width of individual quasimodes is much larger than the spacing of adjacent modes. Even a monochromatic light impinging on the sample, with the frequency equal to that of a particular quasimode, cannot excite only that mode. Instead it excites several modes which overlap with the incident frequency. These modes are excited with constant phase relation, leading to the formation of speckle pattern inside the system. In contrast, the lasing modes in a diffusive system can be well separated in frequency. To check the relation between the lasing modes and the quasimodes of weakly scattering systems, Vanneste *et al* simulate lasing in 2D random systems under uniform pumping.[111] They carefully adjust the pumping rate so that only one mode lases. A slight increase of pumping rate would lead to multimode lasing. Their calculation results illustrate that the first lasing mode above the threshold corresponds to the quasimode of the passive system. The reason that lasing can occur in a single quasimode is that optical amplification greatly enhances the interference of scattered light. In a naive picture, a quasimode is formed by the constructive interference of scattered waves returning to the same coherence volume via different closed paths (loops). The longer the path, the lower the amplitude of returning field. In a weakly scattering system, the amplitude of scattered field returning via a long loop is much lower than that via a short loop. The significant amplitude difference weakens the interference effect. In the presence of gain, light traveling along a long path is amplified more than that along a short path. The preferable amplification of the scattered field over a long loop makes its amplitude approach that over a short loop, which greatly enhances the interference effect. Consequently, the spectral width of the mode decreases dramatically. At the lasing threshold, the mode width is reduced to zero, if the spontaneous emission is neglected.

3.3.6. *Spatial confinement of lasing modes by absorption*

In a diffusive sample, the decay rate distribution of quasimodes is narrow, thus many modes have similar lasing threshold[112]. When optical gain is introduced to the entire sample, many modes can lase simultaneously, and they are closely packed in frequency. If the spectral width of individual lasing modes is larger than the frequency spacing of adjacent lasing modes, the lasing modes cannot be resolved, instead they merge to a single broad peak. In order to resolve the lasing modes, local pumping is often used. Namely, the pump beam is focused to a tiny spot on the sample, so that only a small part of the random system has gain. The number of lasing modes is usually reduced, leading to an increase of mode spacing. The reduction in the lasing mode number is not expected for the diffusive system whose quasimodes are usually spread over the entire volume. Under local pumping, all modes with frequencies within the gain spectrum experience similar gain. Shrinking the gain volume should not decrease the number of potential lasing modes, but only increase their lasing threshold. Imaging of laser light on the sample surface reveals that the lasing modes are not extended over the entire random medium, instead they are localized in the vicinity of excitation region.

The initial explanation for the above experimental observation was that the lasing modes are anomalously localized states.[80] They are analogous to the prelocalized electronic states in diffusive conductors that are responsible for the long-time asymptotics of the current relaxation.[113,114] Such states exhibit an anomalous buildup of intensity in a region of space. If they are located inside the excitation volume, they tend to be the lasing modes because they experience more amplification and less leakage. The anomalously localized states should be rare in the diffusive samples.[114] Experimentally, no matter where on the sample the pump beam is focused, the lasing modes are always confined in the pumped region. Moreover, the lasing threshold does not fluctuate much as the pump spot is moved across the random medium. These experimental observations contradict the theory of anomalously localized states. This discrepancy originates from the assumption that the lasing modes are equal to the eigenmodes of the passive random system. As first illustrated numerically by Yamilov *et al* and later confirmed experimentally by Wu *et al*, this assumption no longer holds if the absorption at the emission wavelength is significant outside the gain volume.[115,116]

Many laser dyes used in the random lasing experiment have significant overlap between the absorption band and emission band. Therefore, the photons that are emitted by the excited dye molecules inside the pumped region may diffuse into the surrounding unpumped region and be absorbed by the unexcited dye molecules there. Such absorption reduces the probability of light returning to the pumped region, thus suppresses the feedback from the unpumped region. Hence, the lasing modes differ from the quasimodes of the total random system.[115,116] Even if all the quasimodes are extended across the entire random system, the lasing modes are confined in the gain volume with an exponential tail outside it. This is because the absorption in the unpumped region effectively reduces the system size to the size of the pumped region plus the absorption length l_{abs}.

In a 3D random medium, the reduction of the effective system size L_{eff} leads to a decrease of the Thouless number $\delta \propto L_{eff}$, as $\delta\nu \propto L_{eff}^{-2}$ and $d\nu \propto L_{eff}^{-3}$. The smaller the value of δ, the larger the fluctuation of the decay rates γ of the quasimodes.[112] The variance of the decay rates is $\sigma_\gamma^2 = \langle\gamma\rangle^2/\delta^{114}$, thus $\sigma_\gamma/\langle\gamma\rangle \propto L_{eff}^{-1}$. The broadening of the decay rate distribution $P(\gamma)$ along with the reduction of the density of states is responsible for the observation of discrete lasing peaks in the tight focusing condition. Despite the reduction in the effective Thouless number, it is still much larger than 1 because of weak scattering. As a result, the lasing modes are the extended states within the effective volume. Because $\sigma_\gamma/\langle\gamma\rangle \ll 1$, the minimum decay rate γ_{min} is still close to $\langle\gamma\rangle$, leading to relatively small fluctuations in the lasing threshold. The threshold gain can be estimated as $g_{th} = \gamma_{min} \approx \langle\gamma\rangle \sim D/L_{eff}^2$.

This estimation gives a threshold equal to that predicted by Letokhov for random lasing with non-resonant feedback[11], $l_{amp} \sim L_{eff}$ or $l_g \sim L_{eff}^2/l_t$. Using this approximation and taking into account the saturation of absorption of pump light, Burin *et al* reproduced the experimentally measured dependence of the lasing threshold pump intensity I_{th} on l_t and the pump area A_p, namely, $I_{th} \propto l_t^{1/2}/A_p^q$, where $0.5 \leq q \leq 1$.[117]

Note that the above theory does not deny the possibility of anomalously localized states, in the sense that it does not eliminate the possibility of rare events. If they happen to be *within* the pumped volume, the anomalously localized states could serve as low-threshold lasing modes. However, even in the absence of such rare events, non-uniform distribution of gain and absorption could result in spatial localization of lasing modes in the pumped region. In other words, local pumping in an absorbing medium creates a "trapping" site for lasing modes.

3.3.7. *Effect of local gain on random lasing modes*

Even without absorption, local pumping may also modify the lasing modes, especially in a weak scattering system. Recently Wu *et al* developed a numerical method based on the transfer matrix to calculate the quasimodes of 1D random systems as well as the lasing modes with arbitrary spatial profile of pumping.[118] The boundary condition is that there are only outgoing waves through the system boundary. In a passive system such boundary condition gives the frequency and decay rate of every quasimode, while in an active system it determines the frequency and threshold gain of each lasing mode. This method is valid for linear gain up to the lasing threshold, with both gain saturation and mode competition for gain neglected. This simplification allows identification of all *potential* lasing modes regardless of the material-specific nonlinear gain.

The quasimodes, as well as the lasing modes, are formed by distributed feedback in the random system. The conventional distributed feedback (DFB) laser, made of periodic structures, operates either in the over-coupling regime or the under-coupling regime.[119] The random laser, which can be considered as randomly distributed feedback laser, also has these two regimes of operation. In the under-coupling regime the system size L is much less than the localization length ξ, while in the over-coupling regime $L > \xi$. In the under-coupling regime the electric field of a quasimode grows exponentially towards the system boundaries, while in the over-coupling regime the field maxima are located inside the random system. The frequency spacing of adjacent modes are more regular in the under-coupling regime, and there is less fluctuation in the mode decay rate. The distinct characteristic of the quasimodes in the two regimes result from the different mechanisms of mode formation. In an over-coupling system, the quasimodes are formed mainly through the interference of multiply scattered waves by the particles in the interior of the random system. In contrast, the feedbacks from the system boundaries become important in the formation of quasimodes in an under-coupling system. The contributions from the scatterers in the interior of the random system to the mode formation are relatively weak but not negligible. They induce small fluctuations in mode spacing and decay rate. As the scattering strength is increased, the feedbacks from those scatterers in the interior of the system get stronger, and the frequency spacing of the quasimodes becomes more irregular.

When the gain is uniformly distributed across the random system, the lasing modes (at the threshold) have a one-to-one correspondence with the quasimodes in both over-coupling and under-coupling systems. However, the lasing modes may differ slightly from the corresponding quasimodes in frequency and spatial profile, especially in the under-coupling systems. This is because the introduction of uniform gain removes the feedback caused by spatial inhomogeneity of the imaginary part of the wavevector within the random system and creates additional feedback by the discontinuity of the imaginary part of the wavevector at the system boundaries. As long as the scattering is not too weak, the quasimodes are only slightly modified by the introduction of uniform gain to a random system and they serve as the lasing modes. Because of the correspondence between the lasing modes and the quasimodes, the frequency spacing of adjacent lasing modes is more regular in the under-coupling systems with smaller mode-to-mode variations in the lasing threshold.

When optical gain is introduced to a local region of the random system (with no absorption outside the gain region), some quasimodes fail to lase no matter how high the gain is. The other modes can lase but their spatial profiles may be significantly modified. Such modifications originate from strong enhancement of feedback from the scatterers within the pumped region. It increases the weight of a lasing mode within the gain region. Moreover, the spatial variation in the imaginary part of the refractive index generates additional feedback for lasing. As the size of gain region L_g decreases, the number of lasing modes N_l is reduced, and the frequency spacing of lasing modes is increased. The sublinear decrease of N_l with L_g indicates the feedback from the scatterers outside the pumped region are not negligible. In an under-coupling system, the regularity in the lasing mode spacing remains under local excitation. Note that the increase of mode concentration in the vicinity of the gain region by local pumping have distinct physical mechanism from the absorption-induced localization of lasing modes in the pumped region. The former is based on selective enhancement of feedback within the gain region, while the latter on the suppression of the feedback outside the pumped region by absorption.

3.3.8. *One-dimensional photon localization laser*

One problem of the 3D random lasers is that multiple scattering of pump light restricts the excitation to the proximity of sample surface. The emitted photons readily escape through the sample surface, giving a high lasing threshold. This problem is less serious in 2D random lasers, as the pump light incident from the third dimension does not experience scattering.[120] Yet the pump light is not confined inside the

2D random system. Recently, Milner and Genack realized photon localization laser in which the pump light is localized deep inside an one-dimensional (1D) random structure.[121]

The 1D sample is a stack of partially reflecting glass slides of random thickness between 80 and 120μm with interspersed dye films. The stack is illuminated at normal incidence by the second harmonics of a pulsed Nd:YAG laser. In this 1D structure, light is localized by multiple scattering from the parallel layers which returns the wave upon itself. The average transmittance through a stack of glass slides without intervening dye solution decays exponentially with the number of the slides. Even though the average transport of light is suppressed, resonant tunneling through localized states gives spectrally narrow transmission peaks. A large number of narrow peaks are observed in the transmission spectra. When the pump wavelength is tuned into one of the narrow transmission lines, the pump light penetrates deep into the sample's interior via resonant excitation of a long-lived spatially-localized mode. Energy absorbed from this mode is subsequently emitted into long-lived localized modes which fall within the dye emission spectrum. Stimulated emission is enhanced when the spatial energy distributions at both the excitation and emission wavelengths overlap. The deposition of pump energy deep within the sample and its efficient coupling to long-lived emission modes removes a major barrier to achieving low-threshold lasing in the presence of disorder. The threshold is sufficiently low for lasing to be induced with a quasi-continuous illumination by a 3W argon-ion laser.

Another feature of the 1D photon localization laser is the large fluctuation in lasing power when the pump beam is focused onto different parts of the sample which correspond to different realizations of disorder. The pump transmission is strongly correlated with the output lasing power. High pump transmission results from resonant excitation of a localized mode which is spatially peaked near the center of the sample. Therefore, the pump energy is exponentially enhanced within the sample and is efficiently transferred to the gain medium. Since excitation in the center of the sample is likely to escape via emission into localized modes with long lifetime, the opportunity for stimulated emission is enhanced and the laser output is high.

In the localization regime, the sample length L is much larger than the localization length ξ. If the gain is uniform throughout the 1D structure, the first lasing mode is usually the most localized mode in the middle of the random structure. Let us consider such a mode located at $x_0 \sim L/2$. The random media on its left side and right side are characterized by the reflection coefficients (r_l, r_r) and transmission coefficients (t_l, t_r). The threshold gain g_{th} is related to the reflection coefficients[122]:

$$|r_l r_r| \exp[g_{th}(d\Phi_l/d\omega + d\Phi_r/d\omega)/2] = 1, \tag{25}$$

where $r = |r|e^{i\Phi}$. The frequency dependence of $|r|$ can be neglected because $L \gg \xi$ and $1 - |r| \ll 1$. The factor in the exponent of Eq.(25) represents a product of photon amplification rate and the trapping time of photons inside the system $\tau_0 = d\Phi_l/d\omega + d\Phi_r/d\omega$. Since $|r_l|$ and $|r_r|$ are very close to 1, g_{th} can be expressed in terms of the transmission coefficients with the linear expansion:

$$g_{th} \approx \frac{|t_l|^2 + |t_r|^2}{d\Phi_l/d\omega + d\Phi_r/d\omega}.$$ (26)

In the localization regime, $t_l \sim \exp(-x_0/\xi)$, $t_r \sim \exp[-(L - x_0)/\xi]$. Hence, $g_{th} \sim \{\exp(-2x_0/\xi) + \exp[-2(L - x_0)/\xi]\}/\tau_0$. Since $x_0 \sim L/2$, $g_{th} \propto \exp(-L/\xi)$, the lasing threshold depends exponentially on the system length L. The photons emitted inside the random system need an exponentially long time to escape from it, due to exponentially small transmission in the localization regime. Therefore, an exponentially small gain is enough to initiate lasing oscillation.

3.4. *ASE spikes vs. lasing peaks*

One fundamental difference between a random laser with resonant feedback and a random laser with nonresonant feedback is that the lasing frequency of the former is determined by the spatial resonance of random structure and the latter by the maximum of gain spectrum. The emergence of discrete narrow peaks in the emission spectrum, whose frequencies depend on the spatial distribution of refractive index, is a distinct feature of random laser with resonant and coherent feedback. In addition to the lasing peaks, stochastic spikes are also observed in the single-shot spectra of amplified spontaneous emission (ASE) from colloidal solutions over a wide range of scattering strength.[124] They are attributed to single spontaneous emission events which happen to take long open paths inside the amplifying random medium and pick up large gain. One distinction between the ASE spikes and lasing peaks is that, even when the random structure is fixed, the ASE spikes change completely from pulse to pulse, while the lasing peaks remain constant in frequency with only amplitude fluctuation. Next we show that the spectral correlation and intensity statistics of random lasing peaks are very different from those of the ASE spikes, underlying their distinct physical mechanisms.

The experiment was performed on the diethylene glycol solutions of stilbene 420 dye and TiO_2 microparticles.[118] The motion of particles in the solution provides different random configurations for each pump pulse, allowing the ensemble measurement under identical conditions. The dye molecules are excited by the third harmonic of a pulsed Nd:YAG laser which is focused into the solution. The stilbene 420 has well-separated absorption and emission bands, thus the absorption of emitted light outside the pumped region is negligible. At a particle density $\rho = 3 \times 10^9 \mathrm{cm}^{-3}$, the scattering mean free path l_s is on the order of 1mm, exceeding the penetration depth of the pump light. The excitation volume has a cone shape with a length of a few hundred micron and a base diameter of 30μm. Because its length is smaller than l_s, the excitation cone is almost identical to that in the neat dye solution. On one hand, the transport for the emitted light is diffusive in the entire colloidal solution whose dimension is much larger than l_s. On the other hand, light amplification occurs only in the pumped region with size smaller than the mean free path.

     Optical Processes in Microparticles and Nanostructures

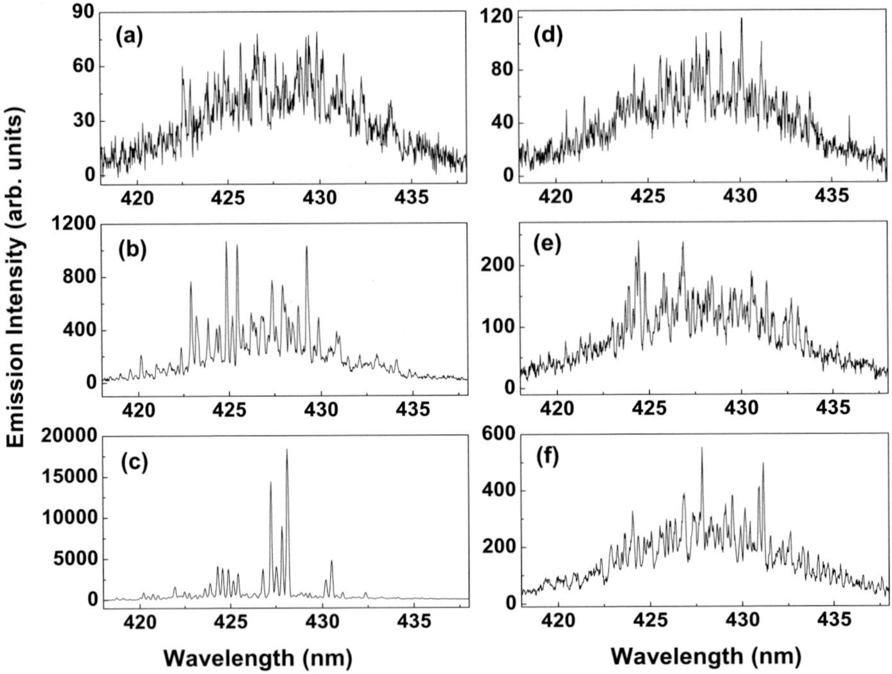

Fig. 7.   Single-shot spectra of emission from the 8.5 mM stilbene 420 dye solutions with (a)-(c) and without TiO$_2$ particles. The particle density $\rho = 3 \times 10^9$cm^{-3} in (a)-(c). The pump pulse energy $E_p = 0.05\mu$J for (a) & (d), 0.09μJ for (b) & (e), 0.13μJ for (c) & (f).

The single-shot emission spectra from the colloidal solution are shown in Figs. 7(a)-(c) with increasing pump pulse energy E_p. At $E_p = 0.05\mu$J [Fig. 7(a)], the spectrum exhibits sharp spikes on top of a broad ASE band. The spikes change completely from shot to shot. The typical linewidth of the spikes is about 0.07nm. The neighboring spikes often overlap partially. As pumping increases, the spikes grow in intensity. When E_p exceeds a threshold, a different type of peaks emerge in the emission spectrum [Fig. 7(b)]. These peaks grow much faster with pumping than the spikes, and dominate the emission spectrum at $E_p = 0.13\mu$J [Fig. 7(c)]. These peaks, with a typical width of 0.13nm, are notably broader than the spikes. Unlike the spikes, the spectral spacing of adjacent peaks is more or less regular. The above experiment was repeated with the neat dye solution of the same M [Fig. 7(d) - (f)]. Although they are similar at $E_p = 0.05\mu$J, the emission spectra with and without particles are dramatically different at $E_p = 0.13\mu$J. Under intense pumping, the emission spectrum of the neat dye solution has only random and closely-spaced spikes but no strong and regularly-spaced peaks [Fig. 7(f)]. The maximum spike intensity is about 50 times lower than the maximum peak intensity from the colloidal solution at the same pumping [Fig. 7(c)]. While the pump threshold for the appearance of peaks depends on ρ, the threshold for the emergence of spikes in solutions with low ρ is similar to that with $\rho = 0$.

The large peaks represent the lasing modes formed by distributed feedback in the colloidal solution. Although the feedback is weak at low ρ, the intense pumping strongly amplifies the backscattered light and greatly enhances the feedback. In contrast, the feedback from the particles is not necessary for the spikes which also exist in the neat dye solution. Thus the spikes are attributed to the amplified spontaneous emission.

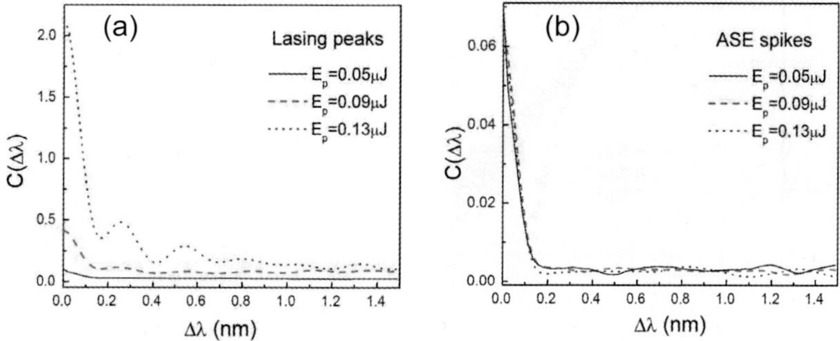

Fig. 8. Ensemble-averaged spectral correlation function $C(\Delta\lambda)$ of single-shot emission spectra. $M = 8.5$ mM. $\rho = 3 \times 10^9 \mathrm{cm}^{-3}$ in (a), and 0 in (b).

The ensemble-averaged spectral correlation function $C(\Delta\lambda) = \langle I(\lambda)I(\lambda + \Delta\lambda)\rangle/\langle I(\lambda)\rangle\langle I(\lambda + \Delta\lambda)\rangle - 1$ is obtained from 200 single-shot emission spectra over the wavelength range 425-431nm within which the gain coefficient has only slight variation. For $\rho = 3 \times 10^9 \mathrm{cm}^{-3}$, $C(\Delta\lambda)$ changes dramatically with pumping [Fig. 8(a)]. Below the lasing threshold, it starts with a small value at $\Delta\lambda = 0$, and decays quickly to zero as $\Delta\lambda$ increases. Above the lasing threshold, the amplitude of $C(\Delta\lambda)$ grows rapidly, and regular oscillations with $\Delta\lambda$ are developed. The oscillation period is about 0.27nm, corresponding to the average spacing of lasing peaks. Despite the change of lasing peaks from shot to shot, the oscillations survive the ensemble average. This result confirms the lasing peaks in a single-shot spectrum are more or less regularly spaced and the average peak spacing is nearly the same for different shots. The periodicity of lasing peaks in similar random samples was also revealed by Polson and Vardeny using the power Fourier transform technique.[124] $C(\Delta\lambda)$ for the ASE spikes at $\rho = 0$ barely changes with pumping [Fig. 8(b)]. It is similar to that of the colloidal solution below the lasing threshold where the spectrum has only ASE spikes. Although ASE spikes produce irregular oscillations in the spectral correlation function of individual single shot spectrum, such oscillations are smeared out after averaging over many shots. This result reflects the stochastic nature of the ASE spikes.

The statistical distribution $P(\delta\lambda)$ of wavelength spacing $\delta\lambda$ between adjacent lasing peaks is distinct from that of ASE spikes. $P(\delta\lambda)$ for the ASE spikes decays exponentially as $\delta\lambda$ increases from zero. It suggests the spectral spacing of ASE spikes satisfies the Poisson statistics, which means the frequencies of individual ASE spikes are uncorrelated. $P(\delta\lambda)$ for the lasing peaks has a value close to zero at $\delta\lambda \sim 0$. It increases with $\delta\lambda$ and reaches the maximum at $\delta\lambda = 0.27$nm, which coincides with the average lasing peak spacing obtained from the oscillation period of $C(\Delta\lambda)$. Such distribution reflects the spectral repulsion of lasing peaks.

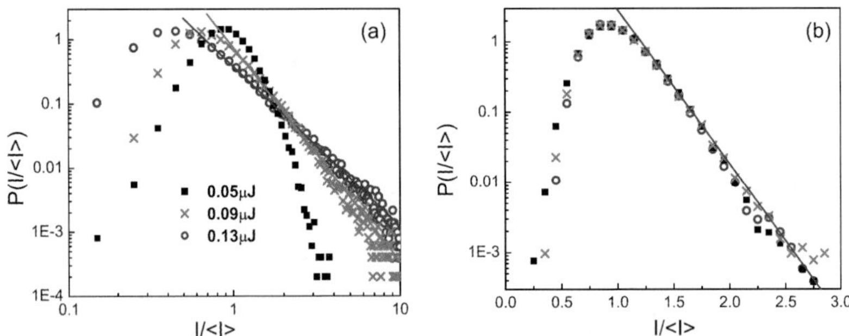

Fig. 9. Statistical distributions of normalized emission intensities $I(\lambda)/\langle I(\lambda)\rangle$ for 425nm $< \lambda <$ 431nm. $\rho = 3 \times 10^9$cm^{-3} in (a), and 0 in (b). $E_p = 0.05\mu$J (squares), 0.09μJ (crosses), 0.13μJ (circles). The solid lines represent curve-fitting. In (a) $P(I/\langle I\rangle) = 0.77(I/\langle I\rangle)^{-3.3}$ (for crosses) and $P(I/\langle I\rangle) = 0.38(I/\langle I\rangle)^{-2.5}$ (for circles). In (b), $P(I/\langle I\rangle) = 467\exp(-5.1\,I/\langle I\rangle)$.

In the study of the statistics of emission intensity, the average intensity $\langle I(\lambda)\rangle$ is first computed from 200 single-shot emission spectra, then the statistical distribution of the normalized emission intensity $I(\lambda)/\langle I(\lambda)\rangle$ is computed. In Fig. 9(a), the log-log plot of $P(I/\langle I\rangle)$ for $\rho = 3 \times 10^9$cm^{-3} clearly reveals a power-law decay at large I above the lasing threshold. The solid lines represent the fit of $(I/\langle I\rangle)^{-b}$ to $P(I/\langle I\rangle)$, with $b = 3.3$ and 2.5 for $E_p = 0.09\mu$J and 0.13μJ respectively. Since only the high lasing peaks contribute to the tail of $P(I/\langle I\rangle)$, the power-law decay reflects the intensity statistics of lasing peaks. Below the lasing threshold, $P(I/\langle I\rangle)$ is similar to that of the neat dye solution, which exhibits an exponential tail. As shown in the log-linear plot of Fig. 9(b), the exponential decay rate is almost the same for different pumping levels at $\rho = 0$. The solid line is an exponential fit $P(I/\langle I\rangle) \sim \exp(-a\,I/\langle I\rangle)$ with $a = 5.1$. In the absence of lasing peaks, the ASE spikes contribute to $P(I/\langle I\rangle)$ at large I. Therefore, the exponential decay describes the intensity statistics of ASE spikes.

The above experimental results demonstrate the fundamental difference between the ASE spikes and lasing peaks. The stochastic structures of the pulsed ASE spectra of neat dye solutions were observed long ago.[125] In the above experiment, the observed ASE spikes originate from photons spontaneously emitted near the excitation cone tip in the direction toward the cone base. As they propagate along the cone, these photons experience the largest amplification due to their longest path length inside the gain volume. The ASE at the frequencies of these photons is the strongest, leading to the spikes in the emission spectrum. Although the spontaneous emission time is a few nanoseconds, the 25ps pump pulse creates the transient gain, and only the initial part of the spontaneous emission pulse is strongly amplified. Hence, the ASE pulse is a few tens of picoseconds long, followed by a spontaneous emission tail. The spectral width of the ASE spikes is determined by the ASE pulse duration. Since different ASE spikes originate from independent spontaneous emission events, their frequencies are uncorrelated. This leads to a Poisson statistics for the frequency spacing of neighboring ASE spikes. Although the occurrence of ASE spikes does not rely on scattering, multiple scattering could elongate the path lengths of spontaneously emitted photons inside the gain volume and increase the amplitudes of some spikes. In the above experiment, however, the size of the gain volume is less than l_s, thus the effect of scattering on the ASE spikes is negligibly small.

In the colloidal solution of low ρ, the large aspect ratio of the excitation cone results in lasing along the cone, confirmed by the directionality of the lasing output. The large gain inside the cone greatly amplifies the feedback from the scatterers within the cone as compared to that from outside the cone. Thus the lasing modes deviate from the quasimodes of the passive system. This explains why the statistical distribution of the lasing peak spacing does not satisfy the Wigner-Dyson distribution which holds for the statistical distribution of quasimode spacing in the diffusive colloidal solution. It also indicates the statistical distribution of decay rate of the quasimodes cannot be applied directly to the calculation of $P(I/\langle I \rangle)$.[126] Moreover, mode competition and gain saturation, as well as the initial spontaneous emission into individual modes, must be taken into account to reproduce $P(I/\langle I \rangle)$. The rapid variation of gain in time and space make the calculation of intensity statistics more difficult. An extensive theoretical study is needed to quantitatively understand the statistical distribution of peak intensity in random lasing.

3.5. *Recent developments*

Over the past few years, random lasers with coherent feedback were realized in many material systems such as semiconductor nanostructrues,[127−136] organic films and nanofibers[137−141], hybrid organic-inorganic composites.[142−144] Various schemes have been proposed to improve the performance of random lasers, e.g., application of external feedback to reduce the lasing threshold and control the output direction of laser emission[145], optimal tuning of random lasing modes through collective particle resonances[100], coupled-cavity ZnO thin-film random lasers for high-power one-mode operation[146], one-mirror random laser for quasi-continuous operation[147,148], waveguide random laser for directional output.[149,150] The progress is so rapid that it is impossible to detail all of the advances. Next only a few examples are mentioned briefly.

Partially-ordered random laser. One way of reducing the random laser threshold is to incorporate some degree of order into an active random medium.[151−153] Shkunov *et al* have observed both photonic lasing and random lasing in dye-infiltrated opals.[154] However, random lasing has a higher threshold than photonic lasing. We numerically simulate lasing in a random system with variable degree of order. When disorder is introduced to a perfectly ordered system, the lasing threshold is reduced. At a certain degree of disorder the lasing threshold reaches a minimum. Then it starts rising with further increase of disorder. Therefore, there exists an optimum degree of order for minimum lasing threshold. We map out the transition from full order to complete disorder, and identify five scaling regimes for the mean lasing threshold versus the system size L. For increasing degree of disorder, the five regimes are (a) photonic band-edge, $1/L^3$, (b) transitional super-exponential, (c) bandgap-related exponential, (d) diffusive, $1/L^2$, and (e) disorder-induced exponential. Experimentally, we have fabricated disordered photonic crystal lasers.[155] The most efficient lasing modes are localized defect states near the edge of a photonic band gap. Such defect states are formed by structural disorder in a 2D triangular lattice. Another advantage of the partially-ordered random laser is efficient pumping. For example, in a 1D random stack of resonant dielectric layers, the pump wavelength can be tuned to a pass band while the emission wavelength stays in a stop band.[156] Then the pump light penetrates into the sample, while the emission is confined inside the system. As a result, the lasing threshold can be significantly reduced.

Mode interaction. The interaction of lasing modes in a random medium is interesting but complicated. Gain competition may lead to mode repulsion in real space for homogeneously broadened gain spectrum or in frequency domain for inhomogeneously broadened gain spectrum.[157,158] In a diffusive random medium, the finesse, defined as the ratio of the quasimode spacing to the mode width, is much less than unity. Strong modal interactions through the gain medium lead to a uniform spacing of lasing peaks.[159−161] In addition, the inhomogeneity of dielectric constant $\epsilon(\mathbf{r})$ modifies the ortho-normalization condition for the quasimodes, and introduces a linear coupling between the quasimodes mediated by the polarization of the gain medium.[162] Finally, the overlapping quasimodes may couple via external bath, that generates excess noise and broadens the lasing linewidth.[163−165] The openness of random laser cavities strongly affects mode interaction, which determines the number of lasing modes and their spacing statistics.[166−168]

Nonlinear random laser. Random laser offers an opportunity to study the interplay between nonlinearity and localization. Nonlinearity is strong in a random laser because the nonlinear coefficient is resonantly enhanced at the lasing frequency and the light intensity is high due to spatial confinement in random media. Noginov *et al.* demonstrated second-harmonic generation in a mixture of powders of laser and frequency doubling materials.[169] Our recent study on the dynamic nonlinear effect in a random laser illustrates that the third-order nonlinearity not only changes the frequency and size of the lasing modes, but also modifies the laser emission intensity and laser pulse width.[170] How nonlinearity affects random lasing process depends on the speed of the nonlinear response. We find two regimes depending on the relative values of two time scales, one is the nonlinear response time, the other is the lifetime of the lasing state. For slow nonlinear response, collective scattering of many particles determines the buildup of a lasing mode. Nonlinearity changes the lasing output through modification of the spatial size of the lasing mode. However, when the nonlinear response is faster than the buildup of a lasing mode, the lasing mode cannot respond fast enough to the nonlinear refractive index change. Rapid change of the phase of scattered waves undermines the interference effect of multiple scattering. Instead, the nonlinear effect of single particle scattering becomes dominant. Strong nonlinearity could lead to temporal instability. One application of optical nonlinearity is upconversion lasing in random media via two-photon or multi-photon pumping.[171,172] Small two-photon/multi-photon absorption coefficient and weak scattering at long pumping wavelength allow the pump light to penetrate deep into a 3D random medium and excite resonant modes far from the surface which have better spatial confinement and lower lasing threshold.[173]

The latest theoretical and experimental studies provide insight into the physical mechanisms for lasing in random media.[96,57,58,62,174–180] However, our understanding of random lasers is far from complete. New ideas and surprises arise frequently, keeping the momentum of random laser study. For example, Rand and coworkers investigated the electrical generation of stationary light (evanescent wave) in ultrafine laser crystal powder.[181] Dice *et al* report the surface-plasmon-enhanced random laser emission from a suspension of silver nanoparticles in a laser dye.[182]

4. Potential applications of random lasers

A random laser is a non-conventional laser whose feedback mechanism is based on light scattering, as opposed to mirror reflection in a conventional laser. This alternative feedback mechanism has important application in the fabrication of lasers in the spectral regimes where efficient reflective elements are not available, e.g., X-ray laser, γ-ray laser. Furthermore, the low fabrication cost, sample specific wavelength of operation, small size, flexible shape, and substrate compatibility of random laser lead to many potential applications.[29,95,183,184]

One application is document encoding and material labeling, as the lasing frequencies represent the "signature" of a random structure. The micro-random laser, described in the last section, may be used as an optical tag in biological and medical studies. When the nanoparticle clusters are attached to biological targets, the positions of the targets can be traced by detecting the lasing emission from the clusters. Each nanoparticle cluster has its unique set of lasing frequencies, that allows us to differentiate the targets.

The multidirectional output of a random laser makes it suitable for display application. For example, a thin layer of a random medium doped with emitters can be coated on an arbitrarily-shaped display panel. Compared to light emitting diodes that are often used for display, random lasers have a much shorter turn-on/off time and would be useful for high-speed displays. A multicolor display may be made by incorporating emitters of different frequencies into a single random medium. The shape flexibility and substrate compatibility of random lasers allow them to be used for machine vision of manufactured parts in assembly lines, search and rescue of downed aircraft or satellites, etc.

So far most random lasers are pumped optically. Some applications such as flat-panel, automotive and cockpit displays require electrical pumping. Recently electrically-pumped continuous-wave laser action was reported in rare-earth-metal-doped dielectric nanophosphors.[185,186] Electrical pumping is much more efficient than optical pumping, because in the latter case most of the pump light is scattered instead of being absorbed by the random medium.

In the medical arena, random lasers may be used for tumor detection and photo-dynamic therapy. Polson and Vardeny have shown that human tissues have strong scattering and can support random lasing when infiltrated with a concentrated laser dye solution.[187] Since cancerous cells grow much faster than the normal cells, they generate much more fragments or waste. The additional disorder in the malignant tissue leads to stronger scattering and more efficient lasing. One could imagine a novel method of probing tumors by scanning a focused laser beam across the human tissue. The laser light pumps the infiltrated dye molecules in a local region. When the pumping is not very strong, lasing can occur only at the location of tumor where scattering is stronger. Since lasing may happen in a region much smaller than 1 mm, it is possible to detect a tumor in the very early stage.

Random lasers could also serve as active elements of photonic devices and circuits. For example, a micro random laser may play a crucial role as a miniature light source in a photonic crystal. The temperature-tunable random laser is expected to find applications in photonics, temperature-sensitive displays and screens, and in remote temperature sensing. Finally, the study of random laser could help us understand the galaxy masers and stellar lasers whose feedback is also caused by scattering.[188,189]

Acknowledgments

I wish to thank my coworkers on the study of random lasers: J. Y. Xu, Y. Ling, X. Wu, Y. G. Zhao, and Prof. Prem Kumar for the experimental work on random lasers; Drs. A. Yamilov, A. L. Burin, B. Liu, S.-H. Chang, Profs. S. T. Ho, A. Taflove, M. A. Ratner and G. C. Schatz for the theoretical investigations of random lasers. Prof. R. P. H. Chang and his students E. W. Seelig, X. Liu fabricated ZnO nanorods and nanoparticles. We enjoyed a fruitful collaboration with Prof. C. M. Soukoulis and Dr. Xunya Jiang on the simulation of random laser. Stimulated discussions are acknowledged with Drs. A. A. Asatryan, A. A. Chabanov, Ch. M. Brikina, L. Deych, M. Giudici, B. Grémaud, F. Haake, G. Hackenbroich, A. Z. Genack, S. John, R. Kaiser, T. Kottos, V. M. Letokhov, C. Miniatura, M. A. Noginov, M. Patra, M.

E. Raikh, S. C. Rand, P. Sebbah, B. Shapiro, C. M. de Sterke, A. D. Stone, J. Tredicce, H. E. Türeci, C. Vanneste, Z. V. Vardeny, C. Viviescas, T. Wellens and D. S. Wiersma. Our research program is partly sponsored by the National Science Foundation through the grants ECS-9877113, DMR-0093949 and ECS-0244457, and by the David and Lucille Packard Foundation, the Alfred P. Sloan Foundation, the Northwestern University Materials Research Center.

References

1. Siegman A, 1986 *Lasers* (Mill Valley: University Science Books).
2. Ambartsumyan R V, Basov N G, Kryukov P G and Letokhov V S 1966 *IEEE J. Quant. Electron.* **QE-2** 442.
3. Cao H 2003 *Waves Random Media* **13** R1.
4. Cao H, Xu J Y, Chang S-H and Ho S T 2000a *Phys. Rev. E* **61** 1985.
5. Cao H, Xu J Y, Ling Y, Burin A L, Seelig E W, Liu X and Chang R P H 2003a *IEEE J. Select. Top. Quant. Electron.* **9** 111.
6. John S 1991 *Phys. Today* **44** 32.
7. Ambartsumyan R V, Kryukov P G, Letokhov V S and Matveets Yu A 1967a *Sov. Phys. JETP* **24** 1129.
8. Ambartsumyan R V, Basov N G and Letokhov V S 1968 *Sov. Phys. JETP* **26** 1109.
9. Ambartsumyan R V, Kryukov P G, Letokhov V S and Matveets Yu A 1967b *Sov. Phys. JETP* **24** 481.
10. Ambartsumyan R V, Bazhulin S P, Basov N G and Letokhov V S 1970 *Sov. Phys. JETP* **31** 234.
11. Letokhov V S 1968 *Sov. Phys. JETP* **26** 1246.
12. Markushev V M, Zolin V F and Briskina CH M 1986 *Sov. J. Quantum Electron.* **16** 281.
13. Markushev V M, Ter-Gabrielyan N E, Briskina Ch M, Belan V R and Zolin V F 1990 *Sov. J. Quantum Electron.* **20** 773.
14. Ter-Gabrielyan N E, Markushev V M, Belan V R, Briskina Ch M, Dimitrova O V, Zolin V F and Lavrov A V 1991a *Sov. J. Quantum Electron.* **21** 840.
15. Ter-Gabrielyan N E, Markushev V M, Belan V R, Briskina Ch M and Zolin V F 1991b *Sov. J. Quantum Electron.* **21** 32.
16. Briskina Ch M, Markushev V M and Ter-Gabrielyan N E 1996 *Quant. Electron.* **26** 923.
17. Briskina Ch M and Li L F 2002 *Laser Physics* **12** 724.
18. Lichmanov A A, Briskina Ch M, Markushev V M, Lichmanova V N and Soshchin N P 1998 *J. Appl. Spectroscopy* **65** 818.
19. Gouedard C, Husson D, Sauteret C, Auzel F and A. Migus 1993 *J. Opt. Soc. Am. B* **10** 2358.
20. Noginov M A, Noginova N E, Egarievwe S U, Caulfield H J, Venkateswarlu P, Thompson T, Mahdi M and Ostroumov V 1996 *J. Opt. Soc. Am. B* **13** 2024.

21. Noginov M A, Noginova N E, Egarievwe S U, Caulfield H J, Cochrane C, Wang J C, Kokta M R and Paitz J 1998a *Opt. Mater.* **10** 297.

22. Zolin V F 2000 *Journal of Alloy. Compd.* **300** 214.

23. Noginov M A, Noginova N E, Egarievwe S U, Caulfield H J, Venkateswarlu P, Williams A and Mirov S B 1997 *J. Opt. Soc. Am. B* **14** 2153.

24. Noginov M A, Egarievwe S U, Noginova N E, Caulfield H J and Wang J C 1999 *Opt. Mater.* **12** 127.

25. Auzel F and Goldner P 2000 *J. Alloy. Compd.* **300** 11.

26. Zyuzin A Yu 1998 *JETP* **86** 445.

27. Zyuzin A Yu 1999 *Europhys. Lett.* **46** 160.

28. Wiersma D S and Lagendijk A 1997a *Phys. Rev. E* **54** 4256.

29. Lawandy N M, Balachandran R M, Gomes A S L and Sauvain E 1994 *Nature* **368** 436.

30. Sha W L, Liu C-H and Alfano R R 1994 *Opt. Lett.* **19** 1922.

31. Lawandy N M and Balachandran R M 1995 *Nature* **373** 204.

32. Zhang W, Cue N and Yoo K M 1995a *Opt. Lett.* **20** 961.

33. Balachandran R M and Lawandy N M 1995 *Opt. Lett.* **20** 1271.

34. Wiersma D, van Albada M P and Lagendijk A 1995 *Nature* **373** 203.

35. Balachandran R M and Lawandy N M 1997 *Opt. Lett.* **22** 319.

36. Siddique M, Alfano R R, Berger G A, Kempe M and Genack A Z 1996 *Opt. Lett.* **21** 450.

37. van Soest G, Tomita M, Sprik R and Lagendijk A 1999 *Opt. Lett.* **24** 306.

38. Beckering G, Zilker S J and Haarer D 1997 *Opt. Lett.* **22** 1427.

39. Totsuka K, van Soest G, Ito T, Lagendijk A and Tomita M 2000 *J. Appl. Phys.* **87** 7623.

40. Noginov M A, Caulfield H J, Noginova N E and Venkateswarlu P 1995 *Opt. Commun.* **118** 430.

41. Zhang W, Cue N and Yoo K M 1995b *Opt. Lett.* **20** 1023.

42. Sha W L, Liu C-H, Liu F and Alfano R R 1996 *Opt. Lett.* **21** 1277.

43. Balachandran R M, Perkins A E and Lawandy N M 1996a *Opt. Lett.* **21** 1603.

44. John S and Pang G 1996 *Phys. Rev. A* **54** 3642.

45. Berger G A, Kempe M and Genack A Z 1997 *Phys. Rev. E* **56** 6118.

46. van Soest G, Poelwijk F J, Sprik R and Lagendijk A 2001 *Phys. Rev. Lett.* **86** 1522.

47. van Soest G and Lagendijk A 2002, *Phys. Rev. E* **65** 047601.

48. Zacharakis G, Heliotis G, Filippidis G, Anglos D and Papazoglou T G 1999 *Appl. Opt.* **38** 6087.

49. Lawandy N M 1994 *Photon. Spectra* **28** 119.

50. Wiersma D S and Lagendijk A 1997b *Physics World* **10** 33.

51. de Oliveira P C, McGreevy J A and Lawandy N M 1997 Opt. Lett. **22**, 895.

52. Balachandran R M, Perkins A E and Lawandy N M 1996b *Opt. Lett.* **21** 650.

53. Martorell J, Balachandran R M and Lawandy N M 1996 Opt. Lett. **21** 239.

54. Wiersma D and Cavalieri S 2001 *Nature* **414** 708.

55. Lee K and Lawandy N M 2002 *Opt. Commun.* **203** 169.

56. Gottardo S, Cavalieri S, Yaroshchuk O and Wiersma D S 2004 *Phys. Rev. Lett.* **93** 263901.

57. Noginov M A, Novak J and Williams S 2004a *Phys. Rev. A* **70** 043811.

58. Noginov M A, Novak J and Williams S 2004b *Phys. Rev. A* **70** 063810.

59. Noginov M A, Zhu G, Frantz A A, Novak J, Williams S N, and Fowlkes I 2004d *J. Opt. Soc. Am. B* **21**, 191.

60. Bahoura M, Morris K J, Zhu G, and Noginov M A 2005 *IEEE J. QUANTUM ELECTRON.* **41**, 677.

61. Noginov M A, Zhu G, Small C, and Novak J 2006 *Appl. Phys. B* **84**, 269-273.

62. Noginov M A, Zhu G, Fowlkes I and Bahoura M 2004c *Laser Phys. Lett.* **1** 291.

63. Noginov M A, Novak J, Grigsby D, Zhu G, and Bahoura M 2005b *Opt. Express* **13**, 8829.

64. Noginov M A, Fowlkes, I N, and Zhu G 2005a *Appl. Phys. Lett.* **86**, 161105.

65. Polson R C, Raikh M E and Vardeny Z V 2002 *Comptes Rendus de l'Academie des Sciences, Serie IV (Physique, Astrophysique)* **3** 509.

66. Frolov S V, Vardeny Z V, Yoshino K, Zakhidov A and Baughman R H 1999a *Phys. Rev. B* **59** R5284.

67. Polson R C, Huang J D and Vardeny Z V 2001a *Syn. Metals* **119** 7.

68. Frolov S V, Vardeny Z V, Yoshino K, Zakhidov A and Baughman R H 1999b *Opt. Commun.* **162** 241.

69. Yoshino K, Tatsuhara S, Kawagishi Y and Ozaki M 1999 *Appl. Phys. Lett.* **74** 2590.

70. Polson R C, Chipouline A Vardeny Z V 2001b *Adv. Mater.* **13** 760.

71. Apalkov V M, Raikh M E and Shapiro B 2003 in *The Anderson Transition and its Ramifications - Localization, Quantum Interference, and Interactions*, vol. 630 of Lecture Notes in Physics, edited by T. Brandes and S. Ketermann (Springer Verlag, Berlin, 2003) p. 119.

72. Apalkov V M, Raikh M E and Shapiro B 2004a *J. Opt. Soc. Am. B* **21** 132.

73. Apalkov V M, Raikh M E and Shapiro B 2002 *Phys. Rev. Lett.* **89** 016802.

74. Patra M 2003b *Phys. Rev. E* **67** 065603(R).

75. Apalkov V M, Raikh M E and Shapiro B 2004b *Phys. Rev. Lett.* **92** 066601.

76. Apalkov V M and Raikh M E 2005 *Phys. Rev. B* **71** 054203.

77. Liu X, Wu X, Cao H, and Chang R P H 2004 *J. Appl. Phys.* **95** 3141.

78. Seelig E W, Chang R P H, Yamilov A and Cao H 2003 *Mat. Chem. Phys.* **80** 257.

79. Cao H, Zhao Y G, Ong H C, Ho S T, Seelig E Q, Wang Q H and Chang R P H 1999a *Phys. Rev. Lett.* **82**, 2278.

80. Cao H, Ling Y, Xu J Y and Burin A L 2002 *Phys. Rev. E* **66** 25601(R).

81. Soukoulis C M, Jiang X, Xu J Y and Cao H 2002 *Phys. Rev. B* **65** R041103.

82. Markushev V M, Ryzhkov V M, Briskina Ch M, Cao H, Zadorozhnayac L A, Li L E, Gevargizovc E I, Demianets L N 2005 *Laser Physics* **15**, 1611.

83. Cao H, Ling Y, Xu J Y, Cao C Q and Kumar P 2001 *Phys. Rev. Lett.* **86** 4524.

84. Mandel L and Wolf E 1995, *Optical Coherence and Quantum Optics* (Cambridge University Press, Cambridge).

85. Ambartsumyan R V, Kryukov P G, Letokhov V S and Matveets Yu A 1967c *JETP Lett.* **5** 378.

86. Beenakker C W J 1998 *Phys. Rev. Lett.* **81** 1829.

87. Misirpashaev T Sh and Beenakker C W J 1998 *Phys. Rev. A* **57** 2041.

88. Patra M and Beenakker C W J 1999 *Phys. Rev. A* **60** 4059.

89. Mishchenko E G, Patra M and Beenakker C W J 2001 *Euro. Phys. J. D* **13** 289.

90. Patra M 2002 *Phys. Rev. A* **65** 043809.

91. Florescu L and John S 2004a *Phys. Rev. Lett* **93** 013602.

92. Florescu L and John S 2004b *Phys. Rev. E* **69** 046603.

93. Cao H, Xu J Y, Chang S-H, Ho S T, Seelig E W, Liu X and Chang R P H 2000b *Phys. Rev. Lett.* **84** 5584.

94. Cao H, Xu J Y, Seelig E W and Chang R P H 2000c *Appl. Phys. Lett.* **76** 2997.

95. Wiersma D 2000 *Nature* **406** 132.

96. Kretschmann M and Maradudin A A 2004 *J. Opt. Soc. Am. B* **21** 150

97. Taniguchi H, Tanosaki S, Tsujita K and Inaba H 1996 *IEEE J. Quant. Electron.* **32** 1864.

98. Burin A L, Ratner M A, Cao H and Chang R P H 2001 *Phys. Rev. Lett.* **87** 215503.

99. Wu X, Yamilov A, Noh H, Cao H, Seelig E W and Chang R P H 2004a *J. Opt. Soc. Am. B* **21** 159.

100. Ripoll J, Soukoulis C M and Economou E N 2004 *J. Opt. Soc. Am. B* **21** 141.

101. Vanneste C and Sebbah P 2005 *Phys. Rev. E* **71** 026612.

102. Jiang X and Soukoulis C M 2000 *Phys. Rev. Lett.* **85** 70.

103. Taflove A and Hagness S C 2000 *Computational Electrodynamics* (Boston: Artech House).

104. Ling Y, Cao H, Burin A L, Ratner M A, Liu X and Chang R P H 2001 *Phys. Rev. A* **64** 063808.

105. Anni M, Lattante S, Stomeo T, Cingolani R, Gigli G, Barbarella G and Favaretto L 2004 *Phys. Rev. B* **70** 195216.

106. Vanneste C and Sebbah P 2001 *Phys. Rev. Lett.* **87** 183903.

107. Sebbah P and Vanneste C 2002 *Phys. Rev. B* **66** 144202.

108. van der Molen K L, Tjerkstra R W, Mosk A P, and Lagendijk A 2007 *Phys. Rev. Lett.* **98**, 143901.

109. Jiang X and Soukoulis C M 2002, *Phys. Rev. E* **65** 025601(R).

110. Li Q, Ho K M and Soukoulis C M 2001 *Physica B* **296** 78.

111. Vanneste C, Sebbah P, and Cao H 2007 *Phys. Rev. Lett.* **98**, 143902

112. Chabanov A A, Zhang Z Q and Genack A Z 2003 *Phys. Rev. Lett.* **90** 203903.

113. Altshuler B L, Kravtsov V E and Lerner I V 1991 in *Mesoscopic Phenomena in Solids* eds. B L Altshuler, P A Lee and R A Webb (Amsterdam: North-Holland)

114. Mirlin A D 2000 *Phys. Rep.* **326** 260.

115. Yamilov A, Wu X, Cao H, and Burin A L 2005 *Opt. Lett.* **30**, 2430.

116. Wu X, Fang W, Yamilov A, Chabanov A A, Asatryan A A, Botten L C,and Cao H 2006 *Phys. Rev. A* **74**, 053812.

117. Burin A L, Cao H and Ratner M A 2003a *Physica B* **338** 212.

118. Wu X and Cao H 2007 arXiv:physics/0703255.

119. Kogelnik H, and Shank C V 1972 *J. Appl. Phys.* **43**, 2327.

120. Stassinopoulos A, Das R N, Giannelis E P, Anastasiadis S H and Anglos D 2005 *App. Surf. Sci.* **247** 18.

121. Milner V and Genack A Z 2005 *Phys. Rev. Lett.* **94** 073901.

122. Burin A L, Ratner A M, Cao H and Chang S-H 2002 *Phys. Rev. Lett.* **88** 093904.

123. Mujumdar S, Ricci M, Torre R and Wiersma D S 2004b *Phys. Rev. Lett.* **93** 053903.

124. Polson R and Vardeny Z V 2005 *Phys. Rev. B* **71** 045205.

125. Sperber P, Spangler W, Meier B, and Penzkofer A 1988 *Opt. Quantum Electron.* **20**, 395.

126. van der Molen K L, Mosk A P, Lagendijk A 2006 *Phys. Rev. A* **74**, 053808.

127. Cao H, Zhao Y G, Ong H C, Ho S T, Dai J Y, Wu J Y and Chang R P H 1998 *Appl. Phys. Lett.* **73** 3656.

128. Mitra A and Thareja R K 1999 *Mod. Phys. Lett. B* **23** 1075.

129. Thareja R K and Mitra A 2000 *Appl. Phys. B* **71** 181.

130. Sun B Q, Gal M, Gao Q, Tan H H, Jagadish C, Puzzer T, Ouyang L and Zou J 2003 *J. Appl. Phys.* **93** 5855.

131. Yu S F, Yuen C, Lau S P, Park W I and Yi G-C 2004a *Appl. Phys. Lett.* **84** 3241.

132. Yu S F, Yuen C, Lau S P and Lee H W 2004b *Appl. Phys. Lett.* **84** 3244.

133. Leong E, Chong M K, Yu S F and Pita K 2004 *IEEE Photon. Tech. Lett.* **16** 2418.

134. Hsu H-C, Wu C-Y and Hsieh W-F 2005 *J. Appl. Phys.* **97** 064315.

135. Yuen C, Yu S F, Leong E S P, Yang H Y and Hng H H 2005a *IEEE J. Quant. Electron.* **41** 970.

136. Lau S P, Yang H Y, Yu S F, Li H D, Tanermura M, Okita T, Hatano H and Hng H H 2005 *Appl. Phys. Lett.* **87** 013104.

137. Anni M, Lattante S, Cingolani R, Gigli G, Barbarella G and Favaretto L 2003 *Appl. Phys. Lett.* **83** 2754.

138. Quochi F, Cordella F, Orrù R, Communal J E, Verzeroli P, Mura A, Bongiovanni G, Andreev A, Sitter H and Sariciftci N S 2004 *Appl. Phys. Lett.* **84** 4454.

139. Quochi F, Andreev A, Cordella F, Orrù R, Mura A, Bongiovanni G, Hoppe H, Sitter H and Sariciftci N S 2005 *J. Luminescence* **112** 321.

140. Klein S, Crégut O, Gindre D, Boeglin A and Dorkenoo K D 2005 *Opt. Express* **13** 5387.

141. Sharma D, Ramachandran H, and Kumar N 2006 *Opt. Lett.* **31**, 1806.

142. Yokoyama S and Mashiko S 2003 *Jpn. J. Appl. Phys.* **42** L970.

143. Anglos D, Stassinopoulos A, Das R N, Zacharakis G, Psyllaki M, Jakubiak R, Vaia R A, Giannelis E P, Anastasiadis S H 2004 *J. Opt. Soc. Am. B* **21** 208.

144. Song Q, Wang L, Xiao S, Zhou X, Liu L and Xu L 2005 *Phys. Rev. B* **72** 035424.

145. Cao H, Zhao Y G, Liu X, Seelig E W and Chang R P H 1999b *Appl. Phys. Lett.* **75** 1213.

146. Yu S F and Leong E S P 2004 *IEEE J. Quant. Electron.* **40** 1186.

147. Feng Y and Ueda K 2003 *Phys. Rev. A* **68** 025803.

148. Feng Y, Bisson J-F, Lu J, Huang S Takaichi K, Shirakawa A, Musha M and Ueda K 2004 *Appl. Phys. Lett.* **84** 1040.

149. Yuen C, Yu S F, Leong E S P, Yang H Y, Lau S P, Chen N S and Hng H H 2005b *Appl. Phys. Lett.* **86** 031112 (2005).

150. Watanabe H, Oki Y, Maeda M and Omatsu T 2005 *Appl. Phys. Lett.* **86** 151123.

151. Chang S-H, Cao H and Ho S T 2003 *IEEE J. Quantum Electron.* **39** 364.

152. Yamilov A and Cao H 2004 *Phys. Rev. A* **69** 031803(R).

153. Burin A L, Cao H, Schatz G C and Ratner M A 2004 *J. Opt. Soc. Am. B* **21** 121.

154. Shkunov M N, DeLong M C, Raikh M E, Vardeny Z V, Zakhidov A A and Baughman R H 2001 *Syn. Metals* **116** 485.

155. Wu X, Yamilov A, Liu X, Li S, Dravid V P, Chang R P H and Cao H 2004b *App. Phys. Let.* **85** 3657.

156. Feng Y and Ueda K 2004 *Opt. Express* **12** 3307.

157. Cao H, Jiang X, Ling Y, Xu J Y and Soukoulis C M 2003b *Phys. Rev. B* **67** 161101(R).

158. Jiang X, Song F, Soukoulis C M, Zi J, Joannopoulos J D and Cao H 2004 *Phys. Rev. B* **9** 104202.

159. Türeci H E, Stone A D, and Collier B 2006 *Phys. Rev. A* **74**, 043822.

160. Türeci H E, Stone A D, and Li G 2007 *Phys. Rev. A* **76**, 013813.

161. Türeci H E, Stone A D, and Li G 2008 *Science* **320**, 643.

162. Deych L I 2005 *Phys. Rev. Lett.* **95** 043902.

163. Patra M, Schomerus H and Beenakker C W J 2000 *Phys. Rev. A* **61** 023810.

164. Frahm K M, Schomerus H, Patra M and Beenakker C W J 2000 *Europhys. Lett.* **49** 48.

165. Schomerus H, Frahm K M, Patra M and Beenakker C W J 2000 *Physica A* **278** 469.

166. Hackenbroich G 2005 *J. Phys. A: Math. Gen.* **38** 10537.

167. Zaitsev O 2006 *Phys. Rev. A* **74** 063803.

168. Zaitsev O 2007 arXiv:cond-mat/0703783.

169. Noginov M A, Egarievwe S U, Noginova N E, Wang J C and Caulfield H J 1998b *J Opt. Soc. Am B* **15** 2854.

170. Liu B, Yamilov A, Ling Y, Xu J Y and Cao H 2003 *Phys. Rev. Lett.* **91** 063903.
171. Zacharakis G, Papadogiannis N A and Papazoglou T G 2002 *Appl. Phys. Lett.* **81** 2511.
172. Fujiwara H and Sasaki K 2004 *Jpn. J. App. Phys.* **43** L 1337.
173. Burin A L, Cao H and Ratner M A 2003b *IEEE J. Select. Top. Quant. Electron.* **9** 124.
174. Patra M 2003a *Phys. Rev. E* **67** 016603.
175. Florescu L and John S 2004c *Phys. Rev. E* **70** 036607.
176. Polson R and Vardeny Z V 2004 *Appl. Phys. Lett.* **85** 1289.
177. Mujumdar S, Cavalieri S and Wiersma D S 2004a *J. Opt. Soc. Am.* **21** 201.
178. Li S, Wang Z-J, Chen L-S, X Sun and George T F 2005 *Appl. Phys. Lett.* **86** 171109.
179. Lubatsch A, Kroha J and Busch K 2005 *Phys. Rev. B* **71** 184201.
180. Vasa P, Singh B P and Ayyub P 2005 *J. Phys. Condens. Matter* **17** 189.
181. Redmond S M, Armstrong G L, Chan H-Y, Mattson E, Mock A, Li B, POtts J R, Cui M, Rand S C, Oliveira S L, Marchal J, Hinklin T and Laine R M 2004 *J. Opt. Soc. Am. B* **21** 214.
182. Dice G D, Mujumdar S and Elezzabi A Y 2005 *App. Phys. Lett.* **86** 131105.
183. Rand S 2003 *Opt. Photon. News* 33.
184. Cao H 2005 *Opt. Photon. News* **16** 24.
185. Williams G R, Bayram S B, Rand S C, Hinklin T and Laine R M 2001 *Phys. Rev. A* **65** 013807.
186. Li B, Williams G R, Rand S C, Hinklin T, and Laine R M 2002 *Opt. Lett.* **27**, 394.
187. Polson R C and Vardeny V 2000 *Appl. Phys. Lett.* **85**, 70.
188. Letokhov V S 1972 *IEEE J. Quant. Electron.* **QE-8** 615.
189. Letokhov V S 1996 *Amazing Light* ed R Y Chiao (Berlin: Springer) 409.

CHAPTER 13

OPTICAL PROPERTIES OF ZINC OXIDE QUANTUM DOTS

WEN-FENG HSIEH,* HSU-CHENG HSU, WAN-JIUN LIAO, HSIN-MING CHENG, KUO-FENG LIN, WEI-TZE HSU and CHIN-JIU PAN

Department of Photonics & Institute of Electro-Optical Engineering,
National Chaio Tung University,
1001 Tahsueh Rd., Hsinchu, 30050, Taiwan.
wfhsieh@mail.nctu.edu.tw

Size-dependence of efficient UV photoluminescence (PL) and absorption spectra of various sizes of zinc oxide (ZnO) quantum dots (QDs) give evidence for the quantum confinement effect. Bandgap enlargement is in agreement with the theoretical calculation based on the effective mass model for the size of ZnO QDs being comparable to the Bohr radius of bulk exciton. By using the modified spatial correlation model to fit the measured Raman spectra, we reveal that the Raman spectral shift and asymmetry for E_2(high) mode are caused by localization of optical phonons. Furthermore, we present temperature-dependent PL of different sizes of ZnO particles. The unobvious LO-phonon replicas of free exciton (FX) were observed when the ZnO particle sizes were under 12 nm in diameter. The increasing exciton energy (E_b) with the decreasing quantum dot size can be obtained from temperature-dependent PL. From the temperature-dependent change of FX emission energy, we deduce that the exciton-LO phonon coupling strength reduces as the particle size decreases. The reduced exciton Bohr radius a_B with particle size obtained from E_b and PL spectrum confirms that the exciton becomes less polar in turn reducing the Fröhlich interaction and the exciton-LO phonon interaction is reduced with decreasing size of the ZnO QDs. In addition, the nearly unchanged spectral shape in power dependent PL of ZnO quantum dots reveals stable exciton states without formation of biexcitons and exciton-exciton scattering.

1. Introduction

Wurtzite zinc oxide (ZnO) has a hexagonal structure (space group C_{6mc}) with lattice parameters a = 0.3296 nm and c = 0.52065 nm. The structure of ZnO can be simply described as a number of alternating planes composed of tetrahedrally coordinated O^{2-} (open circles in Fig. 1) and Zn^{2+} ions (solid circles), stacked alternately along the c-axis. The tetrahedral coordination in ZnO results in non-central symmetric structure and consequently possesses piezoelectricity and pyroelectricity. The wide bandgap (3.37 eV), large exciton binding energy (60 meV), and radiation hardness of ZnO make it an excellent candidate of room temperature (RT) ultraviolet (UV) light-emitter for use in lasers and light-emitting diodes (LEDs).[1,2] Some optoelectronic applications of ZnO overlap with that of wurtzite gallium nitride (GaN), another wide-gap semiconductor (E_g ~ 3.4 eV at RT), which is widely used for production of green, blue-ultraviolet, and white light-emitting devices. However, ZnO has some advantages over GaN among which are the availability of fairly high-quality ZnO bulk single crystals and the larger exciton binding energy over that of the GaN E_b of 25 meV.

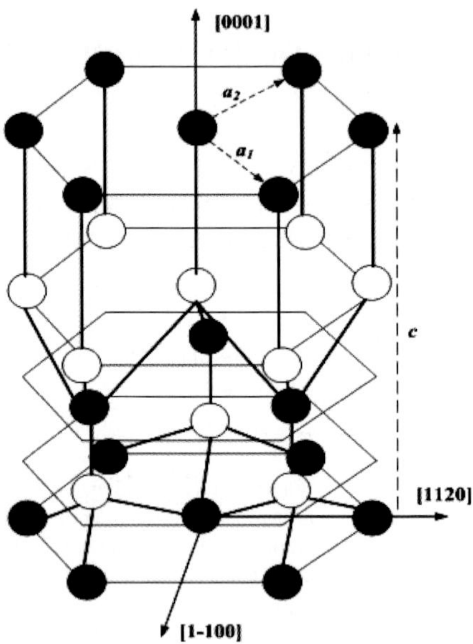

Fig. 1.   The wurtzite structure of ZnO.

The synthesis of semiconductor nanomaterials has aroused worldwide interest in the last few years. Given their large surface area to volume ratios, size effects and possible quantum confinement effects, nanomaterials are predicted to exhibit new and enhanced properties relative to those of the corresponding bulk materials and offer routes to fabricating novel nanodevices. The size-tunable optical properties of quantum confined semiconductor nanocrystals have motivated further investigations into the luminescence of semiconductor quantum dots (QDs), especially prepared under proper surface passivation showing bandgap engineered emission over wide wavelength range with high quantum yields.

Nanostructured ZnO materials have received broad attention due to their distinguished performance in electronics and photonics. From the 1960s, synthesis of ZnO thin films has been an active field because of their applications as sensors, transducers and catalysts. In the last few decades, a variety of ZnO nanostructure morphologies, such as nanowires,[3-5] nanorods,[6-9] tetrapods,[10-12] nanoribbons/belts,[13-15] and nanoparticles[16,17] have been reported. ZnO nanostructures have been fabricated by various methods, such as thermal evaporation,[11-14] metal–organic vapor phase epitaxy (MOVPE),[8] laser ablation,[9] hydrothermal synthesis,[6,7] sol-gel method[16,17] and template-based synthesis.[5] Recently, novel morphologies such as hierarchical nanostructures,[18] bridge-/nail-like nanostructures,[19] tubular nanostructures,[20] nanosheets,[21] nanopropeller arrays,[22,23] nanohelixes and nanorings[22,24] have, amongst others, been demonstrated and applied to dye sensitized solar cells.[25] Several recent review articles have summarized progress in the growth and applications of ZnO nanostructures.[26-28]

However, as the dimensions of semiconductors are reduced to the nanometer scale, the optical properties of semiconductors are much different from their bulk counterparts materials.[29-31] There are two incompatible physical mechanisms in modifying the energy band structure of nanostructures, i.e., the quantum confinement effect (QCE) and surface states.[32] These two mechanisms compete with each other to influence the photoluminescence (PL) spectra. For nanodots or nanostructures in ZnO system with diameters less than 10 nm, the QCE plays a dominant role as has been much reported.[33,34] On the other hand, the surface-to-volume ratio also brings much influence on the system's Hamiltonian when the material size is reduced to the nanometer scale.[35,36] The predominance of surface states is responsible for many novel physical features of nanomaterials.[37,38] Recently, Guo et al.[39] demonstrated significantly enhanced UV luminescence, diminished visible luminescence and excellent third-order nonlinear optical response with polyvinyl pyrrolidone (PVP) modified surface of ZnO nanoparticles. Norberg and Gamelin[40] noted that changes in nanocrystal size, shape, and luminescence intensities have been measured for nanocrystals capped by dodecylamine (DDA) and trioctylphosphine (TCP) oxide after different growth times. They explained that the green trap emission intensities show a direct correlation with surface hydroxide concentrations. Contrary to expectations, there is no direct correlation between excitonic emission quenching and surface hydroxide concentrations. The nearly pure excitonic emission observed after heating in DDA is attributed to the removal of surface defects from the ZnO nanocrystal surfaces and to the relatively high packing density of DDA on the ZnO surfaces. Furthermore, Shaish et al.[41] showed that intensity relations of below-bandgap and band-edge luminescence in ZnO nanowires depend on the wire radius. The relative intensity of this surface luminescence increases as the wire radius decreases at the expense of the band-edge emission. Pan et al.[42] also predicated a significant increase in the intensity ratio of the deep-level to the near-band-edge emission is observed with ever-increasing nanorod surface-aspect ratio. Thus, in quantum-size nanostructures, surface-recombination may entirely quench band-to-band recombination, presenting an efficient sink for charge carriers that unless deactivated may be detrimental for electronic devices.

The interaction between exciton and longitudinal-optical (LO) phonon has a great influence on the optical properties of polar semiconductors. Ramvall, et al.[43] reported a diminishing temperature-dependent shift of the PL energy with decreasing GaN QD size caused by a reduction of the LO-phonon coupling. Chang, et al.[44] reported on the exciton LO-phonon interaction energy $|E_{ex\text{-}ph}|$ theoretically evaluated as functions of electric field strength and the size of the quantum dots. The field enhanced by reducing the separation between electron and hole would increase $|E_{ex\text{-}ph}|$. Whereas, the decrease of dot size leads to delocalization of the wave functions of both electron and hole in turn decreasing $|E_{ex\text{-}ph}|$.

Although there were many experiments describing the influence of surface states and electronic behavior in ZnO nanostructures, there is still a lack of experimental studies of the influences of crystalline size on electronic structure and exciton-LO-phonon coupling in ZnO. In this chapter, we first review our recent reports on increasing quantum confinement effect[45-47] and decreasing exciton-LO phonon coupling, while decreasing the size of the ZnO quantum dots.[48] We then report a power dependent PL of ZnO quantum dots that reveals stable exciton states with unchanged spectral shape in which formation of biexcitons and exciton-exciton scattering that occurs in the ZnO powder[49] were not observed in the ZnO quantum dots.

This chapter is organized as follows. Section 2 presents the synthesis of the ZnO QDs and illustrates the characterization techniques. In Section 3 we present (A) the morphology and crystal structures, (B) quantum confinement effect and (C) lattice dynamics with different crystallize size of ZnO QDs, (D) the increase of exciton binding energy resulting from the decrease of exciton Bohr radius making the exciton less polar and thereby reducing the coupling to LO phonons, and (E) Coulomb blockade in QDs using the power-dependent PL. Finally, we conclude our recent study on optical properties of ZnO QDs in Section 4.

2. Experimental details

ZnO QDs have been obtained by metal organic chemical vapor-phase deposition (MOCVD) epitaxy,[50,51] pulsed laser deposition (PLD),[52] and vapor phase transport (VPT) deposition process[53,54] techniques. However, these methods are expensive and require high vacuum and formation controlling conditions. Compared with these methods, the sol-gel process is an attractive technique for compound semiconductors preparation because of its simplicity, low cost, and ease of composition control.[55,56] In particular, the sol-gel technique has the potential to produce samples with large areas and complicated forms on various substrates. In this research, the sol-gel method was used to fabricate ZnO QDs. The detailed growth mechanisms and characterization techniques of the ZnO QDs are discussed as follows.

2.1. *Sample preparation*

We produce monodisperse ZnO colloidal spheres by sol-gel method. Sol-gel method was chosen due to its simple handling and narrow size distribution. The ZnO colloidal spheres were produced by an one-stage reaction process. All chemicals used in this study were reagent grade and employed without further purification. Zinc acetate dihydrate (99.5% $Zn(OAc)_2$, Riedel-deHaen) was added to diethylene glycol (99.5% DEG, EDTA). Our first observation is that we can control the quantum dots size with concentration of zinc acetate in the solvent (DEG). Then the temperature of reaction solution was increased to 160°C and maintained for different aging time. White colloidal ZnO was formed in the solution that was employed as the primary solution. The product was then placed in a centrifuge. The supernatant (DEG, dissolved reaction products, and unreacted ZnAc and water) was decanted off and saved, and the polydisperse powder was discarded. Finally, the supernatant was then dipped on substrates (SiO_2/Si (001) or SiO_2) and dried at 150°C. A flow chart of the detailed fabricating ZnO quantum dots (QDS) by sol-gel method can be found in Ref. 45.

2.2. *Characterization of ZnO QDs*

The crystal structures of the as-grown powder were inspected by Bede D1 diffractometer at Industrial Technology Research Institute, Taiwan using a CuK X-ray source (λ = 1.5405Å). We used small angle diffraction. The ω was fixed at 5°, the scanning step was 0.04°, scanning rate was 4 degree/min and count time was 1.00 second. The shapes and sizes of ZnO QDs were analyzed using JEOL JEM-2100F field emission transmission electron microscope (FETEM) operated at 200 KeV.

Raman scattering is a very powerful probe for investigating the vibration properties of materials. It is also important in understanding problems as diverse as the structure of amorphous insulators, and the conduction mechanisms in ionic conductors. The micro-Raman spectroscopy was performed in the backscattering geometry with a confocal Olympus (BX-40) optical microscope. The scattered light was dispersed through the triple-monochromator and detected by a liquid-nitrogen-cooled charge-coupled device (CCD) camera. The 515-nm line of a frequency-doubled Yb^{3+}:YAG laser was used as the Raman excitation source. The best spatial resolution during Raman measurements was 1.0 μm with a spectral resolution of 0.2 cm^{-1}. Similarly, micro-Raman spectroscopy was measured with an Ar-ion laser (Coherent INNOVA 90) as an excitation source emitting at a wavelength of 488 nm. The scattered light was collected by a camera lens and imaged onto the entrance slit of the Spex spectrometer.

PL provides a non-destructive technique for the determination of certain impurities in semiconductors. The shallow-level and the deep-level of impurity states were detected by PL. The PL showed that radiative recombination events dominate non-radiative recombination. In the PL measurements, the 325-nm line from a He–Cd laser was used as the excitation light. Light emission from the samples was collected into the TRIAX 320 spectrometer and detected by a photomultiplier tube (PMT). The PL detection system includes mirror, focusing and collecting lens, the sample holder and the cooling system. The excitation laser beam was directed normally and focused onto the sample surface with power being varied with an optical attenuator. The spot size on the sample is about 100 μm. Spontaneous and stimulated emissions were collected by a fiber bundle and coupled into a 32-cm focal-length monochromator (TRIAX 320) with a 1200 lines/mm grating, then detected by either an electrically cooled CCD camera (CCD-3000) or a PMT (PMT-HVPS) detector. The temperature-dependent PL measurements were carried out using a closed cycle cryostat between 15 K and 300 K.

3. Results and discussion

3.1. *Morphology and crystal structures*

Figure 2(a) shows a typical high-resolution transmission electron microscope (HRTEM) image of the ZnO nanoparticles prepared from solution concentration of 0.06M aged at 160 °C for 1 hr. The particles shape are predominantly spherical, many also exhibit surface facet, as shown in the inset of Fig. 2(a) where a step of one atomic layer can be seen. The nanoparticles are clearly well-separated and essentially have some aggregation.

Figure 2(b) shows the size distribution of particles obtained from analysis of more than 35 particles in this sample. The average diameter was determined to be 4.36 ± 0.30 nm. Presumably due to the viscosity of DEG, the solvent may have modified the Ostwald ripening kinetics such that the growth rate decreases with the size of the ZnO QDs. This would narrow the size distribution of ZnO QDs effectively.

The XRD patterns and the interplane spacing of the prepared sample (ZnO solution dip on SiO_2 glass) can be well-matched to the standard diffraction pattern of wurtzite ZnO with lattice constants a = 3.253Å and c = 5.219Å, which are consistent with the value in the standard card (JCPDS 36-1451), demonstrating the formation of wurtzite ZnO nanocrystals. The mean diameter of the ZnO nanocrystallites is evaluated from the full-width half-maximum (FWHM) of the (110) peak at $\theta = 47.56°$ to be 3.5 to 12 nm for the solution concentrations from 0.06 M to 0.32 M by the Debye-Scherer formula: $D = 0.89\lambda\, /\, B\cos\theta$, where D is the average diameter of the nanocrystallite, $\lambda = 1.5406$Å is the wavelength of the X-ray source, and B is the FWHM of X-ray diffraction peak at the diffraction angle θ ranging from 0.75° to 2.45°. The particle size is basically linearly proportional to the solution concentration under 160 °C aging for 1.5 hrs, and the statistical result is consistent with the observation from HRTEM.

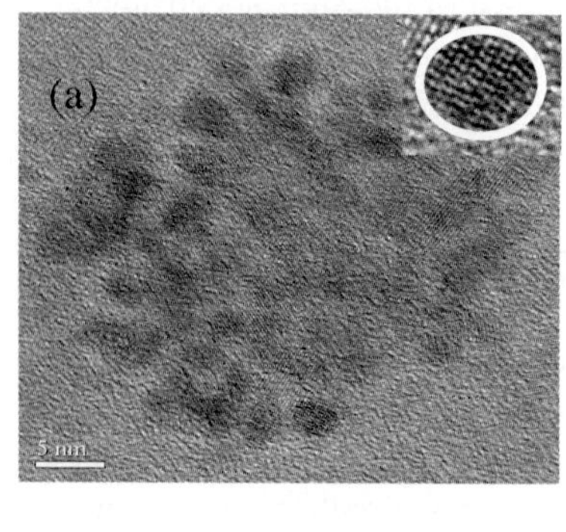

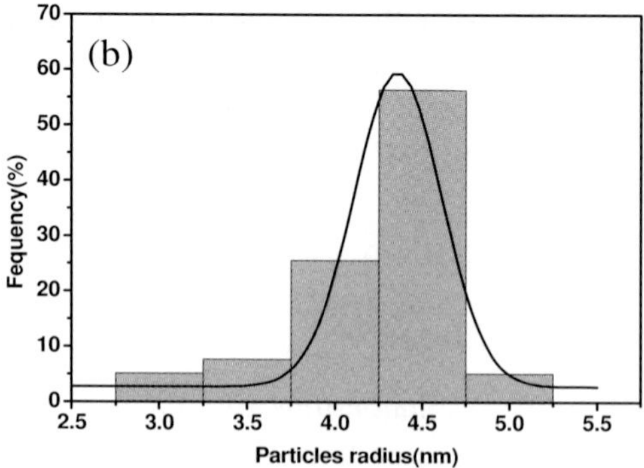

Fig. 2.   HRTEM image (a) and size distribution (the scale bar is 5 nm) (b) of the ZnO QDs using 0.06M $Zn(OAc)_2$.

3.2. *Quantum confinement effect*

Figure 3 shows typical PL and absorption spectra of the samples with different average QD sizes at room temperature. The UV emission represents a relaxed state of the exciton near the band edge in the ZnO QDs. The highly efficient UV emission near the band edge is attributed to confined exciton emission with high density of states that shifts to the higher energies from 3.30 to 3.43 eV as the size of QDs decreases from 12 to 3.5 nm, which are comparable to or smaller than the diameter (4.68 nm) of the exciton (Bohr radius of bulk ZnO is 2.34 nm).[57] In general, quantum confinement widens the energy bandgap and gives rise to a blueshift in the transition energy as the crystal size decreases. Such a phenomenon is also revealed in the absorption spectra, although the faint excitonic absorption peaks are due to the moderate size distribution of ZnO QDs. From this figure it can clearly be seen that the absorption onset exhibits a progressive blueshift from 3.43 to 3.65 eV as the size of ZnO QD decreases from 12 to 3.5 nm. We calculated the bandgaps by using the effective-mass model[58] for different sizes of ZnO QDs and the results showed good agreement with the experimental data.

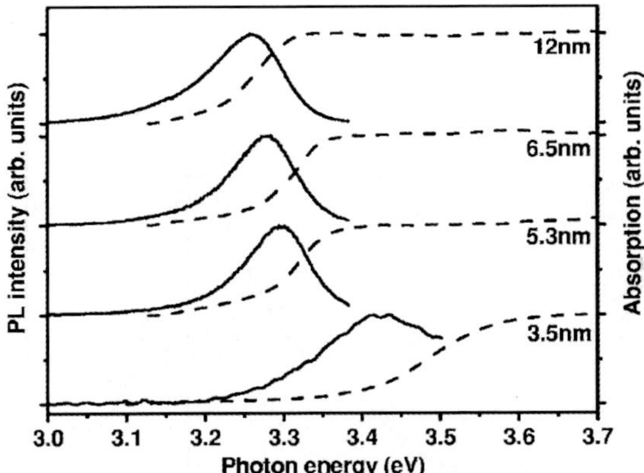

Fig. 3.   PL (solid line) and absorption (dashed line) spectra near the band edge of various ZnO QD sizes.

3.3. *Spatial phonon confinement effect*

In order to observe the optical phonon confinement effect, the Raman spectra with different sizes of ZnO QDs were measured using micro-Raman spectroscopy under a fixed excitation laser power of 3.1 mW. Because the phonon wave function is partially confined to the volume of the crystallite and if a spherical shape of finite-size ZnO QDs is assumed, the first-order Raman spectrum $I(\omega)$ can be described by the following equation:[59]

$$I(\omega) \propto \int_0^1 \frac{4\pi q^2 \exp(-q^2 L^2 / 4)\, dq}{[\omega - \omega(q)]^2 + (\Gamma / 2)^2} , \qquad (1)$$

where q is expressed in unit of $2\pi/a$, a is the lattice constant, $\omega(q)$ is the phonon dispersion relation, Γ is the linewidth of the LO phonon of the ZnO bulk, and L is spatial correlation length corresponding to grain size. Furthermore, Islam et al.[60,61] reported on the influence of crystallite size distribution (CSD) on the shifts in Raman scattering frequencies and lineshapes in silicon nanostructures. They modified the Raman intensity expression of Eq. (1) to $I(\omega, L_0, \sigma)$ by using Gaussian CSD of an ensemble of spherical crystallites with mean crystallite size L_0 and standard deviation σ,

$$I(\omega, L_0, \sigma) \propto \int_0^1 \frac{f(q)q^2 \exp(-q^2 L_0^2 / 4)dq}{[\omega - \omega(q)]^2 + (\Gamma / 2)^2} \tag{2}$$

where $f(q) = 1/\sqrt{1 + q^2\sigma^2/2}$ is the characteristic of the CSD. By matching the calculated normalization Raman profiles of E_2(high) mode at ~ 436 cm^{-1} from an ensemble of ZnO QDs with the measured Raman spectra by varying σ to illustrate the effect of σ on the Raman lineshape (as shown in Fig. 4) we found $L_0 = 6.5$ nm for matching the results having a mean crystallite size of 6.5 nm for 0.16-M sample. It is clear that a single-crystalline component with $\sigma = 0.27$ describes the Raman spectra of 6.5-nm ZnO QDs quite well. The CSD of all samples were about 27%, which agrees with the TEM result, e.g., the obtained crystal size is 4.3 nm $\pm$ 1.1 nm.

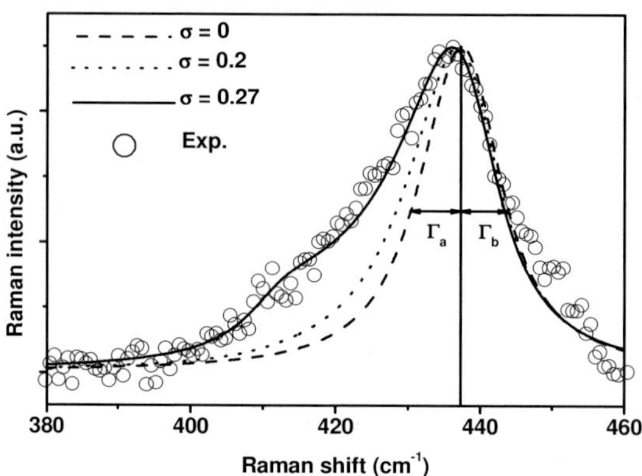

Fig. 4. Fitting of the modified spatial correlation model with $\sigma = 0$, 0.2, and 0.27, respectively, to the measured result for average size of 6.5 nm ZnO QDs.

Furthermore, the frequency shift $\Delta\omega$ and the asymmetry, Γ_a/Γ_b, of E_2(high) mode from the ZnO bulk (439 cm^{-1}) as a function of diameter or correlation length with $\sigma = 0.27$ were plotted in Fig. 5, in which the solid curves are the calculated results of the modified spatial correlation model and hollow circles are the measurements. We found that the measured frequency shift and asymmetry agrees very well with the calculations by the modified SC model and the mean values of crystallite sizes obtained from our fitting are also in good agreement with the XRD results.

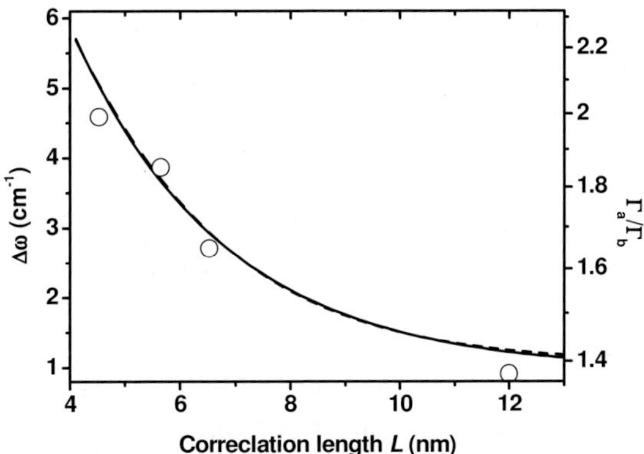

Fig. 5. Raman shift $\Delta\omega$ (solid curve) and asymmetric broadening Γ_a/Γ_b (dashed curve overlays with the solid curve) of E_2(high) phonon as a function of correlation length L or average size of nanocrystal.

4. Reducing exciton-phonon coupling due to quantum confinement effect

Figure 6 shows the PL spectra of different ZnO sizes at 13 K. The spectrum of ZnO powders consists of the free exciton (FX) and the donor-bound exciton (D^0X) emission peaks along with three obvious LO-phonon replicas. The FX emission of ZnO powders is 3.377 eV which behaves as ZnO bulk. The energy shift (dashed line) from 3.377 eV to 3.475 eV due to QCE can be observed.[45] The FWHM which increases as the dot size decreases may be caused by the contribution of surface-optical phonon,[62] surface-bound acceptor exciton complexes,[63] and size distribution. Accordingly, we observed that LO-phonon replicas are obvious in ZnO powders but are missing in other QD-samples. Duke et al.[64] interpreted that the intensities of LO-phonon replicas depend strongly on their exciton-phonon coupling strengths.

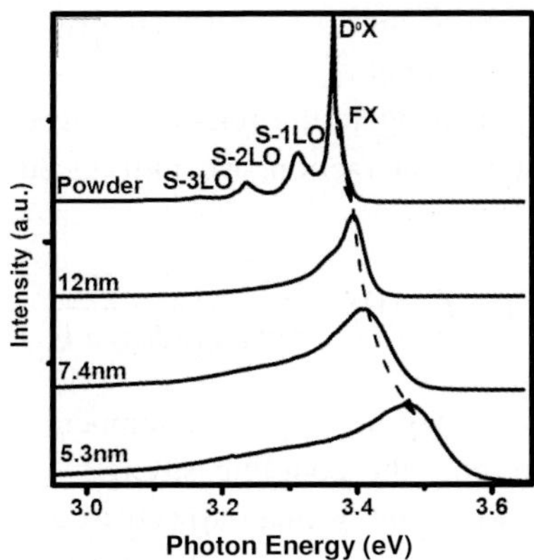

Fig. 6. PL spectra of different ZnO particle sizes at 13K. The dashed line indicates the free exciton peak energy shift.[48]

From the temperature-dependent PL, we can obtain the exciton binding energy (E_b), from the following relation:[65]

$$I(T) = \frac{I(0)}{1 + A\exp(-E_b / k_B T)} \tag{3}$$

where I(T) is the integrated intensity of the peak at specific temperature, I(0) is the integrated intensity at absolute zero, A is a constant, and k_B is Boltzmann constant. The fitting results reveal that E_b of the ZnO powder is 60 meV, which is close to that of ZnO bulk, and E_b = 67, 87 and 132 meV, respectively, for 12-, 7.4-, and 5.3-nm QDs. The decreasing particle size would raise the electron-hole interaction as a result of the compressing boundary to cause increasing Coulomb energy. Therefore, the binding energy increases as the particle size decreases.

The increasing E_b gives an indication for reduction of exciton-LO phonon interaction α_0. The enhancement of E_b (or Coulomb potential) indicates a reduction of exciton Bohr radius a_B. It makes the exciton less polar capable for efficiently interacting with LO-phonon through the Fröhlich interaction.[66] To find out the relation between a_B and α_0, we calculated a_B from our PL spectra including the FX emission energy and E_b for different dot sizes based on the weak confinement model:[67]

$$E_g^*(R) \cong E_g^{bulk} + \frac{\hbar^2 \pi^2}{2eR^2}\left(\frac{1}{m_e} + \frac{1}{m_h}\right) - \frac{1.8e^2}{4\pi\varepsilon\varepsilon_0 R} \tag{4}$$

and $a_B^2 = \hbar^2/(2\mu^* E_b)$,[68] where $E_g(R)$ is the measured FX emission energy plus E_b, E_g = 3.43 eV is the bandgap energy of bulk ZnO, e is the charge of electron, $\hbar$ is Planck's constant divided by 2π, R is the particle radius, μ^* is the reduced mass of exciton, ε = 3.7 is the relative permittivity, and ε_0 is the permittivity of free space. The calculated exciton Bohr radii a_B-QD for 5.3-nm, 7.4-nm and 12-nm QDs are 0.977 nm, 1.038 nm and 1.328 nm respectively. The ratios of a_B QD to the exciton Bohr radius for bulk ZnO of a_B powders = 2.34 nm are 0.42, 0.46 and 0.57, respectively, which agree well with 0.42, 0.49 and 0.59 obtained by Senger, et al .[69]

In order to quantitatively investigate the relation between the quantum confinement size and the exciton-LO phonon interaction, we introduced the temperature-dependent exciton energy.[70]

$$E_{ex}(T) = E_{ex}(0) - \sum_i \frac{\alpha_{0i}}{\exp(\hbar\omega_i / k_B T) - 1} \tag{5}$$

where $E_{ex}(T)$ is the exciton energy at a specific temperature T, $E_{ex}(0)$ is the exciton energy at 0 K, and α_{0i} represents the coupling strength with the optical phonon with energy $\hbar\omega_i$. Our previous Resonant Raman Spectroscopy (RRS)[46] and PL[47] results indicate the most promising LO-phonon involved in RRS and PL with an energy of 72 meV, we therefore take only a single one of the summation terms with $\hbar\omega$ = 72 meV into account to discuss exciton-LO phonon coupling. Then α_0 represents the weighting of exciton-LO-phonon coupling. The fitting to the temperature-dependent exciton transition energies, we obtained α_0 = 0.59, 0.40, 0.21, and 0.19 for powders, 12-nm, 7.4-nm, and 5.3-nm QDs, respectively. These results are consistent with the observations of PL spectra in Fig. 5, where the coupling strength of exciton-LO phonon weakens as the particle size decreases.

Figure 7 shows similar trends of $\alpha_{0\ QD}/\alpha_{0\ Powders}$ and $a_{B\ QD}/a_{B\ powders}$ against the dot size. The exciton formation is attained by Coulomb interaction, as the particle sizes decrease, the quantum confinement effect causes an increase of E_b and a decrease of a_B. The electric dipole, which is proportional to the distance of electron-hole pair, is then reduced. The exciton formation thus becomes less polar, in turn reducing the coupling strength with the polar lattice via the Fröhlich interaction. Consequently, exciton-LO phonon interaction is reduced as the particle size decreases in ZnO-QD system.

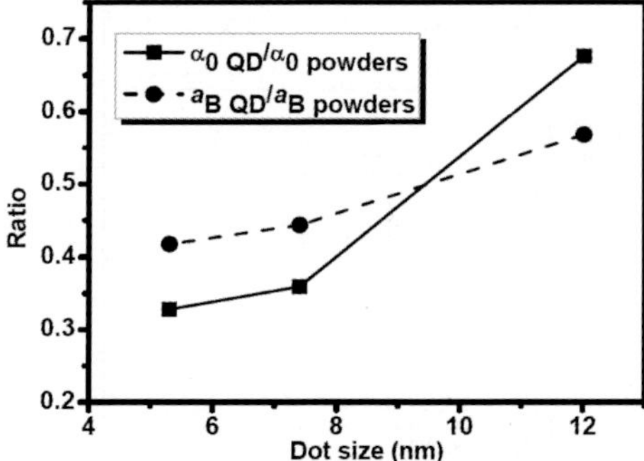

Fig. 7.   The relation of $\alpha_{0\ QD}/\alpha_{0\ Powders}$ and $a_{B\ QD}/a_{B\ Powders}$ with different dot sizes.

5. Power-dependent photoluminescence

Figure 8 shows the power-dependent PL spectra of ZnO QDs with diameter of 12 nm taken at 10K. The spectral lineshape is basically unchanged with the excitation power varying over three orders of magnitude. The spectral lineshape is quite different from those of micrometer-sized ZnO powder taken at both 10 K and 80 K,[49] as shown in Fig. 9, in which the spectral shapes dramatically change as increasing the excitation power from 2 to 40 mW. At T =10 K, the lowest curve (2-mW excitation) presents two peaks at the lower energy side of the biexciton (BX) peak at 3.358 eV, denoted as P_2 and P_∞, which originate from inelastic scattering between excitons.[71]   As a result of such scattering, one exciton is scattered into one of the higher states (n = 2, 3, 4, . . . , ∞), while the other exciton loses its kinetic energy to occupy the lower polariton branch, which is roughly located at 3.326 and 3.312 eV for its counterpart exciton being scattered to n = 2 and n = ∞, respectively. In addition, the peak due to 1LO phonon-assisted radiative recombination of the neutral donor-bound exciton (D^0X) is observed at 3.295 eV denoted by D^0X-1LO. These three bands shift towards the lower energy side and finally merge into a broad P line as further increasing the excitation power. Similar change of spectral shape due to increasing formation of biexciton and exciton-exciton scattering can be found at 80K. The spectral shape of near-band-edge emission measured at 80 K is decomposed into BX, D^0X, $FX^A_{n=1}$, and $FX^B_{n=1}$ by Lorentzian functions. The typical fitting results are shown in the inset of Fig. 9(b) with dashed lines denoting the various emissions and the solid line corresponding to the sum of the theoretical fits, which shows good agreement with the experimental data denoted as the open dots. The integrated intensity of the free-exciton peak exhibits linear dependence on the excitation power, while that of the biexciton follows a superlinear dependence as $I_{ex}^{1.86}$. The results support our assignment of $FX^A_{n=1}$ and BX peaks. The unchanged spectral shape in power-dependent PL of ZnO quantum dots (Fig. 8) reveals no formation of biexcitons or exciton-exciton scattering in ZnO QDs that may be due to hardly more than one single exciton co-existence inside a QD within the exciton.

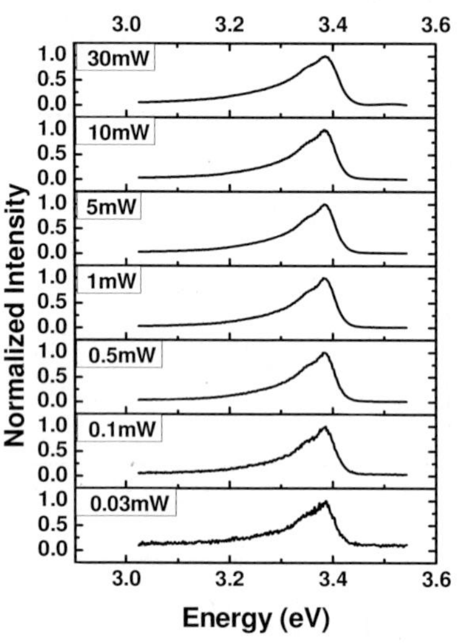

Fig. 8.   Power-dependent PL of ZnO QDs with diameter of 12 nm taken at 10K.

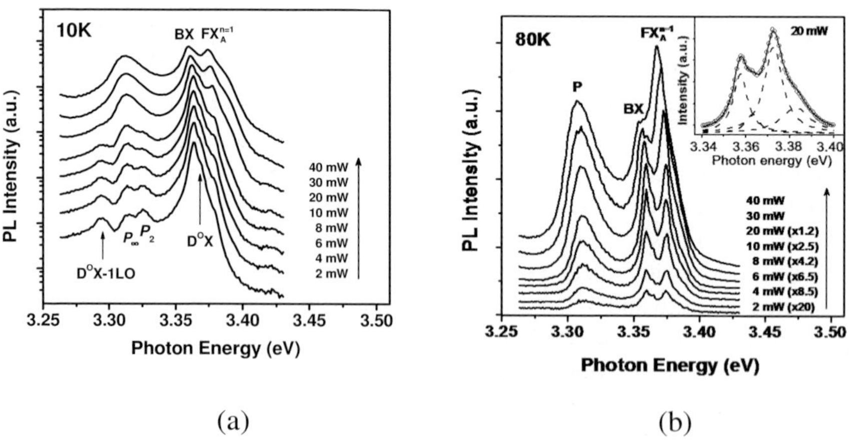

(a)   (b)

Fig. 9.   Dependence of PL spectra on excitation power measured at (a) 10 K and (b) 80 K. The inset of (b) shows typical theoretical fit to the PL spectrum for excitation power of 20 mW. Solid lines correspond to the fit and open dots represent the data. The fitted lineshapes are also shown separately in dashed lines.

6. Conclusion

Size-dependence of efficient UV photoluminescence and absorption spectra of various QD sizes give evidence for the quantum confinement effect. Bandgap enlargement is also in agreement with the theoretical calculation based on the effective mass model. By using the modified spatial correlation model to fit the measured Raman spectra, we reveal that the Raman spectral shift and asymmetry for the E_2(high) mode are caused by localization of optical phonons. Furthermore, we presented temperature-dependent PL of different sizes of ZnO particles. The unobvious LO-phonon replicas of FX were observed when the ZnO particle sizes were under 12 nm in diameter. The increasing exciton E_b with the decreasing quantum dot size can be obtained from temperature-dependent PL. From the temperature-dependent change of FX emission energy, the exciton-LO phonon coupling strength reduces as the particle size decreases. The reduced a_B with particle size obtained from E_b and PL spectrum confirms that the exciton becomes less polar in turn reducing the Fröhlich interaction and the exciton-LO phonon interaction is reduced with decreasing ZnO QDs.

In addition, from the nearly unchanged spectral shape in power-dependent PL of ZnO quantum dots, we conclude that no more than one exciton co-existence inside a ZnO QD within the exciton lifetime as compared with formation of biexcitons or exciton-exciton scattering in the ZnO powder. Further investigation of quantum confinement effect in quantum well structures is under way.

References

1. D. Özgür, Ya. I. Alivov, C. Liu, A. Teke, M. A. Reshchikov, S. Dogan, V. Avrutin, S. J. Cho, and H. Morkoç, J. Appl. Phys. **98**, 041301 (2005).
2. R. Triboulet, J. Perriere, Prog. Cryst. Growth Charact. Mater. **47**, 65 (2003).
3. M. H. Huang, S. Mao, H. Feick, H. Yan, Y. Wu, H. Kind, E. Weber, R. Russo, P. Yang, Science **292**, 1897 (2001).
4. M. H. Huang, Y. Wu, H. Feick, N. Tran, E. Weber, P. Yang, Adv. Mater. **13**, 113 (2001).
5. C. Liu, J. A. Zapien, Y. Yao, X. Meng, C. S. Lee, S. Fan, Y. Lifshitz, S. T. Lee, Adv. Mater. **15**, 838 (2003).
6. B. Liu, H. C. Zeng, J. Am. Chem. Soc. **125**, 4430 (2003).
7. M. Guo, P. Diao, S. Cai, J. Solid State Chem. **178**, 1864 (2005).
8. W. I. Park, Y. H. Jun, S.W. Jung, G.-C. Yi, Appl. Phys. Lett. **82**, 964 (2003).
9. A. B. Hartanto, X. Ning, Y. Nakata, T. Okada, Appl. Phys. **A78**, 299 (2003); W. -R. Liu, Y. -H. Li, W. F. Hsieh, C. -H. Hsu, W. C. Lee, Y. J. Lee, M. Hong, J. Kwo, Crystal Growth & Design **9(1)**, 239-242 (2009).
10. Y. Dai, Y. Zhang, Z. L. Wang, Solid State Commun. **126**, 629 (2003).
11. V. A. L. Roy, A. B. Djurisic, W. K. Chan, J. Gao, H. F. Lui, C. Surya, Appl. Phys. Lett. **83**, 141 (2003).
12. H. Yan, R. He, J. Pham, P. Yang, Adv. Mater. **15**, 402 (2003).
13. Z.W. Pan, Z. R. Dai, Z. L. Wang, Science **291**, 1947 (2001).
14. H. Yan, J. Johnson, M. Law, R. He, K. Knutsen, J. R. McKinney, J. Pham, R. Saykally, P. Yang, Adv. Mater. **15**, 1907 (2003).
15. Y. B. Li, Y. Bando, T. Sato, K. Kurashima, Appl. Phys. Lett. **81**, 144 (2002).
16. K. F. Lin, H. M. Cheng, H. C. Hsu, L. J. Lin, and W. F. Hsieh, Chem. Phys. Lett. **409**, 208 (2005).

17. C. J. Pan, K. F. Lin, W. T. Hsu, and W. F. Hsieh, J. Appl. Phys. **102**, 123504 (2007).
18. J. Y. Lao, J. G. Wen, Z. F. Ren, Nano Lett. **2**, 1287 (2002).
19. J. Y. Lao, J. Y. Huang, D. Z. Wang, Z. F. Ren, Nano Lett. **3**, 235 (2003).
20. Y. J. Xing, Z. H. Xi, X. D. Zhang, J. H. Song, R. M. Wang, J. Xu, Z. Q. Xue, D. P. Yu, Solid State Commun. **129**, 671 (2004).
21. J.-H. Park, H.-J. Choi, Y.-J. Choi, S.-H. Sohn, J.-G. Park, J. Mater. Chem. **14**, 35 (2004).
22. Z. L. Wang, X. Y. Kong, Y. Ding, P. Gao, W. L. Hughes, R. Yang, Y. Zhang, Adv. Funct. Mater. **14**, 943 (2004).
23. P. X. Gao, Z. L. Wang, Appl. Phys. Lett. **84**, 2883 (2004).
24. X. Y. Kong, Z. L. Wang, Nano Lett. **3**, 1625 (2003).
25. W. –H. Chiu, C.-H. Lee, H.-M. Cheng, H.-F. Lin, S.-C. Liao, J.-M. Wu and W. F. Hsieh, Energy Environ. Sci. **2**, 694 (2009).
26. Z. Y. Fan, J. G. Lu, J. Nanosci. Nanotechnol. **5**, 1561 (2005).
27. G. C. Yi, C. R. Wang, W. I. Park, Semicond. Sci. Technol. **20**, S22 (2005).
28. Y.W. Heo, D. P. Norton, L. C. Tien, Y. Kwon, B. S. Kang, F. Ren, S. J. Pearton, J. R. LaRoche, Mater. Sci. Eng. **47**, 1 (2004).
29. X.Y. Kong, Y. Ding, R. Yang and Z. L. Wang, Science **303**, 1348 (2004).
30. S. Nakamura, M. Senoh, N. Iwasa, T. Yamada, T. Matsushita, Y. Sugimoto, H. Kiyoku, Appl. Phys. Lett. **69**, 1568 (1996).
31. L. Bergman, X.B. Chen, J.L. Morrison, J. Huso, A.P. Purdy, J. Appl. Phys. **96**, 675 (2004).
32. L.T. Canham, Appl. Phys. Lett. **57**, 1046 (1990).
33. Y. Kayanuma, Phys. Rev. **B38**, 9797 (1988).
34. M.D. Mason, G.M. Credo, K.D. Weston, S.K. Buratto, Phys. Rev. Lett. **80**, 5405 (1998).
35. F. Koch, V. Petrova-Koch, T. Muschit, J. Lumin. **57**, 271 (1993).
36. J.B. Xia, K.W. Cheah, Phys. Rev. **B59**, 14876 (2003).
37. J.C. Tsang, M.A. Tischler, R.T. Collins, Appl. Phys. Lett. **60**, 2279 (1992).
38. I. Shalish, H. Temhin, V. Narayanamurti, Phys. Rev. **B69**, 245401 (2004).
39. L. Guo, S. Yang, C. Yang, P. Yu, J. Wang, W. Ge, and G. K. L. Wong, Appl. Phys. Lett. **76**, 2901 (2000).
40. N. S. Norberg and D. R. Gamelin, J. Phys. Chem. **B109**, 20810 (2005).
41. I. Shalish, H. Temkin, and V. Narayanamurti, Phys. Rev. **B69**, 245401 (2004).
42. N. Pan, X. Wang, M. Li, F. Li, and J. G. Hou, J. Phys. Chem. **C111**, 17265 (2007).
43. P. Ramvall, P. Riblet, S. Nomura, and Y. Aoyagi, J. Appl. Phys. **87**, 3883 (2000).
44. R. Chang, and S. H. Lin, Phys. Rev. **B68**, 045326 (2003).
45. K. F. Lin, H. M. Cheng, H. C. Hsu, and W. F. Hsieh, Chem. Phys. Lett. **409**, 208–211 (2005).
46. H. M. Cheng, H. C. Hsu, K. F. Lin, and W. F. Hsieh, Appl. Phys. Letts. **88**, 263117 (2006).
47. K. F. Lin, H. M. Cheng, H. C. Hsu, and W. F. Hsieh, Appl. Phys. Lett. **88**, 263117 (2006).
48. W. T. Hsu, K. F. Lin, and W. F. Hsieh, Appl. Phys. Lett. **91**, 181913 (2007).
49. C. J. Pan, K. F. Lin, and W. F. Hsieh, Appl. Phys. Lett. **91**, 111907 (2007).
50. L. M. Yang, Z. Z. Ye, Y. J. Zeng, W. Z. Xu, L. P. Zhu and B. H. Zhao, Solid State Commun. **138**, 577 (2006).
51. S. T. Tan, X. W. Sun, X. H. Zhang, B. J. Chen, S. J. Chua, A. Yong, Z. L. Dong and X. Hu, J. Crystal Growth **290**, 518 (2006).
52. S. Barik, A. K. Srivastava, P. Misra, R. V. Nandedkar and L. M. Kukreja, Solid State Commun. **127**, 463 (2003).
53. J. G. Lu, Z. Z. Ye, J. Y. Huang, L. P. Zhu, B. H. Zhao, Z. L. Wang, and Sz. Fujita, Appl. Phys. Lett. **88**, 063110 (2006).
54. J. G. Lu, Z. Z. Ye, Y. Z. Zhang, Q. L. Liang, Sz. Fujita, and Z. L. Wang, Appl. Phys. Lett. **89**, 023112 (2006).
55. A. E. Jimenez-Gonzalez, J. A. Soto Urueta, and R. Sua´rez-Parra, J. Cryst.Growth **192**, 430 (1998).
56. D. Bao, H. Gu, and A. Kuang, Thin Solid Films **312**, 37 (1998).

57. R. T. Senger and K. K. Bajaj, Phys. Rev. **B68**, 045313 (2003).

58. S.A. Studenikin, N. Golego, M. Cocivera, J. Appl. Phys. **84**, 2287 (1998).

59. H. Richter, Z. P. Wang, and L. Ley, Solid State Commun. **39**, 625 (1981).

60. Md. N. Islam and S. Kumar, Appl. Phys. Lett. **78**, 715 (2001).

61. Md. N. Islam, A. Pradhan, and S. Kumar, J. Appl. Phys. **98**, 024309 (2005).

62. Z. D. Fu, Y. S. Cui, S. Y. Zhang, J. Chen, D. P. Yu, S. L. Zhang, L. Niu and J. Z. Jiang, Appl. Phys. Lett. **90**, 263113 (2007).

63. V. A. Fonoberov and A. A. Balandin, Appl. Phys. Lett. **85**, 5971 (2004).

64. C. B. Duke and G. D. Mahan, Phys. Rev. **139**, 1965 (1965).

65. S. J. Sheih, K. T. Tsen, D. K. Ferry, A. Botchkarev, B. Sverdlov, A. Salvador, and H. Morkoç, Appl. Phys. Lett. **67**, 1757 (1995).

66. D. S. Jiang, H. Jung, and K. Ploog, J. Appl. Phys. **64**, 1371 (1988).

67. J. J. Shiang, S. H. Risbud, and A. P. Alivisatos, J. Chem. Phys. **98,** 8432 (1993).

68. L. E. Brus, J. Chem. Phys. **80**, 4403 (1984).

69. R. T. Senger and K. K. Bajaj, Phys. Rev. **B68**, 045313 (2003).

70. L. Vina, S. Logothetidis, and M. Cardona, Phys. Rev. **B30**, 1979 (1984).

71. C. F. Klingshirn, *Semiconductor Optics*, edited by C. F. Klingshirn (Springer, Berlin, 1997), Chap. 19, pp. 284–290.

CHAPTER 14

CONSTANT FLUX STATES AND THEIR APPLICATIONS

MARTIN CLAASSEN and HAKAN E. TÜRECİ

Institute for Quantum Electronics,
Eidgenössische Technische Hochschule-Zürich,
CH-8093 Zürich, Switzerland
tureci@phys.ethz.ch

Constant flux states describe the steady state response of a photonic medium with an arbitrary, possibly frequency dependent index of refraction $n(x,\omega)$ to a harmonically oscillating source. Constant flux states were initially introduced to describe steady state oscillations of complex lasers. Here, we describe their application to various phenomena in photonics and quantum optics.

1. Introduction

The description of many phenomena in photonics relies on the idea of normal modes. The concept of normal modes is so powerful because it provides conceptually and numerically the most economical way to describe light-matter interactions. For instance, both the semiclassical laser theory and quantum optics rely heavily on the expansion of the radiation field in normal modes that are suitable for the given optical structure, be it a "cavity" or a continuous medium. Despite the ubiquitousness of normal mode description, when it comes to the description of open photonic structures, there are a number of alternative descriptions of what normal modes of the system should be. In open structures, the radiation can escape to infinity from the cavity, requiring either a non-hermitian description via (discrete) quasi-modes of a finite system[1] or a hermitian description via the (continuous) modes of the universe of an infinite system[2]. Both descriptions have their advantages and disadvantages: While modes of the universe provide a mathematically consistent framework thanks to the complete understanding of the spectral problem of hermitian operators, they fail to focus on the details of the cavity; quasi-modes on the other hand use the cavity structure as a point of departure but in general yield a problematic mathematical framework. We will focus here on quasi-modes, because they provide the most effective description and show that a consistent mathematical and computational framework can be reached with Constant Flux (CF) modes. Our aim in this chapter is to illustrate how seeming mathematical pathologies of quasi-modes are related to the time-independent (frequency-domain) language that is employed many times, and that everything falls into place when one considers the underlying problem that is *fundamentally* time-dependent, due to omnipresence of matter (sources) which emits or absorbs photons.

The structure of this chapter is as follows: In Section 2, we will analyze in depth the source-field problem in the most simple setting for a one-dimensional (1D) dielectric cavity to illustrate the underlying physics. This system has the advantage that it contains a number of the generic features of open systems, while being mathematically transparent. In Section 3 we will discuss two applications of CF states, in semiclassical laser theory and quantum optics, respectively.

2. Source-field solutions for 1D cavity

Consider the field measured at x at time t, $e(x,t)$, in response to a point source oscillating harmonically at the frequency Ω located at x' inside a one-dimensional dielectric cavity

$$\left[\partial_x^2 - n^2(x)\,\partial_t^2\right] e(x,t) = f(x,t) \tag{1}$$

with $f(x,t) = \delta(x - x')\mathrm{e}^{-i\Omega t}$. The "cavity" is defined by the discontinuity of the index of refraction (for simplicity we assume here a dispersionless medium) $n(x) = n_0$ for $0 < x < a$, and $n(x) = 1$ for $x > a$. Without a loss in generality, we assume that the cavity is terminated by a perfectly reflecting mirror at $x = 0$, thus $e(x = 0, t) = 0$. The dielectric continuity conditions at $x = a$ are given by the continuity of $e(x,t)$ and $\partial_x e(x,t)$. Finally, the radiation generated by the source has to be flowing to spatial infinity, a condition expressed by outgoing boundary conditions for $x \to \infty$, given by $\partial_x e(x,t) = -\partial_t e(x,t)$. We set $c = 1$, so that outside the cavity the field $e(x,t)$ has to oscillate *in space* at the wavevector $k = \Omega$ due to linear dispersion of vacuum. Fourier transforming Eq. (1) in time, we obtain

$$\left[\partial_x^2 + n^2(x)\,\Omega^2\right] \tilde{e}(x,\Omega) = \delta(x - x') \tag{2}$$

Consider the *auxiliary* generalized eigenvalue problem of the Laplace operator $\mathcal{L} = -\partial_x^2$

$$-\partial_x^2\, \varphi_m(x,\Omega) = n^2(x)\, \omega_m^2(\Omega)\, \varphi_m(x,\Omega) \tag{3}$$

Note that we do not have to assign any physical meaning to these eigensolutions at this stage; they have to be calculated to solve the inhomogeneous problem (2), and hence by Fourier transform Eq. (1), which is the actual physical problem at hand.

The Ω-dependence of the eigenfrequencies $\omega_m(\Omega)$ and eigenmodes $\varphi_m(x,\Omega)$ are due to the outgoing boundary conditions $\partial_x \varphi_m(x = \infty) = i\Omega\, \varphi_m(x = \infty)$ the eigenfunctions have to satisfy. The solutions are simply trigonometric functions

$$\varphi_m(x,\Omega) = \sin(n_0\omega_m(\Omega)x) \qquad\qquad x < a \qquad\qquad (4)$$

$$= \sin(n_0\omega_m(\Omega)a)e^{i\Omega(x-a)} \qquad\qquad x > a \qquad\qquad (5)$$

The eigenfrequencies $\omega_m(\Omega)$ are found through the characteristic equation

$$\tan(n_0\omega_m a) = -i\frac{n_0\omega_m}{\Omega} \qquad\qquad (6)$$

and always have a non-zero imaginary part. In Fig. 1 we show the parametric dependence on Ω of the eigenfrequencies $\omega_m(\Omega)$.

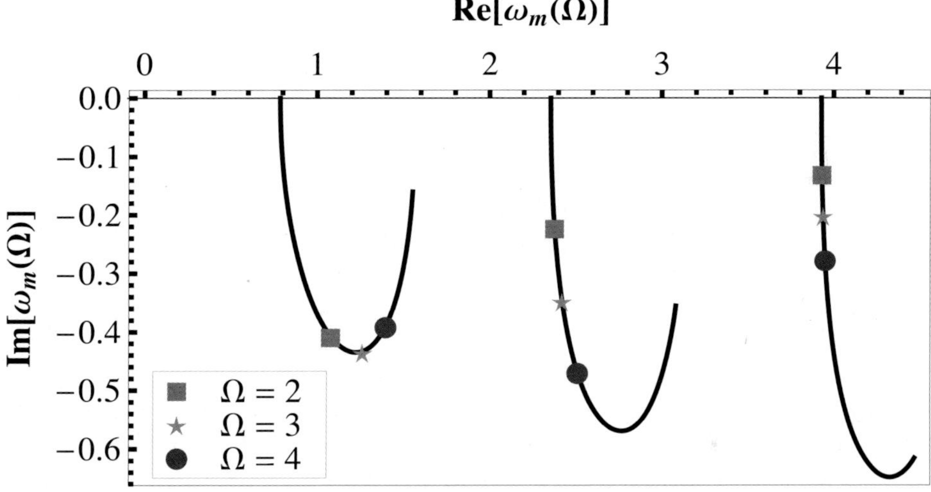

Fig. 1. Parametric dependence of the CF eigenvalues $\omega_m(\Omega)$. Solid lines show the variation of $\omega_m(\Omega)$ for $m = 1, 2, 3$ as Ω is varied in the interval $(0, 5)$. Note that all the eigenvalues are in the lower half of the complex plane. We sample three of the points with the three dots (rectangle, circle, star) on each ω_m trace. For this plot, $a = 1$ and $n_0 = 2$.

In order to construct the solution to Eq. (1), we also need to solve the adjoint spectral problem. This is defined by the adjoint operator $\mathcal{L}^\dagger = \mathcal{L} = -\partial_x^2$ and the replacement $n(x) \to n^*(x)$ in the generalized eigenvalue equation, defining an adjoint function space $\{\bar\varphi(x)\}$. Additionally, the boundary condition has to be replaced by the adjoint boundary condition $\partial_x\bar\varphi(x = \infty, \Omega) = -i\Omega\bar\varphi(x = \infty, \Omega)$. The adjoint eigenfunctions are found to be

$$\bar\varphi_m(x,\Omega) = \sin(n_0^*\bar\omega_m(\Omega)x) \qquad\qquad x < a \qquad\qquad (7)$$

$$= \sin(n_0^*\bar\omega_m(\Omega)a)e^{-i\Omega(x-a)} \qquad\qquad x > a \qquad\qquad (8)$$

The dual eigenfrequencies are found through the adjoint characteristic equation

$$\tan(n_0^*\bar\omega_m a) = i\frac{n_0^*\bar\omega_m}{\Omega} \qquad\qquad (9)$$

By comparing to Eq. 6, one can easily see that $\bar{\omega}_m = \omega_m^*$ and $\bar{\varphi}_m = \varphi_m^*$. It can then be straightforwardly shown that

$$\langle\langle \varphi_m | \varphi_n \rangle\rangle = \int_0^a dx\, n^2(x) \bar{\varphi}_m^*(x, \Omega) \varphi_n(x, \Omega) = \delta_{mn} \eta_n(\Omega) \tag{10}$$

Thus the usual orthogonality relation only exists between the eigenfunctions of adjoint spaces.

The solution of (2) is the Green's function $\tilde{e}(x, \Omega) = G(x, x'; \Omega)$, which can then be constructed by the solutions of the auxiliary eigenvalue problem and its adjoint:

$$G(x, x'; \Omega) = \sum_m \frac{\varphi_m(x, \Omega) \bar{\varphi}_m^*(x', \Omega)}{\eta_m(\Omega)(\Omega^2 - \omega_m^2(\Omega))} \tag{11}$$

The normalization factor is given by

$$\eta_m(\Omega) = \frac{an_0^2}{2}\left(1 - \frac{\sin(2n_0\omega_m(\Omega)a)}{2n_0\omega_m(\Omega)a}\right) \tag{12}$$

Finally, the solution $e(x, t)$ for an *arbitrary* time-dependent source term $f(x, t)$ can be calculated using the Green's function found above:

$$e(x, t) = \int_0^a dx' \int_{-\infty}^{\infty} d\Omega' \int_{-\infty}^{\infty} dt'\, e^{-i\Omega'(t-t')} G(x, x'; \Omega')\, f(x', t') \tag{13}$$

To understand the physical meaning of the modes φ_m consider a source distribution in Eq. (1) of the form $f(x, t) = \varphi_m(x, \Omega)e^{-i\Omega t}$ inside the cavity. The response to such a source distribution, using Eq. (13), is

$$e(x, t) = \frac{\varphi_m(x, \Omega)}{\Omega^2 - \omega_m^2(\Omega)} e^{-i\Omega t} \tag{14}$$

showing that the states $\varphi_m(x, \Omega)$ correspond to special spatial distributions of harmonically oscillating sources at frequency Ω in the cavity ($0 < x < a$) that produce a field proportional to themselves. Another important feature of these modes is that the modes $\varphi_m(x, \Omega)$ defined here carry a constant flux to infinity ($j \sim \text{Im}\,[\varphi_m(x)\partial_x \varphi_m^*(x)] = \Omega\,|\sin(n_0\omega_m(\Omega)a)|^2$ for $x > a$). These modes are therefore termed *Constant Flux (CF) modes*.

Note that the definition of a harmonic source implies that the sources are switched on in the dim past, at $t \to -\infty$, so that there is no transient response. Consider now a pulse-source of the form $f(x, t) = \delta(x-x')\delta(t)$, the resulting impulse-response can be obtained via Eq. (13):

$$e_\delta(x, t) = \int_{-\infty}^{\infty} d\Omega\, e^{-i\Omega t}\, G(x, x'; \Omega) \tag{15}$$

Inspecting the Green's function (11) we find two sets of poles. First, the factor $[\Omega^2 - \omega_m^2(\Omega)]^{-2}$ has poles at $\Omega = +\omega_m(\Omega)$, which we denote by ω_m^{QB}; these are the *Quasi-bound (QB) modes* of the system[1]. These poles lie in the lower half

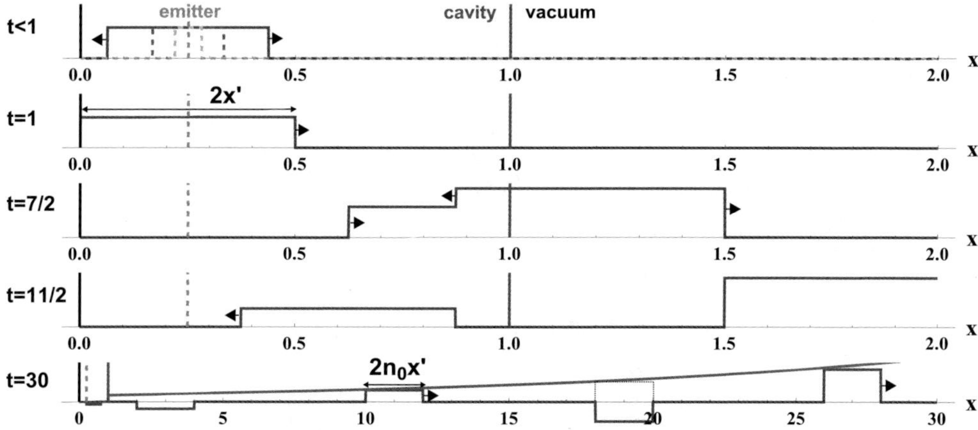

Fig. 2. Pulse propagation excited by an impulse of the form $f(x,t) = \delta(x - x')\delta(t)$. Vertical line at $x = 0$ is a perfectly reflecting mirror, vertical line at $x = a = 1$ denotes the dielectric interface and the vertical dashed line denotes position of the source at $x' = 1/4$. Thick arrows attached to pulse fronts denote their respective propagation directions. Thick asymptote in the lowest panel depicts the exponential decay envelope. Note that the lowest panel is rescaled to display all the pulses and the dotted boxes show the absolute value of the pulse amplitudes. For this plot, $a = 1$ and $n_0 = 2$.

plane - note that no solution exists for $\Omega = -\omega_m(\Omega)$. The poles ω_m^{QB} can be calculated analytically and are given by[3]:

$$\omega_m^{QB} \equiv \nu_m - i\kappa = \frac{1}{n_0 a}\left[\pi(m + 1/2) - \frac{i}{2}\ln\left(\frac{n_0 + 1}{n_0 - 1}\right)\right] \tag{16}$$

Second, the Green's function has additional poles at the zeros of the normalization factor $\eta_m(\Omega)$ which are in the lower half-plane and symmetrically distributed around the imaginary Ω-axis.

One widely encountered curiosity is that the QB modes $\varphi_m^{QB}(x) = \varphi_m(\omega_m^{QB}, x)$ blow up at $x = \infty$ as $e^{\kappa x}$; naively, one might expect that this divergence reflects the expectation that in the long-time limit all energy will have accumulated at $x = \infty$ and that it doesn't make sense to consider $e(x,t) \propto e^{\kappa(x-t)}$ (for $x > a$) beyond the 'causality cone' $x > t$. However, a more careful analysis of the integral (15) shows that through a delicate destructive interference of the terms in Eq. (11), the divergence at $x = \infty$ is removed and that the causality need not be imposed by hand.

Calculating the integral (15) by using appropriate contour integration and ignoring the poles resulting from zeros of $\eta_m(\Omega)$, we obtain for $x > a$

$$e_\delta(x,t) = -\theta(t - x + a - n_0(a - x'))\frac{2\pi i}{a}\sum_m \frac{\varphi_m^{QB}(x)(\bar{\varphi}_m^{QB}(x'))^*}{\omega_m^{QB}}e^{-i\omega_m^{QB}t}$$

$$= \theta(t - x + a - n_0(a - x'))\frac{2\pi}{a}\int_{-\infty}^{t} d\tau \sum_m \varphi_m^{QB}(x)(\bar{\varphi}_m^{QB}(x'))^* e^{-i\omega_m^{QB}\tau}$$

$$= \theta(t - x + a - n_0(a - x'))\frac{2\pi n_0}{n_0 + 1}\sum_k (-1)^k \left(\frac{n_0 - 1}{n_0 + 1}\right)^k$$

$$\left[\theta\left(t - (x - a) - n_0(a + x') - 2kn_0 a\right)\right.$$

$$\left. - \theta\left(t - (x - a) - n_0(a - x') - 2kn_0 a\right)\right] \tag{17}$$

Here, $\theta(x)$ is the Heaviside step-function. The step-function outside the sum results from the choice of closing the integration contour either in the lower or upper-half of the complex Ω-plane. Note that due to this prefactor, the entire function is non-zero only for $t > L(x, x')$ and that $L(x, x') = (x - a) + n_0(a - x')$ is the *direct* (without intermediate reflections) optical path from the source to the observation point. The factor $r = (n_0 - 1)/(n_0 + 1)$ in the summation is immediately recognized as the Fresnel reflection coefficient, its powers r^k yielding the correct amplitude for a multiply reflected pulse. Figure 2 displays snapshots of the propagating pulse for different times. We find a pulse that initially spreads in both directions from the source at x'. Upon the left front reaching the reflecting boundary at $x = 0$ (assuming $x' < a/2$), the rectangular pulse remains at a width $2x'$ and starts to propagate towards the interface at $x = a$. Reaching it, the pulse partially reflects back into the cavity, and partially transmits a pulse of width $2n_0 x'$ to infinity. The reflected pulse propagates back towards the perfect mirror, at which the left front reflects and annihilates the pulse exactly at position x'. The field stays not zero at this point however, as the first derivative of the field amplitude will cause a new pulse to start expanding at x' but with negative amplitude. A camera in the far field would observe a train of pulses of width $2n_0 x'$ of decaying amplitudes and alternating sign (assuming a phase-sensitive detector). The envelope of the pulse-amplitudes decays as $e^{-\kappa t}$ starting with the first pulse that arrives at x a time $\Delta t = x - a + n_0(a - x')$ after it is emitted at x'.

This at first sounds somewhat odd when we think of pulse propagation in uniform media with a frequency independent index of refraction, which we expect to be non-dispersive. The initial pulse $\delta(x - x')\delta(t)$ spreads and seems to stop spreading once the reflecting boundary is reached, reflecting back and forth after that. Such an expectation is based on our intuition about wave-propagation in three-dimensional (3D) space. In reduced dimensions, because of the effective sources being extended (i.e. a point source in two dimensions is physically a line-source in 3D and so on), the fundamental solutions are different than those in 3D. While the 3D solution is an expanding spherical *shell* of electric field with its center at the source, the 2D solution is a front that has a trailing tail that decays as a power-law all the way back to the source, and the 1D solution is simply a uniform field with fronts on both ends expanding[4]. That a pulse started in a cavity described by Eq. (17) stops spreading once the reflecting wall is reached can be simply understood by utilizing the method of images: image sources located periodically at $x = -x'$, $x = -x' - 2a$, $x = -2x' - 2a$, $x = -2x' - 4a$, and so forth, do also emit spreading pulses that cancel the physical pulse at $x = 0$ to satisfy the boundary condition there. The net effect is that the pulse width after reflection from $x = 0$ remains constant.

In conclusion, we see that QB modes are relevant only in describing transient response while CF states form a more general basis to describe any type of source; in fact we have derived (17) starting with the CF representation of the Green's function (11). In the Appendix, we show that the impulse response calculated above fully agrees with what is obtained starting from the exact Green's function. The latter can be calculated in a closed form in the 1D case.

Note that the CF modes in Eq. (4) all peak at the open interface. This can be easily understood in a semiclassical description: We want the ray amplitude and phase to be conserved through a roundtrip (the phase up to 2π) and there is an amplitude loss at the open interface. This is only possible if the amplitude increases towards that interface. If we had two open interfaces, we would have a reflection symmetric distribution with respect to the center of the cavity with higher amplitude at both interfaces.

Finally, we would like to highlight a physically appealing and useful interpretation of the CF modes. The quantization condition Eq. 6 can alternatively be thought in the following way. Define a frequency dependent effective index of refraction $n_m(\Omega) = n_0 \omega_m / \Omega$ which is complex. Then the quantization condition can be rewritten as

$$\tan(n_m(\Omega)\,\Omega\,a) = -i n_m(\Omega) \tag{18}$$

This is the equation for the quasi-bound modes of the system, with a frequency dependent index of refraction. Starting with the discrete quasi-bound modes enumerated by m for a constant index of refraction n_0, imagine tuning the real and imaginary part of the index of refraction to $n_m(\Omega) = n_m^R(\Omega) + i n_m^I(\Omega)$ so that we impart precisely the amplification and dispersion necessary for the mode m to oscillate at the stationary frequency Ω. The corresponding solutions are exactly the CF modes $\varphi_m(x, \Omega)$. In this manner, at each Ω we can construct a discrete set of modes which are continuously related to the original quasi-bound modes of the cavity with a constant index of refraction.

3. Applications of CF states

We will focus in this section on two applications of CF states in the semiclassical laser theory and quantum optics. For transparency of presentation we will focus in this section again on 1D cavities.

3.1. *Steady-state semiclassical laser theory*

The starting point of semiclassical laser theory is the set of Maxwell-Bloch (MB) equations, which describe the electric and magnetic fields using Maxwell's equations, with a source term (polarization) that is generated by a set of uniformly distributed two-level emitters, described quantum mechanically. Using the rotating wave and stationary inversion approximations[3] the MB equation can be reduced to the following source-field problem:

$$\left[\partial_x^2 - n^2(x)\partial_t^2\right] e(x,t) = f(x,t) \tag{19}$$

where $e(x,t) = \sum_\mu \Psi_\mu(x)\mathrm{e}^{-i\Omega_\mu t}$ and

$$f(x,t) = -\frac{D_0(x)}{1 + \sum_\nu \Gamma_\nu |\Psi_\nu(x)|^2} \sum_{\mu=1}^{N} \left(\frac{\Omega_\mu}{\omega_a}\right)^2 \frac{\Psi_\mu(x)\,\mathrm{e}^{-i\Omega_\mu t}}{-i(\Omega_\mu - \omega_a) + \gamma_\perp}$$

Here, $D_0(x)$ measures the pumped energy in terms of the inversion generated per unit volume in the absence of gain saturation ($e = 0$), $\Gamma_\nu = \Gamma(\Omega_\nu) = \gamma_\perp^2 / \left[(\Omega_\nu - \omega_a)^2 + \gamma_\perp^2\right]$ is the laser gain-curve, ω_a and $\gamma_\perp$ are the two-level emitter resonance frequency and the associated homogeneous broadening, respectively. This problem is very similar to the problem given by Eq. (1) discussed in Section 2, with one notable difference: the source term is dependent on the field $e(x,t)$ itself via the unknown non-linear lasing modes $\Psi_\mu(x)$ and the (real) laser frequencies Ω_μ. This makes the problem Eqs. (19)-(20) non-linear requiring a self-consistent solution technique that we describe below to determine the solutions.

The formal solution is obtained by using the Green's function methods of Section 2 for harmonically oscillating sources, Eq. (1), via Eq. (13):

$$\Psi_\mu(x) = \frac{i\gamma_\perp}{-i(\Omega_\mu - \omega_a) + \gamma_\perp} \frac{\Omega_\mu^2}{\omega_a^2} \int_D dx' \frac{D_0(x')G(x,x';\Omega_\mu)\Psi_\mu(x')}{1 + \sum_\nu \Gamma_\nu |\Psi_\nu(x')|^2} . \tag{20}$$

This set of non-linear integral equations for lasing modes $\Psi_\mu(x)$ and the laser frequencies Ω_μ constitute the *Steady-state Ab-initio Laser Theory* (SALT) first described in Ref. [3]. Note that the integral is limited to the domain D, within which the gain medium is located; for simplicity we assume that the pump is uniform across the entire cavity $0 < x < a$ i.e. $D_0(x) = D_0\theta(a - x)\theta(x)$. In contrast to standard approaches, cavity losses enter this equation through the Green's function and ultimately, as we have seen in Section 2, through the boundary conditions at infinity. Below we briefly outline the solution technique and a powerful numerical algorithm to determine the multimode lasing solutions at any pump-rate D_0.

The relations found in Section 2 make it possible to expand an arbitrary lasing solution in the form

$$\Psi_\mu(x) = \sum_{m=1}^{\infty} a_m^\mu \varphi_m^\mu(x) , \tag{21}$$

so that each $\Psi_\mu(x)$ is defined by the vector of complex coefficients $\boldsymbol{a}^\mu$ in the space of CF states. In what follows we will use the shorthand $\varphi_m^\mu(x) \equiv \varphi_m(x, \Omega_\mu)$ and

$\omega_m^\mu \equiv \omega_m(\Omega_\mu)$. Because only CF states with frequencies near the center of the gain curve contribute to the lasing state, it is possible to truncate the sum in Eq. (21) to a finite number (N) of components, making $\boldsymbol{a}^\mu$ a finite dimensional vector. By substitution of Eq. (21) into Eq. (20) and use of the biorthogonality relations (10) one finds:

$$a_m^\mu = \frac{iD_0\gamma_\perp}{(\gamma_\perp - i(\Omega_\mu - \omega_a))} \frac{(\Omega_\mu/\omega_a)^2}{(\Omega_\mu^2 - \omega_m^{\mu\,2})} \int_\mathcal{D} dx' \frac{\bar{\varphi}_m^{\mu*}(x') \sum_p a_p^\mu \varphi_p^\mu(x')}{1 + \sum_\nu \Gamma_\nu |\Psi_\nu(x')|^2} . \tag{22}$$

This is the form of Eq. (20) that is employed in SALT for finding the lasing modes and frequencies. As discussed above, it reduces the problem to finding the complex vector of coefficients $\boldsymbol{a}^\mu$ and the frequency Ω_μ for each lasing mode, which depends non-linearly on all the other lasing modes and itself through the infinite order non-linearity evident in the denominator of equation (22). Here we will focus on the near-threshold solutions of this set of equations. The details of the technique for determining the multimode lasing solutions for any pump rate D_0 is described in Ref. 5.

For a pump rate below the first lasing threshold $D_0 < D_{th}$, only the trivial solution exists: $\Psi_\mu = 0$, for all μ. This is the non-lasing solution i.e. a standard lamp, which because of the neglect of the quantum fluctuations appears to have zero amplitude. Very close to but above the first threshold $D_0 = D_{th} + \epsilon$, we may neglect the term $\sum_\nu \Gamma_\nu |\Psi_\nu(x')|^2$ in the denominator of Eq. (22). The resulting set of equations can then be compactly written in the form of an eigenvalue equation

$$\boldsymbol{\mathcal{T}}^{(0)}(\Omega)\boldsymbol{a}^\mu = (1/D_0)\boldsymbol{a}^\mu . \tag{23}$$

where $\Lambda_m(\Omega) = i\gamma_\perp(\Omega^2/\omega_a^2)/[(\gamma_\perp - i(\Omega - \omega_a))(\Omega^2 - \omega_m^2(\Omega))]$ and $\boldsymbol{\mathcal{T}}^{(0)}(\Omega)$ is a matrix

$$\mathcal{T}_{mn}^{(0)}(\Omega) = \Lambda_m(\Omega) \int dx' \bar{\varphi}_m^*(x', \Omega) \varphi_n(x', \Omega) \tag{24}$$

parametrically dependent on Ω and acting on the properly truncated N-dimensional vector space of complex amplitudes $\boldsymbol{a}^\mu = (a_1^\mu, a_2^\mu, \ldots, a_N^\mu)$. This equation can in general not be satisfied for a *real value* of D_0 except at discrete values of Ω. $\boldsymbol{\mathcal{T}}^{(0)}(\Omega)$ is a non-hermitian matrix and has N complex eigenvalues $\lambda_n(\Omega)$ for general values of Ω. As Ω is varied, the eigenvalues $\lambda_n(\Omega)$ flow in the complex plane (see Fig. 3), each one crossing the positive real axis at a specific Ω_n, determined by $\text{Im}[\lambda_n(\Omega = \Omega_n)] = 0$. The modulus of the eigenvalue defines the "noninteracting" lasing threshold corresponding to that eigenvalue, $D_{th}^{(n)} = 1/\lambda_n(\Omega_n)$, the real frequency Ω_n is the non-interacting lasing frequency, and the eigenvector $\boldsymbol{a}^n$ gives the "direction" of the lasing solution in the space of CF states. Among these solutions, the smallest $D_{th}^{(1)}$ (i.e., the largest of the real eigenvalues λ_n) gives the actual threshold for the first lasing mode; the frequency Ω_1 is the lasing frequency at threshold and the eigenvector $\boldsymbol{a}^1$ defines "direction" of the lasing solution at threshold. The "length" of $\boldsymbol{a}^1$ cannot be determined from the linear equation (23) but rises continuously from zero at threshold and is determined by the non-linear equation (22) infinitesimally above threshold. As noted, the remaining real eigenvalues of $\boldsymbol{\mathcal{T}}^{(0)}(\Omega)$

define the non-interacting thresholds for other modes, however the actual thresholds of all higher modes will differ substantially from their non-interacting values due to the non-linear term in (22) which now comes into play. The actual lasing frequencies of higher modes have a relatively weak dependence on D_0 and differ little from their non-interacting values. In Fig. 3 we show the flow of the eigenvalues of $\mathcal{T}^{(0)}(\Omega)$ for the case of a 1D cavity, with the non-interacting thresholds and lasing frequencies indicated.

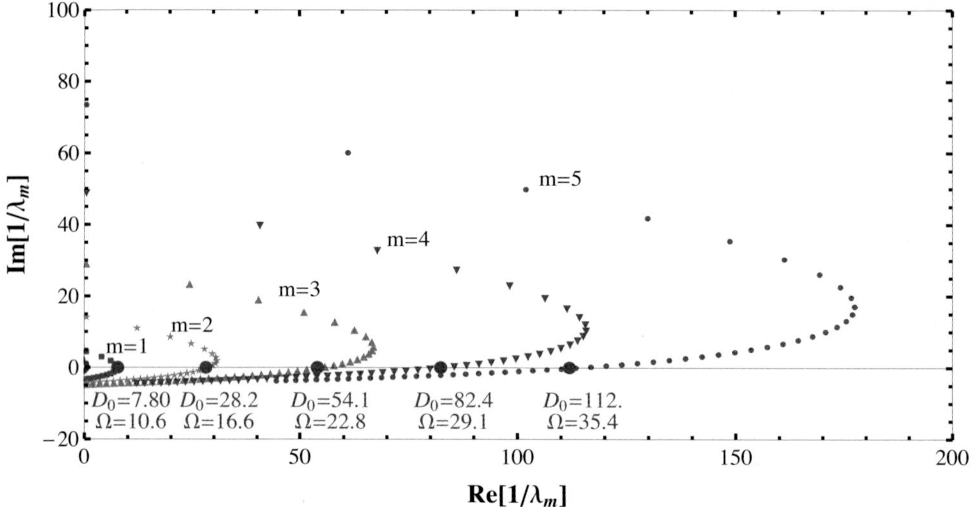

Fig. 3. Linearized lasing thresholds. Solution of the linearized eigenvalue problem (23) displays the flow of eigenvalues D_0 in the complex plane. The smallest *real* D_0 corresponds to the actual lasing threshold. For this plot, $\gamma_\perp = 0.1$, $\omega_a = 1$, $a = 1$ and $n_0 = 2$.

SALT has been successfully used to calculate the steady-state lasing characteristics of arbitrary 1D cavities, 2D uniform dielectric cavities of general shape, and 2D disordered media embedded in a disk-shaped gain medium[6].

3.2. *Calculation of Purcell factors for open cavity structures*

Spontaneous emission is a ubiquitous phenomenon that is responsible for most of the light around us. Any optically active emitter that is prepared in high energy states decays to lower energy states by emitting a photon through the action of various sources of noise. The most basic model of spontaneous emission considers the case of a two-level system (TLS) where the noise is caused by quantum fluctuations of electromagnetic vacuum that causes transitions between the two levels. In the perturbative regime, the spontaneous emission rate Γ_{sp} can be calculated via the Fermi Golden rule and related to the local density of states (LDOS) of photons at the emitter position x_e, $\rho(x_e, \omega_a)$

$$\Gamma_{sp} = \frac{\pi \omega_a \mathcal{P}_{12}^2}{\hbar \epsilon_0} \rho(x_e, \omega_a) \tag{25}$$

Here $\mathcal{P}_{12}$ is the electric dipole moment and ω_a is the transition frequency of the TLS. In free space (in three dimensions) $\rho = \omega_a^2/3\pi^2$ and we get the vacuum spontaneous emission rate

$$\Gamma_{sp}^{vac} = \frac{\omega_a^3 \mathcal{P}_{12}^2}{3\pi \hbar \epsilon_0} \tag{26}$$

It was Purcell[7] who pointed out that the spontaneous emission rate of an emitter can be dramatically modified in the presence of a cavity. The action of cavity can simply be thought of as modifying the LDOS of photons. Note that although spontaneous emission itself is due purely to the quantum nature of light, the spontaneous emission rate can be related to a purely classical quantity, the LDOS, which in turn can be related to the Green's function of the wave equation that we calculated in Section 2:

$$\rho(x_e, \omega_a) = \frac{2\omega_a}{\pi} \mathrm{Im}\left[G(x_e, x_e; \omega_a)\right] \tag{27}$$

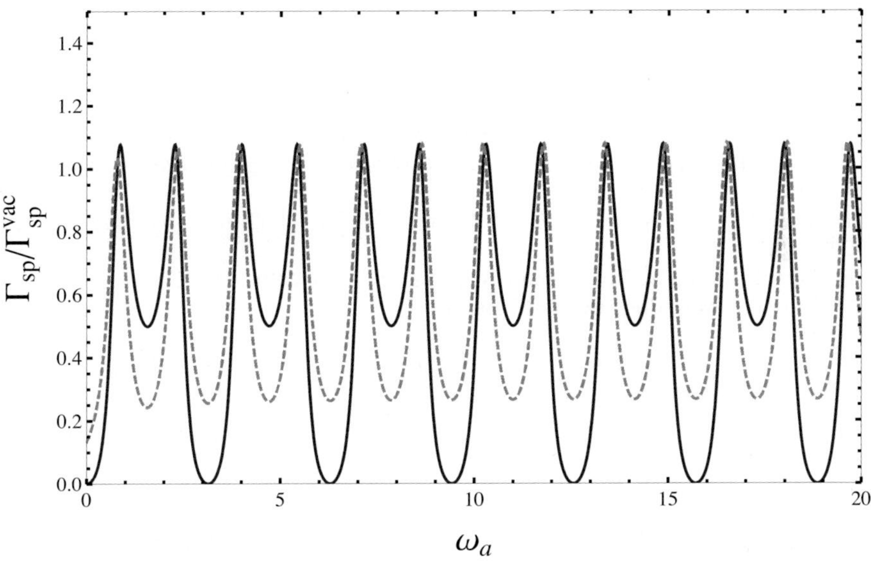

Fig. 4. Purcell enhancement factor for an emitter with a transition frequency at ω_a. The enhancement factor, plotted as a solid line, is shown for ω_a varying within a range $(0, 20)$. The emitter is located at $x_e = a/2$. The dashed line shows a comparison to a fit by a series of Lorentzian functions $\sum_m \kappa_m/((\omega_a - \nu_m)^2 + \kappa_m^2)$ with central frequencies and decay rates given by those of the QB modes, Eq. (16). For this plot, $a = 1$ and $n_0 = 2$.

Once the CF modes of a cavity are determined, the spontaneous emission rate can be calculated readily using this expression. For illustration, we calculate in

Fig. 4 the Purcell enhancement, $\Gamma_{sp}/\Gamma_{sp}^{vac}$ for a one-dimensional dielectric cavity. One striking feature in this plot is the strongly non-Lorentzian form of each of the peaks (to avoid confusion, we emphasize that what is calculated and plotted is *not* the spectrum). This *does not* derive from interference between various terms of the spectral expansion in Eq. (27); it's present in each individual term contributing to (27). More significantly, individual terms in the sum typically change sign as ω_a is varied (due to the complex valued CF functions in the numerator), however the sum total remains positive.

In conclusion, Purcell enhancement factors for arbitrary photonic structures, including plasmonic systems, can straightforwardly be calculated once the CF modes are determined.

Acknowledgments

We would like to acknowledge discussions with Matthias Liertzer, Stefan Rotter, Christian Vanneste and support from Swiss NSF under Grant No. PP00P2-123519/1.

References

1. E. S. C. Ching, P. T. Leung, A. M. van den Brink, W. M. Suen, S. S. Tong, and K. Young, *Rev. Mod. Phys.* **70**, 1545 (1998).
2. R. Lang, M. O. Scully, and W. E. Lamb, *Phys. Rev* **A7**, 1788 (1973).
3. H. E. Türeci, A. D. Stone, and B. Collier, *Phys. Rev.* **A74**, Art. No.043822 (2006).
4. P. M. Morse and H. Feshbach, *Methods of Theoretical Physics, Part I* (McGraw-Hill, New York, NY, USA, 1953).
5. H. E. Türeci, A. D. Stone, L. Ge, S. Rotter, and R. J. Tandy, *Nonlinearity* **22**, C1 (2009).
6. H. E. Türeci, L. Ge, S. Rotter, and A. Stone, *Science* **320**, 643 (2008).
7. E. M. Purcell, *Phys. Rev.* **69**, 681 (1946).

Appendix

Here, we compare the CF Green's function Eq. (17) to the exact Green's function

$$\left[\frac{\partial^2}{\partial x^2} + n^2(x)\Omega^2\right] G(x, x'; \Omega) = \delta(x - x') \tag{28}$$

that can be obtained in analytical form for the 1D case. In addition to the conditions at $x = 0$, $x = a$ and $x = \infty$, we have the additional conditions $G(x' - 0^+, x'; \Omega) = G(x' + 0^+, x'; \Omega)$, $\partial G(x, x'; \Omega)/\partial x|_{x=x'-0^+} - \partial G(x, x'; \Omega)/\partial x|_{x=x'+0^+} = 1$. The resulting Green's function is given by

$$G(x, x'; \Omega) = -\frac{1}{\Omega} \frac{\sin(n_0 \Omega x') e^{i\Omega(x-a)}}{\Delta(\Omega)} \tag{29}$$

outside the cavity ($x > a$). Here

$$\Delta(\Omega) = n_0 \cos(n_0 \Omega a) - i \sin(n_0 \Omega a) \tag{30}$$

The field response to an impulse-source is given by Eq. (15). To calculate this integral, we analyze the analytic structure of the integrand. There is no pole at $\Omega = 0$, because the singularity is canceled by the trigonometric functions in the numerator. All the poles are the zeros of $\Delta(\Omega)$ and are given by $\tan(n_0 \Omega a) = -i n_0$; these are identical to Eq. (16). The final expression for the impulse-response is given by

$$\begin{aligned}
e_\delta(x, t) = \theta(t - x + a - n_0(a - x')) \frac{2\pi n_0}{n_0 + 1} \sum_k (-1)^k \left(\frac{n_0 - 1}{n_0 + 1}\right)^k \\
\left[\theta\left(t - (x - a) - n_0(a + x') - 2kn_0 a\right)\right. \\
\left. -\theta\left(t - (x - a) - n_0(a - x') - 2kn_0 a\right)\right]
\end{aligned} \tag{31}$$

This is exactly the CF result Eq. (17).

CHAPTER 15

ELECTRO-OPTICAL APPLICATIONS OF HIGH-Q CRYSTALLINE WGM RESONATORS

VLADIMIR S. ILCHENKO, ANDREY B. MATSKO, ANATOLIY A. SAVCHENKOV, and LUTE MALEKI

OEwaves Inc., 2555 E. Colorado Blvd., Ste. 400, Pasadena, California, 91107, U.S.A.
vladimir@oewaves.com

We review results of our recent research and development efforts devoted to application of electro-optical crystalline whispering gallery mode resonators in the area of microwave photonics. In particular, we describe deployment of these resonators in functional radio frequency signal processing devices operating at frequencies laying in microwave and millimeter-wave bands. We discuss demonstrations of quadratic photonic detectors, coherent quadratic receivers, terahertz receivers, optical pulse generators and parametric converters, as well as examples of small footprint packaging prototypes of these devices.

1. Introduction

Studies of high-Q optical whispering gallery modes (WGMs) (morphology-dependent resonances) in liquid droplets, pioneered by Professor Richard K. Chang,[1–9] initiated the development of a new field in optics and photonics devoted to WGM resonators (for review see Refs. 10–15). Later, studies of optical WGMs in solidified droplets of amorphous materials, such as fused silica,[16–20] have proven that the optical whispering gallery modes are not only a subject of laboratory studies, or a tool to describe and explain peculiar scattering phenomena in aerosol media, but can also be deployed in practical applications in optics, and as photonic devices.[21,22]

Ultra–high-Q WGM microresonators are commonly used for enhancement of efficiency of nonlinear optical processes such as three–[23,24] and four–wave mixing.[25–28] Strong spatial confinement of light, as well as a long interaction times of light with the nonlinear medium of the resonator host material, results in a substantial reduction of thresholds of lasing,[29,30] stimulated Raman scattering,[31–33] and stimulated Brillouin scattering.[34,35]

Spectacular new possibilities have been opened with successful demonstration of WGM resonators made with electro-optic materials. WGM resonators made out of

$LiNbO_3$ and $LiTaO_3$ have been successfully used in prototypes of functional devices for optical and microwave photonic applications. These resonators are characterized by bandwidth in the hundred kilohertz (weakly coupled) to gigahertz (fully loaded) range. Optical transparency and nonlinearity of lithium niobate and tantalate help to realize a number of high performance devices: tunable and multipole filters,[36] resonant electro-optic modulators, photonic microwave receivers, opto-electronic microwave oscillators, and parametric frequency converters. An approach to induce coupling between light and microwave fields in a WGM resonator was initially proposed in Refs. 37 and 38. Mode-matched resonant interaction of two (and more) optical WGMs with a microwave or millimeter-wave mode was achieved by optimal engineering the geometry of a micro-strip microwave resonator coupled to an optical WGM resonator. Based on this interaction, electro-optic modulators, as well as photonic microwave receivers have been realized.[39–49] Resonant electro-optic modulators with narrow modulation bandwidth that can be tuned over a wide microwave frequency range, have recently been introduced and demonstrated.[50–54]

Fabrication of optical WGM resonators with lithium niobate[36] has led to demonstration of high-Q tunable filters with a linewidth less than a megahertz. A crystalline WGM resonator has been used as a Lorentzian microwave filter with ten gigahertz tuning range.[55] A miniature resonant electro-optically tunable third-order Butterworth filter based on cascaded lithium niobate WGM resonators has also been demonstrated.[56] Optical parametric frequency conversion has been realized in a periodically poled lithium niobate (PPLN) WGM resonator.[24] A daisy-shaped poling of $LiNbO_3$ domains, optimal for WGM resonators, was also proposed[57] and realized.[58]

Solid state WGM resonators are also an important tool of fundamental studies, e.g. in material science. An advantage of crystalline WGM resonators in material studies stems from the fact that the resonator is a mechanically shaped sample carved out of a nonlinear material. Adaptation of machining and polishing techniques, known in optical industry, allows to obtain axi-symmetrical dielectric resonators with nearly molecular surface roughness — and minimal invasion in chemistry, doping and stoichiometry of source material. This provides the opportunity to investigate the intrinsic properties of the resonator material while avoiding the undesirable changes in morphology and doping content (such as those inevitable in reflow of amorphous materials). It is also important that the resonator comprises no complementary optical components — except for the dielectric itself — such as mirrors in conventional Fabry-Perot resonators. In this way, measurement of frequencies and quality-factors of the WGM gives direct information about the value of refractive index and optical attenuation in the material. As an example of such an application, resonators have been used in high resolution studies of photorefractive phenomena in lithium niobate in the infrared.[59–61]

The traditional, and ever challenging, benchmark of the state-of-the-art with fabrication of WGM resonators with intrinsic material properties is the unloaded quality-factor. In fact, with development of angstrom level surface polishing techniques (virtually eliminating scattering losses, down to values below 10^{-12}), Q-factor measurement in WGM resonators has become one of the most sensitive method of detecting and measuring small optical loss in crystals and other materials, where large propagation lengths and/or large sample dimensions are unattainable.

With the choice of available highest purity grades of commercial material, we were able to achieve Q-factors approaching 10^9 in both lithium niobate and lithium tantalate — our choice materials in electro-optic device integration. Presented in Fig. 1 is the typical absorption resonance of a lithium tantalate resonator (diameter 1.5 mm, thickness 0.1 mm), obtained with a single diamond prism coupler using a Koheras 1550 nm wavelength fiber laser with the intrinsic linewidth < 15 kHz.

In this Chapter we review some of our recent studies devoted to microwave photonic applications of WGM resonators. We focus on devices that utilize electro-optical modulation based on WGM resonators fabricated from optical crystalline materials possessing quadratic nonlinearity.

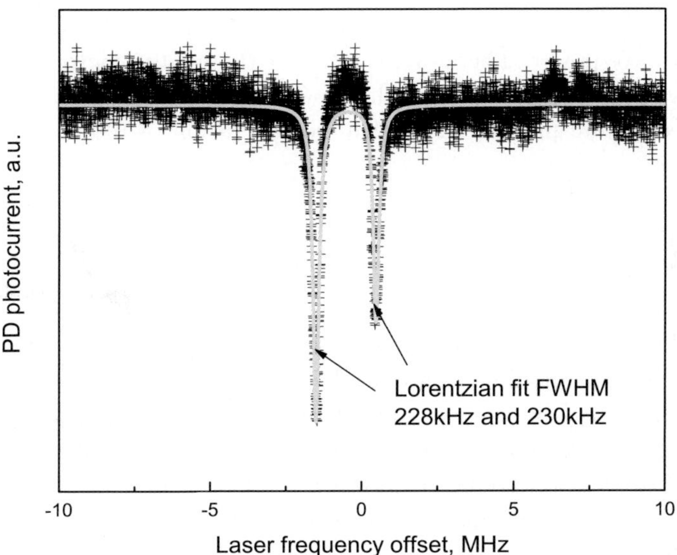

Fig. 1.   An absorption resonance of a high-Q WGM as observed in the resonator with a single prism coupler. The resonator has 1.5 mm diameter and 0.1 mm thickness. Data taken at 1550 nm with a Koheras Adjustik laser (linewidth smaller than 10 kHz), at power level <50 μW (to avoid nonlinear loss due to Raman scattering). The deduced quality factor is $Q = 0.85 \times 10^9$. Frequency splitting of the resonance (1.9 MHz) is due to the intracavity Rayleigh backscattering.

2. Theory of all-resonant electro-optical modulators

Both microwave and optical resonators are used to increase efficiency of electro-optical modulators (EOMs) within limited frequency bands around the respective microwave and optical carrier frequencies.[62–69] Electro-optically active WGM resonators are attractive for this application because they feature high quality factors at any optical frequency within the transparency window of the resonator host material.[37–47,70–76] These modulators generally operate either at the baseband (see, e.g., Ref. 73) or at bands detuned from the baseband by several free spectral ranges (FSRs) of the WGM resonator. The latter versions of the modulator have the highest efficiency at higher microwave frequencies (see, e.g., Refs. 39, 44 and 76).

Modulation in all-resonant EOMs occurs because the microwave field in an externally pumped microwave resonator interacts with modes of a nonlinear optical WGM resonator. All but one optical fields are the modulation sidebands with frequencies equal to a multiple of the microwave pump frequency. The microwave frequency is selected to correspond to the FSR of the WGM resonator so that the optical modulation sidebands coincide with WGMs. The modulation coefficient defined as the ratio of the output power of the first optical harmonic and the optical pump power is proportional to $P_{mw}Q^2 Q_M$,[45] where Q and Q_M are the loaded quality factors of the optical and microwave modes respectively, and P_{mw} is the applied microwave power. Therefore, the higher the quality factors are, the smaller the microwave power has to be for achieving the same modulation efficiency.

In this section we present theoretical consideration of resonant electro-optical modulators.

2.1. *EOM as an example of a three-wave mixer*

Let us study the interaction of three optical and a single microwave resonator modes via $\chi^{(2)}$ nonlinearity.[77] We start from a consideration of a purely theoretical conservative problem of a nonlinear interaction of modes having infinite Q-factors, and consider generation of Stokes and anti-Stokes photons occurring as a result of interaction of optical and microwave photons initially stored in the modes.

The Hamiltonian describing the interaction of three optical ($\hat{a}$, $\hat{b}_-$, and $\hat{b}_+$) and one microwave ($\hat{c}$) modes is presented as a sum of free ($\hat{H}_0$) and interaction ($\hat{V}$) parts $\hat{H} = \hat{H}_0 + \hat{V}$:

$$\hat{H}_0 = \hbar\omega\hat{a}^\dagger\hat{a} + \hbar\omega_-\hat{b}_-^\dagger\hat{b}_- + \hbar\omega_+\hat{b}_+^\dagger\hat{b}_+ + \hbar\omega_c\hat{c}^\dagger\hat{c}, \tag{1}$$

$$\hat{V} = \hbar g(\hat{b}_-^\dagger\hat{c}^\dagger\hat{a} + \hat{b}_+^\dagger\hat{c}\hat{a}) + \text{adjoint}, \tag{2}$$

where ω and $\omega_\pm$ are the eigenfrequencies of the optical modes, ω_c is the eigenfrequency of the microwave mode, $\hat{a}$, $\hat{b}_\pm$, and $\hat{c}$ are the annihilation operators for these modes respectively, g is the coupling constant given by the system geometry and nonlinearity.

We derive equations of motion for the field operators using the Hamiltonian

$$\dot{\hat{a}} = -i\omega\hat{a} - ig^*(\hat{b}_-\hat{c} + \hat{c}^\dagger\hat{b}_+), \tag{3}$$

$$\dot{\hat{b}}_- = -i\omega_-\hat{b}_- - ig\hat{c}^\dagger\hat{a}, \tag{4}$$

$$\dot{\hat{b}}_+ = -i\omega_+\hat{b}_+ - ig\hat{c}\hat{a}, \tag{5}$$

$$\dot{\hat{c}} = -i\omega_c\hat{c} - ig\hat{b}_-^\dagger\hat{a} - ig^*\hat{a}^\dagger\hat{b}_+. \tag{6}$$

Set (3-6) was solved in the case of undepleted pump in Ref. 78. However, such a solution does not show the maximum number of microwave photons that could be generated in the absence of the microwave signal.

Set (3-6) is integrable in the case of equidistant modes. Assuming that $\omega - \omega_- = \omega_+ - \omega = \omega_c$, and introducing operators $\hat{S}_+ = (g\hat{b}_-^\dagger\hat{a} + g^*\hat{a}^\dagger\hat{b}_+)\exp(-i\omega_c t)/|g|$ and $\hat{C} = \hat{c}\exp(-i\omega_c t)$ we obtain several integrals of motion

$$\hat{n}_{b+} + \hat{n}_{b-} + \hat{n}_a = \hat{N}, \tag{7}$$

$$\hat{n}_{b-} - \hat{n}_{b+} - \hat{n}_c = \hat{N}_S, \tag{8}$$

$$\hat{S}_+^\dagger\hat{S}_+ + \hat{S}_+\hat{S}_+^\dagger - (\hat{n}_{b-} - \hat{n}_{b+})^2 = \hat{N}_C, \tag{9}$$

$$\hat{S}_+^\dagger\hat{C} + \hat{C}^\dagger\hat{S}_+ = \hat{N}_I; \tag{10}$$

where $\hat{n}_\xi \equiv \hat{\xi}^\dagger\hat{\xi}$ is the photon number operator.

It is easy to obtain a closed set of two equations

$$\dot{\hat{C}} = -i|g|\hat{S}_+ \tag{11}$$

$$\dot{\hat{S}}_+ = -i|g|[\hat{n}_c + \hat{N}_S]\hat{C}, \tag{12}$$

which is transformed to a second order nonlinear equation with respect to operator $\hat{C}$ as

$$\ddot{\hat{C}} + |g|^2[\hat{n}_c(t) + \hat{N}_S]\hat{C} = 0. \tag{13}$$

and an equation with respect to the operator $\hat{n}_c$ as

$$\ddot{\hat{n}}_c + |g|^2[\hat{n}_c^2 + \hat{n}_c - \hat{N}_C - \hat{N}_S^2 + \hat{N}_S] = 0. \tag{14}$$

Eq. (14) has a solution in terms of elliptic integrals. For simplicity reasons, we do not discuss this solution here, and only give an approximate solution for the expectation value of $\hat{n}_c$ in two cases. In the first case, we assume that $\langle\hat{n}_c(0)\rangle \gg \langle\hat{n}_a(0)\rangle \gg 1$ and sideband modes $b_\pm$ are initially in the vacuum state and obtain

$$\langle\hat{n}_c\rangle \approx \langle\hat{n}_c(0)\rangle + 2\frac{\langle\hat{n}_a(0)\rangle}{\langle\hat{n}_c(0)\rangle}\sin^2\left[|g|t\sqrt{\frac{\langle\hat{n}_c(0)\rangle}{2}}\right] \approx \langle\hat{n}_c(0)\rangle. \tag{15}$$

Therefore, the photon number in the mode c is nearly constant.

In the second case, if the pump mode a is in the coherent state and $\langle\hat{N}\rangle \gg 1$ and sideband modes $b_\pm$ are initially in the vacuum state, we derive

$$\langle\hat{n}_c\rangle \approx \sqrt{6\langle\hat{N}\rangle}\sin^2\left[|g|t\left(\frac{\langle\hat{N}\rangle}{3}\right)^{1/4}\right]. \tag{16}$$

This equation correctly describes the maximum of the photon number as well as the oscillation period, however it gives $\sqrt{2}$ faster growth at $t \to 0$. The correct expression is

$$\langle \hat{n}_c \rangle |_{t \to 0} \simeq \langle \hat{N} \rangle |g|^2 t^2. \tag{17}$$

Eq. (17) corresponds to the solution of a linearized problem for the photon number.[78] The general linearized solution is

$$\hat{b}^\dagger_-(t) = \hat{b}^\dagger_-(0) + ig^* a^*(\hat{c}(0)t + \tilde{c}t^2/2), \tag{18}$$

$$\hat{b}_+(t) = \hat{b}_+(0) - iga(\hat{c}(0)t + \tilde{c}t^2/2), \tag{19}$$

$$\hat{c} = \hat{c}(0) + \tilde{c}t, \quad \tilde{c} = -i(gab^\dagger_-(0) + g^* a^* \hat{b}_+(0)). \tag{20}$$

This solution results in

$$\langle \hat{n}_{b-} \rangle |_{t \to 0} \simeq \langle \hat{N} \rangle |g|^2 t^2 + \langle \hat{N} \rangle^2 |g|^4 t^4/4, \tag{21}$$

$$\langle \hat{n}_{b+} \rangle |_{t \to 0} \simeq \langle \hat{N} \rangle^2 |g|^4 t^4/4. \tag{22}$$

The basic conclusions of this idealized analysis are (i) that one can neglect the change of microwave field in the resonator and assume that the field is constant when both strong microwave and optical fields are present; (ii) that the optical sidebands as well as the microwave field can be excited from the ground state — if some photons are present in the pump mode. This, in turn, leads us to the conclusion that the presence of an optical pump will result in additional noise in the Stokes and anti-Stokes sidebands in a photonic microwave receiver based on wave interaction. To estimate the significance of this effect, we have to study an open system. This noise can be neglected in the usual resonator-based EOMs pumped with classical fields.

2.2. *An open system*

In reality, all optical and microwave resonators are open systems, pumped externally, and the pumping and decay terms do not follow from the Hamiltonian approach. To describe analytically the interaction among different modes inside an open resonator, we use the quasi-mode approach (see, for example, Ref. 79). We assume that each mode in the resonator can be considered independently. This assumption is valid when frequency splitting between the modes significantly exceeds the mode bandwidth, which is usually the case for any high-Q mode of a cavity.

Let us consider a lossless resonator connected to an ideal transmission line via a lossless coupler with energy transmission coefficient T ($1 > T$). The value $1/T$ determines the resonator finesse (FWHM/FSR). Electromagnetic field $E_{in}(t)$ in the transmission line enters the resonator through the coupler, and field $E_{out}(t)$ exits the resonator and travels in the opposite direction to the field $E_{in}(t)$. We introduce the resonator field via two electromagnetic waves propagating inside the cavity and going out of the coupler ($E_1(t)$) and into the coupler ($E_2(t)$).

Assuming that the coupler has zero response time, we write boundary conditions on the coupler surface

$$E_1(t) = E_{in}(t)\sqrt{T} - E_2(t)\sqrt{1-T}, \tag{23}$$

$$E_{out}(t) = -E_{in}(t)\sqrt{1-T} - E_2(t)\sqrt{T}. \tag{24}$$

Fields E_1 and E_2 are connected by condition

$$E_2(t) = -E_1(t-\tau), \tag{25}$$

where τ is the round-trip time in the resonator ($\tau = 2\pi Rn/c$ for the case of a whispering gallery mode, where R is the mode radius, n is the medium index of refraction, c is the speed of light in the vacuum).

The set of equations (23-25) is general. Let us simplify the problem and consider the case when the field inside the resonator can be presented as a product of a fast oscillating part $\exp(-i\omega t)$ and a slow oscillating part $\widetilde{E}(t)$, i.e. $E(t) = \widetilde{E}(t)\exp(-i\omega t)$. The carrier frequency ω coincides with one of the resonant frequencies of the resonator, $\omega\tau = 2\pi l$ (l is a real number).

We consider the case when the slow field amplitude inside the resonator does not change significantly during the round-trip time. Then expression (25) can be decomposed into a Taylor series keeping linear term in τ only

$$\widetilde{E}_2(t) \simeq -\widetilde{E}_1(t) + \tau\dot{\widetilde{E}}_1(t). \tag{26}$$

We also assume that

$$1 - \sqrt{1-T} \simeq \frac{T}{2}. \tag{27}$$

Substituting (26) and (27) into (23) we get an equation that allows us to calculate the field inside the resonator if we know the pump field

$$\dot{\widetilde{E}}_1(t) + \frac{T}{2\tau}\widetilde{E}_1(t) = \frac{E_{in}(t)}{\sqrt{\tau}}e^{i\omega t}\sqrt{\frac{T}{\tau}}. \tag{28}$$

It is convenient to introduce decay $\gamma = T/(2\tau)$ and external "force" $F = E_{in}\sqrt{T/\tau^2}$, then Eq. (28) can be rewritten in the form

$$\dot{E}_1(t) + (i\omega + \gamma)E_1(t) = F(t). \tag{29}$$

This is an equation that describes the evolution of the amplitude of the field inside the linear resonator. For exact resonant tuning and time independent pumping field E_{in} the ratio of light powers inside the resonator P_1 and outside the resonator P_{in} is $2/(\gamma\tau)$. The total energy accumulated in the resonator is $P_1\tau = 2P_{in}/\gamma$.

To find the field that exits the resonator we use Eq. (24) that can be rewritten as

$$E_{out}(t) = -E_{in}(t) + E_1(t)\sqrt{T}. \tag{30}$$

It is worth noting that Eq. (29) describes the evolution of a single resonator mode with carrier frequency ω. This equation is valid under conditions (26) and

(27). Moreover, we have to demand the pump field E_{in} to be nearly resonant with the mode.

Let us now return to the modulator and consider open resonators. For such an open symmetric system as well as the exact resonant tuning of the optical and the microwave pumps Eqs. (3-6) transform to

$$\dot{\hat{A}} = -\gamma\hat{A} - ig^*(\hat{B}_-\hat{C} + \hat{C}^\dagger\hat{B}_+) + \hat{F}_A, \tag{31}$$

$$\dot{\hat{B}}_- = -\gamma\hat{B}_- - ig\hat{C}^\dagger\hat{A} + \hat{F}_{B-}, \tag{32}$$

$$\dot{\hat{B}}_+ = -\gamma\hat{B}_+ - ig\hat{C}\hat{A} + \hat{F}_{B+}, \tag{33}$$

$$\dot{\hat{C}} = -\gamma_M\hat{C} - ig\hat{B}_-^\dagger\hat{A} - ig^*\hat{A}^\dagger\hat{B}_+ + \hat{F}_C. \tag{34}$$

where $\hat{A}$, $\hat{B}_\pm$, and $\hat{C}$ are the slowly-varying amplitudes of the operators $\hat{a}$, $\hat{b}$, and $\hat{c}$ respectively; γ and γ_M are the optical and microwave decay rates respectively (we assume that the modes are overloaded so the intrinsic losses can be neglected and γ and γ_M are the decay rates of the loaded modes); $\hat{F}_A$, $\hat{F}_{B\pm}$, and $\hat{F}_C$ are the Langevin forces with properties given by

$$\langle\hat{F}_A\rangle = \sqrt{\frac{2P\gamma}{\hbar\omega}}, \tag{35}$$

$$\langle\hat{F}_C\rangle = \sqrt{\frac{2P_{mw}\gamma_M}{\hbar\omega_c}}, \tag{36}$$

$$\langle\hat{F}_A(t)\hat{F}_A^\dagger(t')\rangle = 2\gamma\delta(t - t'), \tag{37}$$

$$\langle\hat{F}_{B\pm}(t)\hat{F}_{B\pm}^\dagger(t')\rangle = 2\gamma\delta(t - t'), \tag{38}$$

$$\langle\hat{F}_C(t)\hat{F}_C^\dagger(t')\rangle = 2\gamma_M(\bar{n}_{th} + 1)\delta(t - t'), \tag{39}$$

$$\langle\hat{F}_C^\dagger(t)\hat{F}_C(t')\rangle = 2\gamma_M\bar{n}_{th}\delta(t - t'), \tag{40}$$

where $\langle\ldots\rangle$ stands for the reservoir averaging, P is the power of the external optical pump of mode A, P_{mw} is the power of the external microwave pump of mode C, $\bar{n}_{th} = [\exp(\hbar\omega_c/k_BT) - 1]^{-1}$ is the average number of thermal photons in the microwave mode, k_B is the Boltzmann constant, and T is the temperature. The other expectation values, quadratic deviations, and correlations are equal to zero.

Let us solve the linearized set (31-34) in steady state, assuming an undepleted classical pump $\hat{A} \to A = \langle\hat{F}_A\rangle/\gamma$ =const. We present the force and the field operators as a sum of an expectation and fluctuational parts like

$$\hat{F}_{B\pm} = \int_{-\infty}^{\infty} \hat{f}_{B\pm}(\tilde{\omega})e^{-i\tilde{\omega}t}\frac{d\tilde{\omega}}{2\pi}, \tag{41}$$

$$\hat{F}_C = \langle\hat{F}_C\rangle + \int_{-\infty}^{\infty} \hat{f}_C(\tilde{\omega})e^{-i\tilde{\omega}t}\frac{d\tilde{\omega}}{2\pi}, \tag{42}$$

where $\langle \hat{f}_A(\omega)\hat{f}_A(\omega)^\dagger \rangle = \langle \hat{f}_{B\pm}(\omega)\hat{f}_{B\pm}(\omega)^\dagger \rangle = 4\pi\gamma\delta(\omega - \omega')$ and $\langle \hat{f}_C(\omega)\hat{f}_C(\omega)^\dagger \rangle = 4\pi\gamma_M(\bar{n}_{th} + 1)\delta(\omega - \omega')$; and

$$\hat{B}_\pm = B_\pm + \int_{-\infty}^{\infty} \delta\hat{B}_\pm(\tilde{\omega})e^{-i\tilde{\omega}t}\frac{d\tilde{\omega}}{2\pi}, \tag{43}$$

$$\hat{C} = C + \int_{-\infty}^{\infty} \delta\hat{C}(\tilde{\omega})e^{-i\tilde{\omega}t}\frac{d\tilde{\omega}}{2\pi}. \tag{44}$$

The solution of the linearized set (31-34) with respect to the fluctuations is given by

$$\delta\hat{B}_-(\omega) = \left(1 + \frac{|g|^2|A|^2}{\Gamma\Gamma_M}\right)\frac{\hat{f}_{B-}(\tilde{\omega})}{\Gamma} + \frac{g^2 A^2}{\Gamma\Gamma_M}\frac{\hat{f}_{B+}^\dagger(-\tilde{\omega})}{\Gamma} - \frac{igA}{\Gamma}\frac{\hat{f}_C^\dagger(-\tilde{\omega})}{\Gamma_M}, \tag{45}$$

$$\delta\hat{B}_+(\tilde{\omega}) = \left(1 - \frac{|g|^2|A|^2}{\Gamma\Gamma_M}\right)\frac{\hat{f}_{B+}(\tilde{\omega})}{\Gamma} - \frac{g^2 A^2}{\Gamma\Gamma_M}\frac{\hat{f}_{B-}^\dagger(-\tilde{\omega})}{\Gamma} - \frac{igA}{\Gamma}\frac{\hat{f}_C(\tilde{\omega})}{\Gamma_M}, \tag{46}$$

$$\delta\hat{C}(\tilde{\omega}) = -\frac{igA}{\Gamma_M}\frac{\hat{f}_{B-}^\dagger(-\tilde{\omega})}{\Gamma} - \frac{ig^* A^*}{\Gamma_M}\frac{\hat{f}_{B+}(\tilde{\omega})}{\Gamma} + \frac{\hat{f}_C(\tilde{\omega})}{\Gamma_M},$$

where we use notations $\Gamma = \gamma - i\tilde{\omega}$ and $\Gamma_M = \gamma_M - i\tilde{\omega}$.

It is easy to see now that in the absence of microwave pumping ($\langle \hat{F}_C \rangle = 0$) and in the steady state, the expectation values of the photon number in the optical modes and the photon number in the microwave mode are

$$\langle \hat{n}_{B-} \rangle = \frac{(2\gamma + \gamma_M)\gamma_M}{2\gamma^2}\left[\frac{|g|^2|A|^2}{\gamma_M(\gamma_M + \gamma)}\right]^2 + \frac{|g|^2|A|^2}{\gamma(\gamma_M + \gamma)}(\bar{n}_{th} + 1), \tag{47}$$

$$\langle \hat{n}_{B+} \rangle = \frac{(2\gamma + \gamma_M)\gamma_M}{2\gamma^2}\left[\frac{|g|^2|A|^2}{\gamma_M(\gamma_M + \gamma)}\right]^2 + \frac{|g|^2|A|^2}{\gamma(\gamma_M + \gamma)}\bar{n}_{th}, \tag{48}$$

$$\langle \hat{n}_C \rangle = \bar{n}_{th} + \frac{|g|^2|A|^2}{\gamma_M(\gamma_M + \gamma)}. \tag{49}$$

Let us find the output signal in the system if the microwave pumping is present. Neglecting the optical saturation of the microwave as well as the pump field, we obtain the expectation values for the fields from (31-34)

$$A \simeq \frac{\langle \hat{F}_A \rangle}{\gamma}, \quad C \simeq \frac{\langle \hat{F}_C \rangle}{\gamma_M}, \quad B_- \simeq \frac{igA}{\gamma}C^*, \quad B_+ \simeq \frac{igA}{\gamma}C. \tag{50}$$

We finally derive the expression for the expectation values

$$E_{out} \simeq E_{in}, \quad E_{out-} = E_{in}\frac{2ig}{\gamma}C^* = E_{in}\frac{2ig}{\gamma\gamma_M}\langle \hat{F}_C^\dagger \rangle,$$

$$E_{out+} = E_{in}\frac{2ig}{\gamma}C = E_{in}\frac{2ig}{\gamma\gamma_M}\langle \hat{F}_C \rangle, \tag{51}$$

as well as fluctuations of the optical fields

$$\hat{e}_{out-}(\tilde{\omega}) = \left(\frac{\Gamma^*}{\Gamma} + \frac{2\gamma|g|^2|A|^2}{\Gamma^2\Gamma_M}\right)\hat{e}_{in-}(\tilde{\omega}) + \frac{2\gamma g^2 A^2}{\Gamma^2\Gamma_M}\hat{e}^\dagger_{in+}(-\tilde{\omega}) - \frac{2igA\sqrt{\gamma\gamma_M}}{\Gamma\Gamma_M}\hat{e}^\dagger_C(-\tilde{\omega}),$$

$$\tag{52}$$

$$\hat{e}_{out+}(\tilde{\omega}) = \left(\frac{\Gamma^*}{\Gamma} - \frac{2\gamma|g|^2|A|^2}{\Gamma^2\Gamma_M}\right)\hat{e}_{in+}(\tilde{\omega}) - \frac{2\gamma g^2 A^2}{\Gamma^2\Gamma_M}\hat{e}^\dagger_{in-}(-\tilde{\omega}) - \frac{2igA\sqrt{\gamma\gamma_M}}{\Gamma\Gamma_M}\hat{e}_C(\tilde{\omega}).$$

$$\tag{53}$$

We used expressions $\langle A\rangle\sqrt{2\gamma\tau} = 2E_{in}$, $F_A = E_{in}\sqrt{2\gamma/\tau}$, $E_{out\pm} = B_\pm\sqrt{2\gamma\tau}$; τ is the resonator round trip time $\tau = 2\pi R/c$, where R is the radius of the resonator. We assume, for the sake of simplicity, that the resonator is phase matched (or empty), so that $\tau = \tau_M$. Eqs. (51-53) represent the solution of the problem and describe the EOM in the undepleted case.

2.3. *Saturation of the EOM*

To find saturation of the modulator, we have to take into account depletion of the optical pump. This has been done in the analysis presented in Ref. 45. We rewrite Eqs. (31-34) as

$$\dot{A} = -\Gamma_A A - ig^*(B_-C + C^*B_+) + F_A, \tag{54}$$

$$\dot{B}_- = -\Gamma_{B-}B_- - igC^*A, \tag{55}$$

$$\dot{B}_+ = -\Gamma_{B+}B_+ - igCA, \tag{56}$$

$$\dot{C} = -\Gamma_C C - igB_-^* A - ig^*A^*B_+ + F_M, \tag{57}$$

where

$$\Gamma_A = i(\omega - \omega_0) + \gamma,$$
$$\Gamma_{B\mp} = i(\omega_\mp - \omega_0 \pm \omega_M) + \gamma,$$
$$\Gamma_C = i(\omega_c - \omega_M) + \gamma_M,$$

where ω_0 and ω_M are the frequencies of the optical and microwave pumping respectively. We neglect the quantum effects at this point.

Let us solve set (54-57) in the steady state. Neglecting the optical saturation of the microwave oscillations, we obtain from (54) and (57)

$$A = -i\frac{g^*}{\Gamma_A}(B_-C + C^\dagger B_+) + \frac{F_A}{\Gamma_A}, \tag{58}$$

$$C \simeq \frac{F_M}{\Gamma_C}. \tag{59}$$

Substituting (58) and (59) into (55) and (56) in the steady state we get

$$B_- = -\frac{igF_A F_M^*\Gamma_{B+}\Gamma_C}{\Gamma_{B-}\Gamma_{B+}\Gamma_A|\Gamma_C|^2 + |g|^2|F_M|^2(\Gamma_{B+} + \Gamma_{B-})}, \tag{60}$$

$$B_+ = -\frac{igF_A F_M\Gamma_{B-}\Gamma_C^*}{\Gamma_{B-}\Gamma_{B+}\Gamma_A|\Gamma_C|^2 + |g|^2|F_M|^2(\Gamma_{B+} + \Gamma_{B-})}. \tag{61}$$

Finally, for the pumping mode we derive

$$A = \frac{\Gamma_{B-}\Gamma_{B+}|\Gamma_C|^2 F_A}{\Gamma_{B-}\Gamma_{B+}\Gamma_A|\Gamma_C|^2 + |g|^2|F_M|^2(\Gamma_{B+} + \Gamma_{B-})}. \tag{62}$$

To find expressions for the optical output we use Eq. (30) and derive

$$E_{out-} = -E_{in}\frac{2ig\gamma F_M^*\Gamma_{B+}\Gamma_C}{\Gamma_{B-}\Gamma_{B+}\Gamma_A|\Gamma_C|^2 + |g|^2|F_M|^2(\Gamma_{B+} + \Gamma_{B-})}, \tag{63}$$

$$E_{out+} = -E_{in}\frac{2ig\gamma F_M\Gamma_{B-}\Gamma_C^*}{\Gamma_{B-}\Gamma_{B+}\Gamma_A|\Gamma_C|^2 + |g|^2|F_M|^2(\Gamma_{B+} + \Gamma_{B-})} \tag{64}$$

$$E_{out} = E_{in}\frac{\Gamma_{B-}\Gamma_{B+}(\Gamma_A - 2\gamma)|\Gamma_C|^2 + |g|^2|F_M|^2(\Gamma_{B+} + \Gamma_{B-})}{\Gamma_{B-}\Gamma_{B+}\Gamma_A|\Gamma_C|^2 + |g|^2|F_M|^2(\Gamma_{B+} + \Gamma_{B-})}. \tag{65}$$

Let us now calculate the power of the output light for the carrier wave (P_0) and the harmonics ($P_\pm$) with respect to the pump power (P_{in}). For the sake of simplicity, let us consider the entirely resonant case, i.e. $\Gamma_A = \Gamma_{B\pm} = \gamma$, $\Gamma_C = \gamma_M$. Introducing quality factors as $Q = \omega_0/(2\gamma)$ and $Q_M = \omega_M/(2\gamma_M)$, and recalling $|C|^2 = |F_M|^2/\gamma_M^2 = 4P_{mw}Q_M/(\hbar\omega_M^2)$, where P_{mw} is the input microwave power, we derive from (63-65)

$$\frac{P_\pm}{P_{in}} = \left(\frac{2\mathcal{S}}{1 + 2\mathcal{S}^2}\right)^2, \quad \frac{P_0}{P_{in}} = \left(\frac{1 - 2\mathcal{S}^2}{1 + 2\mathcal{S}^2}\right)^2, \tag{66}$$

where

$$\mathcal{S} = \frac{4|g|Q}{\omega_0}\sqrt{\frac{P_{mw}Q_M}{\hbar\omega_M^2}} \tag{67}$$

is the saturation parameter. Equations (66) show that there is an optimum value of the microwave power for which the conversion of our carrier frequency to the Stokes and anti-Stokes sidebands is the most efficient. In principle, a complete conversion is possible for $2\mathcal{S}^2 = 1$. In practice, however, the three mode model works only far from saturation of the modulator. Other optical modes should be taken into account closer to the saturation point. Moreover, in the realistic cases one should not neglect the absorption in the system - unless all the modes are significantly overcoupled.

It is useful to derive the expression for the coupling constant g now. We took into account that the microwave electric field E_z changes the index of refraction of the material by

$$\Delta n_e = \frac{1}{2}n_e^3 r_{33}E_z, \tag{68}$$

where r_{33} is the electro-optic constant for the material of the dielectric resonator (for the sake of simplicity we assume that all the fields are polarized along the lithium niobate crystalline axis), n_e is the refractive index for the optical mode. The refractive index change results in the shift of the frequency of the optical mode

$$\Delta\omega = -\frac{1}{2}\omega n_e^2 r_{33}E_z. \tag{69}$$

The microwave electric field in the microwave resonator is quantized as[80]

$$E_z = \sqrt{\frac{2\pi\hbar\omega_c}{n_c^2 \mathcal{V}_c}}(\hat{c} + \hat{c}^\dagger),$$

(70)

where $\mathcal{V}_c$ is the volume of the microwave field, n_c is the index of refraction at the microwave frequency.

On the other hand, the interaction Hamiltonian for the light and electric field can be presented in form

$$V_{int} = \hbar g \hat{a}^\dagger \hat{a}(\hat{c}^\dagger + \hat{c}),$$

(71)

so that the optical frequency shift is equal to

$$\Delta\omega = g(\hat{c}^\dagger + \hat{c}).$$

(72)

It is easy to find now, keeping in mind partial overlap of the optical and microwave modes,

$$g = -\frac{1}{2}\omega n_e^2 r_{33}\sqrt{\frac{2\pi\hbar\omega_c}{n_c^2 \mathcal{V}_c}}\eta,$$

(73)

where

$$\eta = \frac{1}{\mathcal{V}}\int_{\mathcal{V}} d\mathcal{V}\Psi_a\Psi_b\Psi_c$$

(74)

is the overlap integral of the optical and microwave fields, $|\Psi_a|^2$, $|\Psi_b|^2$, and $|\Psi_c|^2$ are the spatial distributions of the power of the optical and microwave fields respectively, $(1/\mathcal{V})\int|\Psi_a|^2 dV = 1$, $\mathcal{V}$ is the whispering gallery mode volume. We assume here that the whispering gallery modes are nearly identical, i.e. $\omega \simeq \omega_\pm \gg \omega_c$, $\mathcal{V} \simeq \mathcal{V}_\pm$, $\int d\mathcal{V}\Psi_a\Psi_{b-}\Psi_c = \int d\mathcal{V}\Psi_a\Psi_{b+}\Psi_c$.

Substituting (73) into (67) we get

$$S = 2Q n_e^2 r_{33}\sqrt{\frac{2\pi P_{mw}Q_M}{n_c^2 \mathcal{V}_c \omega_M}}\eta.$$

(75)

2.4. *Generation of multiple harmonics*

We have studied the interactions of light and microwave fields that lead to the generation of two optical harmonics. This generally is not the case. With increasing pump, multiple harmonics start to occur because WGMs are nearly equidistant in frequency.

To conclude this theoretical consideration of the all-resonant modulator, below we consider the multimode modulator.[45]

To describe the harmonic generation we rewrite the interaction Hamiltonian as

$$\hat{V} = \hbar g \sum_{n=-\infty}^{\infty}(\hat{a}_{n-1}^\dagger\hat{c}^\dagger\hat{a}_n + \hat{a}_{n+1}^\dagger\hat{c}\hat{a}_n) + \text{adjoint},$$

(76)

where $\hat{a}_n$ is annihilation operator for $\mathrm{n^{th}}$ resonator mode. We assume that modes are completely identical with respect to their quality factors and coupling to the microwave field.

Using (76) we derive equations of motion for the modes. For the sake of simplicity, we will consider the case of exact resonance for all the modes. In slowly varying amplitude and phase approximation equations for the expectation values of the field amplitudes are

$$\dot{A}_n = -\gamma A_n - ig(A_{n-1}C + C^* A_{n+1}) + F_A \delta_{n,0}, \tag{77}$$

$$\dot{C} = -\gamma_M C - ig \sum_{n=-\infty}^{\infty} (A_{n-1}^* A_n + A_n^* A_{n+1}) + F_M, \tag{78}$$

where F_A stands for the pumping of the modes, $\delta_{i,j} = 1$ if $i = j$ and $\delta_{i,j} = 0$ if $i \neq j$. In other words, we assume that only mode with $n = 0$ is pumped. Then $A_{\pm 1}$ corresponds to $B_\pm$ in our previous consideration.

We assume that set (77,78) has a steady state solution,

$$\gamma A_n + ig(A_{n-1}C + C^* A_{n+1}) = F_A \delta_{n,0}, \tag{79}$$

$$\gamma_M C + ig \sum_{n=-\infty}^{\infty} (A_{n-1}^* A_n + A_n^* A_{n+1}) = F_M, \tag{80}$$

This set of equations can be solved using Fourier transformation

$$A(t) = \sum_{n=-\infty}^{\infty} A_n e^{-i\omega_M n t}. \tag{81}$$

In the case of exact resonance the amplitude of the microwave field does not depend on the light intensity (see Eq. (59)) $C = |F_M| \exp(i\phi_M)/\gamma_M$. Then, multiplying each term of Eq. (79) on $\exp(-i\omega_M n t)$ (n corresponds to the index of term γA_n), and summing them over all n we derive

$$A(t) = \frac{F_A \gamma_M}{\gamma \gamma_M + 2ig|F_M| \cos(\omega_M t + \phi_M)}. \tag{82}$$

The solution for each mode A_n can be written as

$$A_n = \frac{1}{2\pi} \int_0^{2\pi/\omega_M} A(t) e^{i\omega_M n t} dt. \tag{83}$$

To find expressions for the output light we use Eq. (30)

$$E_{out} = \frac{1 - 2i\mathcal{S}\cos(\omega_M t + \phi_M)}{1 + 2i\mathcal{S}\cos(\omega_M t + \phi_M)} E_{in}(t) = E_{in} \exp\left\{-2i \arctan\left[2\mathcal{S}\cos(\omega_M t + \phi_M)\right]\right\}.$$

Therefore, in the case of multiple harmonics, the behavior of the system is different from the case of three harmonics. An increase of the microwave power leads to the increase in the number of optical harmonics instead of saturation, as well as a decrease in the field amplitude, as was shown in the previous section.

In the more general case of absorptive resonator Eq. (82), the electric field of the light escaping the resonator can be presented as follows

$$E_{out} = \frac{\gamma - \gamma_{cp} - i(\omega - \omega_0) + 2i\tilde{g}\cos(\omega_M t + \phi_M)}{\gamma + \gamma_{cp} - i(\omega - \omega_0) + 2i\tilde{g}\cos(\omega_M t + \phi_M)} E_{in} \tag{84}$$

where γ and γ_{cp} are the intrinsic and coupling half widths at half maximum of the WGM resonances ($\gamma + \gamma_{cp}$ is the loaded half width at half maximum), ϕ_M is the phase of the microwave signal, and $\tilde{g}$ is the modulation parameter given by

$$\tilde{g} = \omega r_{33} n_e^2 \sqrt{\frac{\pi P_{mw} Q_M}{n_c^2 \omega_M \mathcal{V}_c}} \, \eta, \tag{85}$$

where we assume that the microwave resonator is critically coupled and Q_M is the loaded quality factor.

3. Photonic receivers for microwave and THz radiation

Photonic microwave receivers hold promise for achieving high detection sensitivity at high microwave frequencies. Direct processing of high frequency microwave signals with the conventional electronic approaches is hindered by the absence of efficient microwave amplifiers and detectors. Microwave photonic receivers enable transfer of microwave signals to an intermediate frequency (IF) or to baseband, where subsequent efficient detection is feasible. Examples of such receivers based on all-resonant interaction of light and microwaves in solid-state WGM resonators have been recently demonstrated.[39–75] The sensitivity of photonic receivers based on these resonance structures can be made very high within relatively narrow reception bandwidth. However, unlike conventional receivers in which noise properties, dielectric and Joule losses, energy consumption, and even component availability deteriorate with increasing frequency, the sensitivity of photonic devices fundamentally does not degrade with the increasing microwave frequency. (Practically, the sensitivity does degrade because of the reduction of efficiency and power handling in ever smaller active area photodiodes that are used at very high modulation frequencies. However recent advances in travelling-wave photodiodes, the so called uni-travelling-carrier photodiodes, and advanced heat sinking solutions in small size, small capacity photodiodes, have made possible demonstrations of multi-milliWatt microwave returns at frequencies approaching 100GHz and beyond. Microwave components that are part of the photonic receivers also suffer from increasing loss. Nevertheless, microwave photonic receivers remain very compact and consume low power, even at high frequencies. They also allow separating the detection and processing of the microwave signals in space when they are combined with high quality optical links used to transmit the upconverted microwave signals. $X-$, $K_\mathrm{u}-$band, and $K_\mathrm{a}-$band WGM-based receivers have been demonstrated.

Resonant electro-optic modulators are in the heart of photonic receivers. The core physical basis for the operation of the receiver is the phenomenon of three-wave mixing occurring under proper conditions in a medium possessing a quadratic nonlinearity (c.f the operation principles of the WGM modulators). A microwave signal is sent into the medium that is simultaneously pumped optically. The pump light interacts with the microwaves creating optical harmonics. Each harmonic is spectrally separated from the optical carrier by a value equal to an integer number of the microwave carrier frequency. Information carried by the microwave signal is retrieved by means of processing and detecting the optical harmonics and/or the optical carrier leaving the medium.

To enhance the efficiency of wave mixing the interaction length of the optical and microwave photons can be increased using resonant interaction, as was shown in the previous section. By properly engineering the morphology of the resonator we achieve strong interaction of light and microwaves resulting in the generation of multiple optical harmonics separated by the FSR of the WGM resonator. Obviously, the operational bandwidth of the receiver — in terms of information coded as modulation on microwave carrier — is limited by the lesser of the spectral widths of optical and microwave modes.

3.1. *Initial experiments: linear X-band photonic link*

WGM-based X-band receiver/photonic link with 2.5 nW minimal detectable microwave power, about 14 dB signal-to-noise ratio, corresponding to a noise floor of ~ 0.1 nW, and ~ 5 kHz analog bandwidth has been demonstrated in Ref. 44. While the bandwidth of this device is narrower than those of electronic receivers,[81] it has other advantages due to small power consumption and the simplicity associated with optical signal processing. This performance may be further enhanced by increasing the quality factors, and improving the microwave and optical field overlap.

The optical pump has to be provided by a laser that is either passively tuned, or actively locked to the frequency of operational optical WGM, by means of either electronic or optical feedback. In the experiment light from a distributed feedback (DFB) laser having 1550 nm wavelength is launched into Z-cut $LiNbO_3$ spheroidal optical resonator using a diamond prism (Fig. 2). The light is tuned to be near a particular resonator mode. The input optical power is about 2 mW, the output optical power detected is 10 dB less than the input power.

The resonator is a thin disk with radius 2.4 mm, thickness 150 μm, and with its perimeter edge polished into a spheroidal geometry with effective transverse curvature diameter 180 μm. The crystal C-axis of $LiNbO_3$ coincides with the symmetry axis of the resonator within 0.1 degree uncertainty. The loaded optical quality factor of WGMs is $Q \simeq 5 \times 10^6$ (optical resonance bandwidth 30 MHz).

The field confinement volume of the WGMs overlaps with that of a microwave resonator that is excited by a strip-line coupler delivering the input microwave signal. We used a half-wave microstrip resonator arranged by placing a half-circular electrode along the rim of the resonator to tailor the microwave field structure for optimal nonlinear-optic interaction. The quality factor of this microwave resonator is $Q_M = 120$ and the bandwidth ~ 80 MHz. By tuning the length of the strip-line

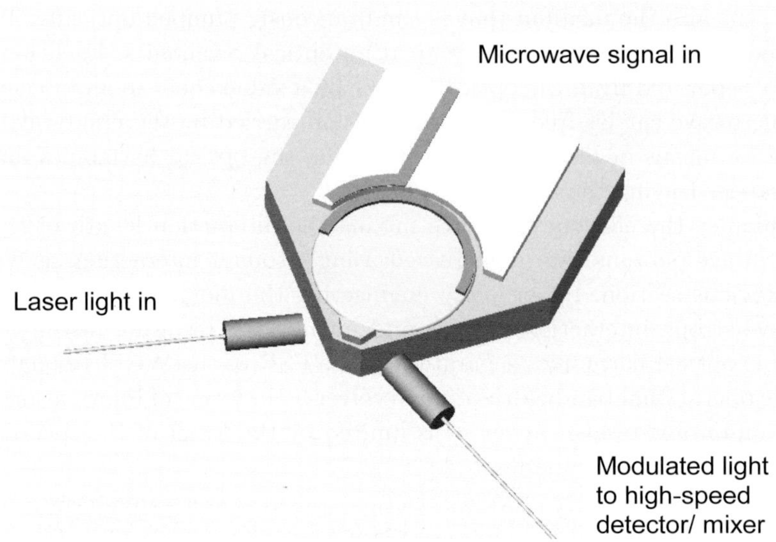

Fig. 2. Experimental setup for the linear X-band receiver based on a lithium niobate WGM resonator. (Reprinted from Ref. 44 by permission of IEEE).

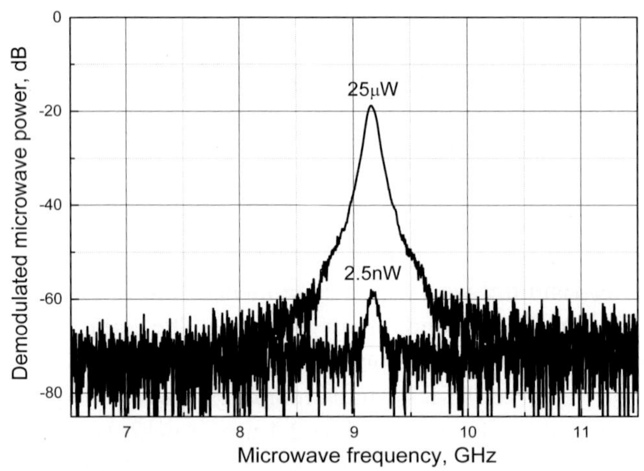

Fig. 3. Demodulated microwave power vs frequency of the microwave pumping. Zero level corresponds to the saturation power. (Reprinted from Ref. 44 by permission of IEEE).

electrodes, the microwave resonator resonance frequency can be tuned to be equal to the optical free spectral range (9.155 GHz in our case).

We studied the dependence of light modulation as a function of frequency of the input microwave signal power in this device. The typical frequency response of

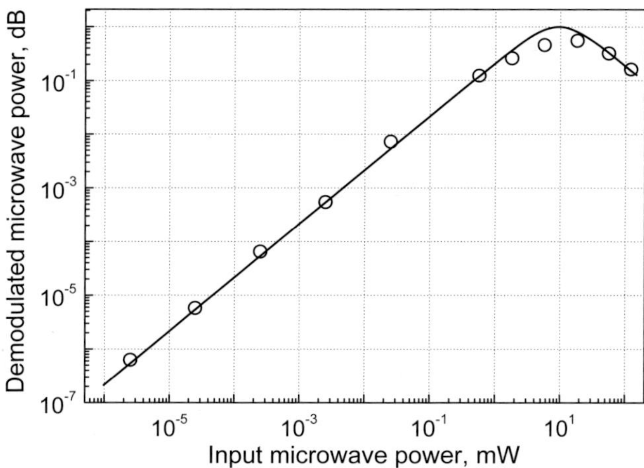

Fig. 4. Normalized demodulated microwave power vs power of the microwave pump. Experimental data are shown by circles. Theoretical simulations shown by the solid line. (Reprinted from Ref. 44 by permission of IEEE).

the device is shown in Fig. 3. Here the laser is constantly kept at the slope of the resonance curve of an optical mode, with the microwave frequency scanned, and the demodulated microwave power is obtained by a high-speed photodetector. This signal is recorded by a network analyzer. For the 2 mW laser used in the experiment, the absolute value of the demodulated signal power is 31 dB less than the input microwave signal power. The curves in Fig. 3 are well under the saturation limit shown by the zero dB level. As follows from Fig. 3, the operational 3 dB bandwidth of our receiver is ~ 85 MHz.

The amplitude characteristics of our modulator/receiver was studied as the dependence of the demodulated microwave power, obtained by a high-speed photodetector at the optical output of the modulator, on the input microwave power. Results are presented in Fig. 4. The saturation point at ~ 10 mW corresponds to the limit imposed by harmonic multiplication, as well as power broadening, of the optical resonances. The optimal power within the linear regime of the modulator is about 1 mW, and the dynamic range of the receiver is ~ 70 dB.

This system may produce both phase and amplitude modulated signals. When the laser is tuned to the slope of the optical resonance, the modulation is mostly of the amplitude-type. The modulation is of phase-type for the laser tuned to the center of the resonance. To demonstrate the dependence of amplitude modulation (AM) response on optical detuning, we have monitored demodulated microwave power while scanning the laser frequency around the WGM resonance; see Fig. 5. Amplitude modulation is maximal when the laser is tuned to the slope of the resonance curve, and turns to zero at exactly the center of optical resonance.

It has to be mentioned that other types of nonlinear interactions could occur,

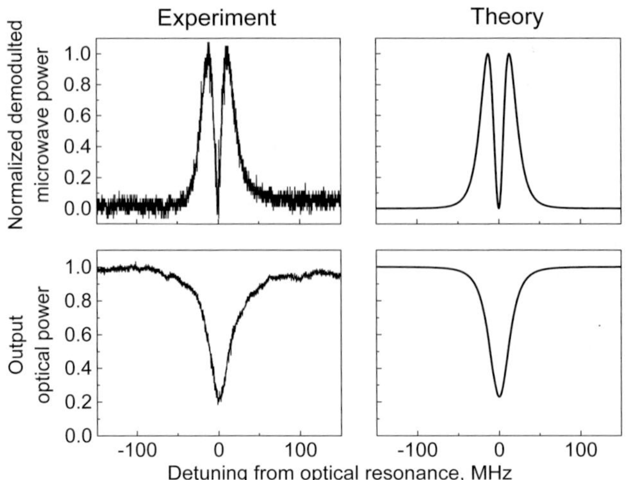

Fig. 5.   Top: Demodulated microwave power vs detuning of the pump light from the whispering gallery mode resonance. Bottom: Whispering gallery mode resonance. No signal is found for the resonant tuning. (Reprinted from Ref. 44 by permission of IEEE).

which may be potentially detrimental to the functionality of the modulator in a small $LiNbO_3$ resonator such as ours. For example, parasitic acoustic excitations by piezoelectric effect have been reported in conventional electro-optic modulators. Nevertheless, we have not observed any indication of these phenomena with our modulator within the full range of the applied microwave power. The unconventional geometry of our modulator prevents effective interaction of light and acoustic waves: in the area of modal overlap (perimeter of the resonator) acoustic modes are orthogonal to optical WGMs, and even if excited, the strain modes induced in the material cannot influence the light. The same conclusion is applicable to the Brillouin interaction: while usually limiting the performance of fiber-optic devices, it does not impose any serious restrictions in our case of WGM optical resonator — unless its FSR exactly coincides with the Brillouin frequency, or if the resonator is highly overmoded (i.e. has a very dense spectrum of WGM).

Strictly speaking, the receiver demonstrated in Ref. 44 was not a complete receiver, but rather a transducer (an optical link). The microwave frequency demodulated on the photodiode had the same frequency as the frequency of the signal. Real receiver should have the ability to shift the signal frequency to an intermediate frequency (IF) or base band. An example of such a receiver is discussed below.

3.2. *Quadratic K_a-band receiver*

The modulator and photonic link (Fig. 2) discussed in the previous section also can operate as a quadratic receiver. We have demonstrated a quadratic receiver that downconverts microwave pulses into the base-band.[48] The high Q lithium niobate

WGM resonator was excited with 1319 nm laser light at the resonant frequency of one of the resonator TE modes (the electric field is parallel to the resonator symmetry axis which coincides with the C-axis of $LiNbO_3$). When a pulse-modulated radio signal with repetition frequency from 10 kHz to 100 kHz impressed onto the microwave carrier was applied to the resonator, the light output detected at a low frequency photodetector directly reproduced the envelope shape of the pulse train. This direct signal recovery did not require a local oscillator, and was the result of nonlinear response of the material of the WGM resonator modulator, combined with the automatic filtering produced by the low-pass response function of the photodetector.

A quadratic receiver based on a lithium niobate WGM resonator was proposed and demonstrated by M. Hossein-Zadeh and A. F. J. Levi.[46,47] The basic innovations of our work include i) a high operational frequency, 35 GHz vs previously demonstrated 8.7 GHz and 14.6 GHz; and ii) the ultra-high quality factor of the lithium niobate resonator (loaded optical Q-factor was on the order of 10^8) that resulted in much higher sensitivity (by three orders of magnitude) of reception. While extending the range of operation from X-band to K_a- band is theoretically straightforward, the basic microwave electrode structures is substantially more complicated. The reduced size of the resonator for achieving the K_a- band FSR combined with the required precision electrode structures presented unique challenges absent in the X-band devices. We note that a resonator with multiples of its FSR frequency equal to K_a- band, in principle, could also be used. Our study further includes a theoretical characterization of the receiver, and we present analytical expressions for its sensitivity and dynamic range. The dynamic range restriction results from nonlinear microwave saturation of the receiver.

The basic operation principle of the receiver is related to the decrease of the resonant absorption of light when the microwave signal is on (see Fig. 6). A critically coupled resonator absorbs all the input resonant light, and the spectrum of the modes of a critically coupled resonator has 100% contrast. In our case the coupling efficiency was approximately 60%, and the insertion loss was less than 3 dB. When a microwave signal is applied to the resonator the excited WGM resonance broadens and the light coupling efficiency decreases.[45] The change of the DC optical power transmitted through the resonator is proportional to the power of the microwave signal. In fact, the receiver resembles an opto-electronic transistor-like device. The small amount of the microwave power applied (a few tens of microwatts) changes the transmission of a milliwatt optical signal. Similarly a modulated microwave input results in the microwave modulation of light passing through the resonator. The microwave modulation frequency of the microwave carrier cannot exceed the spectral width of both the optical and microwave resonances.

It is convenient to characterize the baseband response of the receiver with respect to the voltage applied to the microwave resonator $V_{RF} = (\rho P_{mw})^{1/2} \cos(\omega_M t + \phi_M)$,

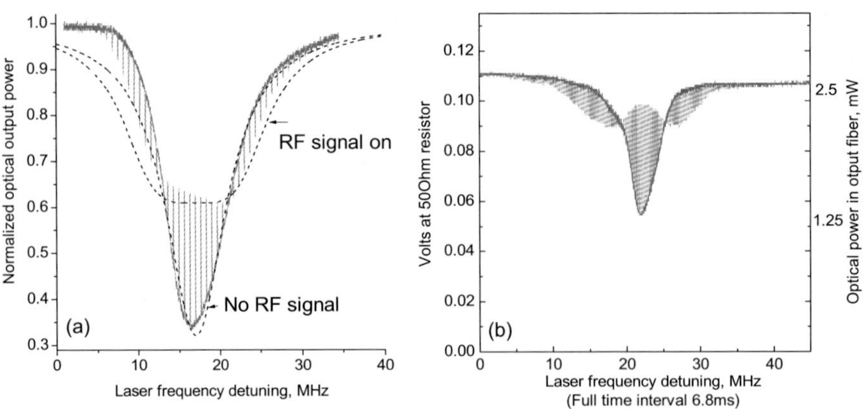

Fig. 6.   Typical shapes of the WGM resonance with and without applied microwave field. (a) Dashed curves stand for the theoretical fit of the experimental parameters. To measure the change of the optical resonance in the experiment we sent a train of microwave pulses into the microwave resonator and slowly swept the optical frequency through the WGM resonance. We observed the resulting change of the optical transmission to observe the change in the resonance curve. In the particular experiment shown in the figure microwave pulses have 3 μs duration and 30 kHz repetition rate. The scan of the optical frequency has 1.5 ms duration. The carrier microwave frequency is 32.92 GHz. Microwave Q is equal to 59. Microwave peak power is 2 mW. The optical passband is 9 MHz, and the photodiode responsivity is 0.37 A/W. Optical power at the photodiode is 2 mW if the laser is moved out of the resonance. (b) Observation similar to (a). The optical mode has 5.2 MHz optical passband. The increase in the optical Q results in stronger RF modulation for the same parameters of the experiment.

where ρ is the resistance. The output optical power P_{out} can be presented as

$$\frac{P_{out}}{P_{in}} = \frac{N_2 V_{RF}^2 / P_{in}}{1 + N_2 V_{RF}^2 / P_{in}} \tag{86}$$

in the case of the critical coupling ($\gamma = \gamma_c$, $Q = \omega/4\gamma$) of resonant monochromatic pump light ($\omega = \omega_0$) to a WGM (see Eq. (84)), and

$$N_2 = 16\pi\eta^2 \frac{P_{in} Q^2 Q_M r_{33}^2 n_e^4}{\rho V_M n_M^2 \omega_M}. \tag{87}$$

The characterization approach for the photonic receiver is different from the approach for linear receivers. The performance of the latter type is generally given by the sensitivity, spurious-free dynamic range, and third-order intercept point. Spurious-free dynamic range and third-order intercept point depend on the residual nonlinearity of the receiver. The method for characterizing a linear receiver is based on injection of equal levels of two closely spaced carriers into the device and the study of generated spurious sidebands at product frequencies. Resonant photonic receivers, by contrast, are fundamentally nonlinear. A major problem with them is the generation of second-order products of signals from two nearby microwave communication channels. One component of these second-order products is located at baseband, and thus interferes with the desired signal, degrading performance. We assume that in our case the receiver is used to detect a signal coming from a single channel. The nonlinearity of the receiver is given by its saturation power estimated as

$$P_{mw\ sat} = \frac{4P_{in}}{3N_2\rho},\tag{88}$$

where we took into account an approximate expression for the base band (BB) optical signal

$$\frac{P_{out}}{P_{in}} = \frac{g^2\cos^2(\omega_M t + \phi_M)}{\gamma^2 + g^2\cos^2(\omega_M t + \phi_M)}\Big|_{BB} \approx \frac{2g^2}{4\gamma^2 + 3g^2},\tag{89}$$

valid for $g < 2\gamma$.

The dynamic range (DR) of the resonant photonic receiver is given by the ratio of its saturation power and the noise floor equivalent power. The noise factor is equal to the ratio of the signal-to-noise ratios at the input and output of the receiver. The noise floor equivalent power is

$$\delta P_{mw\ out} = \left[\left(1 + 2\frac{P_{mw\ out}}{P_{mw}}\right)k_B T + 2e\mathcal{R}\rho_{pd}P_{out}\right]\Delta F,\tag{90}$$

where ρ_{pd} is the resistivity of the photodiode, $\mathcal{R}$ is the responsivity of the photodiode, k_B is the Boltzmann constant, T is the temperature, e is the charge of the electron, and ΔF is the video-bandwidth. Three terms on the right hand side of Eq. (90) are the temperature noise of the receiver, the amplified temperature noise of the input signal, and the shot noise. We have assumed that the input noise power is $\Delta P_{mw\ in} = k_B T\Delta F$. It is easy to see that $\delta P_{mw\ out} = \Delta P_{mw\ in}$ for a small microwave power. The noise floor equivalent power should be found from equation $P_{mw\ out} = \delta P_{mw\ out}$ under condition $P_{mw} \ll P_{mw\ sat}$. Using relationships

$$P_{out} \simeq N_2 V_{RF}^2 = \frac{\rho}{2}N_2 P_{mw} = P_{in}\frac{2P_{mw}}{3P_{mw\ sat}},\tag{91}$$

$$P_{mw\ out} = \rho_{pd}\mathcal{R}^2 P_{out}^2.\tag{92}$$

we find the dynamic range

$$DR = \frac{2}{3}\left(\frac{\rho_{pd}\mathcal{R}^2 P_{in}^2}{k_B T\Delta F}\right)^{1/2}\tag{93}$$

Let us compute the receiver performance parameters. We assume that $Q = 7 \times 10^7$, so that FWHM of the optical resonance is 3 MHz, $\Delta F = 3$ MHz, and $\gamma = \gamma_c = 2\pi \times 10^6$ s^{-1}. We further assume $Q_M = 80$, $\omega_M = 2\pi \times 35$ GHz, $P_{in} = 50$ μW,

$r_{33} = 30$ pm/V, $n_e = 2.1$, $n_M = 5.4$, $\mathcal{V}_M = 5 \times 10^{-6}$ cm^3, $\rho = 50$ Ohm, $\rho_{pd} = 10$ kOhm, $\mathcal{R} = 0.8$ A/W, $k_B = 1.38 \times 10^{-23}$ J/K. Using these values we obtain $N_2 = 0.5$ W/V^2, $P_{mw\ sat} = 2.5$ μW, and $DR = 44$ dB. The noise floor equivalent power is $P_{noise} = P_{mw\ sat}/DR = 0.1$ nW. It is easy to see that the theoretically calculated numbers approximately correspond to our experimental results.

The quadratic receiver has almost two orders of magnitude improved performance as compared to electronic receivers operating at frequencies lower than 5 GHz[82] (we have measured $\mathcal{R}N_2 = 0.4$ A/V^2); and it is more than three orders of magnitude more sensitive as compared to earlier demonstrations of similar photonic devices.[46,47] We also note that sensitivity values quoted here will further improve when the input optical power is increased. The performance of a photonic receiver based on this approach does not degrade for higher microwave frequencies. Hence, the photonic receiver is an attractive alternative for improving the performance of conventional electronic receivers.

3.3. *Quadratic coherent receiver*

The quadratic receiver can also be operated in a coherent mode that allows retrieving the phase of a narrowband microwave signal with increased signal-to-noise ratio. Let us assume that one pumps the microwave resonator with a microwave local oscillator wave (LO) in addition to the microwave signal. The LO power must be below the receiver saturation threshold ($P_{mw\ sat} \gg P_{LO}$). The electric field of the light escaping the critically coupled WGM resonator can be presented in form

$$E_{out} \simeq \frac{ig}{\gamma} \left[\cos(\omega_M t + \phi_M) + \sqrt{\frac{P_{LO}}{P_{mw}}} \cos(\omega_{LO} t + \phi_{LO}) \right] E_{in},$$

if frequencies of both LO (ω_{LO}) and microwave signal (ω_{mw}) nearly coincide with the FSR of the WGM resonator. The time averaged power of the radio-frequency signal ($P_{RF\ BB}$, carrier frequency $\omega_{LO} - \omega_{mw}$) at the exit of the photodiode is given by parameter

$$G = \frac{P_{RF\ BB}}{P_{mw}} = \frac{16}{9} \frac{\rho_{pd} \mathcal{R}^2 P_{LO} P_{in}^2}{\left[P_{mw\ sat} + 2P_{LO} \right]^2}, \tag{94}$$

where P_{mw} is the power of the input microwave signal.

The noise of the receiver is given by

$$\delta P_{RF\ BB} = \left[(1 + G)k_B T + 2e\mathcal{R}\rho_{pd} P_{out\ dc} \right] \Delta F, \tag{95}$$

where the first term in the right hand side corresponds to the thermal noise of the receiver, the second term (proportional to G) stands for the amplified thermal noise of the signal, and the last, term describes the shot noise; ΔF is the reception bandwidth.

In the most interesting case $G \gg 1$, and the noise floor (the maximum sensitivity of the receiver corresponding to the signal that makes signal-to-noise ratio equal to unity) of the receiver is

$$\delta P_{RF\ BB\ nf} = \frac{\delta P_{RF\ BB}}{G} \approx k_B T \Delta F. \tag{96}$$

The dynamic range is

$$DR \simeq \frac{P_{mw\ sat}}{2 k_B T \Delta F}. \tag{97}$$

We have realized such a receiver and an example of its response is shown in Fig. 7.

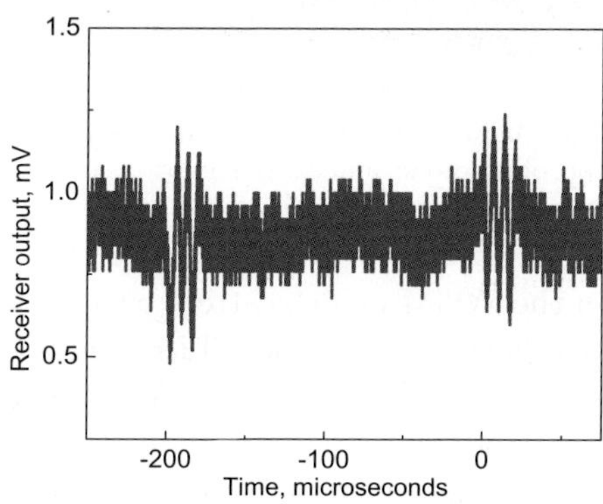

Fig. 7.   Response of the coherent quadratic receiver with no averaging performed. The RF signal pulses are characterized with 0.4 nW peak power, 5 kHz repetition rate, 20 μs duration, and 35 GHz carrier frequency. The microwave local oscillator has 10 μW power and the carrier frequency 200 kHz away from the carrier frequency of the signal.

A K_a−band coherent photonic receiver based on all-resonant interaction of light and RF radiation in solid-state WGM resonators have been recently packaged at OEwaves (Fig. 8).[54] The core of the receiver, the mixer based on material nonlinearity of the electro-optical WGM resonator, operates well for RF frequencies ranging from several GHz to a hundred GHz. The coherent photonic receiver has spurious free dynamic range in excess of 55 dB and sensitivity in excess of –100 dBm in 10 MHz reception band. The sensitivity does not degrade with increasing RF frequency. These devices are compact and have low power consumption. They also enable detecting and processing the RF signals in separate points in space, when combined with high quality optical links used to transmit the upconverted RF signals.

As mentioned above, the laser providing the optical pump must be frequency-locked to the operational WGM mode. Successful tight packaging of a WGM-resonator-based receiver at OEwaves was made possible thanks to deployment of optical injection locking technique of the semiconductor laser to the WGM resonator. The locking mechanism is very fast and it does not require any electronics.

Fig. 8.　Packaged OEwaves Ka-band photonic receiver prototype on RF test board. The entire optical bench (laser, resonator, photodiode, and coupling optics) is incorporated into surface mount RF package inside the interposer (center, lid removed).

The light propagating in the WGM mode scatters back to the laser due to residual scattering in the resonator host material as well as residual surface roughness. The scattering locks the laser frequency to the mode. As a result of this feedback, laser linewidth collapses and the optical frequency remains "parked" at WGM frequency within a practically useful locking range, up to several GHz in units of free-running laser frequency, or up to tens of milliAmperes in terms of driving current of utilized semiconductor laser. We measured the linewidth and the residual optical frequency instability by beating two prototype lasers injection-locked to two CaF_2 resonators Fig. (9). The free-running laser linewidth was measured to be approximately 2MHz. The linewidth of the locked laser was found to be less than 1 kHz.

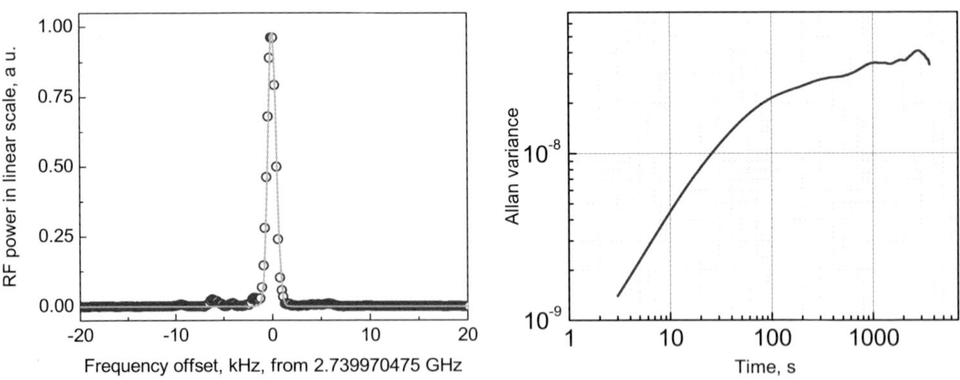

Fig. 9.　Left: RF spectrum of 2.74 GHz beatnote of two 1560 nm DFB lasers injection locked to two CaF_2 WGM resonators, independently packaged and temperature stabilized, optically isolated and pigtailed. Measured by Discovery DSC30 high speed photodiode and Agilent 8564 RF spectrum analyzer. Gaussian fit yields FWHM at 830 Hz; at nominal instrument resolution bandwidth 1 kHz. Right: Allan variance of the beatnote.

3.4. *Fundamental limitations of sensitivity of a photonic receiver*

To realize the potential of photonic microwave receivers let us discuss their fundamental noises in this section. We will first analyze the receiver based on the phase-type WGM modulator and then consider advantages of usage of a single-sideband WGM modulator for the reception purposes.

3.4.1. *Receiver based on usual phase EOM*

Let us assume that one measures $\hat{E}^{\dagger}_{out+}\hat{E}_{out+}$ (see Section 2.2) to detect a monochromatic microwave signal with averaged power $P_{in\,mw}$. The optical power that corresponds to photon number in the " $+$ " mode is given by

$$P_{out+} = P_{in}\frac{4g^2}{\gamma^2}\frac{2(P_{mw} + \Delta P_{mw})}{\hbar\omega_c\gamma_M}. \tag{98}$$

The microwave power P_{mw} is the averaged microwave signal and ΔP_{mw} is the noise part of the signal we calculate using Eq. (53). We find the following expressions for the averaged signal

$$
\begin{aligned}
P_{mw} &= P_{in\,mw} + \int_{-\infty}^{\infty} S(\tilde{\omega})\frac{d\tilde{\omega}}{2\pi} \\
&= P_{in\,mw} + \frac{\gamma\gamma_M}{\gamma + \gamma_M}\frac{k_B T}{2} + \frac{P|g|^2}{(\gamma + \gamma_M)^2}\frac{\omega_c}{\omega}\left(1 + \frac{\gamma_M}{2\gamma}\right),
\end{aligned}
\tag{99}
$$

$$S(\tilde{\omega}) = k_B T\frac{\gamma^2\gamma_M^2}{(\gamma^2 + \tilde{\omega}^2)(\gamma_M^2 + \tilde{\omega}^2)} + P\frac{\omega_c}{\omega}\frac{2|g|^2\gamma^2\gamma_M}{(\gamma^2 + \tilde{\omega}^2)^2(\gamma_M^2 + \tilde{\omega}^2)}, \tag{100}$$

and the noise

$$\langle\Delta P_{mw}^2\rangle = 2\bar{n}_{th}(\bar{n}_{th} + 1)\left[\frac{\hbar\omega_c}{2}\frac{\gamma\gamma_M}{\gamma + \gamma_M}\right]^2 + P_{in\,mw}\int_{-\infty}^{\infty}\left[\frac{\hbar\omega\gamma}{2P}\frac{\gamma\gamma_M}{4|g|^2}\hbar\omega_c + 2S(\tilde{\omega})\right]\frac{d\tilde{\omega}}{2\pi}.$$

It is worth noting that the detection bandwidth for the noise floor (calculated for $P_{in\,mw} \to 0$) is given by the geometrical average of the bandwidths of the microwave and optical resonators $(\gamma^{-1} + \gamma_M^{-1})^{-1}$.

It is possible to introduce the noise power density for the case $P_{in\,mw} \neq 0$:

$$\Delta P(\tilde{\omega}) = \hbar\omega_c\left[\frac{\hbar\omega\gamma}{2P}\frac{\gamma\gamma_M}{4|g|^2} + \frac{1}{2}\frac{2P}{\hbar\omega\gamma}\frac{4|g|^2}{\gamma\gamma_M}\frac{\gamma^4\gamma_M^2}{|\Gamma|^4|\Gamma_M|^2}\right] + 2k_B T\frac{\gamma^2\gamma_M^2}{|\Gamma|^2|\Gamma_M|^2}. \tag{101}$$

This noise power density determines the signal-to-noise ratio of the receiver

$$\frac{S}{N} = \frac{P_{in\,mw}}{\int \Delta P(\tilde{\omega})d\tilde{\omega}/(2\pi)}.$$

Eq. (101) contains three terms. The first term, inversely proportional to P, is due to the photon shot noise. The second term, proportional to P, comes from the spontaneous emission noise. Finally, the third term, independent of P, is from the thermal noise.

We are interested in the case $|\tilde{\omega}| < \gamma, \gamma_M$. The receiver has the smallest noise power density

$$\Delta P(\tilde{\omega})|_{min} = \sqrt{2}\hbar\omega_c + 2k_B T, \tag{102}$$

if

$$\frac{2P_{opt}}{\hbar\omega\gamma}\frac{4|g|^2}{\gamma\gamma_M} = \sqrt{2}. \tag{103}$$

Let us represent the optimal power in measurable units. We introduce a directly measurable modulation coefficient m

$$m = \frac{P_{out+}}{P_{out}} = 8\frac{|g|^2}{\gamma^2}\frac{P_{mw}}{\hbar\omega_c\gamma_M}. \tag{104}$$

It is easy to measure experimentally the input microwave power $P_{mw\,sat}$ corresponding to $m \simeq 1$. The optimum optical power is given then by

$$P_{opt} = \sqrt{2}\frac{P_{mw}}{m}\frac{\omega}{\omega_c} \approx P_{mw\,sat}\frac{\omega}{\omega_c}. \tag{105}$$

The value of the optimal power shows if the nonlinear noise is important. In the case of small optical power $P \ll P_{opt}$ the noise is primarily given by the photon shot noise and $\Delta P(\tilde{\omega})$ equals to

$$\frac{\Delta P(\tilde{\omega})}{P_{mw}} \simeq \frac{2\hbar\omega}{P_{out+}} + \frac{2k_B T}{P_{mw}}. \tag{106}$$

In the case of large optical power $P > P_{opt}$ the nonlinear noise exceeds the shot noise.

3.4.2. *Receiver based on a single sideband EOM*

An advantage of microwave photonic receivers is their ability to detect high frequency RF, microwave, and even THz signals. In what follows we discuss noise properties of photonic THz receiver based on WGM resonators.[83]

The nearly ideal THz reception is possible if spontaneous emission processes are suppressed in the receiver. Spontaneous noise is absent in the case of the anti-Stokes conversion, when the generated optical photon has higher frequency than the carrier frequency of the optical pump. We have considered such a case and studied the interaction of two optical mode and a single THz mode via $\chi^{(2)}$ nonlinearity in a WGM resonator, focusing on the interaction of externally pumped optical and THz modes with finite Q-factors.[83] We study generation of an anti-Stokes sideband in another optical mode occurring as a result of the interaction and show that the anti-Stokes light carries information about THz radiation. The frequency of the anti-Stokes sideband, $\omega + \omega_M$, is given by frequencies of optical (ω) and THz (ω_M) pumping respectively. We assume that the neighboring optical mode is nearly resonant with the sideband and that the process is phase matched in the sense of orbital numbers. The phase matching can be achieved by properly changing the morphology of WGM resonators.

The parametric interaction takes place if the phase matching condition is fulfilled:

$$\eta = \frac{1}{\mathcal{V}} \int_{\mathcal{V}} d\mathcal{V} \Psi_a \Psi_b \Psi_c \neq 0, \tag{107}$$

where $\mathcal{V}$ is the volume of the optical mode, Ψ_a, Ψ_b, and Ψ_c are the normalized dimensionless spatial distributions of the modes. We assume that the optical WGMs are nearly identical, $\mathcal{V} \simeq \mathcal{V}_+$ and that the volume of the THz mode is much larger than the volume of the optical modes, so the integration can be taken only in the volume occupied by the light.

Let us show that the phase matching can be achieved by selecting the proper size and morphology of the resonator. We assume that $\omega = 10^{15}$ s^{-1} and $\omega_M = 6.28 \times 10^{12}$ s^{-1}. The indexes of refraction of the material are 2.2 in optical and 5 in THz domains. We assume that the resonator is a cylinder with radius R and thickness L and select all the WGMs to belong to the basic TE mode family. The frequencies of the modes obey the equation

$$\omega_l \simeq \frac{c}{n} \left\{ \left[l + \alpha_1 \left(\frac{l}{2} \right)^{1/3} - \frac{n}{\sqrt{n^2 - 1}} \right]^2 + \left[\frac{\pi}{L} \right]^2 \right\}^{1/2}, \tag{108}$$

where l is the mode number, $\alpha_1 \simeq 2.338$, and n is the refractive index of the mode. By simple algebra we verify that resonator with $R = 0.1$ cm and $L = 34.5$ μm has modes with given frequencies $\omega_a \simeq 10^{15}$ s^{-1} ($l_a = 7350$), $\omega_b \simeq 10^{15} + 2\pi \times 10^{12}$ s^{-1} ($l_b = 7396$), and $\omega_M = 6.28 \times 10^{12}$ s^{-1} ($l_c = 46$). Because $l_b - l_a = l_c$ the modes are phase matched and $\eta_{pm} = 1$. This technique is similar to the phase matching technique of nonlinear processes via geometrical confinement of the light.[84]

Interaction of the optical and THz fields is described by equations (c.f. 54-57)

$$\dot{\hat{A}} = -\gamma \hat{A} - ig^* \hat{C}^\dagger \hat{B}_+ + \hat{F}_A, \tag{109}$$

$$\dot{\hat{B}}_+ = -\gamma \hat{B}_+ - ig \hat{C} \hat{A} + \hat{F}_{B+}, \tag{110}$$

$$\dot{\hat{C}} = -\gamma_M \hat{C} - ig^* \hat{A}^\dagger \hat{B}_+ + \hat{F}_C. \tag{111}$$

where $\hat{A}$, $\hat{B}_+$, and $\hat{C}$ are the slowly-varying amplitudes pump (optical), idler(optical), and signal (THz) fields; γ and γ_M are the optical and THz decay rates respectively (we assume that the optical modes are overloaded so the intrinsic losses can be neglected and γ is the decay rate of the loaded modes); g is the coupling constant, $\hat{F}_A$, and $\hat{F}_{B+}$ are the Langevin forces, properties of which are given by

$$\langle \hat{F}_A \rangle = \sqrt{\frac{2P\gamma}{\hbar\omega}}, \tag{112}$$

$$\langle \hat{F}_A(t)\hat{F}_A^\dagger(t') \rangle = 2\gamma\delta(t-t'), \tag{113}$$

$$\langle \hat{F}_{B+}(t)\hat{F}_{B+}^\dagger(t') \rangle = 2\gamma\delta(t-t'), \tag{114}$$

where $\langle \ldots \rangle$ stands for the ensemble averaging, P is the power of the external optical pump of mode A. We assume that the sources of thermal noise are uncorrelated. The Langevin force describing the thermal fluctuations consists of two uncorrelated parts,

$$\hat{F}_C = \hat{F}_{Cs} + \hat{F}_{Cr}, \tag{115}$$

$$\langle \hat{F}_{Cs} \rangle = \sqrt{\frac{2P_{mw}\gamma_{Ms}}{\hbar\omega_c}}, \tag{116}$$

$$\langle \hat{F}_{Cs}(t)\hat{F}_{Cs}^\dagger(t') \rangle = 2\gamma_{Ms}(\bar{n}_{th\ s} + 1)\delta(t-t'), \tag{117}$$

$$\langle \hat{F}_{Cr}(t)\hat{F}_{Cr}^\dagger(t') \rangle = 2\gamma_{Mr}(\bar{n}_{th\ r} + 1)\delta(t-t'), \tag{118}$$

where γ_{Ms} and γ_{Mr} stand for coupling of the thermal fluctuations from the signal channel and receiver channel to the THz mode (the spectral width of the THz resonance is given by $\gamma_M = \gamma_{Ms} + \gamma_{Mr}$), $\bar{n}_{th\ s} = [\exp(\hbar\omega_c/k_B T_s) - 1]^{-1}$ is the averaged number of thermal photons coming from the signal, $\bar{n}_{th\ r} = [\exp(\hbar\omega_c/k_B T_r) - 1]^{-1}$ is the average number of thermal photons coming from the receiver, P_{mw} is the power of the external THz pump of mode C. The other expectation values, quadratic deviations, and correlations are equal to zero.

We solve linearized set of differential equations (109-111) using standard Fourier technique assuming undepleted classical pump: $\hat{A} \to A = \langle \hat{F}_A \rangle/\gamma =$ const. This approximation is valid if $|g|^2|\langle \hat{n}_c \rangle|^2/\gamma^2 \ll 1$. We present the forces and the field operators as a sum of an expectation and fluctuational parts like

$$\hat{F}_{B+} = \int_{-\infty}^{\infty} \hat{f}_{B+}(\tilde{\omega})e^{-i\tilde{\omega}t}\frac{d\tilde{\omega}}{2\pi}, \tag{119}$$

$$\hat{F}_C = \langle \hat{F}_{Cs} \rangle + \int_{-\infty}^{\infty} [\hat{f}_{Cs}(\tilde{\omega}) + \hat{f}_{Cr}(\tilde{\omega})]e^{-i\tilde{\omega}t}\frac{d\tilde{\omega}}{2\pi}, \tag{120}$$

where $\langle \hat{F}_{Cs} \rangle$ stands for the THz signal, $\hat{f}_{B+}(\omega)$ and $\hat{f}_C(\omega)$ are the noise Fourier components, $\langle \hat{f}_{B+}(\omega)\hat{f}_{B+}(\omega)^\dagger \rangle = 4\pi\gamma\delta(\omega - \omega')$ and $\langle \hat{f}_{Cs}(\omega)\hat{f}_{Cs}(\omega)^\dagger \rangle = 4\pi\gamma_{Ms}(\bar{n}_{th\ s} + 1)\delta(\omega - \omega')$ (similar expression for $\hat{f}_{Cr}(\omega)$); and the field operators have presentation similar to that one of the Langevin forces

$$\hat{B}_+ = \int_{-\infty}^{\infty} \delta\hat{B}_+(\tilde{\omega})e^{-i\tilde{\omega}t}\frac{d\tilde{\omega}}{2\pi}, \tag{121}$$

$$\hat{C} = C + \int_{-\infty}^{\infty} \delta\hat{C}(\tilde{\omega})e^{-i\tilde{\omega}t}\frac{d\tilde{\omega}}{2\pi}. \tag{122}$$

The solution of the linearized set (109-111) with respect to the fluctuations is given by the Fourier amplitudes

$$\delta\hat{B}_+(\tilde{\omega}) = \frac{\Gamma_M}{\Gamma\Gamma_M + |g|^2|A|^2}\hat{f}_{B+}(\tilde{\omega}) - \frac{igA}{\Gamma\Gamma_M + |g|^2|A|^2}\hat{f}_C(\tilde{\omega}),$$

$$\delta\hat{C}(\tilde{\omega}) = \frac{\Gamma}{\Gamma\Gamma_M + |g|^2|A|^2}\hat{f}_C(\tilde{\omega}) - \frac{ig^*A^*}{\Gamma\Gamma_M + |g|^2|A|^2}\hat{f}_{B+}(\tilde{\omega}),$$

where we use notations $\hat{f}_C(\tilde{\omega}) = \hat{f}_{Cs}(\tilde{\omega}) + \hat{f}_{Cr}(\tilde{\omega})$, $\Gamma = \gamma - i\tilde{\omega}$ and $\Gamma_M = \gamma_M - i\tilde{\omega}$.

Neglecting the optical saturation of the pump field, we obtain the following expectation values for the fields from (109-111)

$$A \simeq \frac{\langle\hat{F}_A\rangle}{\gamma}, \tag{123}$$

$$C \simeq \frac{\langle\hat{F}_C\rangle}{\gamma_M}\frac{\gamma\gamma_M}{\gamma\gamma_M + |g|^2|A|^2}, \tag{124}$$

$$B_+ \simeq \frac{igA}{\gamma}C. \tag{125}$$

It is important to note at this point that there exists a maximum of the sideband amplitude B_+ with respect to the pump amplitude $|A|$ for a given signal amplitude ($\langle\hat{F}_{Cs}\rangle$). This is similar to the saturation phenomena in Raman lasers occurring due to the power broadening.

We finally derive expressions for the expectation values

$$E_{out} \simeq E_{in}, \quad E_{out+} = E_{in}\frac{2ig}{\gamma}C. \tag{126}$$

as well as fluctuations of the output optical field of the sideband

$$\hat{e}_{out+}(\tilde{\omega}) = \frac{\Gamma^*\Gamma_M - |g|^2|A|^2}{\Gamma\Gamma_M + |g|^2|A|^2}\hat{e}_{in+}(\tilde{\omega}) - \tag{127}$$

$$\frac{2igA\sqrt{\gamma}}{\Gamma\Gamma_M + |g|^2|A|^2}\left[\sqrt{\gamma_{Ms}}\hat{e}_{Cs}(\tilde{\omega}) + \sqrt{\gamma_{Mr}}\hat{e}_{Cr}(\tilde{\omega})\right].$$

We used expressions $\langle A\rangle\sqrt{2\gamma\tau} = 2E_{in}$, $F_A = E_{in}\sqrt{2\gamma/\tau}$, $E_{out\pm} = B_\pm\sqrt{2\gamma\tau}$, τ is the resonator round trip time $\tau = 2\pi R/c$, where R is the radius of the resonator. We assume, for the sake of simplicity, that the resonator is phase matched (or empty), so that $\tau = \tau_M$.

Eq. (127) shows that the fluctuations of the anti-Stokes sideband are given by the optical shot noise fluctuations as well as by the thermal fluctuations in the THz mode and the receiver. It is easy to see that the thermal noise contributions come with coefficients describing the coupling of the thermal sources to the resonator mode. The effective mode temperature is (in the limit of large temperature, when $n_{th} \approx k_B T/\hbar\omega_c$)

$$T_{eff} \approx \frac{\gamma_{Ms}}{\gamma_M}T_s + \frac{\gamma_{Mr}}{\gamma_M}T_r. \tag{128}$$

The contribution from the receiver temperature is significantly suppressed if the resonator is overcoupled, $\gamma_s \gg \gamma_r$.

The fluctuations of the outgoing light do not depend on the optical pumping. The photon number in the anti-Stokes mode is given by the thermal fluctuations and is independent on the pumping power if there is no THz signal. Hence, the excess power fluctuations of the power of the pump laser will not impact the sensitivity in the case of undepleted pump.

Let us consider an important case for THz astronomy and assume that the noise background $\hat{e}_{Cs}$ is our signal ($P_{mw} = 0$). The power noise of the anti-Stokes light is given by expression

$$\langle \Delta P_{mw}^2 \rangle \simeq 2 \frac{\gamma_{Mr}^2}{\gamma_{Mc}^2} n_{th\,r}^2 \left[\frac{\hbar \omega_c}{2} \frac{\gamma \gamma_{Mc} + |g|^2 |A|^2}{\gamma + \gamma_{Mc}} \right]^2 .$$

In other words, the detector temperature is $T_{eff} \simeq T_r \gamma_{Mr} / \gamma_{Mc}$ now.

Let us consider another example and assume that one measures photon number $\hat{E}_{out+}^\dagger \hat{E}_{out+}$ in the anti-Stokes field to detect a monochromatic THz signal with averaged power $P_{in\,mw}$ when $T_r = T_s = T$. One will observe the averaged signal

$$P_{mw} = P_{in\,mw} + \frac{\gamma \gamma_M + |g|^2 |A|^2}{\gamma + \gamma_M} \frac{k_B T}{2} . \tag{129}$$

masked by noise

$$\langle \Delta P_{mw}^2 \rangle = 2 n_{th} (n_{th} + 1) \left[\frac{\hbar \omega_c}{2} \frac{\gamma \gamma_M + |g|^2 |A|^2}{\gamma + \gamma_M} \right]^2 +$$

$$P_{in\,mw} \int_{-\infty}^{\infty} \left\{ \frac{\hbar \omega_c}{4} \left(\frac{\sqrt{\gamma \gamma_M}}{|g||A|} + \frac{|g||A|}{\sqrt{\gamma \gamma_M}} \right)^2 + 2 k_B T \left| \frac{\gamma \gamma_M + |g|^2 |A|^2}{\Gamma \Gamma_M + |g|^2 |A|^2} \right|^2 \right\} \frac{d\tilde{\omega}}{2\pi} .$$

In the case the signal has significant average power $P_{in\,mw} \neq 0$ and $|\tilde{\omega}| < \gamma, \gamma_M$ we have

$$\frac{\langle \Delta P_{mw}^2 \rangle}{P_{in\,mw}^2} \simeq \frac{\int \Delta P_{mw}(\tilde{\omega}) d\tilde{\omega}/(2\pi)}{P_{in\,mw}} \tag{130}$$

The density of power fluctuations $\Delta P_{mw}(\tilde{\omega})$ is given by

$$\Delta P_{mw}(\tilde{\omega}) = \frac{\hbar \omega_c}{4} \left(\frac{\sqrt{\gamma \gamma_M}}{|g||A|} + \frac{|g||A|}{\sqrt{\gamma \gamma_M}} \right)^2 + 2 k_B T . \tag{131}$$

The receiver has the smallest noise floor

$$\Delta P_{mw}(\tilde{\omega})|_{min} = \hbar \omega_c + 2 k_B T, \tag{132}$$

if

$$\frac{2 P_{opt}}{\hbar \omega \gamma} \frac{|g|^2}{\gamma \gamma_M} = 1. \tag{133}$$

This is an obvious result for the maximum sensitivity of a linear coherent detector of THz radiation. Now, if one uses the photonic receiver in the THz photon counting mode, so that $P_{in\,mw} = 0$, the detection noise (130) coincides with the noise of an ideal quadratic THz receiver. Thermal noise in the THz mode determines the maximum measurement sensitivity at room temperature. Shot noise determines the sensitivity in a cooled receiver.

4. Generation of optical pulses using WGM resonators

A short length for the optical resonator is essential for establishing stable generation of optical pulses with high repetition rates. In the case of fiber lasers, for example, the dense mode spectrum and material nonlinearity of a long length of fiber forming the ring ring resonator results in various kinds of instabilities. Here, an optical soliton becomes unstable as the soliton-laser cavity approaches the length of several soliton periods.[85,86] Furthermore, long cavity based harmonically mode-locked oscillators suffer from the supermode noise.[87] Short cavities allow for solutions to these problems. For example, 2 ps pulses at a 16.3 GHz repetition rate were obtained for a 2.5 mm-long actively mode-locked monolithic laser;[88] 420 GHz subharmonic synchronous mode locking was realized in a laser cavity of total length approximately 174 μm.[89] A significant supermode noise suppression was demonstrated by inserting a small high-finesse Fabry-Perot resonator to the cavity of an actively mode-locked laser.[90,91]

In a system that includes a high-Q resonator, the minimum pulse width and timing that characterizes an optical pulse train generated is determined by the resonator dispersion. Depending on the material and geometrical size, WGM resonators can possess either positive, or negative, or zero group velocity dispersion (GVD).[92] Resonators possessing positive group velocity dispersion may be used for GVD compensation in optical fiber lines. Negative GVD resonators sustain nonlinear Schrödinger soliton propagation and may be used for pulse shaping and soliton shortening in conventional mode-locked lasers (see, e.g. Refs. 93–95). Zero GVD resonators may be used as high-finesse etalons to stabilize actively mode-locked lasers (as in Ref. 91).

We have proposed to use WGM electro-optical modulator for generation of short optical pulses by a direct conversion of a continuous laser beam into a stable pulse train.[92] It is known that an electro-optic modulator placed in an optical resonator may generate an optical frequency comb.[96–99] The output of such a device is similar to that of a mode locked laser. However, the system is passive and the pulse duration is not limited by the bandwidth, as is the case in the mode-locked laser. The pulse width decreases with the modulation index increase, and with the overall resonator dispersion decrease.

We have proposed an architecture including an advanced monolithic integrated mode-locked source based on WGM dielectric resonator. This source would generate low jitter picosecond optical pulse trains with repetition rate up to 100 GHz, consume low microwave power, and have low oscillation threshold. The active element would have several millimeter in size and it would not need any active control of the optical cavity length. The idea of this laser is based on two recently realized WGM devices: electro-optical modulator and Er-doped microsphere glass laser.[37,100–103]

Finally, we have shown that the WGM electro-optic modulator can improve performance of a coupled opto-electronic oscillator.[92] The main advantage of the WGM OEO is that it also play a role of a photonic filter that stabilizes the system.

4.1. *Optical comb generation with a WGM EOM*

Optical frequency comb generators are usually based on an intracavity electro-optical modulator. Such devices can produce picosecond or even sub-picosecond optical pulses.[99] The devices can be small in size. The operation frequency is determined by the pump laser so short optical pulses can be produced in a broad range of wavelengths using the same comb generator and changing pump lasers. The repetition rate of the pulses may be very high and, unlike to an active mode-locked laser, it is equal to twice the EOM modulation frequency.

EOM based on a WGM resonator[57] may produce a comb of optical harmonics in the same fashion as conventional EOMs do. Because of the high finesse of WGMs and small geometrical dispersion introduced by the resonator structure compared with the material dispersion (this is generally the case for lithium niobate and large enough resonators) the number of harmonics may be as large as in a Fabry-Perot resonator with a conventional EOM inside. If a WGM resonator modulator is utilized, the system would also benefit from the improved modulation index.

The scheme of the WGM comb generator coincides with the scheme of the WGM EOM (Fig. 2). The resonant frequency of the microwave field should be adjusted to fit the frequency difference between optical modes by change of the microwave resonator shape. Due to quadratic nonlinearity of $LiNbO_3$ the modes of the microwave and optical resonators are effectively coupled. This coupling increases significantly for resonant tuning of the fields due to high quality factors of the modes of optical and microwave resonators as well as small mode volumes.

The spectral width of the comb can be found from general expression for the modulation efficiency (84). The number of the sidebands is restricted by the optical dispersion of the resonator. In approximation $\mathcal{F} \to \infty$ the maximum frequency span of the generated harmonics is (see Ref. 98 and (84))

$$\delta f_{comb} \simeq \sqrt{\frac{2\omega r_{33} n_e^2}{\pi^2 \omega_M |\beta_2| R} \left(\frac{\pi P_{mw} Q_M}{n_c^2 \omega_M V_c}\right)^{1/2} \eta}, \qquad (134)$$

where we assumed that the length of the EOM is approximately πR, β_2 is the second order dispersion of the resonator ($|\beta_2| \sim 100$ ps^2/km in $LiNbO_3$ at 1550 nm), R is the radius of the resonator. The value of δf_{comb} can exceed several THz in WGM resonators.

4.2. *Actively mode-locked WGM lasers*

Generation of high-repetition rate picosecond optical pulses at 1.5 μm is important for modern optical telecommunications and microwave photonic applications. In this regard, various schemes of mode-locked lasers were proposed. Actively mode-locked lasers based on intracavity amplitude or phase modulation are among the most promising here.[104–108]

The repetition rate of the lasers is generally restricted by the pulse roundtrip time in the cavity. The shorter is the round-trip time, the higher is the repetition rate. High repetition rate fiber lasers with long roundtrip time are operated in a high order harmonic configurations and suffer from mode competition. Stabilization of such lasers requires use of intracavity narrow-band filters.[90]

It was shown recently that doped electro-optic materials allow for monolithic integration of an active mode-locker.[109] Erbium doped $LiNbO_3$ is attractive for such applications because of its excellent electro-optic properties on one hand, and Er solubility in $LiNbO_3$ crystals without fluorescence quenching on the other.[110]

Active mode-locking at up to 10 GHz pulse repetition rate have been already demonstrated in $Ti:Er:LiNbO_3$ broad-band Fabry-Perot waveguide cavities.[111–114] The integrated cavity should be long enough to achieve significant frequency modulation necessary for the mode locking for these lasers. Stable pulse generation at high repetition rate with long active cavity needs supermode selection with additional passive cavity. EOM integrated into the cavity requires about 0.5 Watt of applied microwave power to achieve full modulation.

We here discuss an architecture of an advanced monolithic integrated mode-locked source based on WGM nonlinear dielectric resonator.[45] This source would generate low jitter picosecond optical pulse trains with repetition rate up to 100 GHz. It will consume low microwave power and have low oscillation threshold. The active element will have several millimeter in size and it will not need any active control of the optical cavity length, though thermal stabilization would be advantageous.

A configuration of the laser we discuss here nearly coincides with the configuration of the WGM-based EOM. A cw pump laser radiation at $\lambda_p = 1.48$ μm is sent into z-cut $Er:LiNbO_3$ spheroid optical resonator via a coupling prism. Oblate spheroid resonator shape is essential to clean up the cavity spectrum. Modes of the cavity with frequencies $\lambda_s \simeq 1.54$ μm experience amplification due to interaction of the pump and the erbium ions. The system emits coherent radiation at λ_s when the pump power exceeds some threshold value. The optical resonator is placed between two plates of the microwave resonator, like in a usual WGM EOM.

The device becomes a classical example of an actively mode-locked laser when both cw microwave radiation and optical pump are applied. Interaction of the fields results in generation of pulses in the resonator. Because the optical amplification procedure is not phase sensitive, pulses running in both directions around the cavity rim should be observed.

There is a significant difference between construction of phase modulator in our case compared with the modulators described in Refs. 105 and 106 and other studies where time dependent description of mode locking was used. In those works the length of the modulator was much shorter than the length of the pulse round trip in a cavity. Therefore, it was possible to assume that the interaction time of the pulse and the modulator is short enough and use this fact for a decomposition of the expression for the modulator response by a small parameter $\omega_M t$, where ω_M is the modulation frequency, and t is the interaction time.

In our case a pulse travels a half of its round trip time in the modulator, i.e. $\omega_M t \simeq \pi$. The same condition is valid in the experiments with Er:Ti:LiNbO$_3$ mode locked lasers[113] where the length of the modulator is comparable with the path length of a pulse. In the theoretical description of such a mode locking[115] Floquet theory was used. This theory is powerful for numerical simulations, however it does not give a clear analytical expression for the pulse parameters. In what follows we show how to modify the problem of a long modulator mode locked laser to solve it analytically.

One more peculiarity of the device is that the microwave field is excited in a microwave resonator and the amplitude of the field changes along the rim of the cavity, not only in time, as in usual short modulators utilized in mode locked lasers. It is this geometry which allows us to achieve an efficient modulation and interaction among cavity modes.[45]

In augmentation to the conventional theory,[105,106] let us assume that the pulse length is much smaller than the modulator length. Than the round trip transmission through the modulator is given by

$$e^{i\delta(t)} = \exp\left[\frac{i}{2}\omega_0 n_e^2 r_{eff} \int_0^{\pi/\omega_M} E_M(\tau,t)d\tau\right], \qquad (135)$$

where t characterizes the time of entrance of the pulse into the modulator, $E_M(\tau,t)$ is the microwave electric field that the pulse sees, r_{eff} is the electro-optic constant of the material. The electro-optic constant does not necessary coincide with r_{33} because we may use either TE or TM modes, and the electric field in the resonator may be not exactly be aligned with the z direction due to boundary effects for the microwave cavity.

The electric field that short pulse see at time τ in the modulator can be approximated by

$$E_M(\tau,t) = E_{M0}\sin\left(\frac{\pi}{2} - \omega_M\tau\right)\cos\left[\omega_M(\tau+t)\right], \qquad (136)$$

where the "cos" term appears due to temporal modulation of the field, and "sin" results from the pulse motion in the microwave resonator. Eq. (136) shows, for example, that the maximum modulation is on the boundaries of the resonator and there is no modulation in the middle of the resonator. We finally derive

$$e^{i\delta(t)} = e^{2i\delta_e\cos\omega_M t}, \quad \delta_e = 8\pi\frac{\omega}{\omega_M}n_e^2 r_{eff}\left(\frac{\pi P_{mw}Q_M}{n_c^2\omega_M V_c}\right)^{1/2}\eta. \qquad (137)$$

We come to an expression similar to the expression used for the description of short cavity modulators implemented in a mode-locked laser cavities. Therefore, even in the case of long modulator the conventional mode locking theory[105,106] is valid. In our case t is the relative time of entrance of the pulse into the modulator. In accordance with usual FM mode locking theory, a pulse should enter the modulator at the point of maximum field amplitude $\cos \omega_M t = \pm 1$.

The active medium is responsible for lasing itself (amplification of the signal); for the absorption of the pump and, therefore, for the reduction of the effective quality factor of the whispering gallery mode the pump interacts with; and, finally, for the filtering of the signal pulses. The active medium has properties similar to the properties of the usual fiber erbium amplifiers.[116] These amplifiers may be characterized by homogeneous linewidth $\delta\lambda \approx 10$ nm or, in frequency units, $\Gamma \approx 2\pi \times 1.2 \times 10^{12}$ rad/s.

We assume that pulses are much longer than the inversed homogeneous gain linewidth. In this case the amplitude gain may be presented in form[105,106]

$$g(\omega) \simeq e^g \exp\left[2i\frac{g}{\Gamma}(\omega - \omega_0) - \frac{g}{\Gamma^2}(\omega - \omega_0)^2 \right], \tag{138}$$

where g is the saturated amplitude gain through the active medium at the line center for one round trip in the resonator. The direct computation of g are out of scope of the present study and will be presented elsewhere. The lasing threshold pump power, however, can be roughly estimated using the following reasoning. The gain depends on the pump power P as

$$g \sim \frac{P\alpha_{\Sigma\, d}}{P + P_{sat}}, \tag{139}$$

where P_{sat} is the saturation pump power for the pump transition, $\alpha_{\Sigma\, d}$ is the dopant mediated resonant linear absorption of the signal with no pump applied. Eq. (139) means that the gain cannot exceed this linear absorption of the signal and the maximum gain is achieved when the pump creates complete population inversion in the medium. To achieve generation the gain should exceed α_Σ, the residual attenuation of the signal not connected with the dopants. Generally $\alpha_{\Sigma\, d} \gg \alpha_\Sigma$, and therefore $P_{th} \approx (\alpha_\Sigma/\alpha_{\Sigma\, d})P_{sat}$. It is known that typical saturation pump power for an erbium doped fiber amplifier, for example, is approximately $P_{EDFA} = 1$ mW.[116] Therefore, P_{th} should be much less than 1 mW.

We estimate the duration of pulses generated in the system using an expression from[105]

$$\delta t = \frac{\sqrt{2\sqrt{2}\,\ln 2}}{\pi} \left(\frac{g}{\delta_e} \frac{(2\pi)^4}{\Gamma^2 \omega_M^2} \right)^{1/4}. \tag{140}$$

Assuming that $g \simeq \alpha_\Sigma = 0.006$ (α_Σ is the total optical loss in the loop), $\delta_e = 0.001$, $\Gamma \approx 2\pi \times 1.2 \times 10^{12}$ rad/s, and $\omega_M = 2\pi \times 10$ GHz we obtain $\delta t \approx 6$ ps.

It is interesting to note that instead of doping $LiNbO_3$ with erbium, it is possible to use erbium-doped solgel films applied to the surface of lithium niobate resonator to create low-threshold microresonator mode-locked lasers. Silica microsphere lasers with applied solgel films were demonstrated very recently.[103] The solgel does not reduce the quality factor of the microresonator significantly, while the erbium ions are coupled to the cavity modes via the evanescent field.

4.3. *Coupled opto-electronic oscillator with a WGM EOM*

Realization of a compact actively mode-locked laser with high pulse repetition rate is generally hindered by the need for a small high frequency stable microwave source to drive the electro-optical modulator inserted to the laser loop. The problem can be solved if the stable microwave radiation is generated in the same system. Such a device, in which a microwave oscillation and an optical oscillation are generated and directly coupled to each other, is called the Coupled Opto-Electronic Oscillator (COEO).[117,118]

A COEO consists of two photonic loops generating light as well as microwaves. The loops are connected by means of an electro-optic modulator. The transformation of the modulated light power into a microwave signal is achieved via a photodetector. Hence, in an COEO the laser light energy is converted directly to spectrally pure microwave signals, using an electro-optic feedback loop containing a high-Q optical element, at a frequency limited only by the available optical modulation and detection elements. This frequency is the repetition frequency of the optical pulses generated in the system.

The oscillating light loop is a usual ring fiber laser with either erbium or semiconductor optical amplifier. If the microwave photonic loop of the COEO is open, the ring laser generates several independent optical modes. The number of modes is determined by the loop length and the linewidth of the gain of the optical amplifier. If the microwave photonic loop is closed and sufficient microwave amplification is available to ensure microwave oscillations in the system in the manner of usual OEO, the optical modes become phase locked.

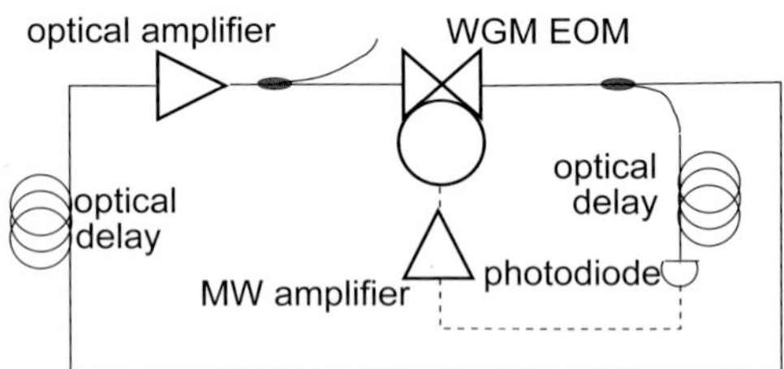

Fig. 10. A schematic of coupled opto-electronic oscillator with whispering gallery mode electro-optic modulator. Solid lines show optical fiber links, while dashed lines show microwave links. (Reprinted from Ref. 92 with permission of SPIE.)

The laser radiation propagates through a modulator and an optical energy storage element (delay line), before it is converted to the electrical energy with a photodetector. The electrical signal at the output of the modulator is amplified and filtered before it is fed back to the modulator, thereby completing a feedback loop with gain, which generates sustained oscillation. Since the noise performance of an oscillator is determined by the energy storage time, or quality factor Q, then the use of optical storage elements allows for the realization of extremely high Q's and thus spectrally pure signals.

The EOM is one of the main sources of power consumption in the COEO because of the large power required to drive the conventional modulators. Broadband Mach-Zender modulators used in COEOs typically require one to a few Watts of microwave power to achieve a significant modulation. This means that either the photocurrent in COEO system should be amplified significantly, or the laser loop of the COEO should operate much above laser threshold to produce enough optical radiation as the source of the drive power for the OEO. If the microwave power sent to the modulator is small, the information about the microwave signal simply will not be transduced to light through the EOM.

By utilizing a high-Q resonance, instead of the zero-order interferometry, as the basis for electro-optic modulation one can reduce the controlling power by many orders of magnitude and reduce the energy consumption of the COEO. For this purpose, the dielectric resonators with whispering gallery modes are useful.

Low timing jitter of generated pulses is one of the main desirable characteristics of a COEO. The timing jitter can be reduced if one uses a harmonically mode locked laser with a short, high-finesse resonator inserted into the long laser resonator.[91] Such a laser may have better performance than a fundamentally mode-locked laser that produces an identical pulse train. Because whispering gallery mode electro-optic modulator simultaneously play the role of a filter, we expect that mode-locked lasers as well as COEOs that utilize WGM EOMs will have low timing jitter in the generated pulses.

5. Conclusion

The small volume and the high quality-factor of whispering-gallery modes result in dramatic enhancement of efficiency of nonlinear optical effects in crystalline WGM resonators. We have described several examples of photonic applications of whispering-gallery mode resonators in functional radio frequency signal processing devices operating at frequencies laying in microwave and millimeter-wave band. We have reviewed proposed schemes and discussed demonstrations of quadratic photonic detectors, coherent quadratic receivers, terahertz receivers, optical pulse generators and parametric converters, as well as examples of small footprint packaging prototypes. Deployment of WGM resonators allows realization of injection-locked lasers with kilohertz linewidth as well as wave-mixing devices with micro-Watt thresholds. Current state-of-the-art in fabrication, optical coupling, and laser locking techniques for WGM crystalline resonators of electro-optic materials allow their successful integration into photonic signal processing devices.

Acknowledgments

Vladimir Ilchenko acknowledges rare but inspiring communications with Professor Chang since 1990-ies, and his support during early years in the United States. All authors acknowledge useful discussions with Dr. David Seidel.

References

1. R. E. Benner, P. W. Barber, J. F. Owen, and R. K. Chang, Phys. Rev. Lett. **44**, 475-478 (1980).
2. J. F. Owen, P. W. Barber, B. J. Messinger, and R. K. Chang, Opt. Lett. **6**, 272-274 (1981).
3. H.-M. Tzeng, K. F. Wall, M. B. Long, J. F. Owen, R. K. Chang, and P. W. Barber, Aerosol Science and Technol. **2**, 193-193 (1983).
4. H. M. Tzeng, K. F. Wall, M. B. Long, and R. K. Chang, Opt. Lett. **9**, 499-501 (1984).
5. H.-M. Tzeng, K. F. Wall, M. B. Long, and R. K. Chang, Opt. Lett. **9**, 273-275 (1984).
6. J. B. Snow, S.-X.Qian, and R. K. Chang, Opt. Lett. **10**, 37-39 (1985).
7. S.-X.Qian, J. B. Snow, and R. K. Chang, Opt. Lett. **10**, 499-501 (1985).
8. S. X. Qian, J. B. Snow, H. M. Tzeng, and R. K. Chang, Science **231**, 486-488 (1986).
9. J.-Z. Zhang, D. H. Leach, and R. K. Chang, Opt. Lett. **13**, 270-272 (1988).
10. R. K. Chang and A. J. Campillo, *Optical Processes in Microresonators*, Advanced Series in Applied Physics v.3 (World Scientific, Singapore, 1996).
11. M. H. Fields, J. Popp, and R. K. Chang, Prog. Opt. **41**, 1-95 (2000).
12. V. V. Datsyuk and I. A. Izmailov, Usp. Fiz. Nauk **171**, 1117-1129 (2001) [Physics-Uspekhi **44**, 1061-1073 (2001)].
13. A. N. Oraevsky, Quant. Electron. **32**, 377-400 (2002).
14. K. J. Vahala, Nature **424**, 839-846 (2003).
15. A. B. Matsko and V. S. Ilchenko, J. Sel. Top. Quant. Electron. **12**, 3-14 (2006).
16. V. B. Braginsky, M. L. Gorodetsky, and V. S. Ilchenko, Phys. Lett. A **137**, 393-397 (1989).
17. L. Collot, V. Lefevre-Seguin, M. Brune, J.-M. Raimond, and S. Haroshe, Europhys. Lett. **23**, 327-334 (1993).
18. M. L. Gorodetsky, A. A. Savchenkov, and V. S. Ilchenko, Opt. Lett. **21**, 453-455 (1996).
19. D. W. Vernooy, V. S. Ilchenko, H. Mabuchi, E. W. Streed, and H. J. Kimble, Opt. Lett. **23**, 247-249 (1998).
20. D. K. Armani, T. J. Kippenberg, S. M. Spillane, and K. J. Vahala, Nature **421**, 925-928 (2003).
21. J. Heebner, R. Grover, and T. Ibrahim, *"Optical Microresonators: Theory, Fabrication, and Applications,"* (Springer Verlag, London, 2007).
22. A. B. Matsko, Editor, *"Practical Applications of Microresonators in Optics and Photonics,"* (CRC Press, Boca Raton, 2009).
23. Z. Yang, P. Chak, A. D. Bristow, H. M. van Driel, R. Iyer, J. S. Aitchison, A. L. Smirl, and J. E. Sipe, Opt. Lett. **32**, 826-828 (2007).
24. V. S. Ilchenko, A. A. Savchenkov, A. B. Matsko, and L. Maleki, Phys. Rev. Lett. **92**, 043903 (2004).
25. T. J. Kippenberg, S. M. Spillane, and K. J. Vahala, Phys. Rev. Lett. **93**, 083904 (2004).
26. A. A. Savchenkov, A. B. Matsko, D. Strekalov, M. Mohageg, V. S. Ilchenko, and L. Maleki, Phys. Rev. Lett. **93**, 243905 (2004).

27. T. Carmon and K. J. Vahala, Nature Physics **3**, 430-435 (2007).

28. P. Del Haye, A. Schliesser, O. Arcizet, T. Wilkins, R. Holzwarth, T.J. Kippenberg, Nature **450**, 1214-1217 (2007).

29. H. B. Lin, A. L. Huston, B. J. Justus, and A. J. Campillo, Opt. Lett. **11**, 614-616 (1986).

30. V. Sandoghdar, F. Treussart, J. Hare, V. Lefevre-Seguin, J. -M. Raimond, and S. Haroche, Phys. Rev. A **54**, R1777-R1780 (1996).

31. H.-B. Lin and A. J. Campillo, Phys. Rev. Lett. **73**, 2440-2443 (1994).

32. S. M. Spillane, T. J. Kippenberg, and K. J. Vahala, Nature **415**, 621-623 (2002).

33. I. S. Grudinin and L. Maleki, Opt. Lett. **32**, 166-168 (2007).

34. I. S. Grudinin, A. B. Matsko, and L. Maleki, Phys. Rev. Lett. **102**, 043902 (2009).

35. M. Tomes and T. Carmon, Phys. Rev. Lett. **102**, 113601 (2009).

36. L. Maleki, V. S. Ilchenko, A. A. Savchenkov, and A. B. Matsko, Chapter 3 in *"Practical Applications of Microresonators in Optics and Photonics"* edited by A. B. Matsko, (CRC Press, Boca Raton, 2009).

37. V. S. Ilchenko, X. S. Yao, and L. Maleki, Proc. SPIE **3930**, 154-162 (2000).

38. V. S. Ilchenko and L. Maleki, Proc. SPIE **4270**, 120-130 (2001).

39. D. A. Cohen and A. F. J. Levi, Electron. Lett. **37**, 37-39 (2001).

40. D. A. Cohen, M. Hossein-Zadeh, and A. F. J. Levi, Electron. Lett. **37**, 300-301 (2001).

41. D. A. Cohen and A. F. J. Levi, Solid State Electron. **45**, 495-505 (2001).

42. D. A. Cohen, M. Hossein-Zadeh, and A. F. J. Levi, Solid State Electron. **45**, 1577-1589 (2001).

43. L. Maleki, A. F. J. Levi, S. Yao, and V. Ilchenko, US papent 6,473,218 (2002).

44. V. S. Ilchenko, A. A. Savchenkov, A. B. Matsko, and L. Maleki, IEEE Photon. Tech. Lett. **14**, 1602-1604 (2002).

45. V. S. Ilchenko, A. A. Savchenkov, A. B. Matsko, and L. Maleki, J. Opt. Soc. Am. B **20**, 333-342 (2003).

46. M. Hossein-Zadeh and A. F. J. Levi, Solid State Electron. **49**, 1428-1434 (2005).

47. M. Hossein-Zadeh and A.F.J. Levi, IEEE MTT **54**, 821-831 (2006).

48. V. S. Ilchenko, A. A. Savchenkov, I. Solomatine, D. Seidel, A. B. Matsko, and L. Maleki, IEEE Photon. Tech. Lett. **20**, 1600-1602 (2008).

49. V. S. Ilchenko, J. Byrd, A. A. Savchenkov, D. Seidel, A. B. Matsko, and L. Maleki, lanl.arXiv.org>physics>arXiv:0806.3239.

50. A. A. Savchenkov, W. Liang, V. S. Ilchenko, A. B. Matsko, D. Seidel, and L. Maleki, Proc. of 2009 IEEE Radar Conference, pp. 1-6, May 2009.

51. A. A. Savchenkov, W. Liang, A. B. Matsko, V. S. Ilchenko, D. Seidel, and L. Maleki, Opt. Lett. **34**, 1300-1302 (2009).

52. A. A. Savchenkov, W. Liang, A. B. Matsko, V. S. Ilchenko, D. Seidel, and L. Maleki, Proc. of 2009 IEEE/LEOS Summer Topical Meeting, pp. 63-64 July 2009.

53. A. A. Savchenkov, A. B. Matsko, V. S. Ilchenko, D. Seidel, and L. Maleki, Proc. of 2009 International Topical Meeting on Microwave Photonics, pp. 1-4 Oct. 2009.

54. A. B. Matsko, V. S. Ilchenko, P. Koonath, J. Byrd, A. A. Savchenkov, D. Seidel, and L. Maleki, Proc. of 2009 IEEE Radar Conference, pp. 1-6, May 2009.

55. A. A. Savchenkov, V. S. Ilchenko, A. B. Matsko, and L. Maleki, Electron. Lett. **39**, 389-391 (2003).

56. A. A. Savchenkov, V. S. Ilchenko, A. B. Matsko, and L. Maleki, IEEE Photonics Technology Lett. **17**, 136-138 (2005).

57. V. S. Ilchenko, A. B. Matsko, A. A. Savchenkov, and L. Maleki, J. Opt. Soc. Am. B **20**, 1304-1308 (2003).

58. M. Mohageg, D. Strekalov, A. A. Savchenkov, A. B. Matsko, V. S. Ilchenko, and

L. Maleki, Opt. Express **13**, 3408-3419 (2005).

59. A. A. Savchenkov, A. B. Matsko, D. Strekalov, V. S. Ilchenko, and L. Maleki, Appl. Phys. Lett. **88**, 241909 (2006).

60. A. A. Savchenkov, A. B. Matsko, D. Strekalov, V. S. Ilchenko, and L. Maleki, Phys. Rev. B **74**, 245119 (2006).

61. A. A. Savchenkov, A. B. Matsko, D. Strekalov, V. S. Ilchenko and L. Maleki, Opt. Commun. **272**, 257-262 (2007).

62. E. I. Gordon and J. D. Rigden, Bell System Tech. J. **42**, 155-179 (1963).

63. R. C. Alferness, IEEE Trans. Microwave Theor. and Techniques **30**, 1121-1137 (1982).

64. K. P. Ho and J. M. Kahn, IEEE Photon. Technol. Lett. **5**, 721-725 (1993).

65. T. Kawanishi, S. Oikawa, K. Higuma, Y. Matsuo, and M. Izutsu, Electron. Lett. **37**, 12441246 (2001).

66. I. L. Gheorma and R. M. Osgood, IEEE Photon. Technol. Lett. **14**, 795-797 (2002).

67. M. Kato, K. Fujiura, and T. Kurihara, Electron. Lett. **40**, 299-301 (2001).

68. N. Benter, R. P. Bertram, E. Soergel, K. Buse, D. Apitz, L. B. Jacobsen, and P. M. Johansen, Appl. Opt. **44**, 6235-6239 (2005).

69. M. Kato, K. Fujiura, T. Kurihara, Appl. Opt. **44**, 1263-1269 (2005).

70. P. Rabiei, W. H. Steier, C. Zhang, and L. R. Dalton, L. Lightwave Technol. **20**, 1968-1975 (2002).

71. M. A. J. Weldon, S. V. Hum, R. J. Davies, and M. Okoniewski, IEEE Photon. Technol. Lett. **16** 1295-1297 (2004).

72. H. Tazawa and W. H. Steier, Electron. Lett. **41**, 12971298 (2005).

73. H. Tazawa, W. H. Steier IEEE Photon. Technol. Lett. **18**, 211-213 (2006).

74. H. Tazawa, Y. H. Kuo, I. Dunayevskiy, J. D. Luo, A. K.-Y. Jen, H. R. Fetterman, and W. H. Steier, J. Lightwave Technology **24**, 3514-3519 (2006).

75. S. Li, F. Yi, X. M. Zhang, and S. L. Zheng, Microwave Opt. Technol. Lett. **49**, 313-316 (2007).

76. B. Bortnik, Y.-C. Hung, H. Tazawa, B.-J. Seo, J. D. Luo, A. K.-Y. Jen, W. H. Steier, and H. R. Fetterman, IEEE J. Sel. Top. Quantum Electron. **13**, 104-110 (2007).

77. A. B. Matsko, A. A. Savchenkov, V. S. Ilchenko, D. Seidel, and L. Maleki, Opt. Express **15**, 17401-17409 (2007).

78. E. A. Mishkin and D. F. Walls, Phys. Rev. **185**, 1618-1628 (1969).

79. D. F. Walls and G. J. Milburn, *"Quantum Optics"* (Springer, New York, 1994).

80. M. E. Crenshaw and C. M. Bowden, Opt. Commun. **203**, 115-124 (2002).

81. K. Ohata, T. Inoue, M. Funabashi, A. Inoue, Y. Takimoto, T. Kuwabara, S. Shinozaki, K. Maruhashi, K. Hosaya, and H. Nagai, IEEE Trans. Microw. Theory Tech. **44**, 2354 (1996).

82. D. M. Pozar, *Microwave engineering.* New York: Wiley, 1998.

83. A. B. Matsko, D. V. Strekalov, and N. Yu, Phys. Rev. A 77, 043812 (2008).

84. A. Yariv, IEEE J. Quantum Electron. **QE-9**, 919-933 (1973).

85. L. F. Mollenauer, J. P. Gordon, and M. N. Islam, IEEE J. Quantum Electron. **22**, 157-173 (1986).

86. H. A. Haus, K. Tamura, L. E. Nelson, and E. P. Ippen, IEEE J. Quantum Electron. **31**, 591-598 (1995).

87. M. F. Becker, D. J. Kuizenga, and A. E. Siegman, IEEE J. Quantum Electron. **QE 8**, 687-693 (1972).

88. K. Sato, K. Wakita, I. Kotaka, Y. Kondo, M. Yamamoto, and A. Takada, Appl. Phys. Lett. **65**, 1-3 (1994).

89. S. Arahira and Y. Ogawa, IEEE Photon. Technol. Lett. **14**, 537-539 (2002).

90. G. T. Harvey and L. F. Mollenauer, Opt. Lett. **18**, 187-189 (1993).

91. C. M. DePriest, T. Yilmaz, P. J. Delfyett, S. Etemad, A. Braun, and J. Abeles, Opt. Lett. **27**, 719-721 (2002).

92. L. Maleki, A. A. Savchenkov, V. S. Ilchenko, and A. B. Matsko, Proc. SPIE **5104**, 1-13 (2003).

93. J. D. Kafka, T. Baer, and D. W. Hall, Opt. Lett. **14**, 1269-1271 (1989).

94. F. X. Kartner, D. Kopf, and U. Keller, J. Opt. Soc. Am. B **12**, 486-496 (1995).

95. T. F. Garruthers and I. N. Duling III, Opt. Lett. **21**, 1927-1929 (1996).

96. M. Kourogi, K. Nakagawa, and M. Ohtsu, IEEE J. Quantum Electron. **29**, 2693-2701 (1993).

97. L. R. Brothers, D. Lee, and N. C. Wong, Opt. Lett. **19**, 245-247 (1994).

98. M. Kourogi, B. Widiyatomoko, Y. Takeuchi, and M. Ohtsu, IEEE J. Quantum Electron. **31**, 2120-2126 (1995).

99. G. M. Macfarlane, A. S. Bell, E. Riis, and A. I. Ferguson, Opt. Lett. **21**, 534-536 (1996).

100. M. Cai, O. Painter, K. J. Vahala, and P. C. Sercel, Opt. Lett. **25**, 1430-1432 (2000).

101. W. von Klitzing, E. Jahier, R. Long, F. Lissillour, V. Lefevre-Seguin, J. Hare, J. M. Raimond, and S. Haroche, J. Opt. B **2**, 204-206 (2000).

102. H. Fujiwara, K. Sasaki, Japan. J. Appl. Phys. II **41** L46-L48 (2002).

103. L. Yang and K. J. Vahala, Opt. Lett. **28**, 592-594 (2003).

104. S. E. Harris and O. P. McDuff, IEEE J. Quant. Electron. **QE-1**, 245-262 (1965).

105. D. J. Kuizenga and A. E. Siegman, IEEE J. Quant. Electron. **QE-6**, 694-708 (1970).

106. J. T. Darrow and R. K. Jain, IEEE J. Quant. Electron. **27**, 1048-1060 (1991).

107. A. E. Siegman, *"Lasers"* (University Science Books, Mill Valley. Calif. 1996).

108. G. H. C. New, Rep. Prog. Phys. **46**, 877-971 (1983).

109. E. Lallier, J. P. Pocholle, M. Papuchon, M. de Micheli, M. J. Li, Q. He, and C. Grezes-Besset, Electron. Lett. **27**, 936-937 (1991).

110. I. Baumann, R. Brinkmann, M. Dinand, W. Sohler, L. Beckers, C. Buchal, M. Fleuster, H. Holzbrecher, H. Paulus, K. H. Muller, T. Gog, G. Materlik, O. Witte, H. Stolz, and W. von der Osten, Appl. Phys. A **64**, 33-44 (1997).

111. H. Suche, L. Baumann, D. Hiller, and W. Sohler, Electron. Lett. **29**, 1111-1112 (1993).

112. H. Suche, R. Wessel, S. Westenhofer, W. Sohler, S. Bosso, C. Carmannini, and R. Corsini, Opt. Lett. **20**, 596-598 (1995).

113. H. Suche, A. Greiner, W. Qiu, R. Wessel, and W. Sohler, IEEE J. Quant. Electron. **33**, 1642-1646 (1997).

114. C. Becker, T. Oesselke, J. Pandavenes, R. Ricken, K. Rochhausen, G. Schreiber, W. Sohler, H. Suche, R. Wessel, S. Balsamo, I. Montrosset, and D. Sciancalepore, IEEE J. Sel. Top. Quant. Electron. **6**, 101-113 (2000).

115. D. Sciancalepore, S. Balsamo, and I. Montrosset, IEEE J. Quant. Electron. **35**, 400-409 (1999).

116. E. Desurvire, *"Erbium-doped fiber amplifiers: principles and applications"* (Wiley, New York, 1994).

117. X. S. Yao and L. Maleki, Opt. Lett. **22**, 1867-1869 (1997).

118. X. S. Yao, L. Davis, and L. Maleki, J. Lightwave Technol. **18**, 73-78 (2000).

CHAPTER 16

VLSI PHOTONICS: A STORY FROM THE EARLY STUDIES OF OPTICAL MICROCAVITY MICROSPHERES AND MICRORINGS TO PRESENT DAY AND ITS FUTURE OUTLOOK

EL-HANG LEE

Graduate School of Information Technology, Inha University, Incheon 402-751 South Korea
ehlee@inha.ac.kr

This chapter is intended to tell a story, in a small way, how the early studies of the optical microcavity microspheres by the author, Prof. Richard K. Chang, and Prof. John B. Fenn in the 1970's and subsequent studies of the microdisks and microrings by others in Prof. Richard K. Chang's group evolved to affect and shape the present day very large scale integrated photonics technology and its future outlook. The chapter first describes how the concept of VLSI photonics was born out in the early 2000's out of the need for high-speed and high-capacity information technology and discusses the scientific and technological issues and challenges to be worked out for its future. The chapter then reviews how the early studies on the optical microcavity microspheres in the 1970's progressed to the understanding of the optical whispering gallery modes in Prof. Richard K. Chang's group and how they later evolved to the studies of microcylinders, microdisks, and microrings, which found their applications in photonic integration and very large scale integrated photonics of today. It then discusses on the outlook of the very large scale integrated photonics in the future.

1. Introduction

The title of this chapter is "VLSI Photonics: A Story from the Early Studies of Optical Microcavity Microspheres and Microrings to Present Day and its Future Outlook." The title contains three essential key words: VLSI (very large scale integration), photonics, and optical microcavity. The primary theme of this chapter is VLSI Photonics, but it looks into this theme particularly from the perspective of the optical microcavity structures and discusses how the two themes have met and merged in their own way. "VLSI photonics" is a bold proposition at the present stage, both in concept and technology, considering the level of integration density and complexity of what we now know as "VLSI electronics." But the evolution of information technology and its prospects for the future demand that this subject should be considered seriously. Since this particular field is relatively new, there are still more questions than answers and more problems than solutions. This chapter is intended to lay out the issues and challenges in VLSI photonics technology of today and to examine its outlook in the future. While there are several different kinds of candidate building blocks for VLSI photonics like optical microrings, photonic crystals, plasmonics, and quantum dots that are being investigated by many researchers around the world, this chapter particularly focuses on the aspect of the optical microcavity structures of the microspheres, microcylinders, microdisks, and microrings. It attempts to tell a story how the early studies of the optical microcavity microspheres by the author, Prof. Richard K. Chang, and Prof. John B. Fenn in the 1970's and subsequent studies of the microcylinders, microdisks and microrings by others in Prof. Richard K. Chang's group evolved and affected the present day VLSI photonics technology. It then discusses on the future outlook of the VLSI photonics technology.

2. Trends and prospects in information technology

The whole world is wired and networked for the information processing and transmission. "Welcome to the Wired World" is the title of a special report by the Time magazine.[1] The progresses of information technology owes a great deal to the advances in electronic technology, but it started showing its limitations in speed and capacity due to limitations in electronics technology. To overcome these limitations, information technology started using photonics based on the lasers and optical fibers invented in the latter part of the 20th century. Thereafter, the world has witnessed an unprecedented growth of photonics in the information technology, especially with the optical fiber cables replacing electrical cables around the world.

Terabits of information can be stored in and retrieved from tiny pieces of silicon chips in an ever-increasing speed and capacity. Terabits of information are flowing through the mega-scale optical networks. Optical fiber cables are now penetrating every home and office. Fiber optics has become another essential utility for information transmission in addition to the existing utilities of electrical power, water, and natural gas. The dream and drive to store ever-larger amounts of information on the smallest possible chips and the dream and drive to transmit the ever-larger amounts of information to the longest possible distance in the shortest possible period of time through the wired network will persist as long as the demand for increased information speed and capacity increases. Further, there also have been increasing demands in recent years for the information technology to be more sensitive to the rising global concerns of the energy and environment of the 21st century.

While mega-scale optical cables are replacing metal based electrical cables, the information technology is also witnessing a rapid growth of micro/nano-scale photonics technology, where optical wires and optical devices are replacing the metal-based electrical wires and devices on a board scale or chip-scale. In the 20th century, micro/nano-scale electrical wires and electronics penetrated every corner of the world, and, now at the threshold of a new century, many people are wondering how the micro/nano-scale optical wire and devices would shape the information technology of the 21st century.

Ever since the challenges on the miniaturization and integration of devices were made some half-century ago by Richard Feynman,[2] the micro- and nano-scale utilization of space for information technology has made great progresses. In the 20th century, the miniaturization was accomplished by way of micro/nano-scale electrical wires. Recently, we begin to see many attempts, where optical wires are used to open up their ways to the micro/nano-scale electronic boards and chips in order to complement or overcome the limitations of the electrical wires. Micro/nano-scale optical channels and optical devices are in increasing need to meet the new challenges and many researches are being conducted around the world to meet these goals.[3] This is where, the need for micro/nano-optical channels and devices arises, and this is where, the concept of VLSI photonics comes in.[4-11]

3.　VLSI photonics and its background

It might be useful to consider the nature of "VLSI Photonics." First, the acronym "VLSI" represents "very large scale integration," like the one used through the 20th century "VLSI electronics." The term "photonics" refers to the study of lightwaves and photons for practical applications or technology and is now a well established field. We thus have become familiar with both the terminologies and their implications, but we are not very familiar with the combined terminology, "VLSI Photonics." [4-11]

VLSI photonics may be described in many possible ways. However, it may be generically described as "a process of high-density integration of photonic wires, devices, and circuits in small sizes reaching down to the scale of micron, submicron, and quantum dimensions." VLSI photonics requires two essential building elements: optical micro/nano-scale channels and micro/nano-scale devices. The optical channels are used to carry information and the optical devices are to perform functions. The channels can be formed by optical wires and the devices can be made out of optical functional structures. Devices made of optical microcavities are, for example, good candidate elements in VLSI photonics. The two functional elements then need to be integrated to carry out more complicated functions, either in monolithic integration or in hybrid integration.

The concept of VLSI photonics has been born out largely out of the need for the progress in the information technology but it also has its roots in the historical development of miniaturization and integration. Through the 20th century of electronics technology, there are two most pivotal inventions that made the electronic revolution possible: first, the electrical printed circuit board (E-PCB) technology and, second, the VLSI electronic technology. The E-PCB technology was grown out of the need to integrate increasing number of discrete electrical and electronic components on a board in the late 1940s. The increasing complexity of the electrical and electronic systems, connected to the copper cables, needed much simplified and integrated array of electrical and electronic devices on a board, which can be used in compact modular forms in place of entangled copper cables and wires. The VLSI electronics technology was also grown out of the impending need to integrate increasing number of semiconductor transistors on a chip-scale.

Although the concepts of "optical printed circuit board (O-PCB)" and "VLSI photonic chip" have been largely inspired by the concepts of the E-PCBs and VLSI electronics, they can be also regarded as a continuation of the photonic integration efforts of the past several decades. While most of the copper cables were being replaced by optical cables in the past several decades, there have been increasing needs for miniaturization and integration of photonic devices and components at the systems level. These efforts can find their roots in the fields of micro-optics or integrated optics which started from the early days of 1970's. With the rapid development of semiconductor optoelectronics, a variety of fields named optoelectronic integration, integrated optoelectronics, optical integration, photonic integration, and integrated photonics evolved to be continuing subjects of study to date.[12-17] Now, the process of photonic integration is approaching much smaller scale with much more complexity.

An O-PCB may be defined as a flat modular board which consists of a planar array of optical wires, circuits, and devices to perform the functions of information technology. O-PCBs use micro/nano-scale optical wires or waveguides for optical interconnection and integration of the discrete micro/nano-photonic devices. VLSI photonics is proposed as a platform for high density integration of photonic devices and circuits on micron or submicron scale on a chip.[4-11]

The advantages of the VLSI photonics technology may be partly found from the advantages of the VLSI electronic technology. The advantages of making electron devices small are that the components can be made compact, light-weight, energy-efficient, resource-saving, cost-efficient, flexible and portable. Now, in photonics, too, these advantages will be similarly effective. Further, the concept of VLSI photonics can fully embrace the concerns of the rising global issues of green technology, environmental technology and biotechnology of the new century. VLSI photonics is an area of new frontier and exploration. This entails technological and scientific issues of both evolutionary and revolutionary nature into the future. Many new potential applications of VLSI photonics can also be expected.

With a brief review of the background of the VLSI photonics technology given above, we now turn to review how the studies of optical microcavity structures, such as microspheres, microdisks, and microrings, from the early years of the 1970's evolved over the next 30 some years and how they affected and shaped the VLSI photonics technology today.

4. Optical microcavity structure

Optical microcavities are defined as the microstructures where the photons can be confined by resonance. Traditionally an optical cavity refers to a three dimensional structure formed by two light-reflecting surfaces on both sides of an optical medium, like that of a laser. An optical microcavity is a resonator that has the cavity dimensions on the scale of a single optical wavelength or smaller, sometimes reaching down to the nanometer ranges, and the fact that the devices can be made small makes the optical microcavities attractive as a candidate component for VLSI photonics. Due to the small nature of the microcavities, quantum effects of the electromagnetic lightwaves can also be observed. This is a situation where no photon is emitted or only a very narrow spectrum of light is emitted. The quality factor of an optical microcavity thus depends on how narrow this emitted spectrum appears with respect to the wavelength of the light. The narrower the spectral emission is, the better the quality of the microcavity is.[18]

Optical microcavities can be used to make devices in many ways depending on their geometrical and resonant properties. They can control the optical emission properties, the spatial distribution of the radiation, and the spectral width of the emitted light. Dependence of the optical properties and characteristics on the geometrical and resonant properties offers the optical microcavities promises for new functional optical devices. In addition to the traditional examples of optical microcavities, such as distributed Bragg reflectors (DBRs) and vertical cavity surface emitting lasers (VCSELs), more recent examples include the microring structures and photonic crystal structures. The reasons that the more recent microcavity structures can be considered as good candidates for VLSI photonics is because they can not only offer a variety functions but also they can be made small for high-density integration.

5. Early studies on optical microcavity: case of microspheres

Of the many microstructures and microcavities mentioned above, it will be particularly interesting to note that the evolution of VLSI photonics technology has strong historical roots and threads that can be traced back to the early days of studies on microspheres. In the 1970's, while Prof. Richard K. Chang first pioneered and led the investigations on the linear and nonlinear optical processes in microparticles, El-Hang Lee, under the guidance of Prof. John B. Fenn and Prof. Richard K. Chang, started investigating the optical scattering processes from microspheres, such as elastic and inelastic light scattering, including the Lorenz-Mie scattering, Raman scattering, fluorescence scattering, and other novel light scattering phenomena. These studies used both homogeneous plane waves and inhomogeneous plane waves (or evanescent waves) for the excitation of the microspheres, as described well in the summary report published in the Faraday Discussions[19] and in other earlier literatures.[20-22]

Fluorescence scattering measurements were made using monodispersed spherical particles of polystyrene latex microspheres containing fluorescent dye molecules uniformly distributed either within the particle volume or over its surface. At non-absorbing excitation frequencies the particles behaved like dielectric, and only elastic Mie scattering occurred. At absorbing excitation frequencies, the particles contributed both elastic Mie scattering and fluorescence emission signals. Thus, one could do both the Mie scattering and the fluorescence emission measurements using the same particles by simply selecting the proper detection wavelength for each measurement.[20,21]

When a dipole emits radiation at a frequency, which is different from the frequency of the incident radiation, the scattering is said to be inelastic. Fluorescence emission and Raman scattering are examples of such inelastic scattering. The characteristics of the inelastic scattering from small spherical particles, that is, the angular distribution of the scattered light intensity, its dependence on polarization, particle size and refractive index had not been well known then. Fluorescence and Raman scattering from molecules embedded in dielectric particles became of practical importance, especially in the study of biological cells and in the study of particulate constituents of the atmosphere. For example, when the fluorescent molecules are attached to a biological cell, deoxyribonucleic acid (DNA), cytoplasm, or the cell membrane, fluorescence scattering can be used in monitoring specific cell functions or in cell identification and sorting systems. Another example of the use of inelastic scattering was in the remote sensing of particulate constituents of the atmosphere, in particular, where there was a need for chemical species identification and concentration information of the ambient aerosols in real time and in situ, including light detection and ranging. In this context, the experimental researches were conducted on the inelastic scattering from microspherical dielectric particles. Fluorescence and Raman scattering processes are quite different from each other. Usually fluorescence emission intensity, whenever it exists, overwhelmingly dominates the Raman scattering intensity. Therefore, the detection of fluorescence emission is orders of magnitude easier than the detection of Raman scattering. For this reason, it is convenient and appropriate to study the fluorescence emission first. Knowledge and experience gained by the fluorescence measurements are applicable to the study of Raman scattering.[19-22]

The fluorescence emission from monodispersed spherical particles was studied in terms of its dependence on particle size, on the polarization, and on the medium refractive index. It was found that the interference patterns of the Lorenz-Mie scattering were absent in the case of fluorescence emission. This was explained in terms of incoherent and coherent sums of the field contributions within the particle. Further, it was found that the fluorescence from particles showed deviations from the fluorescence of dye molecules in an alcohol solution. The strong deviations in the backward direction were explained in terms of the focusing effect of the field in the forward portion of the particle. This effect was consistent with another observation that matching the medium refractive index with that of particle decreased the deviation.[20, 21]

Effectively, this study, which was conducted for the first time on microspheres, later led to the study of optical lightwaves traveling inside the microspheres by way of total internal reflection (TIR) and that was the first time that the concept of the whispering gallery mode (WGM) in the optical microcavities was introduced. The story is well accounted in the Faraday Discussions by Richard K. Chang and it goes as follows. While the angular distribution of fluorescence emission from an ensemble of polystyrene latex microspheres dyed with fluorescent molecules showed broad emission profile, diluted sample solution having only a few spheres per milliliter started showing many sharp peaks on top of a broad fluorescent dye profile. With a large number of spheres, the sharp peaks merged into a broader peak because the average smoothed out the sharp peaks. Because the monodispersed spheres had a size distribution and since no two spheres were exactly of the same size, the fluorescence from an ensemble of microspheres showed fluorescence emission profile that smears out the resonance peaks of the individual microcavity or polystyrene latex microspheres. Further refined studies with the help of electromagnetic theories led to the understanding that, there were some enhanced internal fields inside the microspheres, which increased the fluorescence emission and that within the wide fluorescence spectra there were some wavelengths that were resonant with the polystyrene spheres. These wavelengths were the ones which were in optical resonance within the microsphere cavities with varying high Q values. This is how the study of optical lightwaves traveling inside the microspheres led to the first time discovery and formulation of the concept of WGM in the optical domain.[19]

While the above studies were mainly conducted using homogeneous plane waves for the excitation of the microcavity spheres, the study of optical microcavities and WGM was further strengthened and expanded by the use of inhomogeneous plane waves (evanescent waves) for the excitation of the microspheres. To the best of the author's knowledge, this was the first attempt to report on the effect of resonance in a microsphere cavity as excited by evanescent waves. Evanescent waves have been known since the time of Newton. Evanescent waves are inhomogeneous plane waves generated in a medium of low refractive index when homogeneous plane waves traveling in an adjacent medium of high refractive index strike the interface and undergo TIR. This evanescent wave is considered inhomogeneous because the planes of constant phase do not coincide with the planes of constant amplitude. An evanescent wave has its propagation vector directed parallel to the interface while its amplitude falls off exponentially with distance away from the interface. In homogeneous plane waves, the planes of constant phase coincide with the planes of constant amplitude.[20,22]

In this study, the dielectric polystyrene latex microspheres containing efficient fluorescent dye molecules were made to come in contact with the flat surface of a cylindrical sapphire prism and the dielectric spheres were excited by the evanescent waves generated at the interface by a laser light entering a cylindrical sapphire prism at varying angles. Scattering patterns were observed over the surface of the polystyrene microcavity spheres. The detailed analysis of the optical scattering processes in this configuration and a more detailed experimental observations and theoretical analysis of the optical resonant processes within the spheres are well described in the Lee's doctoral dissertation and in the paper entitled "Angular distribution of fluorescence from liquids and monodispersed spheres by evanescent wave excitation." [20,22] The interaction of evanescent waves with matter in terms of absorption and emission had been recognized since the beginning of the 20th century, but practical application did not begin until the mid-sixties, when the concept of internal reflection spectroscopy (IRS) began to emerge as a powerful technique in studying surface phenomena and the spectra of materials which would be otherwise difficult to prepare for reflectance or transmission spectrometry in conventional, free-space spectroscopy.[23] The IRS technique can obtain absorption spectra of very thin films or even powdered samples without significant scattering losses. In the spectrometry of reflectance or transmittance, elastic scattering would alter the spectral shape considerably. This required a systematic understanding of the elastic and inelastic light scattering from the fluorescent or non-fluorescent microspheres excited by the evanescent waves and the study by Lee, Chang, and Fenn further enlightened the understanding of the elastic and inelastic light scattering in the microsphere cavities. This study, and the underlying principles of this study, was further deepened and expanded by many other researches, including applications in biosensing, environmental sensing and others.[24-29] Thus, while the elastic and inelastic light scattering studies of the microcavity spheres using the homogeneous plane wave excitation paved the way to the first time understanding of the WGM in the optical microcavities, the understanding was further strengthened and expanded by the use of inhomogeneous plane waves (evanescent waves) for the excitation of the microcavities. Again, to the best of the author's knowledge, this was the first study ever to report on the effect of optical resonance in a microsphere cavity as excited by evanescent waves that led to the study of WGM phenomena in the optical microcavities.[22] The study of WGM phenomena in microspheres later opened up the studies of optical processes in the microcylinders, microdisks, and microrings and this will be discussed in the following section.

But, before we move on to the next section, it should be noted that the basic studies on the optical microcavity studies on the microspheres discussed above were later followed upon for their application in VLSI photonics by Ali Serpengüzel and others in the 1990's and recently in the 2000's. These studies include the basic studies on the analysis of the whispering gallery modes and emission patterns, the efforts to increase the resonance quality of the microspheres, various means to couple the light in and out of the microspheres by way of optical fibers, and studies on the use of fiber-coupled microspheres for wavelength drop filters and sensors. These studies came to be of important values especially in light of the use of silicon microspheres for photonic application.[30-40]

6. Subsequent studies of the optical microcavity: microcylinders, microdisks and microrings

The early studies on the optical processes in the microspheres, described above, were extended to the study of optical processes in microcylinders, microdisks, and microrings. While the optical processes in the microspheres are three-dimensional (3-D), the optical processes in microcylinders, microdisks, and microrings can be treated in two-dimensional (2-D) frame of reference. The well-accepted 3-D spherical Lorenz–Mie theory reduces the problem to a 2-D problem. Again, the early studies on the WGM effects in the 3-D and 2-D microcavity structures are well-described in the Faraday Discussions by Richard K. Chang.[19] The studies on WGM later evolved not only to deeper scientific studies of the microcavities of today but also to the use of the resonance characteristics of the microcylinders, microdisks, and microrings for a variety of applications in information technology. Studies on microcylinders made the understanding on WGM simpler. Microdisks were first used to make microcavity lasers. And, the microrings were made for various applications, such as wavelength filters and optical sensors.[41-43]

The studies on WGM were later extended to the microstructures of diverse morphological shapes and designs, including hexagonal structures, microtoroids, microracetrack structures, micropillar structures and others. The WGM phenomena in such structures were later developed extensively to applications in devices for information technology, like micro-optical light sources, switches, modulators, optical wavelength filters, wavelength converters, channel add-drop filters, detectors, attenuators, and optical sensors. Further, in concert with the development of micro/nano-scale optical integration technology, these devices are now considered highly viable candidates as basic building blocks for future VLSI photonics technology. Silicon is transparent in the telecommunication wavelength bands and using the morphology-dependent high-quality-factor resonances for diverse device applications and VLSI photonics.[44-52]

More recent studies report, as reviewed well by Andrew W. Poon and others, on the developments in cascaded microresonator-based matrix switches for silicon photonic interconnection networks in many-core computing applications. They use, for example, a matrix switch comprising two-dimensionally cascaded microring resonator-based electro-optic switches coupled to a waveguide cross-grid on a silicon chip. Such a microresonator-based matrix switch offers non-blocking interconnections among multiple inputs and multiple outputs, with the merits of tens to hundreds of micrometers-scale footprint, gigabit/second-scale data transmission, nanosecond-speed circuit-switching, and large-scale integration (LSI) for networks-on-chips applications.[51,52]

These studies collectively show that the advantage of optical microcavity resonator devices is that, it can be fabricated for VLSI photonic systems by cascading the resonators in various ring sizes. As is the case in most of the scientific and technological evolution, these studies described above for the microspheres and the microresonators show another example that the new visions and outlook toward VLSI photonics has historical roots that can be effectively traced back to the fundamental principles of the microcavities.

7. Photonic devices for VLSI photonic integration

Photonic devices for VLSI photonic integration need to be made in the form of miniaturized micro/nano-channels and micro/nano-cavities.[53] From the technological point of view, the miniaturization and integration of photonic devices entails the need to understand the behavior of photons traveling or confined in micro/nano-scale structures. The photonic devices that are being developed in recent years include: microring resonator devices,[41-52] photonic crystals,[54-59] surface plasmon polaritons[60-66] and metamaterials.[67-68]

One of the candidate devices for VLSI photonics is the microring resonator.[41-52] A microresonator device, in its basic form, consists of a ring-type curved waveguide and one or two straight waveguides coupled adjacently to the curved ring waveguide. Microring resonators are based on the optical resonance in the ring, which is determined by the phase matching between the lightwave entering the resonator and the lightwave traveling around the resonator. Microring resonators can be made into a variety of optical functional components like filters, switches, routers, and lasers and can be fabricated for very large scale integrated photonic systems by cascaded integration of the ring resonators in various sizes and shapes.[41-52]

Another candidate device for VLSI photonics is are photonic crystals.[54-59] Photonic crystals are composed of periodic arrays of dielectric or metallo-dielectric nanostructures of regularly repeating high and low dielectric constants, and they can affect the propagation of electromagnetic waves as the periodic potential in a semiconductor crystal affects the electron motion by way of allowed and forbidden electronic energy bands. When the periodicity of dielectrics is closer to the wavelength of a lightwave, lightwaves cannot propagate and photonic bandgap develops. However, when one introduces a defect in the periodic structure, one can control the characteristics of the lightwave propagation such as path, speed, dispersion, and nonlinearity. As such, photonic crystals have been utilized for many micro/nano-scale optical functional devices and have shown promises of very large scale integrated photonic systems. Also, by controlling and modifying the optical characteristics of the lightwaves at the photonic band edge, lightwave propagation can be made to slow down or even to stop, leading to the possibilities of realizing new functional devices including optical switches and modulators.[54-59]

Another candidate for VLSI photonics is the surface plasmonic waves.[60-66] Plasmonic waves result from the collective motion of electrons at the interface between metals and dielectrics. Despite large propagation losses, plasmonic waves show promises of light confinement in the regime below the diffraction limit for miniaturization and integration of photonic devices for VLSI photonics. Many progressive studies have been made in recent years.[60-66]

8. Technical issues and challenges of VLSI photonics: miniaturization, integration, and scaling rules

Unlike VLSI electronics, VLSI photonics is still at an early stage with most of the questions and problems still open for further study. However, the fundamental question to ask is whether "VLSI photonics" can achieve the level of integration of photonic devices comparable to that of "VLSI electronics?" The physical nature and characteristics of electrons and those of photons radically differ from each other. Electrons have mass and obey the Fermi-Dirac statistics, while photons have no mass and obey the Bose-Einstein statistics. Electron devices have been miniaturized and integrated to the level that we are very well aware. Simply, if photonics uses the wavelengths of the light in the visible and infrared for information and telecommunication applications, the question is whether "VLSI photonics" can go over the diffraction limit of the light and achieve the same level of miniaturization and integration as that of VLSI electronics?[4-11]

The process of miniaturization requires attention to various scientific and technological issues arising from the variations of the size, lightwave confinement, optical interference, high field, nonlinearity, and others such as chaotic noise. These properties and characteristics, however, can be utilized for various functional devices for VLSI photonics.[4-11]

The process of integration requires attention to various scientific and technological issues arising from the interface mismatches of different types, shapes, sizes, modes, polarizations, powers, materials, functions, proximities, mechanics, and others such as path and alignment. Other issues will include the design and packaging of a system for reliable function and operation. Many of these integration issues have been the subjects of study for mega-scale optical fiber communication technology and now these issues need to be addressed in the integration of the micro/nano-scale photonic devices.[4-11]

In terms of materials, silicon is becoming an especially important material for photonics because silicon has high refractive index that can allow good confinement of light and it is compatible with the existing silicon foundry for electronics.[69-76] Despite such advantages, silicon has shortcomings as photonic material. Since silicon is an indirect transition material, light emission from silicon is not an efficient process.

However, silicon has recently been made to emit light by way of Raman processes.[77-79] Silicon photonics in recent years utilizes the so called silicon-on-insulator (SOI) wafers. SOI wafers constitute a thin layer of single crystal silicon isolated from the main silicon substrate by way of a thin oxide insulation layer formed between the upper silicon layer and the bottom main silicon substrate. It is an interesting historical evolution that SOI was invented in the late 1970's and in the early 1980's for the purpose of making silicon devices formed in the upper layer faster by isolating them from the parasitic resistance and capacitance elements causing large RC time constants in the main substrate. The SOI wafers are now considered ideal for silicon photonics because the three layers, that is, the upper silicon layer, middle oxide layer, and the bottom silicon substrate, can be concertedly or independently utilized for photonics. The upper silicon layer and the oxide layer can be used for photonics while the bottom layer can still be used for electronics. The high contrast in the refractive index between the silicon and the oxide, for example, is another highly attractive aspect because the light in the silicon can be very effectively confined to small dimensions even leading to the studies of nonlinear optical phenomena in micro/nano-structures. Further, the silicon based photonics offer great potential for low cost manufacturing of the photonics because standard electronic device fabrication tools and processes can also be used for the photonic devices and circuits. Silicon has some limitations because silicon has extremely low electro-optic coefficient. The low electro-optic coefficient usually makes the device size too large. However, recent efforts to use silicon for modulation have been successfully demonstrated.[80]

In addition, VLSI photonics needs some design rules, or scaling rules, for in miniaturization and integration. The two most essential rules that govern the VLSI microelectronics are the scaling rules for device miniaturization (known as the K-law) and for device integration (known as the Moore's Law). Some efforts have been made in recent years to find the scaling rules for the miniaturization and integration of the most basic photonic devices such as directional coupler.[5,9] Scaling rules for photonic devices will be needed for the future development of VLSI photonics, as the scaling rules for electron devices did for the development of VLSI electronics.

There have been efforts in recent years to achieve VLSI-level photonic chips and systems. Some of these efforts include: the optoelectronic VLSI chip with 540-element receiver and transmitter arrays; optoelectronic VLSI switching chip with greater than Tbit/s optical input and output bandwidth; optical interconnects for neural and reconfigurable VLSI architectures; high-speed optoelectronic VLSI switching chip using quantum-well modulators and detectors. These exploratory studies collectively offer various solutions to the issues and challenges of VLSI design and fabrication.[81-88]

One of the critical technological issues that need to be eventually addressed in the development and application of VLSI photonics is the study of photonic circuit theories and network theories on micro/nano-scale like the circuit theories of the VLSI electronics technology and large-scale network theories. In recent years, there have been some attentions given to the development of photonic circuit theories and this is an area to be addressed as the level of integration diversity and complexity rises for diverse photonic applications in the coming years.[89-92]

9. Applications of VLSI photonics

The application of VLSI photonics, when fully developed, can be as diverse as it can be. Like the VLSI electronics technology in the 20th century, VLSI photonics shows promises for many potential applications. VLSI photonics can provide a platform for integrated photonics as applicable for telecommunication, data communication, aerospace, transportation, environment, biomedical, and manufacturing systems.

Typical applications for information technology may include: monitoring, sensing, storage, processing, high-speed and high-capacity transmission, switching, and routing. Applications for environmental and biomedical technologies may include green technology, green energy, sensing, detecting, monitoring, exploration, diagnosis, analysis, imaging, control, and distribution.

With many potential advantages of making photonic devices compact, light-weight, energy-efficient, resource-saving, cost-efficient, flexible, and portable, VLSI photonics can further embrace the rising global concerns of the green technology, environmental technology and biotechnology of the new century. VLSI photonics is a widely open area of frontier study and exploration and is awaiting great challenges of both evolutionary and revolutionary nature in the years to come.

10. Summary

This chapter has briefly introduced and described the concept of VLSI photonics technology that has been born out from the historical and global needs in information technology. The VLSI photonics in the 21st century can be regarded as the analogue of the VLSI electronics in the 20th century. The chapter discussed and described why the high-density photonic integration in micro/nano-scale is becoming increasingly important. It explained ways to approach the technology of micro/nano-photonic integration, especially using the optical microcavity structures as one particular example. It also discussed the need of developing scaling rules for the design of photonic wires and devices' and discussed the scientific and engineering issues and challenges for the miniaturization and integration of photonic wires and devices in micro/nano-scale. The chapter also discussed on the building block photonic devices of VLSI photonic integration and related materials thereof. It has been particularly noted that VLSI photonics technology has strong historical roots and threads that can be related to the early days of studies on microspheres, especially by El-Hang Lee, Richard K. Chang, and John B. Fenn, in the late 1970's. This study later evolved not only to the deeper scientific studies of the microcavities but also to the use of their unique characteristics for applications in the microdisks and microrings, which are considered highly viable candidates as basic building blocks for VLSI photonics technology. And, it reviewed some examples of recent studies that led to the photonic integrated circuits and VLSI photonic circuits using microsphere cavities and microring cavities. Finally, it discussed on the potential applications of the VLSI photonics. Like VLSI electronics, which has penetrated almost every corner of the modern day electronic civilization, VLSI photonics shows a strong potential to make its way into the information technology of the future. This chapter has attempted to lay out the scientific and technological issues and challenges to that goal and to examine the steps to that end. With many potential advantages, VLSI photonics can further embrace the rising global concerns of the green technology, environmental technology and biotechnology of the new century. VLSI photonics is a widely open area of frontier study and exploration and is awaiting great challenges of both evolutionary and revolutionary nature in the years to come.

Acknowledgments

The author expresses deep gratitude to Prof. John B. Fenn (Professor Emeritus, Yale University, Nobel Laureate, Chemistry, 2002) and Prof. Richard K. Chang (Henry Ford II Professor, Emeritus, Yale University, Applied Physics and Electrical Engineering, former student of Prof. Nicolaas Bloembergen, Harvard University, Nobel Laureate, Physics, 1981) who, as great advisors and teachers, gave the author a firm foundation of education and inspiration for a lifetime pursuit of science. The story given in this chapter is part of a small fruit of that education and teaching.

References

1. "Welcome to the Wired World – What the networked society means to you, your business, your country, and the globe." TIME, Special Report, February 3, 1997.
2. Richard Feynman, "There is plenty of room at the bottom," *American Physical Society, Annual Meeting*, California Institute of Technology, Pasadena, CA, USA, December 29, 1959.
3. David Miller, "Device Requirements for Optical Interconnects to Silicon Chips," *Proceedings of the IEEE*, Vol. 97, No. 7, p.1166, July 2009; and the references therein.
4. E. H. Lee, S. G. Lee, B. H. O, "Microphotonics: physics, technology, and an outlook toward the 21st Century," *SPIE Proceedings*, Vol. 4580, p. 263, 2001.
5. E. H. Lee, S. G. Lee, B. H. O, "VLSI microphotonics: issues, challenges, and prospects," *SPIE Proceedings*, Vol. 4652, p. 1, 2002.
6. E. H. Lee, S. G. Lee, B.H.O, S.G. Park, "Miniaturization and integration of micro/nano-scale photonic devices: scientific and technical issues," *SPIE Proceedings*, Vol. 5356, 2004.
7. E. H. Lee, S. G. Lee, B. H. O, S. G. Park, "Optical Printed Circuit Board (O-PCB): A New Platform toward VLSI Micro/Nano-Photonics?" *IEEE/LEOS, Summer Topical Meeting, Optical Interconnects and VLSI Photonics*, June 29-30, 2004, San Diego, CA, USA.
8. E. H. Lee, S. G. Lee, B. H. O, S. G. Park, "Optical Printed Circuit Board (O-PCB): A New Platform toward VLSI Micro/Nano-Photonics?" *IEEE-LEOS Newsletter*, Vol. 18, No. 5, October, 2004.
9. E.-H. Lee, S.-G. Lee, B. H. O, M.-Y. Jeong, K.-H. Kim, and S.-H. Song, "Fabrication and integration of VLSI micro/nano-photonic circuit board," *Microelectronic Engineering*, Vol. 83, p. 1767, 2006.
10. E. H. Lee, "VLSI photonics" Tutorial Lecture, Optoelectronic and Optical Communication Conference (OECC), Hong Kong, July 13-28, 2009.
11. E. H. Lee, "Optical printed circuit board (O-PCB) and VLSI photonics," *IEEE Photonics Society, Distinguished Lecture Series*, 2007-2009.
12. S. E. Miller, "Integrated Optics: An Introduction," *Bell Sys. Tech. J.*, Vol. 48, p. 2059, 1969.
13. Dietrich Marcuse, Ed., Integrated Optics, *IEEE Press*, New York, 1973.
14. R. G. Hunsperger, Integrated Optics: Theory and Technology, *Springer Verlag, Berlin*, 1984.
15. T. L. Koch, et al., "Semiconductor photonic integrated circuit," *J. Quantum Electron.*, Vol. 27, p. 641, 1991.
16. D. A. B. Miller, "Physical Reasons for Optical Interconnection," *Int. J. Optoelectron.*, Vol. 11, p. 155, 1997.

17. Ellias Towe, Ed., Heterogeneous Optoelectronic Integration, *SPIE Press, Bellingham*, 2000.

18. R. K. Chang and A. J. Campillo, Eds., *Optical Processes in Microcavities*. Singapore: World Scientific, 1996, and all the references therein.

19. R. K. Chang and Y. L. Pan, "Linear and non-linear spectroscopy of microparticles: Basic principles, new techniques and promising applications," *Faraday Discuss.*, Vol. 137, pp. 9–36, 2008, and the references therein.

20. El-Hang Lee, "Elastic and Inelastic Light Scattering by Small Spherical Particles by Plane Wave and Evanescent Wave Excitation," *Ph.D. Thesis, Yale University*, 1978, and all the references therein.

21. E. H. Lee, R. E. Benner, R. K. Chang, and J. B. Fenn, "Angular Distribution of Fluorescence from Monodispersed Particles", *Appl. Opt.*, 17, pp.1980 - 1982 (1978).

22. E. H. Lee, R. E. Benner, R. K. Chang, and J. B. Fenn, "Angular Distribution of Fluorescence from Liquids and Monodispersed Spheres by Evanescent Wave Excitation", *Appl. Opt.*, 18, 862, 1979.

23. N. J. Harrick, Internal Reflection Spectroscopy, *John Wiley & Sons*, Inc., New York, 1967.

24. C. F. Bohren and D. R. Huffman, "Absorption and Scattering of Light by Small Particles," *John Wiley & Sons*, New York, 1983.

25. P. R. Conwell, P. W. Barber, and C. K. Rushforth, "Resonant spectra of dielectric spheres', *J. Opt. Soc. Am. A* 1, 62-66 (1984).

26. S. X. Qian, J. B. Snow, H. M. Tzeng and R. K. Chang, *Science*, 1986, 231, 486.

27. S. X. Qian and R. K. Chang, *Phys. Rev. Lett.*, 1986, 56, 926.

28. P. W. Barber and R. K. Chang, Eds. "Optical Effects Associated with Small Particles," *World Scientific*, Singapore (1988).

29. P. W. Barber and S. C. Hill, "Light Scattering by Particles: Computational Methods," *World Scientific*, Singapore, 1990.

30. M. L. Gorodetsky and V. S. Ilchenko, ''High-Q optical whispering gallery microresonators: precession approach for spherical mode analysis and emission patterns,'' *Opt. Commun.* 113, 133–143 (1994).

31. A. Serpengüzel, S. Arnold, and G. Griffel, "Excitation of Resonances of Microspheres on an Optical Fiber," *Opt. Lett.* 20, 654-656 (1995).

32. M. L. Gorodetsky, A. A. Savchenkov, and V. S. Ilchenko, "Ultimate Q of optical microsphere resonators," *Opt. Lett.* 21, 453-455 (1996).

33. A. Serpengüzel, S. Arnold, G. Griffel, and J. A. Lock, "Enhanced coupling to microsphere resonances with optical fibers," *J. Opt. Soc. Am. B*, 14, 790-795 (1997).

34. H. C. Tapalian, J.P. Laine, and P.A. Lane, "Thermooptical switches using coated microsphere resonators," *IEEE Photon. Technol. Lett.* 14, 1118-1120 (2002).

35. T. Bilici, S. Işçi, A. Kurt, and A. Serpengüzel,, "Microsphere-based channel dropping filter with an integrated photodetector," *IEEE Photon. Tech. Lett.* 16, 476-478 (2004).

36. V. N. Astratov, J. P. Franchak, and S. P. Ashili "Optical coupling and transport phenomena in chains of spherical dielectric microresonators with size disorder," *Appl. Phys. Lett.* 85, 5508-5510 (2004).

37. T. Bilici, S. Işçi, A. Kurt, and A. Serpengüzel, "Microsphere Based Channel Dropping Filter with Integrated Photodetector," *IEEE Photon. Technol. Lett.* 16, 476-478 (2004).

38. Y. O. Yilmaz, A. Demir, A. Kurt, and A. Serpengüzel, "Optical Channel Dropping with a Silicon Microsphere," *IEEE Photon. Technol. Lett.* 17, 1662-1664 (2005).

39. Y. O. Yilmaz, A. Demir, A. Kurt, and A. Serpengüzel, "Optical Channel Dropping with a Silicon Microsphere," *IEEE Photon. Technol. Lett.* 17, 1662-1664 (2005).

40. V. N. Astratov and S. P. Ashili, "Percolation of light through whispering gallery modes in 3D lattices of coupled microspheres," *Opt. Express* 15, 17351-17361 (2007)

41. B. Little, S. Chu, J. Foresi, G. Steinmeyer, E. Thoen, H. A. Haus, E. P. Ippen, L. Kimerling, and W. Greene, "Microresonators for integrated optical devices," *Opt. Photon. News*, Vol. 9, pp. 32–33, Dec. 1998.

42. A. W. Poon. Optical resonances of two-dimensional microcavities with circular and non-circular shapes. *PhD thesis, Yale University*, 2001.

43. G. D. Chern, H. E. Türeci, A. D. Stone, R. K. Chang, M. Kneissl and N. M. Johnson, *Appl. Phys. Lett.*, 2003, 83, 1710–1712.

44. B. E. Little, S. T. Chu, W. Pan, and Y. Kokubun, "Microring resonator arrays for VLSI photonics," *IEEE Photon. Technol. Lett.*, 12, 323-325 (2000).

45. C. Li, N. Ma, and A. W. Poon, "Waveguide-coupled octagonal microdisk channel add-drop filters," *Opt. Lett.* 29, 471-473 (2004).

46. B. E. Little, S. T. Chu, P. P. Absil, J. V. Hryniewicz, F. G. Johnson, F. Seiferth, D. Gill, V. Van, O. King, and M. Trakalo, "Very high-order microring resonator filters for WDM applications," *IEEE Photon. Technol. Lett.*, 16, 2263-2265 (2004).

47. K. J. Vahala, Ed. "Optical Microcavities," *World Scientific*, Singapore (2004).

48. A. B. Matsko and V. S. Ilchenko. Optical resonators with whispering-gallery modes—part I: basics. *IEEE J. Select. Topics Quantum Electron.* 12, 3–14, 2006.

49. V. S. Ilchenko and A. B. Matsko. Optical resonators with whispering-gallery modes—part II: applications. *IEEE J. Select. Topics Quantum Electron.* 12, 15–32, 2006.

50. G. D. Chern, G. E. Fernandes, R. K. Chang, Q. Song, L. Xu, M. Kneissl, and N. M. Johnson, "High-Q-preserving coupling between a spiral and a semicircle μ-cavity," *Opt. Lett.* 32, 1093-1095 (2007).

51. A. W. Poon, X. Luo, F. Xu, and H. Chen, "Cascaded Microresonator-Based Matrix Switch for Silicon On-Chip Optical Interconnection," *Proceedings of the IEEE,* Vol. 97, No. 7, p. 1216, July 2009; and the references therein.

52. A. W. Poon, X. Luo, L. Zhou, C. Li, J. Y. Lee, F. Xu, H. Chen, and N. K. Hon, "Microresonator-based devices on a silicon chip: novel shaped cavities and resonance coherent interference," in Practical Applications of Microresonators in *Optics and Photonics,* Andrey B. Matsko, Ed., Taylor and Francis, 2009.

53. V. R. Almeida, "Building Blocks for Silicon Nanophotonics", *Ph.D. Dissertation, Cornell University*, 2005.

54. E. Yablonovitch and T.J. Gmitter, "Photonic band structure: The face-centered-cubic case", *Phys. Rev. Lett.*, Vol. 63, p. 1950, 1989.

55. V. Berger, "Nonlinear Photonic Crystals," *Physical Review Letters*, vol. 81, pp. 4136-4139, 1998.

56. T. D. Happ, M. Kamp, and A. Forchel, "Photonic crystal tapers for ultracompact mode conversion," *Optics Letters*, Vol. 26, No. 14, pp. 1102-1104, 2001.

57. A. Talneau, P. Lalanne, M. Agio, and C. M. Soukoulis, "Low-reflection photonic-crystal taper for efficient coupling between guide sections of arbitrary widths," *Optics Letters*, Vol. 27, No. 17, pp. 1522-1524, 2002.

58. M. Soljacic and J. D. Joannopoulos, "Enhancement of nonlinear effects using photonic crystals," *Nature Materials*, Vol. 3, pp. 211-219, 2004.

59. Y. A. Vlasov, M. O'Boyle, H. F. Hamann and S. J. McNab, "Active control of slow light on a chip with photonic crystal waveguides," *Nature*, vol. 438, pp. 65-69, 2005.

60. R. H. Ritchie, "Plasma losses by fast electrons in thin films," *Phys. Rev.*, Vol. 106, pp. 874–881, 1957.

61. E. Betzig, J. K. Trautman, T. D. Harris, J. S. Weiner, and R. L. Kostelak, "Breaking the Diffraction Barrier: Optical Microscopy on a Nanometric Scale," *Science*, Vol. 251, p. 1468, 1991.

62. William L. Barnes, Alain Dereux, Thomas W. Ebbesen, "Surface plasmon subwavelength optics," *Nature*, Vol. 424, pp. 824-829, 2003.

63. M. Hochberg, T. Baehr-Jones, C. Walker and A. Scherer, "Integrated plasmon and dielectric waveguides," *Optics Express*, Vol. 12, pp. 5481-5486, 2004.

64. J. R. Krenn and J. C. Weeber, "Surface plasmon polaritons in metal stripes and wires," *Philos. Trans. R. Soc. London Ser. A* 362, 739, 2004.

65. W. Nomura, M. Ohtsu, T. Yatsui, "Nanodot coupler with a surface plasmon polariton condenser for optical far/near-field conversion," *Appl. Phys. Lett.*, Vol. 86, 181108, 2005.

66. L. Yin et al., "Subwavelength Focusing and Guiding of Surface Plasmons, " *Nano Lett.*, Vol. 5, p. 1399, 2005.

67. V. M. Shalaev, "Optical negative-index metamaterials," *Nature Photonics*, Vol. 1, pp. 41 – 48, 2006.

68. S. Linden, C. Enkrich, G. Dolling, M. W. Klein, J. Zhou, T. Koschny, C. M. Soukoulis, S. Burger, F. Schmidt, and M. Wegener, "Photonic Metamaterials: Magnetism at Optical Frequencies," *IEEE Journal of Selected Topics in Quantum Electronics*, Vol. 12, pp. 1097-1105, 2006.

69. R. Soref and B. Bennett, "Electrooptical effects in silicon," *Journal of Quantum Electronics*, Vol. 23, pp. 123–129, 1987.

70. R. A. Soref, "Silicon-based optoelectronics", in *Proceedings of the IEEE,* v. 81, n. 12, pp. 1687-706 (1993).

71. H. Zimmermann, *"Integrated Silicon Optoelectronics," Springer Verlag, Berlin* (2000).

72. G. T. Reed and A. P. Knights, *Silicon photonics: an introduction.* John Wiley & Sons, 2004.

73. L. Pavesi and D. J. Lockwood, *Silicon photonics.* Berlin; New York: Springer, 2004.

74. L. C. Kimerling, L. Dal Negro, S. Saini, Y. Yi, D. Ahn, D., S. Akiyama, D. Cannon, J. Liu, J. G. Sandland, D. Sparacin, J. Michel, K. Wada, and M. R. Watts, "Monolithic Silicon Microphotonics," in *"Silicon Photonics,"* L. Pavesi and D. J. Lockwood, Eds. Springer Verlag, Berlin 89-119 (2004).

75. R. A. Soref, "The past, present, and future of silicon photonics", *IEEE Journal of Selected Topics in Quantum Electronics,* vol. 12, pp. 1678-1687 (2006).

76. G. Guillot and L. Pavesi, "Optical Interconnects: The Silicon Approach," Springer Verlag, Berlin (2006).

77. O. Boyraz and B. Jalali, "Demonstration of a silicon Raman laser," *Opt. Express* 12, 5269-5273 (2004).

78. H. Rong, A. Liu, R. Jones, O. Cohen, D. Hak, R. Nicolaescu, A. Fang and M. J. Paniccia, "An All-Silicon Raman Laser," *Nature* 433, 292-294 (2005).

79. H. Rong, R. Jones, A. Liu, O. Cohen, D. Hak, A. Fang, and M. J. Paniccia, "A Continuous-Wave Raman Silicon Laser," *Nature* 433, 725-728 (2005).

80. O. Boyraz and B. Jalali, "Demonstration of directly modulated silicon Raman laser," *Opt. Express* 13, 796-800 (2005).

81. J. W. Goodman, F. J. Leonberger, S. Y. Kung, and R. A. Athale, Optical Interconnects for VLSI Systems, *Proc. of the IEEE*, 72 (7), pp. 850-866 (1984).

82. M. B. Venditti, E. Laprise, J. Faucher, P-O. Laprise, J.E. Lugo, and D.V. Plant. Design and test of an Optoelectronic-VLSI (OE-VLSI) chip with 540 element receiver/transmitter arrays using differential optical signaling. *IEEE Journal of Selected Topics in Quantum Electronics,* 9: 361-379 (2003).

83. A. L. Lentine, K. W. Goossen, J. A. Walker, J. E. Cunningham, W. Y. Jan, T. K. Woodward, A. V. Krishnamoorthy, B. J. Tseng, S. P. Hui, R. E. Leibenguth, L. M. F. Chirovsky, R. A. Novotny, D. B. Buchholz, and R. L. Morrison, "Optoelectronic VLSI switching chip with greater than 1 Tbit/s potential optical I/O bandwidth," *Electron. Lett.*, vol. 33, pp. 894–895 (1997).

84. D. Fey, W. Erhard, M. Gruber, J. Jahns, H. Bartlet, G. Grimm, L. Hoppe, and S. Sinzinger, "Optical interconnects for neural and reconfigurable VLSI architectures," *Proc. IEEE*, vol. 88, pp. 838–848 (2000).

85. M. B. Venditti, "Receiver, transmitter, and ASIC design for Optoelectronic-VLSI applications", *Ph.D. Thesis*, McGill University (2003).

86. A. L. Lentine, K. W. Goossen, J. A. Walker, L. M. F. Chirovsky, L. A. D'Asaro, S. P. Hui, B. J. Tseng, R. E. Leibenguth, J. E. Cunningham, W. Y. Jan, J.-M. Kuo, D. W. Dahringer, D. P. Kossives, D. D. Bacon, G. Livescu, R. L. Morrison, R. A. Novotny, and D. B. Buchholz,

"High-speed optoelectronic VLSI switching chip with >4000 optical I/O based on flip-chip bonding of MQW modulators and detectors to silicon CMOS," *IEEE J. Select. Topics Quantum Electron.*, vol. 2, pp. 77–84 (1996).

87. A. V. Krishnamoorthy *et al.*, "Triggered receivers for optoelectronic VLSI," *Electron. Lett.*, vol. 36, pp. 249–250 (2000).

88. M. B. Venditti and D. V. Plant. On the design of large transceiver arrays for OE-VLSI applications. *IEEE/OSA J. Lightwave Technol.*, 21: 3406-3416 (2003).

89. N. Engheta, A. Salandrino, and A. Alù, "Circuit Elements at Optical Frequencies: Nanoinductors, Nanocapacitors, and nanoresistors," *Phys. Rev. Lett.* Vol. 95, 095504, 2005.

90. M. Popovic, "Theory and design of high-index-contrast microphotonic circuits", *Thesis Ph.D., Massachusetts Institute of Technology,* Department of Electrical Engineering and Computer Science (2008).

91. A. Barkai *et al.*, "Integrated silicon photonics for optical networks", *Journal of Optical Networking,* vol. 6, pp. 25-47 (2007).

92. Y. Vlasov *et al.*, "High-throughput silicon nanophotonic wavelength-insensitive switch for on-chip optical networks", *Nature Photonics,* vol. 2, pp. 242-246, Apr 2008.

CHAPTER 17

INTEGRATING SPHERES

ALİ SERPENGÜZEL

Koç University, Department of Physics,
Microphotonics Research Laboratory, Istanbul 34450 Turkey
aserpenguzel@ku.edu.tr

Spherical microresonators are natural building blocks of three dimensional integrated photonics. The resulting three dimensional device would be a volumetric lightwave circuit or a photonic integrated circuit. The intersphere communication channels can be evanescently coupled optical waveguides. The possible multidimensional integration of these spherical microresonators evanescently coupled with optical waveguides will usher in a new age in integrated photonics and resulting applications and systems.

1. Introduction

I am honored to participate in the global celebration in the honor of Professor Richard Kounai Chang's (RKC) retirement from Yale University and his 70th birthday. I had the good fortune to be a graduate student and a post doctoral research associate in the RKC laboratory. My contribution was on the applications of stimulated Raman scattering (SRS) to microdroplets of sprays.[1,2] After Yale University, I joined the MicroParticle PhotoPhysics Laboratory (MP3L) of Professor Steve Arnold at Polytechnic University,[3] where we performed elastic light scattering experiments from solid (dielectric, i.e., polymer and glass) microspheres evanescently coupled with optical fiber couplers.[4] Later on at Bilkent University, we performed photoluminescence (PL) experiments on distributed Bragg reflector (DBR) based Fabry-Perot microresonators.[5] Currently, at Koç University, we are working on various solid (glass, sapphire, ruby, and silicon) microspheres for optoelectronic device applications. In this chapter, I will briefly review our recent results on solid microsphere resonators, and outline the future of multidimensional, i.e., one-dimensional (1D), two-dimensional (2D) and three-dimensional (3D) optoelectronics integration, that results in linear lightwave circuits (L^2C), planar lightwave circuits (PLC), and volumetric lightwave circuits (VLCs), respectively.

2. Single microsphere evanescent coupling to waveguides in 1D

The evanescent coupling geometry of the dielectric, i.e., polymer or glass, microspheres on an optical fiber coupler is shown in the following figure.

Fig. 1.   The perpendicular (senkrecht) evanescent coupling geometry of a waveguide to a microsphere resonator.

Because of their geometry, the evanescent light coupling to the microsphere and recently the microtoroid resonators[6] has been challenging for practical device applications. Light can be coupled into the microsphere geometry using different types of coupling devices: side-polished optical fibers, prisms,[7] and tapered optical fibers.[8] The tapered fiber geometry is the preferred coupling method for microtoroids.

The principle of these devices is based on providing efficient energy transfer to the microspheres with their high-quality-factor morphology-dependent resonances (MDR's), i.e., whispering gallery modes (WGMs), through the evanescent fields. Efficient coupling can be expected, if phase synchronism and significant overlap of the MDR and the optical coupler evanescent fields occurs.[9] Fiber optic add-drop filters based on a silica microsphere system on a taper-resonator-taper coupler and channel-dropping filter using a microsphere and integrated waveguides have been developed.[10]

At Koç University, we worked with glass microspheres for optical channel dropping[11] in the near-infrared telecommunication bands. Later on, we evanescently coupled to the MDR's of silicon microspheres with optical fiber couplers.[12] As silicon is a semiconductor, silicon microspheres respond both to electrons and photons.[13] Recently, we observed optical modulation in silicon microspheres coupled with optical fibers.[14]

Progress in silicon photonics integration is continuing with racetrack silicon Raman lasers,[15] silicon microring modulators,[16] and silicon microring wavelength converters.[17] Passive and active silicon racetrack resonators,[18] microring resonators,[19,20] microdisk resonators,[21,22] microspiral resonators,[23] micropolygon resonators,[24] and silicon waveguides with feedback[25,26] are some of the geometries pursued for silicon photonic integration.

Electrophotonic integrated circuit (EPIC) or alternatively, optoelectronic integrated circuit (OEIC) is the natural evolution of the "electronic" integrated circuit (EIC) with the added benefit of photonic capabilities. EPIC/OEIC technology requires the integration of microphotonic circuit elements to the already well integrated microelectronic circuit elements. Traditionally, the EIC industry has been based on group IV element silicon, whereas the photonics industry on group III-V semiconductors. The novel EPIC circuits can be realized with the evanescent compling of dielectric or semiconductor microspheres and waveguides.

The following figures show possible lateral evanescent coupling geometries to microsphere resonatrors. Figure 2 shows the in plane (parallel) shows the out of plane (senkrecht) coupling geometry to a microsphere resonator.

Fig. 2. The in plane (parallel) evanescent coupled waveguide geometry to a microsphere resonator.

Figure 3 shows the out of plane (senkrecht) coupling geometry to a microsphere resonator.

Fig. 3. The out of plane (senkrecht) evanescent coupled waveguide geometry to a microsphere resonator.

3. Multiple microsphere evanescent coupling to optical waveguides in 1D

With the precious coupling geometries, the microsphere resonators can be used in wavelength-division multiplexing (WDM) applications such as high-quality-factor resonant filters, detectors, modulators, switches, generators, converters, and attenuators. However, these are zero-dimensional (0D) discrete devices. Integration has to happen in 1D, 2D and 3D. The 1D integration leads to a linear lightwave circuit (L^2C). The following figure shows the possible geometry of two laterally evanescently coupled microsphere resonators that can form a linear wavelight circuit. Band formation has been observed in such coupled wave structures.[27] Microrings,[28] microdisks,[29] microtoroids,[30] microspirals[31] have been integrated in 1D.

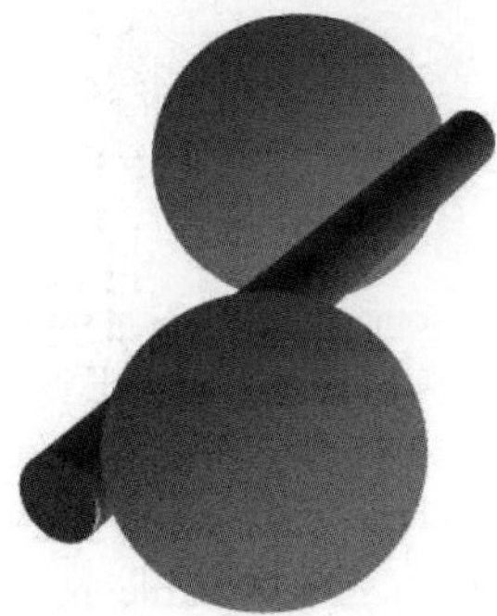

Fig. 4. The senkrecht evanescent coupled optical waveguide to microsphere resonators.

4. Multiple microsphere evanescent coupling to optical waveguides in 2D

2D integration of microresonators leads to photonic lightwave circuits (PLC's). Light coupling is either performed from microcavity to microcavity or through the intermediation of optical waveguides. Microrings,[32] microdisks,[33] and microspheres[34] have been coupled in 2D.

Fig. 5. The lateral evanescent coupled waveguides to microsphere resonators.

5. Multiple microsphere evanescent coupling to optical waveguides in 3D

Spherical microshells and solid microspheres are the natural extensions of microrings and microdisks. Those microspheres, as the building blocks for 3D volumetric lightwave circuits (VLCs), can be integrated in 3D.[35]

Fig. 6. The vertical and lateral evanescent coupled optical waveguides to microsphere resonators.

6. Summary

The microsphere presented to us by geometry and nature is the most symmetric multidimensional microcavity, which is integrable in multiple spatial dimensions. The microspherical resonator is the photonic unit, i.e., cast, atom, of our age and is poised to fulfill all discrete and integrated potential photonic device roles, such as signal generator, converter, modulator, mediator, and detector, for harnessing light in the photonic century.

Acknowledgments

I would like to express my sincere gratitude to Professor Richard Kounai Chang, who as my Ph.D. thesis advisor, initiated me along an exciting journey in scientific research and its technological applications. I would also thank the following researchers, all of whom continued on the bright path laid for us by Professor Chang and join me in wishing many happy returns: Temel Bilici, Şenol İşçi, Ibrahim Inanç, Yiğit Ozan Yılmaz, Abdullah Demir, Onur Akatlar, Ulaş Kemal Ayaz, Emre Yüce, Mohammed Sharif Murib, and Adnan Kurt. We gratefully acknowledge research support from Türkiye Bilimsel ve Teknolojik Araştırma Kurumu (TÜBİTAK), the European Commission (EC), the European Office of Aerospace Research and Development (EOARD), the British Council (BC), and the North Atlantic Treaty Organization (NATO).

References

1. A. Serpengüzel, J. C. Swindal, R. K. Chang, and W. P. Acker, "Two-Dimensional Imaging of Sprays with Fluorescence, Lasing, and Stimulated Raman Scattering," Appl. Opt. 31, 3543 (1992).
2. A. Serpengüzel, S. Küçükşenel, and R. K. Chang, "Microdroplet identification and size measurement in sprays with lasing images," Opt. Express 10, 1118 (2002).
3. A. Serpengüzel, S. Arnold, and G. Griffel, "Excitation of Resonances of Microspheres on an Optical Fiber", Opt. Lett. 20, 654 (1995).
4. G. Griffel, S. Arnold, D. Taskent, A. Serpengüzel, J. Connolly, and N. Morris, "Morphology Dependent Resonances of a Microsphere/Optical Fiber System," Opt. Lett. 21, 695 (1996).
5. A. Serpengüzel, A. Aydinli, A. Bek, and M. Güre, "Visible photoluminescence from planar amorphous silicon nitride microcavities," J. Opt. Soc. Am. B 15, 2706 (1998).
6. S. M. Spillane et al, "Ultrahigh-Q toroidal microresonators for cavity quantum electrodynamics," Phys. Rev. A 71, 013817 (2005).
7. M. L. Gorodetsky and V. S. Ilchenko, "High-Q optical whispering gallery microresonators: precession approach for spherical mode analysis and emission patterns," Opt. Commun. 113, 133 (1994).
8. J. C. Knight, G. Cheung, F. Jacques, and T. A. Birks, " Phased matched excitation of whispering gallery modes by a fiber taper," Opt. Lett. 22, 1129 (1997).
9. A. Serpengüzel, S. Arnold, G. Griffel, and J. A. Lock, "Enhanced coupling to microsphere resonances with optical fibers," J. Opt. Soc. Am. B 14, pp. 790 (1997).
10. M. Cai, G. Hunziker, and K. Vahala, "Fibre optic add-drop device based on a silica microsphere-whispering gallery mode system," IEEE Photon. Technol. Lett. 11, 686 (1999).
11. T. Bilici, S. Isçi, A. Kurt, and A. Serpengüzel, "Microsphere Based Channel Dropping Filter with Integrated Photodetector," IEEE Photon. Technol. Lett. 16, 476 (2004).
12. Y. O. Yilmaz, A. Demir, A. Kurt, and A. Serpengüzel, "Optical Channel Dropping with a Silicon Microsphere," IEEE Photon. Technol. Lett. 17, 1662 (2005).
13. A. Serpengüzel, A. Kurt, and U.K. Ayaz, "Silicon microspheres for electronic and photonic integration," Photon. Nanostructur.: Fundam. Appl. 6, 179 (2008).
14. E. Yüce, O. Gürlü, and A. Serpengüzel, "Optical Modulation with Silicon Microspheres," IEEE Photon. Technol. Lett. 21, 1481 (2009).
15. H. Rong, S. Xu, O. Cohen, O. Raday, M. Lee, V. Sih, and M. Paniccia, "A Cascaded Silicon Raman Laser," Nature Photon. 3, 170 (2008).
16. Q. Xu, S. Manipatruni, B. Schmidt, J. Shakya, and M. Lipson, "12.5 Gbit/s carrier-injection-based silicon microring silicon modulators," Opt. Express 15, 431 (2007).
17. S. F. Preble, Q. Xu, and M. Lipson, "Changing the colour of light in a silicon resonator," Nature Photon. 1, 293 (2007).

18. B.D. Timotijevic, F. Y. Gardes, W. R. Headley, G. T. Reed, M. J. Paniccia, O. Cohen, D. Hak, G. Z. Masanovic, et al., "Multi-stage racetrack resonator filters in silicon-on-insulator," Journal of Optics a-Pure and Appl. Opt. 8, S473 (2006).
19. Q. F. Xu and M. Lipson, "All-optical logic based on silicon micro-ring resonators," Opt. Express 15, 924 (2007).
20. J Niehusmann, et al, "Ultrahigh-quality-factor silicon-on-insulator microring resonator," Opt. Lett. 29, 2861 (2004).
21. A. Morand, Y. Zhang, B. Martin, K. P. Huy, D. Amans, P. Benech, J. Verbert, E. Hadji, and J.-M. Fédéli, "Ultra-compact microdisk resonator filters on SOI substrate," Opt. Express 14, 12814 (2006).
22. T. J Johnson et al, "Self-induced optical modulation of the transmission through a high-Q silicon microdisk resonator," Opt. Express 14, 817 (2006).
23. M. Kneissl et al, "Current-injection spiral-shaped microcavity disk laser diodes with unidirectional emission," Appl. Phys. Lett. 84, 2485 (2004).
24. C. Li et al, "Waveguide-coupled octagonal microdisk channel add-drop filters," Opt. Lett. 29, 471 (2004).
25. B. Jalali, "Teaching silicon new tricks," Nature Photon. 1, 193 (2007).
26. A. Alduino and M. Paniccia, "Interconnects - Wiring electronics with light," Nature Photon. 1, 153 (2007).
27. B. M. Möller et al "Band Formation in Coupled-Resonator Slow-Wave Structures," Opt. Express, 15, 17362 (2007).
28. B. E. Little, et al, "Very high-order microring resonator filters for WDM applications," IEEE Photon. Technol. Lett. 16, 2263 (2004).
29. A. Nakagawa, et al "Photonic molecule laser composed of GaInAsP microdisks," Appl. Phys. Lett. 86, 041112 (2005).
30. M. Hossein-Zadeh et al, "Free ultra-high-Q microtoroid: a tool for designing photonic devices," Opt. Express 15, 166 (2007).
31. G. D. Chern et al, "High-Q-preserving coupling between a spiral and a semicircle μ-cavity," Opt. Lett. 32, 1093 (2007).
32. B. E. Little et al, "Microring resonator arrays for VLSI photonics," IEEE Photon. Technol. Lett. 12, 323 (2000).
33. C. P. Michael et al, "An optical fiber-taper probe for wafer-scale microphotonic device characterization," Opt. Express 15, 4745 (2007).
34. V. N. Astratov et al, "Optical coupling and transport phenomena in chains of spherical dielectric microresonators with size disorder," Appl. Phys. Lett. 85, 5508 (2004).
35. V. N. Astratov, "Percolation of light through whispering gallery modes in 3D lattices of coupled microspheres," Opt. Express 15, 17351 (2007).

CHAPTER 18

MICROSPIRAL AND DOUBLE-NOTCH-SHAPED RESONATORS FOR INTEGRATED PHOTONICS

XIANSHU LUO and ANDREW W. POON

Photonic Device Laboratory, Department of Electronic and Computer Engineering, The Hong Kong University of Science and Technology, Hong Kong, China
eeawpoon@ust.hk

This chapter exemplifies in a small way how some of Prof. Richard Kounai Chang's lasting contributions to the science and technology of optical microcavities are finding promising applications in integrated photonics. Specifically, we review the development of microspiral and double-notch-shaped microdisk resonator-based passive devices. Microspiral resonators as uni-port directional emission microcavity lasers was proposed and demonstrated by Chang's group in 2003. We review how such novel-shaped microresonator design featuring gapless coupling has been bearing fruits as a new breed of silicon photonic passive devices including channel filters and coupled-resonator optical waveguides.

1. Introduction

Microdisk or microring resonators, with their key merits of high-Q modes due to near total internal reflections (TIRs), compact micrometer-scale size, and accessibility with optical waveguides or tapered fibers, have long been regarded as promising microlaser cavities[1-2] and wavelength-selective filter components.[3-4] In large-scale-integrated photonic circuits (LSIPCs) such microresonators may find technologically important applications including telecommunications, optical interconnect and biological / chemical sensing. However, since as early as the mid-1990's, researchers have realized that the conventional circular-shaped microdisk / microring resonators are not favorable for applications as on-chip laser cavities. Due to the cavity rotational symmetry the whispering-gallery modes (WGMs) grazing along the microcavity circumference only output-couple nearly tangentially and homogeneously along the cavity sidewall. This results in nearly isotropic laser emission instead of uni-port directional emission which is desirable for integrated photonics.

Furthermore, in order to efficiently input/output-couple light to the WGMs, evanescent-field coupling is imposed via a narrow gap (typically ~0.1 - ~0.3 μm for high-index-contrast semiconductor substrates[3]). Nonetheless, the fabrication of such gap separations with high fidelity and high uniformity across a chip is still technologically challenging even by using advanced lithography. For integrated photonics, it is thus relevant to find a novel microcavity design that enables uni-port directional laser emission and ease of light coupling without imposing narrow gap separations while preserving high-Q resonances.

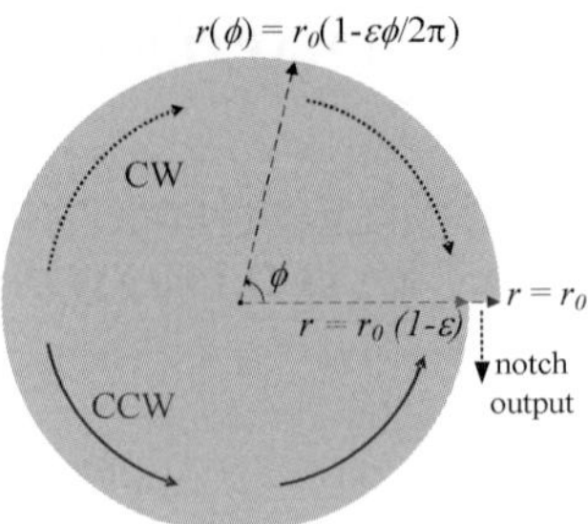

Fig. 1. Schematic of microspiral resonators. r: radius, ε: radius change rate, CW: clockwise, CCW counterclockwise.

In 2003, Prof. Chang's group at Yale University proposed and demonstrated a beautifully simple solution - microspiral resonators.[5] Figure 1 shows the schematic of microspiral resonators. Microspiral resonator is a microdisk or micropillar with the radius linearly varying as a function of the azimuthal angle ϕ as

$$r(\phi) = r_0(1 \pm \frac{\varepsilon\phi}{2\pi}) \tag{1}$$

where r_0 is the radius at $\phi = 0$, and ε is the change rate of the radius. The radius mismatch at $\phi = 2\pi$ gives a notch of width $r_0\varepsilon$, which is typically a small fraction of the radius. Such microspiral shape completely breaks the rotational symmetry. Chang's group experimentally demonstrated unidirectional laser emission from the microspiral notch.[5] Such uni-port directional emission implies that waveguide butt-coupling is feasible and thus microspiral constitutes a potential microcavity laser design for integrated photonics. Various research groups worldwide have since then also realized uni-port directional (or unidirectional) laser emission using microspiral resonators in material systems including III-V semiconductors[6-8] and polymer materials[9-10] using optical pumping and electrical injection.

Inspired by such microspiral resonator design, we have over the years designed and demonstrated various microspiral resonator-based passive devices with waveguide butt-coupling at the microspiral notch[11-13] (see Sec. 2). The notch-coupled waveguide eliminates the evanescent-field coupling. However, as microspiral resonators only have one notch, waveguide side-coupling is still imposed for input/output (I/O) applications. In order to eliminate the waveguide side-coupling, we modified the microspiral design to accommodate two notches (one on each side) for simultaneous input/output gapless coupling as a so-called double-notch-shaped microdisk resonator[12,14-15] (see Sec. 3). Using these two building blocks, we have developed various coupled-microdisk resonator devices (see Sec. 4, 5) including (i) coupled-microspiral resonators,[16-17] (ii) coupled-double-notch-shaped microdisk resonators,[15,18] (iii) coupled microspiral/double-notch-shaped microdisk resonators, and (iv) coupled-resonator optical waveguides (CROWs).[15,18-20] Here we highlight some of these demonstrations, all of which are deeply rooted from Prof. Chang's original work.

2. Waveguide-coupled microspiral resonator devices

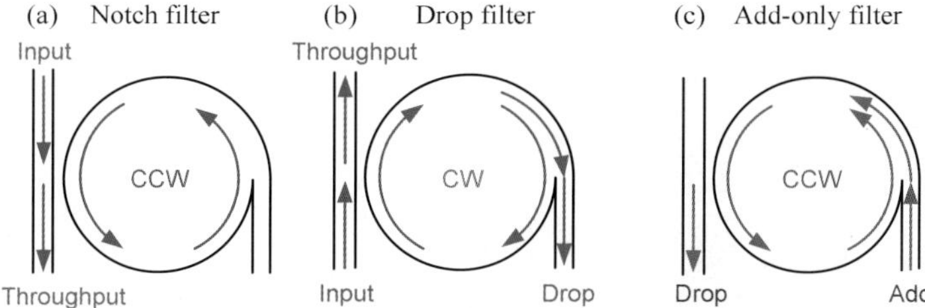

Fig. 2. Schematics of waveguide-coupled microspiral resonator-based filters: (a) notch (channel-rejection) filter, (b) drop filter, and (c) add-only filter. [Reproduced with permission from Optical Society of America [11] © 2007 Optical Society of America].

Figure 2 shows the schematic of waveguide-coupled microspiral resonators.[13] In order to input/output-couple light to the microdisk, we integrate to the cavity an evanescently side-coupled waveguide and a seamlessly butt-coupled waveguide at the microspiral notch. Depending on the light input-coupling port, the microspiral resonator-based filter has three different configurations: notch or channel-rejection filter (Fig. 2(a)), drop filter (Fig. 2(b)) and add-only filter (Fig. 2(c)). For the channel-rejection filter, the evanescently input-coupled counter-clockwise (CCW) traveling wave does not favor output-coupling to the notch-waveguide. For the drop filter, the evanescently input-coupled clockwise (CW) traveling-wave favors partial transmission to the notch-waveguide. For the add-only filter, the gaplessly input-coupled CCW traveling-wave again does not favor output-coupling to the notch-waveguide upon traveling a cavity round-trip, yet the cavity mode field can be evanescently output-coupled to the side-coupled waveguide.

We fabricated the microspiral resonator filters on a silicon nitride (SiN)-on-silica substrate using standard silicon nanoelectronics fabrication processes, which are detailed elsewhere.[11,17] Hereafter, all the device fabrications discussed in the chapter follow the same fabrication process.

Figure 3(a) shows the scanning electron micrograph (SEM) of our fabricated microspiral resonator-based filter. The zoom-in-view SEMs show the evanescent- and notch-coupling regions. The side-coupled waveguide has an interaction length of ~ 18 μm along the microspiral rim in order to enhance the coupling efficiency.

     Optical Processes in Microparticles and Nanostructures

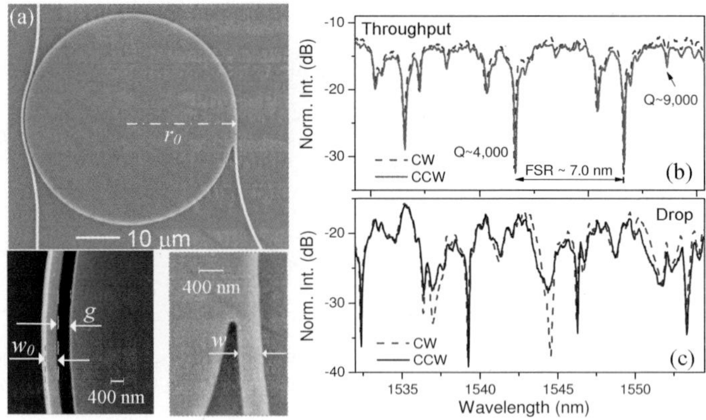

Fig. 3. (a) SEMs of our fabricated microspiral resonator filter on a SiN-on-silica substrate. Zoom-in views: waveguide-resonator evanescent coupling region and gapless notch-coupling region. The fabricated filter is with $r_0 = 25$ μm, $w_0 = w = g = 0.4$ μm, (b) Measured TM-polarized throughput-port transmission spectra for notch filter (CCW) and drop filter (CW). (c) Measured TM-polarized drop-port transmission spectra for drop filter (CW) and add-only filter (CCW).

Figures 3(b)-(c) show the measured TM-polarized (electric field normal to the chip) throughput- and drop-port transmission spectra from the three filter configurations. The free-spectral range (FSR) is ~ 7.0 nm, which correlates well with the microspiral circumference. The measured maximum quality (Q) factor is ~ 9,000, suggesting the microspiral resonators preserve reasonable high-Q modes. The throughput-port transmission spectra of notch filter and drop filter, and the drop-port transmission spectra of drop filter and add-only filter are essentially identical as expected from the reciprocity relation.[11]

The reciprocity relation in microspiral resonators suggests identical cavity losses between the reciprocal transmission pair. However, considering the throughputs of notch filter and drop filter, the drop filter has an additional light output-coupling via the notch-waveguide. Thus, in order to balance the total cavity loss, an additional loss mechanism for notch filter should exist.

In order to gain insights into such loss balancing mechanism, we numerically simulated the microspiral resonator modes using two-dimensional finite-difference time-domain (2-D FDTD) method. We adopt an effective refractive index contrast of 1.9-to-1.4 in order to represent a 0.9-μm-thick SiN-on-silica stack-layered device region and a 0.2-μm-thick SiN slab region.[13] We consider a microspiral with $r_0 = 10$ μm and $w = 0.4$ μm ($\varepsilon = 0.04$). The side-coupled waveguide width is 0.3 μm and the gap separation is 0.2 μm.

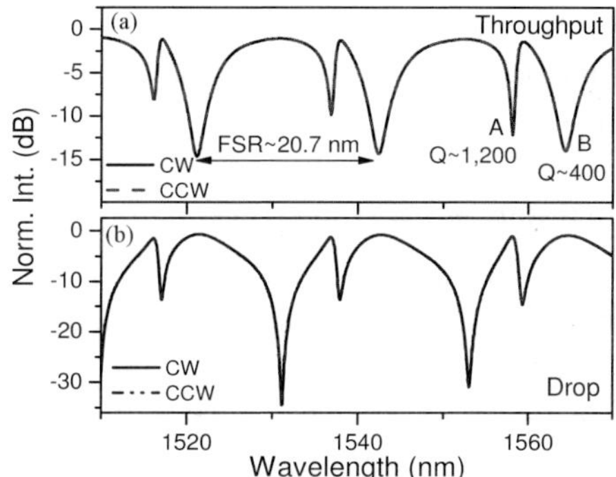

Fig. 4. 2-D FDTD-simulated TM-polarized transmission spectra of a waveguide-coupled microspiral resonator-based filter. (a) Throughput-port transmission spectra of notch filter (CCW) and drop-filter (CW). (b) Drop-port transmission spectra of drop filter (CW) and add-only filter (CCW). In (a) and (b), the solid-line spectrum (CW) and the dashed-line spectrum (CCW) overlap.

Figures 4(a)-(b) show the simulated TM-polarized throughput- and drop-port transmission spectra of a microspiral resonator-based filter. The multimode transmission spectra show a FSR of ~ 20.7 nm, which is consistent with the microspiral disk circumference. The simulated highest Q is ~ 1,200. We clearly observe the reciprocity relation from the overlapped transmission spectra of the corresponding reciprocal transmission pairs.

Figures 5(a)-(f) show the FDTD-simulated resonance mode-field distributions of resonances A and B (see Fig. 4(a)) for all three filter configurations. All the simulated mode-field patterns display WGM-like distributions. Nonetheless, the reciprocity relation does not suggest identical mode-field distributions between reciprocal transmission pairs. Such observation is significant as it is likely the key to balance the cavity round-trip loss between CW and CCW traveling waves, and thus in turn to preserve the reciprocity relation. In our experiments,[11] we observed indeed non-identical resonance mode-field distributions between reciprocal transmission pairs using out-of-plane light scattering collected by a 2-D scanning lensed fiber probe.

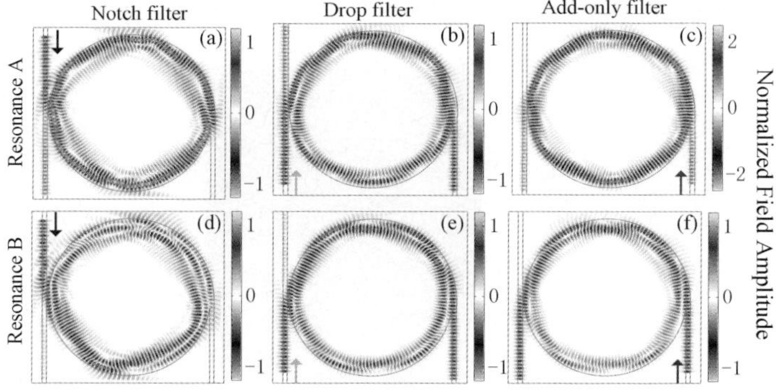

Fig. 5. FDTD-simulated resonance mode-field patterns of resonances A and B for all three filter configurations.

We note that for resonance A, the resonance mode-field pattern in add-only filter exhibits relatively high intensity compared with other filter configurations. This suggests a higher coupling efficiency via the gapless notch input-coupling.

3. Double-notch-shaped microdisk resonators

It is natural to extend the concept of microspiral resonators with a single notch to a microdisk resonator with two notches. Figure 6 shows the schematic of double-notch-shaped microdisk resonators.[14-15] The microdisk shape comprises two jointed non-identical semi-circles with radii of r_1 and r_2. The mismatches between the two semi-circles on both sides along the diameter give two notches with widths of w_1 and w_2. The notch widths are identical in the case that the two semi-circles are concentric. Unlike microspiral resonators, such notch-shaped microdisk with identical notches preserves a two-fold mirror symmetry with the symmetry axis at $\phi = 90°-270°$. Each notch is seamlessly jointed to a waveguide of the same width. Light can be gaplessly input/output-coupled to the microdisk via these notch-waveguides without relying on the evanescent field. It is conceivable that the round-trip cavity light grazing along the cavity circumference can be wavefront-matched (phase-matched) with the input-coupled lightwave and thus attaining WGM-like resonances, with the resonance field only partially output-coupled to the throughput-port. This enables such gaplessly waveguide-coupled microdisk resonator to act as a single-input single-output channel-rejection filter.[14]

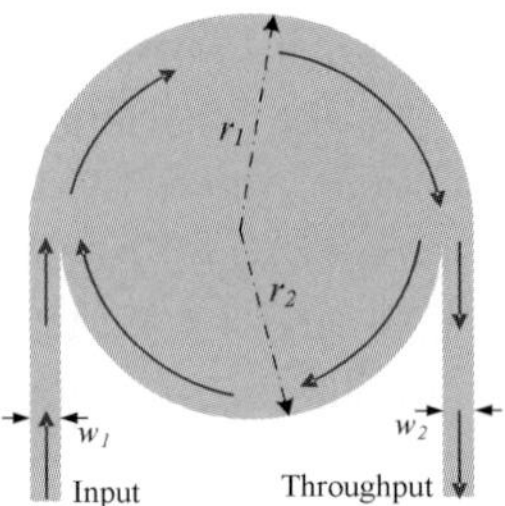

Fig. 6. Schematic of double-notch-shaped microdisk resonators. r_1, r_2: radii of the hemi-circular microdisks; w_1, w_2: notch widths.

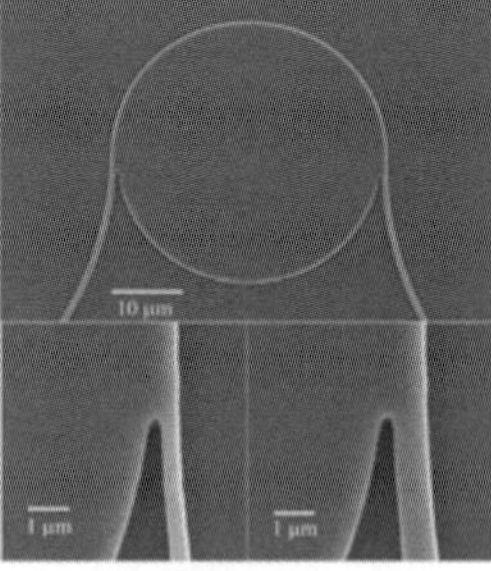

Fig. 7. SEMs of our fabricated double-notch-shaped microdisk resonator-based filter. Zoom-in-views: gapless coupling regions with notch widths of 0.4 μm and 0.8 μm.

We again fabricated the double-notch-shaped microdisk resonator-based filters on a SiN-on-silica substrate. Figure 7 shows the SEMs of a typical device. The microdisks are designed with $r_1 = 20$ μm and different notch widths w. Zoom-in-view SEMs reveal two different designs of the notch-waveguide regions with w of 0.4 μm and 0.8 μm.

Figures 8(a)-(c) show the measured TM-polarized transmission spectra from three double-notch-shaped microdisks with different notch widths. The FSR of ~ 9 nm is consistent with the microdisk circumference, which suggests WG-like mode distributions. For the device with $w_1 = w_2 = 0.4$ μm, we observe more than eight modes (denoted as i to viii) with the highest Q of ~ 15,000 (mode v) and the highest extinction ratio (ER) of ~ 25 dB (mode viii).

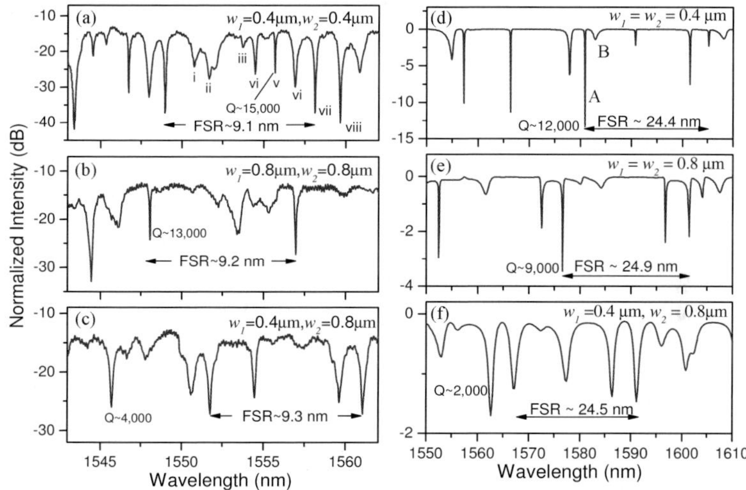

Fig. 8. Measured TM-polarized through-port transmission spectra of a double-notch-shaped microdisk resonator with $r_1 = 20$ μm and various notch widths: (a) $r_2 = 19.6$ μm, $w_1 = w_2 = 0.4$ μm, (b) $r_2 = 19.2$ μm, $w_1 = w_2 = 0.8$ μm, and (c) $r_2 = 19.4$ μm, $w_1 = 0.4$ μm, $w_2 = 0.8$ μm. (d)-(f) 2-D FDTD-simulated TM-polarized throughput-port transmission spectra of a double-notch-shaped microdisk resonator with $r_1 = 10$ μm and various notch widths: (d) $r_2 = 9.6$ μm, $w_1 = w_2 = 0.4$ μm, (e) $r_2 = 9.2$ μm, $w_1 = w_2 = 0.8$ μm, and (f) $r_2 = 9.4$ μm, $w_1 = 0.4$ μm, $w_2 = 0.8$ μm.

As the notch width widens to $w_1 = w_2 = 0.8$ μm given the same r_1, the number of resonances drops to only three dominant modes. We attribute this to the notch-waveguide output-coupling loss, and thereby the wider notch yields a larger cavity loss.[13] Consequently, it is conceivable that modes with a relatively large spatial overlap with the notch-waveguide encounter a relatively large cavity loss and thus are suppressed. While modes with a relatively small spatial overlap with the notch-waveguide encounter a relatively small cavity loss and thus remain as relatively high-Q resonances.

In the case that the notch widths of the input- and output-ports are non-identical (i.e. the two semi-circles are non-concentric), the resonance Q's and ER's are compromised. The measured maximum Q is only ~ 4,000 and the highest ER is only ~ 10 dB (Fig. 8(c)).

Figures 8(d)-(f) show the FDTD-simulated transmission spectra of double-notch-shaped microdisk resonators. The spectra evolution with the various notch width designs is consistent with our measurement, namely relatively high-Q high-ER resonances are obtained for relatively narrow and identical notch widths.

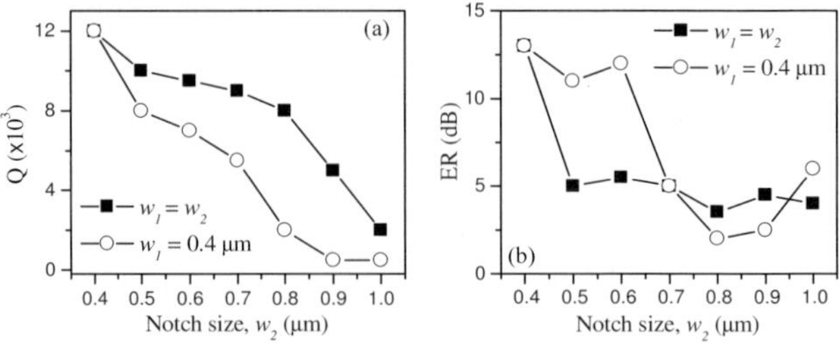

Fig. 9. FDTD-simulated (a) resonance quality factors (Q's) and (b) extinction ratios (ER's) of double-notch-shaped microdisk resonators as functions of notch width w_2.

Figures 9 show the simulated Q's and ER's as functions of the notch width. We fix the radius $r_1 = 10$ μm, while varying r_2 from 9.6 μm to 9.0 μm, giving notch widths of $w_1 = w_2 = 0.4$ μm to 1.0 μm (solid squares). In order to examine the effect of asymmetric notch widths, we also fix $w_1 = 0.4$ μm while varying w_2 from 0.4 μm to 1.0 μm (open circles).

The simulated cavity Q reaches a maximum of ~ 12,000 at $w_1 = w_2 = 0.4$ μm. The Q value drops with the notch width, signaling an increase in the total cavity loss. The Q's for the asymmetric notch designs drop faster than those for the symmetric notch designs. The ER dependence on the notch size is, however, not obvious. As in conventional microdisk resonators, the resonance ER is determined by the input/output-coupling and cavity loss. Hereafter, we adopt double-notch-shaped microdisk resonators with symmetric notch widths.

Figures 10(a)-(b) show the FDTD-simulated on-resonance mode-field patterns of a double-notch-shaped microdisk resonator at the wavelength of 1580.8 nm (resonance A in Fig. 8(d)) and 1582.9 nm (resonance B in Fig. 8(d)). Both resonances reveal WGM-like mode-field distributions. However, the spatial patterns are complicated and may not be readily described in terms of radial and azimuthal orders. More theoretical studies on such double-notch-shaped microdisk modes should be called for.

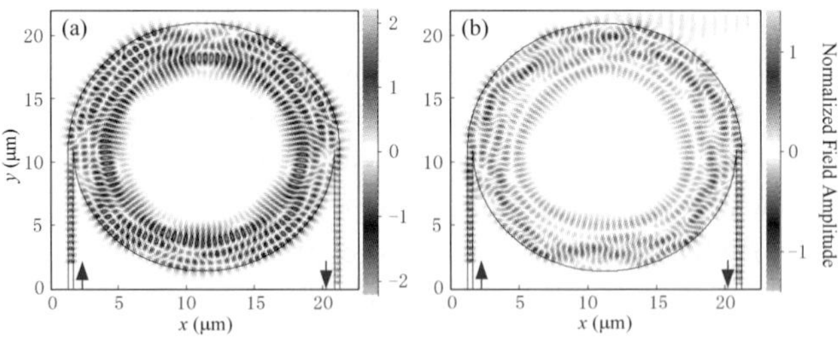

Fig. 10. FDTD-simulated TM-polarized mode-field patterns of a double-notch-shaped microdisk with notch widths of 0.4 μm for on-resonance wavelengths at (a) 1580.8 nm, and (b) 1582.9 nm.

4. Coupled microspiral / double-notch-shaped microdisk resonators

Microspiral and double-notch-shaped microdisk resonators constitute building blocks for coupled or high-order cascaded microdisk resonators. The key is the gapless inter-cavity coupling.[16-17,21-22] Here we show some demonstrated examples of coupled-microdisk resonators including (i) coupled-microspiral resonators, (ii) coupled-double-notch-shaped microdisk resonators, and (iii) coupled microspiral / double-notch-shaped microdisk resonators.

Figure 11(a) shows the SEM of our fabricated coupled-microspiral resonator. Two microspiral resonators are seamlessly jointed via the notches, forming gapless inter-cavity coupling (see the zoom-in-view).[16-17] The microspirals are identically designed with $r_0 = 20$ μm and $w = 0.8$ μm.

Figures 11(b)-(c) show the measured TM-polarized throughput- and drop-port transmission spectra for CCW and CW circulations (light launched from different ends of the side-coupled waveguide). The measured FSR is ~ 8.9 nm, which corresponds to the circumference of a single microspiral resonator. The measured highest Q is ~ 14,000, suggesting that such gapless-coupled microspiral resonators preserve high Q. The CCW and CW throughput transmission spectra are essentially identical, following the reciprocity relation. Nonetheless, the drop-port transmission spectra (not related as reciprocal transmission pairs[17]) only display identical cavity resonance wavelengths with significantly different resonance lineshapes. Furthermore, both our measurements and FDTD-simulations[16-17] suggest that the CW drop-port transmissions always exhibit higher intensities than the CCW drop-port transmissions. This suggests that the gapless intercavity coupling is directional and asymmetric between CCW and CW traveling waves,[16-17] as detailed below.

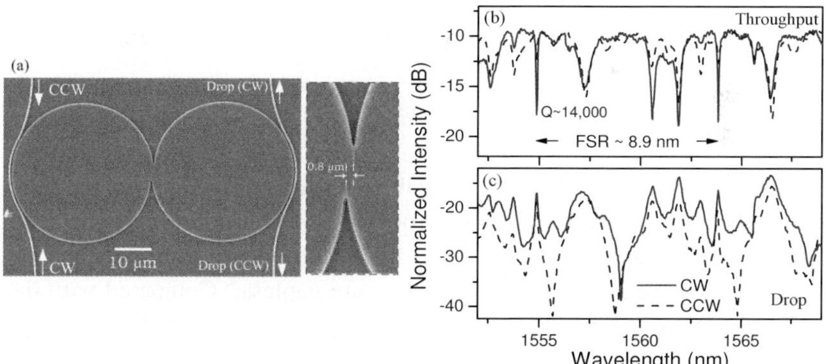

Fig. 11. (a) SEM of the fabricated coupled-microspiral resonator. Zoom-in view: gapless inter-cavity coupling region. (b)-(c) Measured through- and drop-port transmission spectra corresponding to CW and CCW circulations in the first microspiral with different light input-coupling ports.

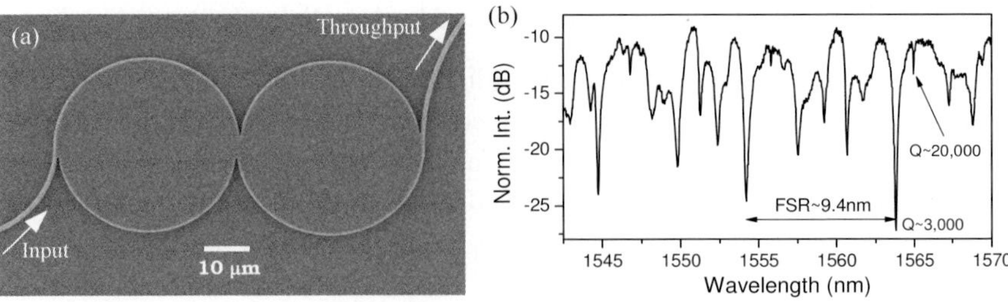

Fig. 12. SEM of our fabricated coupled-double-notch-shaped microdisk resonator filter with r_1 = 20 μm and w_1 = w_2 = 0.4 μm. (b) Measured TM-polarized through-port transmission spectrum.

Figure 12(a) shows the SEM of the fabricated coupled-double-notch-shaped microdisk resonators.[15,18]   Both the input- and output-coupling to the butt-coupled waveguides are gapless.   Figure 12(b) shows the measured TM-polarized throughput transmission spectrum.   The multimode spectrum shows a FSR of 9.4 nm, which corresponds to the single microdisk circumference. The measured maximum Q value is ~ 20,000, again suggesting a high-Q microdisk resonator.

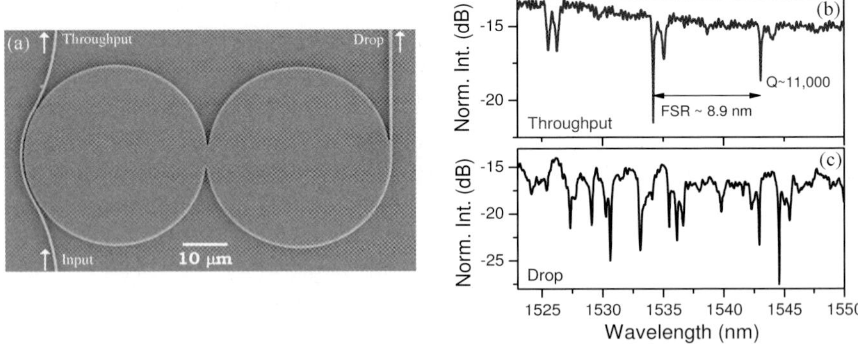

Fig. 13. SEM of the fabricated coupled-microdisk resonator using microspiral and double-notch-shaped microdisks, with r_0 = r_1 = 20 μm and w = 0.4 μm. (b)-(c) Measured TM-polarized throughput- and drop-port transmission spectra.

Figure 13(a) shows the SEM of the fabricated coupled-microdisk resonators using microspiral and double-notch-shaped microdisk resonators. Such combination imposes an evanescent waveguide-coupling to the microspiral resonator while the inter-cavity coupling and the waveguide-to-notch coupling are gapless. Compared with the coupled microspiral resonators (Fig. 11), this structure replaces the evanescently coupled drop-port by the gaplessly coupled drop-port. While compared with the coupled-double-notch-shaped microdisk resonators (Fig. 12), this structure offers an additional evanescently coupled throughput-port. In order to enable seamless integration, we design identical notch width ($w = r_o\varepsilon$) for the microspiral and double-notch-shaped microdisk resonators. Thus, the outer rim circumferences for both microdisks are nearly matched (with a slight mismatch of $3\pi r_0\varepsilon$ using $r_1 = r_0$). The fabricated device is with r_0 = r_1 = 20 μm and identical notch width w = 0.4 μm.

Figures 13(b)-(c) show the measured TM-polarized throughput- and drop-port transmission spectra of the coupled-microdisk resonator using microspiral and double-notch-shaped microdisks, with an FSR of ~ 8.9 nm. The throughput-port transmission spectrum only depicts two dominant modes, with the highest Q value of ~ 11,000. The drop-port transmission spectrum reveals broadened resonance lineshapes. We observe only low side-mode suppression ratio (SMSR) in this case. Enhancement of SMSR requires in general using high-order cascaded microdisk resonators (see Sec. 5).

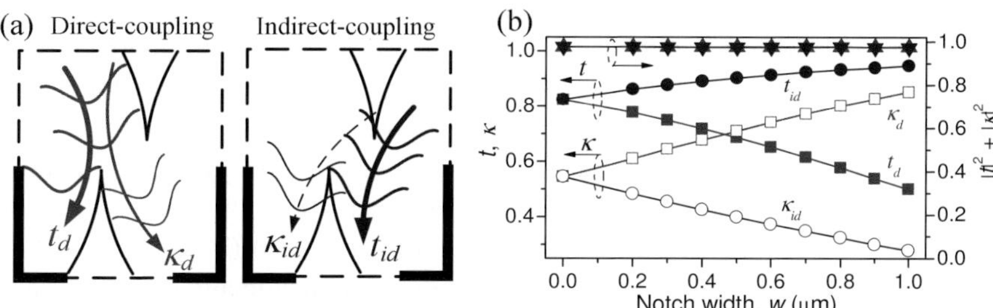

Fig. 14. (a) Schematic of the direct and indirect gapless couplings. (b) FDTD-simulated transmission and coupling coefficients for direct and indirect gapless coupling as functions of notch width. Open squares: κ_d, open circles: κ_{id}, solid squares: t_d, solid circles: t_{id}, solid stars: $|t|^2+|\kappa|^2$.

The gapless inter-cavity coupling is asymmetric in two circulation directions due to the structural asymmetry in the jointed notch region. Figure 14(a) schematically shows the asymmetric inter-cavity coupling. The forward-propagation from the large-hemi-circular disk to the adjacent large-hemi-circular disk is preferred to the backward-propagation from the small-hemi-circular disk to the adjacent small-hemi-circular disk. We term the preferred forward-propagation as direct-coupling and the backward-propagation as indirect-coupling. The transmission and coupling coefficients for direct- and indirect-couplings, t_d, κ_d, t_{id}, κ_{id}, are denoted.

We simulated the inter-cavity coupling using 2-D FDTD simulations with $r_1 = 20$ μm and $w = 0.4$ μm. We assume an effective refractive index n_{eff} of 1.9-to-1.4 for TM mode in a SiN-on-silica substrate.[13] We obtain the transmission and coupling coefficients from the normalized light intensities collected by the corresponding monitors (the solid lines in Fig. 14(a)). Figure 14(b) shows the simulated transmission and coupling coefficients of the direct- and indirect-coupling as functions of the notch width. The direct-coupling coefficient almost linearly scales with the notch width. The indirect-coupling coefficient almost linearly inversely scales with the notch width. The corresponding light transmission coefficients display complementary trends according to energy conservation ($|t|^2+|\kappa|^2 \approx 1$). Thus, our numerical modeling suggests that by designing the notch width, it is possible to tailor the gapless inter-cavity coupling.

5. Coupled-resonator optical waveguides

Gapless inter-cavity coupling is particularly desirable to build coupled-resonator optical waveguides (CROWs). CROWs using microring resonators have long been attracting a lot of research interest due to their potential applications as on-chip high-order optical channel filters[23-27] and optical delay lines.[28-31] However, conventional microring resonators rely on the evanescent-field inter-cavity coupling via many narrow gap spacing. Such coupling gap spacing imposes technologically challenging fabrication even by using advanced lithography. Other researchers also realized this and proposed silicon nanophotonic slow-light waveguides with gapless inter-cavity coupling via a short waveguide segment between adjacent cuboidal microresonators that carry Fabry-Perot-like modes.[32]

Based on the coupled microdisk resonators discussed in Fig. 12 and Fig. 13, we have demonstrated two different CROW structures by cascading many identically designed double-notch-shaped microdisk resonators. One comprises only the double(D)-notch-shaped microdisks with single throughput transmission,[15,18] termed as the D-only structure. The other comprises one microspiral (S) on the periphery and cascaded double-notch-shaped microdisk resonators resulting in both throughput- and drop-port transmissions,[19-20] termed as the S-D structure. Figures 15(a)-(b) show the schematic of D-only and S-D structures. Such designs eliminate the technologically challenging constraints in fabricating many sub-micrometer gap spacing.

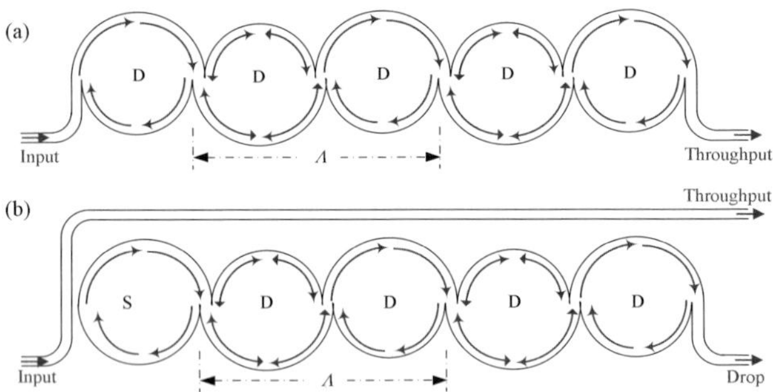

Fig. 15. Schematics of many-element CROWs using (a) double-notch-shaped microdisk resonator and (b) microspiral and double-notch-shaped microdisk resonators. [Reproduced with permission from Optical Society of America [20] © 2009 Optical Society of America].

For D-only and S-D CROWs, the periodicity is the coupled-double-notch-shaped microdisks with period Λ. Based on the CROW theory,[29] such quasi-periodic structures exhibit frequency bands centered at the resonance frequency of the individual resonators, due to the inter-cavity coupling induced mode splitting. These bands exhibit a FSR following that of the individual resonators. As the number of period (the disk number N) increases, the transmission properties (e.g. transmission or rejection band widths) gradually approach those of an ideal periodic structure with infinite length.[33]

We fabricated such CROWs in SiN-on-silica substrates. We designed $r_0 = r_1 = 20$ μm and two sets of notch width values, $w = 0.8$ μm and $w = 0.4$ μm. For S-D CROWs, we designed the evanescently coupled single-mode waveguide width to be 0.4 μm, and the gap spacing between the waveguide and the microspiral resonator is 0.4 μm. We examined the disk number N of 11, 21, 41, 51, 61, 81 and 101.

Figures 16(a)-(b) show the optical micrographs of the fabricated 101-element D-only and S-D CROWs. The 101-element CROWs span a footprint of ~ 4 mm × 40 μm.

Figures 16(c) and (f) show the zoom-in-view SEMs of the input- and output-coupling regions from the S-D CROW. The evanescent-coupled waveguide has an interaction length of ~ 18 µm along the microspiral rim. The input- and output-coupling waveguides are adiabatically tapered to width of 2.5 µm in order to minimize the waveguide propagation loss and also enable better light input-coupling from the lensed fiber (with a spot diameter of ~ 2.5 µm). Figures 16(d)-(e) show the SEMs of a typical jointed notch region.

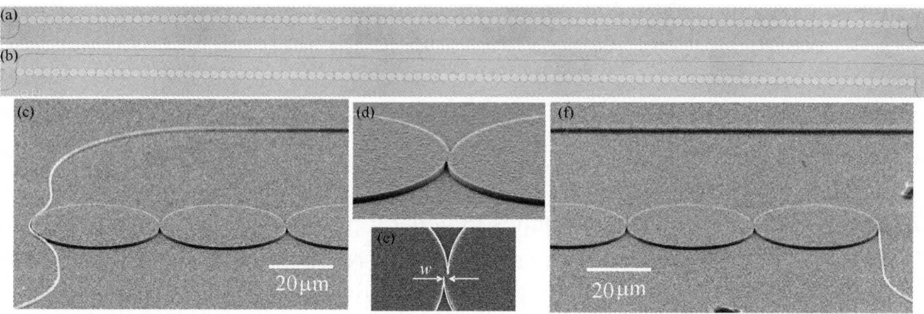

Fig. 16. Fabricated CROWs in a SiN-on-silica substrate. Optical micrographs of (a) D-only and (b) S-D CROWs. (c)-(f) SEMs of the input-coupling, inter-cavity coupling and output-coupling regions of a 101-element S-D CROW. [Reproduced with permission from Optical Society of America [20] © 2009 Optical Society of America].

Figures 17(a)-(f) show the measured TM-polarized multimode transmission spectra of three different device configurations, namely the D-only structure with w = 0.8 µm, the S-D structure with w = 0.8 µm and the S-D structure with w = 0.4 µm. For each configuration, we show the transmission spectra of N = 41 and 101. For each device, the transmission spectra reveal periodic transmission bands. The FSR is ~ 9 nm, which is consistent with the constituent double-notch-shaped microdisk circumference. For N = 41, the side modes due to the highly multimode microdisk resonators are pronounced, while for N = 101, the side-modes are largely suppressed.

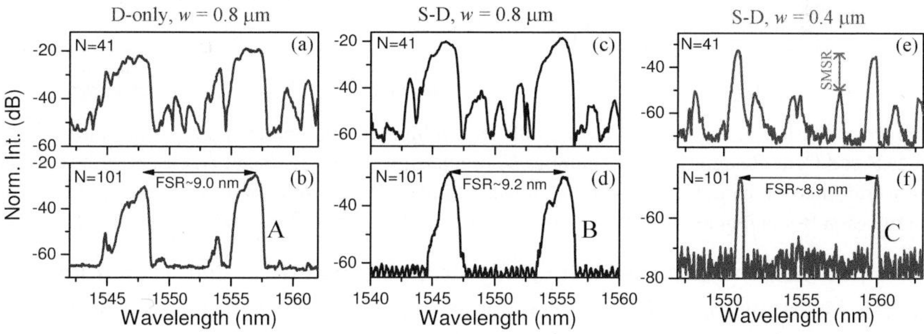

Fig. 17. Measured TM-polarized CROW transmission spectra. (a)-(b) Throughput transmissions of the 41- and 101-element D-only CROWs with w = 0.8 µm. (c)-(d) Drop-port transmissions of the 41- and 101-element S-D CROWs with w = 0.8 µm. (e)-(f) Drop-port transmissions of the 41- and 101-element S-D CROWs with w = 0.4 µm. [Reproduced with permission from Optical Society of America [20] © 2009 Optical Society of America].

In detail, Figs. 17(a)-(b) show the measured throughput transmission spectra for the D-only CROWs. The transmission bands correspond to the off-resonance wavelengths of the constituent microdisk resonators (as D-only CROWs only have throughput transmission as a band-rejection filter).[15,18] For $N = 101$, we observe periodic transmission bands with 3-dB linewidth of ~ 1-nm and SMSR exceeding 20 dB (defined in this case as the intensity difference in dB between the transmission band maximum and the highest side peak).

Figures 17(c)-(f) show the measured drop-port transmission spectra for the S-D CROWs. The transmission bands correspond to the on-resonance wavelengths of the constituent microdisk resonators. For $N = 101$, the side modes are suppressed, leaving only periodic transmission bands. The S-D structures with $w = 0.4$ μm exhibit narrower and lower transmission bands than those with $w = 0.8$ μm. This suggests a weaker inter-cavity coupling via the narrower notch.[30] The microspiral and double-notch-shaped microdisk resonators with small notches in general possess high-Q resonances.[13] Thus, it is conceivable that the design with small notches result in narrow passbands. We note that even with our resolution-limited photolithography (i-line, 365 nm resolution) and the limited fabrication fidelity, the CROWs of identically designed resonators but different N's (fabricated on the same chip) still preserve almost the same center wavelengths at the transmission bands.

For further analysis, we examine transmission bands A, B and C of the three device configurations (denoted in Figs. 17(b), (d), (f)). Figures 18(a)-(c) summarizes the measured transmission band peak intensities, the 3-dB linewidths and the SMSRs as functions of disk number. As N increases, the transmission bands of the three different configurations show three salient features: (i) the transmission band peak intensity (in dB scale) drops near linearly, (ii) the transmission band linewidth narrows until saturated, and (iii) the SMSR enhances until saturated.

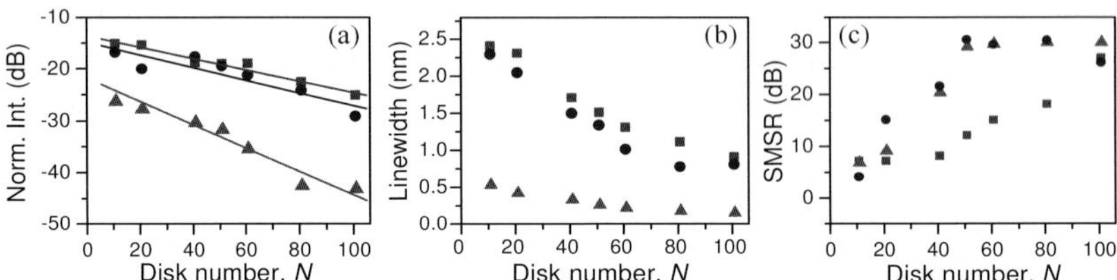

Fig. 18. Transmission band peak intensity, 3-dB linewidth, and side-mode suppression ratio for transmission bands A, B and C, as functions of disk number N. Squares: transmission band A; circles: transmission band B; triangles: transmission band C. Lines in (a): linear fittings. SMSR: side-mode suppression ratio. [Reproduced with permission from Optical Society of America [20] © 2009 Optical Society of America].

The transmission band peak intensity dropping with N is due to additional cascaded microdisk insertion loss. This is similar to that observed in microring CROWs.[30] The linearly fitted disk insertion loss for transmission bands A, B and C are approximately 0.11, 0.13 and 0.24 dB/disk, respectively. The 0.4-μm-width notches (band C) result in a relatively large disk insertion loss possibly due to a relatively large scattering loss at the narrowed jointed notch region.

The linewidth narrowing with N converges to the linewidth of an ideal infinite-length periodic structure.[33] The D-only and S-D CROWs with $w = 0.8$ μm display similar linewidths of ~2.5 nm for 11-element CROWs and ~1 nm for 101-element CROWs. Whereas, the S-D CROWs with $w = 0.4$ μm display overall narrower linewidths of 0.15-0.5 nm. With N increases, the cascaded mode-splitting induced resonance wavelength detuning becomes less spectrally overlapping with the center transmission wavelength, and subsequently limits the overall linewidth. Such linewidth narrowing phenomenon with the increase of the periodic structure length has been calculated for fiber Bragg gating[34] and demonstrated in silicon microring CROWs.[27] In practice, however, complications are prone to arise due to the inevitable fabrication-induced non-uniformities. The fabrication imperfection renders the cascaded microdisk resonators to be non-identical with each other, and thus resulting in spectrally shifted multimode resonances (with possible defects-induced scattering and field localization[32]).

The enhanced SMSR with N is partly attributed to the mode-dependent attenuation along the CROW. The modes with relatively weak inter-cavity coupling tend to be preferentially attenuated as the light propagates along the CROW. Only those modes with sufficiently strong inter-cavity coupling can propagate through the entire CROW. For 0.8-μm-notch D-only CROWs, the maximum SMSR of 101-element is ~ 30 dB. For both 0.8-μm-notch and 0.4-μm-notch S-D CROWs, the SMSRs increase to ~ 30 dB with $N = 50$ and become saturated. We attribute such saturated SMSR to possible measurement-limited detection level.[27]

6. Conclusion

We reviewed the development of microspiral and double-notch-shaped (D) microdisk resonator-based devices for integrated photonics. Rooting from Prof. Richard Kounai Chang's invention of microspiral uniport directional-emission lasers, we proposed and demonstrated a host of microspiral and double-notch-shaped microdisk resonator-based passive devices on silicon chips. Thanks to the notch design, which enables gapless intercavity coupling between coupled microdisks, we demonstrated coupled-resonator optical waveguides with up to 101 microdisks. Such coupled-microdisk arrays promise technologically important applications as spectral filters, optical delay lines and optical biological/chemical sensors.

Acknowledgments

We are greatly indebted to Prof. Richard Kounai Chang for his inspiration and foresight on microspiral resonators. We sincerely acknowledge the significant contributions from Mr Jonathan Y. Lee on the first experimental demonstration of passive microspiral resonators on silicon chips and Dr. Chao Li on proposing the concept of double-notch-shaped microdisk resonators. This work was substantially supported by a grant from the Research Grants Council of The Hong Kong Special Administrative Region, China (Project No. 618506). X. Luo acknowledges the fellowship support from the NANO program of Hong Kong University of Science and Technology (HKUST). We are grateful for the HKUST Nanoelectronics Fabrication Facility for all device fabrication.

References

1. S. L. McCall, A. F. J. Levi, R. E. Slusher, S. J. Pearton, and R. A. Logan, "Whispering-gallery mode microdisk lasers," *Appl. Phys. Lett.* **60**(3), 289-291, January 1992.
2. K. C. Zeng, L. Dai, J. Y. Lin, and H. X. Jiang, "Optical resonance modes in InGaN/GaN multiple-quantum-well microring cavities," *Appl. Phys. Lett.* **75**(17), 2563-2565, October 1999.
3. M.d Soltani, S. Yegnanarayanan, and A. Adibi, "Ultra-high Q planar silicon microdisk resonators for chip-scale silicon photonics," *Opt. Express* **15**(8), 4694-4704, April 2007.
4. B. E. Little, S. T. Chu, H. A. Haus, J. Foresi, and J. P. Laine, "Microring resonator channel dropping filters," *J. Lightwave Technol.* **15**(6), 998-1005, January 1997.
5. G. D. Chern, H. E. Tureci, A. D. Stone, R. K. Chang, M. Kneissl, and N. M. Johnson, "Unidirectional lasing from InGaN multiple-quantum-well spiral-shaped micropillars," *Appl. Phys. Lett.* **83**(9), 1710-1712, September 2003.
6. M. Kneissl, M. Teepe, N. Miyashita, N. M. Johnson, G. D. Chern, and R. K. Chang, "Current-injection spiral-shaped microcavity disk laser diodes with unidirectional emission," *Appl. Phys. Lett.* **84**(14), 2485-2487, April 2004.
7. R. Audet, M. A. Belkin, J. A. Fan, B. G. Lee, K. Lin, and F. Capasso, "Single-mode laser action in quantum cascade lasers with spiral-shaped chaotic resonators," *Appl. Phys. Lett.* **91**(13), 131106, September 2007.
8. C. M. Kim, J. Cho, J. Lee, S. Rim, S. H. Lee, K. R. Oh, and J. H. Kim, "Continuous wave operation of a spiral-shaped microcavity laser," *Appl. Phys. Lett.* **92**(13), 131110, April 2008.
9. T. Ben-Messaoud and J. Zyss, "Unidirectional laser emission from polymer-based spiral microdisks," *Appl. Phys. Lett.* **86**(24), 241110, June 2005.
10. A. Fujii, T. Takashima, N. Tsujimoto, T. Nakao, Y. Yoshida, and M. Ozaki, "Fabrication and unidirectional laser emission properties of asymmetric microdisks based on poly(p-phenylenevinylene) derivative," *Jpn. J. Appl. Phys., Part 2* **45**(29-32), L833-L836, August 2006.
11. J. Y. Lee, X. S. Luo, and A. W. Poon, "Reciprocal transmissions and asymmetric modal distributions in waveguide-coupled spiral-shaped microdisk resonators," *Opt. Express* **15**(22), 14650-14666, October 2007.
12. A. W. Poon, X. Luo, H. Chen, G. E. Fernandes, and R. K. Chang, "Microspiral Resonators for Integrated Photonics," *Opt. Photon. News* **19**(10), 48-53, October 2008.
13. A. W. Poon, X. Luo, L. Zhou, C. Li, J. Y. Lee, F. Xu, H. Chen, and N. K. Hon, "Microresonator-based devices on a silicon chip: novel shaped cavities and resonance coherent interference," In A. Matsko (Ed.), *Practical Applications of Microresonators in Optics and Photonics* (CRC Press, Taylor and Francis, 2009).
14. X. Luo, C. Li, and A. W. Poon, "Double-notch-shaped microdisk resonator-based devices in silicon-on-insulator," *Proc. Conference on Lasers and Electro-Optics 2008 (IEEE and Optical Society of America, 2008)*, paper CTuNN7.
15. X. Luo and A. W. Poon, "Double-notch-shaped microdisk resonator devices with gapless coupling on a silicon chip," *Chin. Opt. Lett.* **7**(4), 296-298, April 2009.
16. X. Luo, J. Y. Lee, and A. W. Poon, "Coupled spiral-shaped microdisk resonators with asymmetric non-evanescent coupling," *IEEE 4th International Conference on Group IV Photonics, Tokyo, Japan*, 18-22 September 2007.
17. X. Luo and A. W. Poon, "Coupled spiral-shaped microdisk resonators with non-evanescent asymmetric inter-cavity coupling," *Opt. Express* **15**(25), 17313-17322, December 2007.
18. X. Luo and A. W. Poon, "50-element cascaded-resonator devices with gapless non-evanescent coupling using double-notch-shaped microdisks on a silicon chip," *IEEE 5th International Conference on Group IV Photonics, Sorrento, Italy*, September 2008.
19. X. Luo and A. W. Poon, "101-element cascaded-microdisk resonators on a silicon chip," *Conference on Lasers and Electro-Optics 2009(IEEE and Optical Society of America, 2009)*, Paper CMAA3, 2009.

20. X. Luo and A. W. Poon, "Many-element coupled-resonator optical waveguides using gapless-coupled microdisk resonators," *Opt. Express* **17**, 23617-23628, December 2009.

21. G. D. Chern, G. E. Fernandes, R. K. Chang, Q. Song, L. Xu, M. Kneissl, and N. M. Johnson, "High-Q-preserving coupling between a spiral and a semicircle μ-cavity," *Opt. Lett.* **32**, 1093-1095, April 2007.

22. G. E. Fernandes, L. Guyot, G. D. Chern, M. Kneissl, N. M. Johnson, Q. Song, L. Xu, and R. K. Chang, "Wavelength and intensity switching in directly coupled semiconductor microdisk lasers," *Opt. Lett.* **33**, 605-607, March 2008.

23. B. E. Little, S. T. Chu, P. P. Absil, J. V. Hryniewicz, F. G. Johnson, F. Seiferth, D. Gill, V. Van, O. King, and M. Trakalo, "Very high-order microring resonator filters for WDM applications," *IEEE Photon. Technol. Lett.* **16**(10), 2263-2265, November 2004.

24. S. J. Xiao, M. H. Khan, H. Shen, and M. H. Qi, "A highly compact third-order silicon microring add-drop filter with a very large free spectral range, a flat passband and a low delay dispersion," *Opt. Express* **15**(22), 14765-14771, October 2007.

25. Q. Li, M. Soltani, S. Yegnanarayanan, and A. Adibi, "Design and demonstration of compact, wide bandwidth coupled-resonator filters on a silicon-on-insulator platform," *Opt. Express* **17**(4), 2247-2254, February 2009.

26. Y. Vlasov, W. M. J. Green, and F. Xia, "High-throughput silicon nanophotonic wavelength-insensitive switch for on-chip optical networks," *Nat. Photonics* **2**(4), 242-246, April 2008.

27. F. N. Xia, L. Sekaric, M. O'Boyle, and Y. Vlasov, "Coupled resonator optical waveguides based on silicon-on-insulator photonic wires," *Appl. Phys. Lett.* **89**(4), 041122, July 2006.

28. F. Xia, L. Sekaric, and Y. Vlasov, "Ultracompact optical buffers on a silicon chip," *Nat. Photonics* **1**, 65-71, January 2007.

29. A. Yariv, Y. Xu, R. K. Lee, and A. Scherer, "Coupled-resonator optical waveguide: a proposal and analysis," *Opt. Lett.* **24**(11), 711-713, June 1999.

30. J. K. S. Poon, J. Scheuer, Y. Xu, and A. Yariv, "Designing coupled-resonator optical waveguide delay lines," *J. Opt. Soc. Am. B* **21**(9), 1665-1673, September 2004.

31. J. K. S. Poon, L. Zhu, G. A. DeRose, and A. Yariv, "Transmission and group delay of microring coupled-resonator optical waveguides," *Opt. Lett.* **31**(4), 456-458, February 2006.

32. S. Mookherjea, J. S. Park, S.-h. Yang, and P. R. Bandaru, "Localization in silicon nanophotonic slow-light waveguides," *Nat. Photonics* **2**, 90-93, February 2008.

33. K. Ohtaka, "Theory I: Basic aspects of photonic bands," in K. Inoue, and K. Ohtaka (Eds.), *Photonic crystals: physics, fabrication, and applications* (Springer-Verlag, Berlin, Hong Kong, 2004).

34. J. W. Goodman, *Introduction to Fourier optics*, 3rd Edition, (ROBERTS & COMPANY, Englewood, Colorado, 2005).

PART V

PHOTOSCAPES OR MULTIDISCIPLINARY APPLICATIONS

CHAPTER 19

TERAHERTZ RADIATION FROM NITRIDE SEMICONDUCTORS

GRACE D. METCALFE, PAUL H. SHEN and MICHAEL WRABACK

Sensors and Electron Devices Directorate, US Army Research Laboratory, 2800 Powder Mill Road, Adelphi, Maryland, USA
grace.metcalfe@us.army.mil

In this chapter we discuss the generation of broadband terahertz radiation from nitride semiconductors by ultrafast optical pulses. We focus in particular on the generation mechanisms related to the photo-Dember effect and acceleration of photogenerated charges in built-in electric fields in-plane as well as normal to the sample surface.

1. Introduction

The terahertz (THz) range of the electromagnetic spectrum, lying between microwave frequencies (100 GHz) and photonic frequencies (30 THz), is a potentially important region for a wide variety of applications, including biomedical imaging, security imaging,[1,2] time-domain spectroscopy,[3] and material identification and characterization. In spite of its potential, the THz regime remains one of the least explored portions of the electromagnetic spectrum partly due to the challenges in efficiently generating and detecting terahertz radiation.

Currently, there are two main approaches to optically generating terahertz radiation from semiconductors. One approach uses ultrafast optical pulses to produce broadband radiation from semiconductors via

(1) optical rectification, a $\chi^{(2)}$ process which depends on crystallographic orientation;[4]
(2) the photo-Dember effect, which generates currents due to the difference between electron and hole mobilities;
(3) or acceleration of photogenerated charge in electric fields created by an external bias, built-in surface fields or internal surface-normal or in-plane fields.

THz radiation from semiconductors is commonly generated by the last mechanism through in-plane carrier acceleration in electric fields created by externally biased photoconductive (PC) switches.[5] Although the geometry of the biased PC antennas is more favorable for coupling out the THz radiation than that for other semiconductor-based THz sources employing transport normal to the surface,[6] these PC switches require electrode processing and an external bias voltage, limited by the dielectric strength of air, to establish an in-plane electric field.

The second optical THz generation approach creates narrow band radiation through the mixing of continuous-wave (CW) lasers separated in frequency by the desired terahertz difference frequency in a semiconductor photomixer.[7] Most photomixers use low-temperature grown GaAs thin films as the active layer due to its short carrier lifetime ($\approx$ 0.5 ps), large breakdown-field threshold (> 300 kV/cm) and high carrier mobility ($\approx$ 200 cm^2/V·s).[8]

With recent advancements in solid-state lasers and III-V materials, especially III-Nitride semiconductors, there are several new approaches to enhancing the performance of THz devices. For instance, unlike conventional III-V semiconductor compounds, wurtzite III-Nitride semiconductors possess a significant spontaneous polarization about an order of magnitude larger than that found in conventional III-V materials.[9] In quantum wells (QWs), the strong spontaneous polarization combined with strain-induced piezoelectric fields lead to large internal electric fields which should generate enhanced THz radiation. THz emission from QWs has been observed from GaAs/AlGaAs single QWs,[10] low-indium-concentration InGaN/GaN multiple quantum wells (MQWs),[11] and high-indium-concentration InN/InGaN MQWs.[12,13]

The strong built-in polarization field can also be exploited in conjunction with stacking faults (SFs) in nonpolar or semipolar nitride semiconductors to enhance THz emission. Evidence of internal electric fields induced by SF-terminated internal polarization in polar crystals has been observed in ZnS,[14] ZnTe,[15] and SiC,[16] and predicted in GaN.[17,18] In 2008, enhanced SF-related THz emission was measured from *m*-plane GaN.[19,20]

Nitride semiconductors can also be applied to fiber laser-based THz systems at 1550 nm. Traditionally, optically generated THz systems utilize femtosecond (fs) mode-locked Ti-sapphire lasers or fiber lasers with a wavelength centered near 800 nm. However, significant advantages associated with the use of telecommunications-grade optoelectronics and compact fiber lasers could be gained in cost, size, weight, and efficiency by shifting the wavelength to 1550 nm if an appropriate narrow bandgap semiconductor THz source were available. In addition, for CW THz generation obtained by photomixing (optical heterodyne conversion), 1550 nm operation is advantageous because the photomixing conversion efficiency, which increases as λ^2,[7] is about four times higher than for conventional GaAs-based THz systems using 800 nm. THz generation using compact fiber laser systems has been demonstrated with InAs at 1040 nm,[21] InSb at 1560 nm,[22] and GaAs at 1560 nm.[23]

InN is another potential material for THz generation with 1550 nm fiber lasers. It has recently been shown to exhibit a bandgap near 0.7 eV,[24] a much smaller value than what was previous accepted[25] (1.9 eV or λ = 650 nm). In 2006, an InN bandgap below 0.65 eV was measured in compressively strained InN films on GaN templates,[26,27] with estimates of the deformation potential pointing to an unstrained bandgap of ~0.63 eV. The uncovering of this smaller bandgap is largely due to a reduction in unintentional doping and concomitant Burstein-Moss shift associated with improvement in crystal quality.

Its small bandgap energy combined with the large internal polarization, high saturation velocity ($>1.5\times10^7$ cm/s),[28] large intervalley spacing to confine carriers within the high mobility gamma valley, short carrier lifetime, and potentially high breakdown voltage makes InN promising as a new material for high efficiency, compact 1550-nm laser-based THz sources.

2. Experimental setup

A typical experimental setup for THz radiation measurements via ultrafast optical excitation is shown in Fig. 1. For the data presented in this chapter, a regenerative amplifier (RegA) system, which operates at 800 nm and has a repetition rate of 250 kHz, is used as the ultrafast laser source. The output of the RegA is split into two beams. For infrared excitation pulses, the stronger beam is frequency-doubled to serve as the pump source for an optical parametric amplifier (OPA), which generates an infrared idler pulse tunable from ~0.9–2.4 μm. The OPA also produces a visible signal beam tunable from ~480–720 nm. The signal beam at 700 nm is doubled for 350 nm excitation. For other ultraviolet pulses, the stronger RegA beam is frequency-doubled (400 nm) or tripled (266 nm).

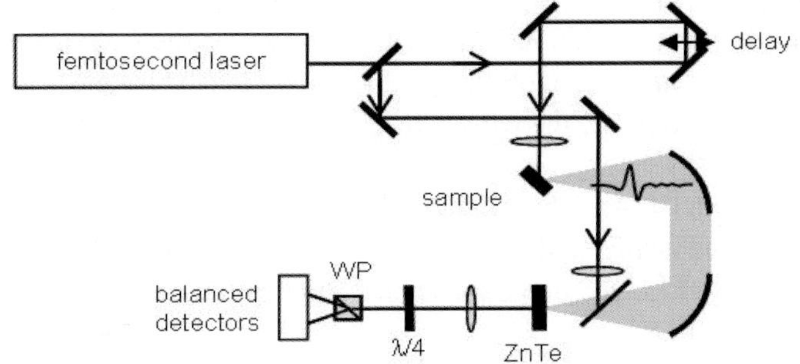

Fig. 1.   Experimental setup for generation and detection of terahertz radiation. WP: Wollaston prism.

The pump pulse, after compression with a prism pair to typical pulse widths of ~150 fs, is then focused onto the semiconductor sample at a 45° incidence angle with a beam diameter of ~1 mm. The pump power is ~4 mW for all excitation wavelengths. The weaker RegA split-off beam is used to probe the THz emission for all excitation wavelengths, as well as to irradiate the samples at 800 nm.

The subsequent THz emission from the semiconductor surface is collected and focused with a pair of off-axis parabolic mirrors onto a 2 mm-thick ZnTe crystal for electro-optic (EO) sampling,[29] which is based on the linear EO effect or Pockels effect and allows for coherent detection of freely propagating THz emission. The THz radiation incident on the EO crystal alters the birefringence of the crystal which results in a phase retardation of the probe beam through the EO crystal. Monitoring of the phase retardation of the probe beam is conducted with a balanced detector system.

3. Terahertz radiation from InN thin films

Although all III-Nitride semiconductors can emit THz radiation, we focus on InN in this section for its applications in 1550nm fiber-laser based systems. The possibility of THz emission from InN has been demonstrated by Ascazubi *et al.*,[30] who attribute the THz generation mechanism to transient photocurrents. In 2006, Chern *et al.*[31] demonstrated that the photo-Dember effect is the dominant THz generation mechanism in *c*-plane *n*-type InN thin films. Similar to InAs, one of the best known THz semiconductor surface emitters, InN has a surface accumulation layer, which has a width that is more than an order of magnitude shorter than the optical absorption length.[32] The high surface-state density leads to a high field near the surface but virtually no field in the bulk of the crystal. The optically induced transient photocurrents in InN should therefore be similar to InAs - that is, due to the photo-Dember effect.

In the photo-Dember effect, THz radiation is generated due to the difference in diffusion between the photogenerated electrons and holes. The diffusion current J_i ($I = e, h$) of the carriers is given by the following equation,

$$J_i \sim \pm q_i D_i \frac{\partial n_i}{\partial x}. \tag{1}$$

where q_i is the carrier charge and n_i is the carrier concentration. The diffusion coefficient D_i is defined by the Einstein relation $D = k_B T \mu / e$, where k_B is the Boltzmann constant, T is the carrier temperature, μ is the mobility, and e is the elementary electric charge. In the presence of an electric field E, the carrier response is described by the drift current $J = ne\mu E$.

As with GaAs[33] and InAs[34] the carrier dynamics due to the photo-Dember effect in InN can be simulated using a drift-diffusion treatment. The sub-ps electron current density J_e is given by the basic one-dimensional momentum conservation and relaxation equation

$$\frac{\partial J_e}{\partial t} = e\left(\frac{k_B T_e}{m*} \frac{\partial n_e}{\partial x} + n_e \frac{eE}{m*} \right) - \frac{J_e}{\tau} \tag{2}$$

where $m*$ is the k-dependent effective mass[35] and τ is the momentum relaxation time. On the right hand side of Eq. (2), the first term describes the diffusion of the photo-generated electrons at elevated electron temperature, the second term expresses the redistribution of the background electrons under the drift effect, and the third term represents the momentum relaxation due to carrier collisions.

Photo-generated and background electron currents must be calculated separately due to their different carrier temperatures. Both the photo-generated and background hole currents are calculated using drift-diffusion equations as well, assuming their temperatures are invariant at room temperature. The drift-diffusion equations in conjunction with carrier conservation equations are solved using a modified Scharfetter-Gummel scheme.[36] The emitted THz signal is then calculated as the volume integration of the time derivative of the total current density.

Figure 2 shows the typical (a) time-domain waveforms and (b) Fourier-transformed amplitude spectra of the THz emission from a 450- μm-thick *p*-type (111) InAs (with a doping level of ~10^{16} cm^{-3}), and a 1- μm-thick *n*-type *c*-plane InN (unintentionally doped with a bulk carrier concentration of $n_{\text{bulk}} = 2.25 \times 10^{17}$ cm^{-3}). The sharp dips in Fig. 2(b) are due to water absorption. Measurements of THz emission as a function of the sample rotation angle (about the surface normal) indicate that the optical rectification component is small (<10 %) for InAs and negligible for InN in the non-saturation regime.[37]

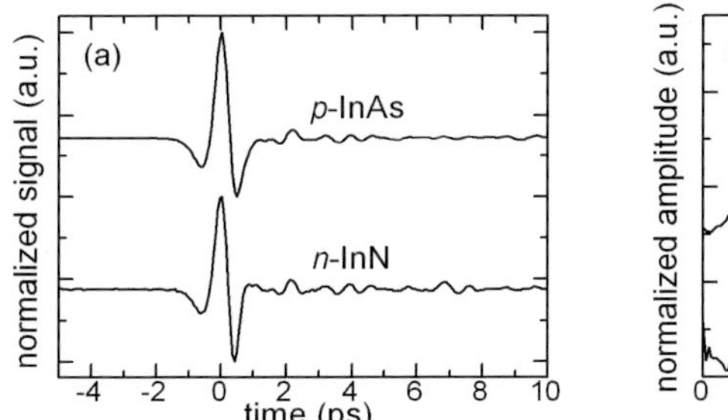

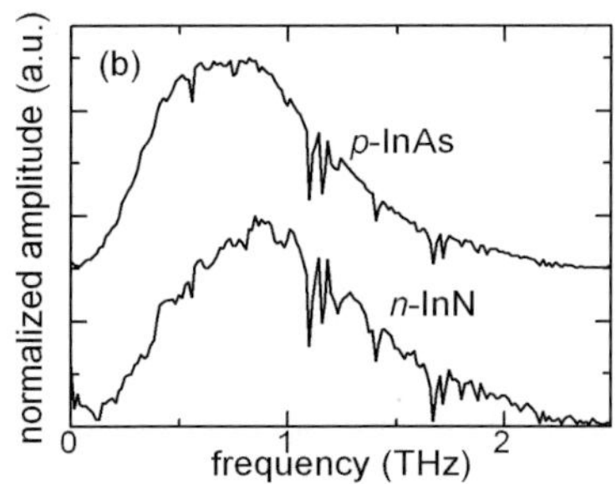

Fig. 2. Typical (a) THz waveform and (b) spectral distribution from *p*-type InAs and *n*-type InN at 800 nm excitation wavelength. Copyright 2006 AIP. Used with permission from Ref. 31 for Fig. 2 (a).

As the excitation wavelength increases, the photo-excited electron temperature decreases and thereby lowers the calculated THz amplitude, in agreement with the experimental data shown in Fig. 3. However, the THz amplitude is also slightly enhanced by an increased photon number with longer excitation wavelengths. As the excitation wavelength goes from 800 nm to 1500 nm for InN, the photo-excited electron temperature decreases by a factor 5 while the photon number increases by a factor of ~1.9, producing an overall decrease of nearly 4 times. The decrease in THz signal for InN at longer excitation wavelengths is only slightly larger than that for *p*-InAs (bandgap energy $E_g = 0.36$ eV), where the photo-excited electron temperature decreases by a factor of 2.5, giving an overall decrease in signal amplitude of nearly 2 times from 800 nm to 1500 nm.

The measured THz amplitude from *n*-InN is about an order of magnitude smaller than that from *p*-InAs due to larger screening from the higher mobility electrons as compared to holes. For example, at the fixed bulk carrier concentration of $n_{\text{bulk}} = 10^{18}$ cm^{-3}, the calculations in Fig. 4 show that the THz amplitude from *p*-InN would be more than an order of magnitude larger than that from *n*-InN. The higher density of background carriers in the InN sample, as compared to InAs, also contributes to the larger screening effect.

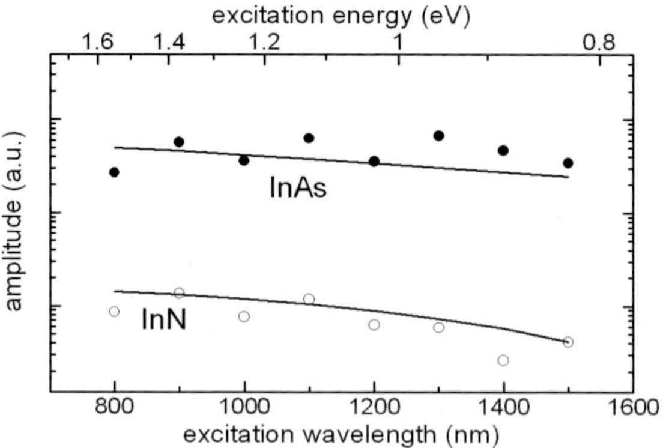

Fig. 3. Plot of THz amplitude, normalized to pump and probe power, as a function of excitation wavelength. Copyright 2006 AIP. Used with permission from Ref. 31.

Although there are several other effects which contribute to the THz amplitude, including photo-excited electron temperature, mobility, absorption, and carrier lifetime, screening from carriers appears to be the dominant effect. It is expected that the THz power from InN will improve with lower n-type bulk carrier concentration or p-type doping, which are the current challenges in the InN growth community.

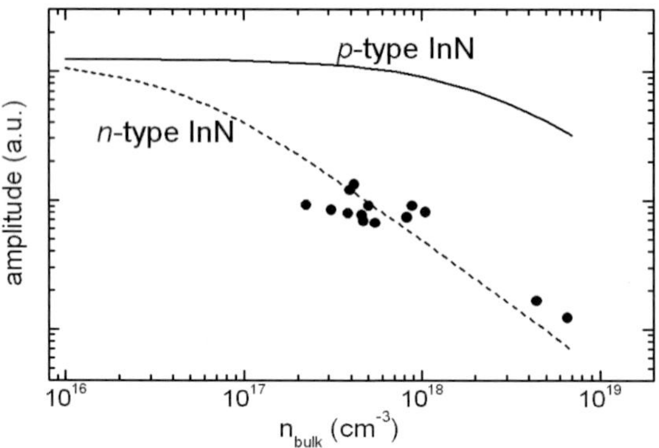

Fig. 4.   THz amplitude from InN as a function of background carrier concentration nbulk. Lines show simulation, dots indicate experimental data points. Copyright 2006 AIP. Used with permission from Ref. 31.

4. Terahertz generation from internal electric fields

Other than improving material quality, the THz output power from nitride semiconductors can also be enhanced by manipulating the strong internal polarization fields inherent in nitride materials. For example, in c-plane quantum well structures, the spontaneous polarization in the highly polar wurtzite crystal and piezoelectric polarization due to the well/barrier lattice mismatch terminate at the well/barrier interface and thereby setup an internal electric field along the c-axis direction.

In nonpolar or semipolar nitride semiconductors, for which there is a projection of the polar axis in-plane, SF-terminated internal polarization fields at wurtzite domain boundaries create strong built-in electric fields along the polar <0001> direction. The high fields within the wurtzite domains terminated by the SFs point in the same direction, effectively creating an array of contactless PC switches. The strong built-in electric field in either MQWs or nonpolar (or semipolar) materials could significantly enhance THz emission from semiconductor surfaces under ultrashort pulse excitation.

4.1. *Normal fields in InN/InGaN multiple quantum wells*

Evidence of a built-in electric field in MQWs can be seen as a blueshift of the photoluminescence (PL) peak with increasing excitation power due to optically induced screening of the quantum-confined Stark effect. This characteristic blueshift has been observed in InGaN/GaN quantum wells,[38] AlGaN/GaN quantum well laser structures,[39] and recently in InN/InGaN MQWs.[40] Although InN/InGaN MQWs have been previously fabricated,[41-43] they have not been as extensively studied as InGaN/GaN and GaN/AlGaN MQWs, in part due to the difficulties in growing high- quality, high-indium-concentration nitride materials with low background carrier concentration.

Recently, Koblmüller *et al.*[27] reported high-quality InN with a bulk carrier concentration n_{bulk} in the low 10^{17} cm^{-3}. Assuming similar material quality in the InN/InGaN MQWs, the carrier concentration became low enough that the partially screened built-in electric field could be optically probed.

Figure 5(a) shows a blueshift of the PL data taken at 30 K from an InN/InGaN MQW structure with increasing excitation power. The MQWs consisted of 25 periods of 1.5-nm InN wells and 15-nm thick In$_{0.88}$Ga$_{0.12}$N barriers. Figure 5(b) demonstrates that at low CW excitation power, the PL peak energy from the MQWs is redshifted relative to that from a 0.5 μm thick bulk InN sample, consistent with the presence of an internal field in the MQWs. As the excitation power increases to over 350 mW, a redshift of the peak energy is seen from the InN material due to heating (Fig. 5(b), empty circles), while a blueshift is observed from the MQWs (Fig. 5(b), solid circles). After correcting for heating effects in the MQWs based on the redshifting of the PL energy of bulk InN, a 44 meV blueshift of the PL peak energy from the MQWs is determined.

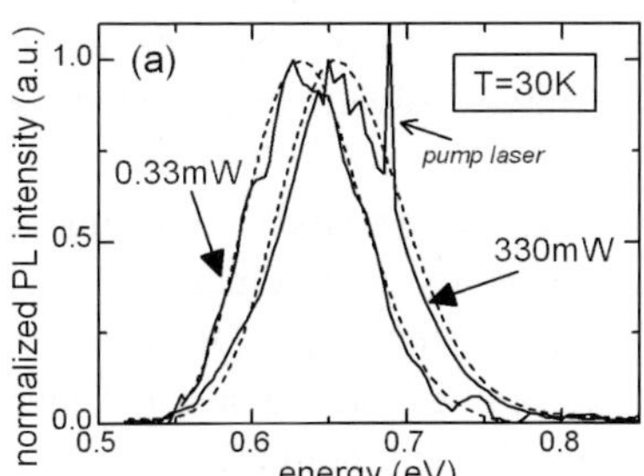

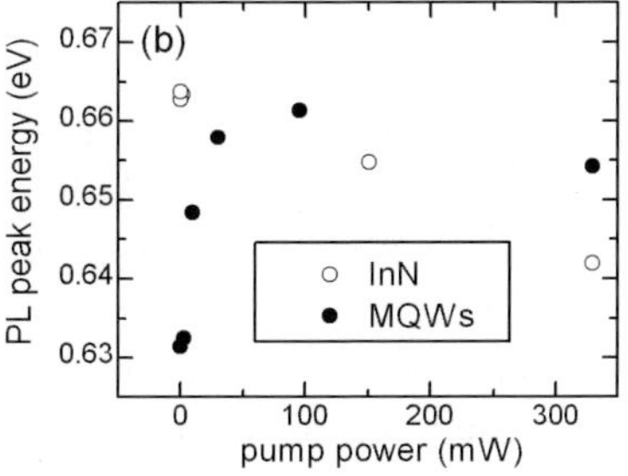

Fig. 5.   (a) Photoluminescence spectra of the InN/InGaN MQW sample at high and low pump powers taken at 30 K. The dotted curves are Gaussian fits. (b) Plot of the PL peak energy as a function of optical pump power of the MQWs (solid circles) and bulk InN (empty circles).

Calculations based on a self-consistent solution of Schrödinger and Poisson equations show the flattening of the conduction and valence band structure of the MQWs as the photo-generated carrier density n_{photo} increases from $n_{\text{photo}} = 0$ [Fig. 6(a), gray line] to $n_{\text{photo}} = 2\times10^{18}$ cm^{-3} [Fig. 6(a), black line]. This effect is due to the screening of the internal field by photo-generated carriers.

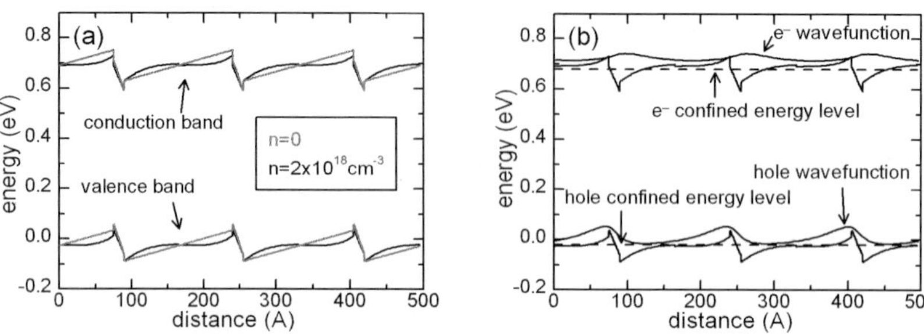

Fig. 6.   Calculations of the (a) MQW band structure as the photo-generated carrier density increases from $n_{\text{photo}} = 0$ to $n_{\text{photo}} = 2\times10^{18}$ cm^{-3} and (b) the electron and hole wavefunctions, and their confined state energy levels for $n_{\text{photo}} = 2\times10^{18}$ cm^{-3}.

Figure 6(b) displays the electron and hole wavefunctions and their confined state energy levels. The shift in the electron ground confined state energy due to a photo-generated carrier density of $n_{\text{photo}} = 2\times10^{18}$ cm^{-3} is ~4 meV while the shift in the hole ground confined state energy is ~24 meV. The majority of the energy shift occurs in the valence band as the hole effective mass is much larger than the electron effective mass. We therefore consider only the hole energy shift, and calculate from the measured 44-meV blueshift the minimum change of the electric field to be ~0.6 MV/cm in the well regions. Using the parameters from literature,[44] the experimental results are consistent with the predicted built-in field in the well region of 0.8 MV/cm for the MQWs.

4.2.   *Terahertz radiation from InN/InGaN multiple quantum wells*

Figure 7(a) shows the enhanced THz emission from InN/InGaN MQWs as compared to bulk InN at a low excitation power of 0.1 mW. With increasing excitation power, the THz signal from the MQW decreases faster than that from the bulk, becoming comparable at the highest pump powers, as shown in Fig. 7(b). The THz signal in Fig. 7(b) is obtained by Fourier transforming the time-resolved THz signal and then integrating the resulting spectra from 0 to 3 THz.

Although there is evidence of a built-in electric field in the MQWs, the enhanced THz signal from the MQWs in this case is primarily due to confinement of background carriers to the quantum wells, reducing the screening of the photo-Dember field in the barriers regions. As mentioned in Section 3, screening due to a large number of background carriers is a dominant contribution to the THz amplitude (Fig. 4).

By simulating the band structure of the InN/InGaN quantum wells (Fig. 6) and assuming a background n-doping of $n_{\text{bulk}} = 10^{17}$ cm^{-3}, the unintentional doping concentration in the bulk InN layers, we find that the majority electrons are localized in triangle quantum wells created by the piezoelectric and spontaneous polarization. The Fermi level is about 60 meV below the maximum of the electron potential barrier. At room temperature, the mobility of the background electrons is therefore greatly restricted, with a concomitant reduction in screening, resulting in enhanced THz radiation at low pump intensity. When the pump intensity increases, photocarrier-induced screening of the internal electric field reduces the potential barrier by flattening the bands, and raises the Fermi level, thus making the background electrons less localized such that their screening effect increases.

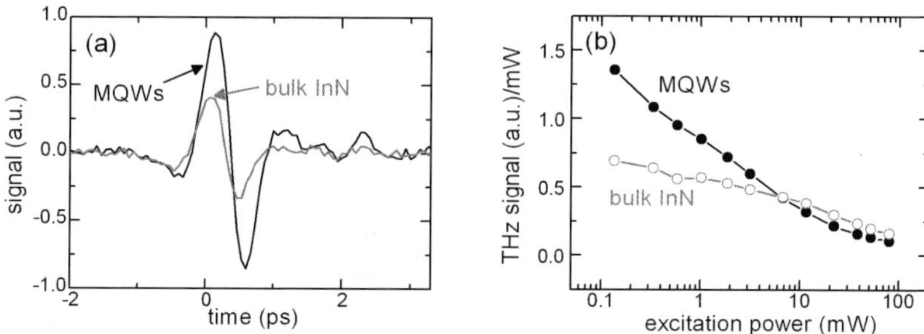

Fig. 7. (a) Time-resolved THz signal from the InN MQWs (black line) and the bulk InN (gray line) when excited with 0.1 mW of 800 nm fs pulses. (b) Plot of excitation power dependence of the THz signal normalized by the pump power from the MQWs (black solid circles) and bulk InN (gray empty circles).

Although localized background carriers is the predominate cause of the enhanced THz signal, the effect of drift and diffusion in the MQWs can still be observed by studying the THz emission as a function of pump wavelength (Fig. 8). At 800 nm, the photoexcited electron thermal energy is much larger in magnitude than the energy from the opposing field in the MQW barrier region, with little effect of the quantum well potential on the electron diffusion. At longer wavelengths, the photoexcited electron thermal energy is smaller in magnitude than the energy supplied by the field in the barrier region, which reduces or provides a negative component to the THz signal from the MQWs (decreases the diffusion current). Also note that at 1700 nm, the absorption length is longer than the thickness of the MQW structure, contributing to the decrease in THz signal. Compared to bulk InN, the THz signal from the MQWs decreases faster with longer wavelength excitation due to the increasing effect of the opposing field in the MQW barrier regions.

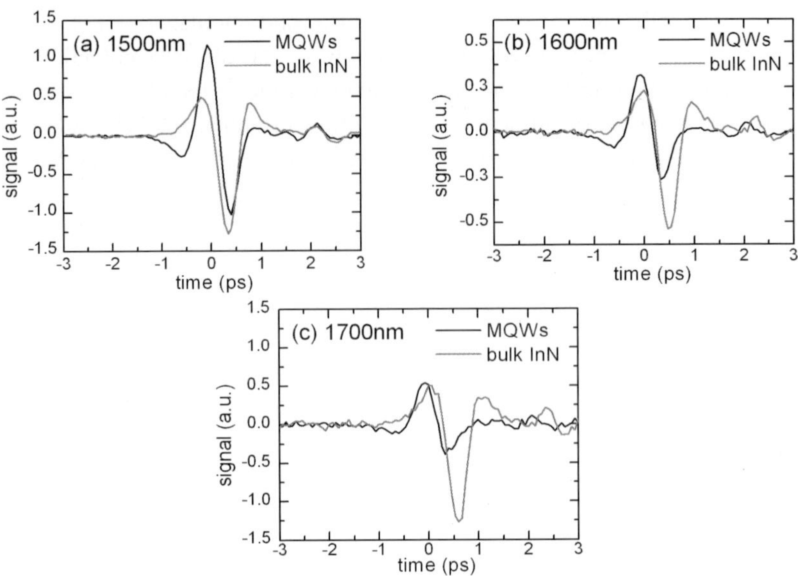

Fig. 8. Time-resolved THz signal from InN MQWs and bulk InN excited with ultrashort pulses at (a) 1500 nm, (b) 1600 nm, and (c) 1700 nm.

4.3. *In-plane fields in nonpolar and semipolar nitride semiconductors*

In nonpolar or semipolar nitrides, THz radiation can be generated due to an internal in-plane electric field induced by stacking fault (SF)-terminated internal polarization at wurtzite domain boundaries.[19] Figure 9 illustrates I_1 type basal plane SFs, which run perpendicular to the c-axis (<0001> direction) and correspond to the stacking sequence of the (0001) basal planes …ABAB**ABC**BCBCB…. The inherent in-plane polarization of the wurtzite crystalline structure is terminated by the SFs, which introduce thin zincblende domains. At the wurtzite/zincblende interface, charge accumulation leading to strong electric fields parallel to the c-axis of the crystal occurs, as illustrated in Fig. 9. In-plane transport of photoexcited carriers proceeds parallel to the electric field, leading to a THz radiation component polarized preferentially along this axis of the sample.

The SFs result primarily from the large density of structural defects associated with heteroepitaxial growth on lattice-mismatched substrates.[45-46] With the appropriate density of SFs, the THz emission can be enhanced by several times relative to that from SF-free material, for which the terahertz signal emanates from surface surge-currents and diffusion-driven carrier transport normal to the surface and is independent of the c-axis orientation.

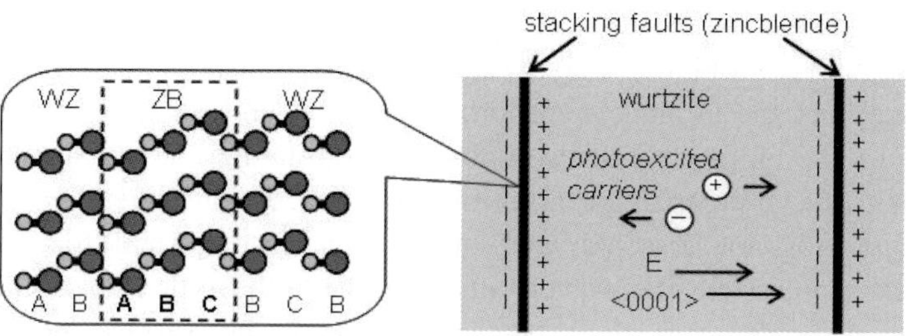

Fig. 9.   Illustration of stacking fault-terminated internal in-plane electric field in *m*-plane wurtzite material.

Figure 10 shows typical THz signals from a ~1 μm-thick *m*-plane ($1\bar{1}00$) GaN epilayer with a SF density of 1×10^6 cm^{-1} for (a) *s*-wave and (b) *p*-wave polarization. The terms *p*- and *s*-polarization refer to the emitted THz polarization parallel and perpendicular to the plane of incidence, respectively. The gray-scaled curves in each plot are measured at different sample rotation angles θ, with the *c*-axis of the nonpolar sample in the plane of incidence at θ = 0°.

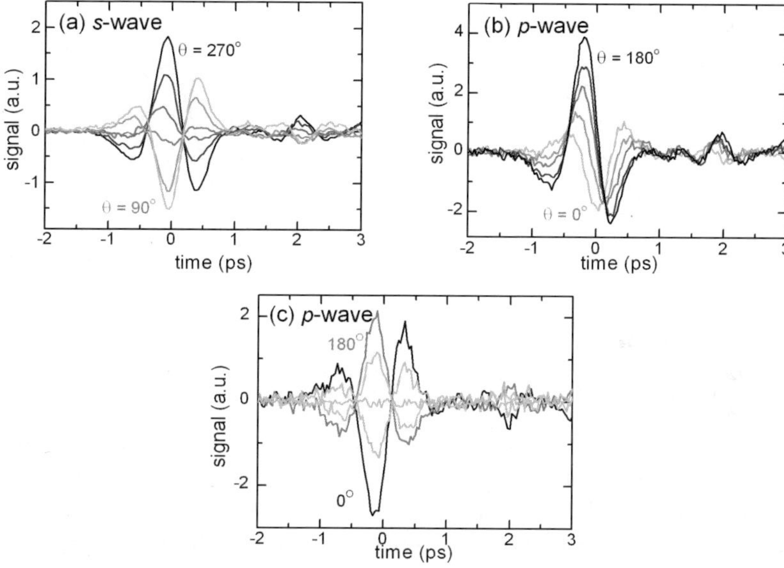

Fig. 10.   Time-resolved (a) *s*-wave and (b) *p*-wave THz signal from a high-stacking-fault-density *m*-GaN sample at different sample rotation angles θ for 266-nm excitation. (c) Figure (b), corrected for the surface normal component. Copyright 2008 AIP. Used with permission from Ref. 19.

As displayed in Fig. 10, the measured *s*-polarized (*p*-polarized) THz waveform polarity flips as the sample rotates, reaching a peak positive signal at θ = 270° (180°) and a peak negative signal at θ = 90° (0°) when the *c*-axis is parallel or antiparallel to the *s*-polarization (*p*-polarization) detection direction. Note that the peak positive signal of the *p*-polarized THz signal [Fig. 10(b)] is larger than the peak negative signal, and the phase appears to shift as the sample rotates. This phase shift and amplitude asymmetry arises from THz radiation contributions due to surface normal transport.

To analyze only the in-plane transport, the normal photocurrent component can be separated from the in-plane drift component by rotating the c-axis of the crystal perpendicular to the p-polarization detection direction. In this orientation, the in-plane drift contribution is not detected, and only the surface normal transport signal is measured. The p-polarized THz waveforms from the high SF-density m-GaN at various sample rotation angles can then be corrected for the surface normal transport contribution, which is independent of the c-axis orientation, to extract only the in-plane drift component. As seen in Fig. 10(c), after removing the normal photocurrent contribution, the phase shift and amplitude asymmetry mentioned above are eliminated and the corrected p-polarized THz waveforms closely resemble the s-polarized THz waveforms in Fig. 10(a).

Analysis of the amplitude asymmetry in Fig. 10(b) relative to the <0001> orientation of the c-axis (which points from the N-face to the Ga-face) shows that the electrons accelerate in the $<000\overline{1}>$ direction, consistent with carrier transport in built-in, in-plane electric fields created by SF-terminated internal polarization at wurtzite domain boundaries (see Fig. 9).

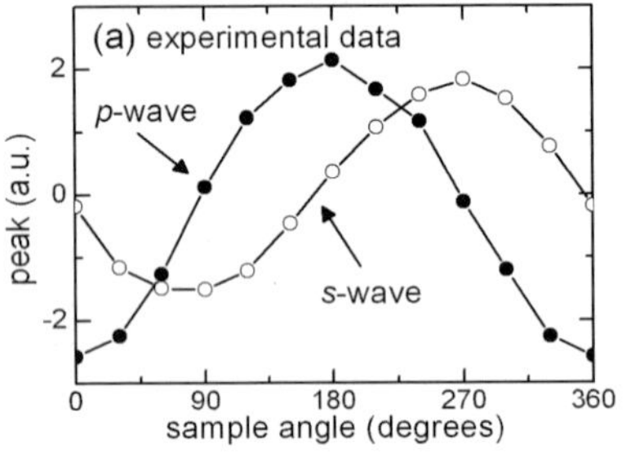

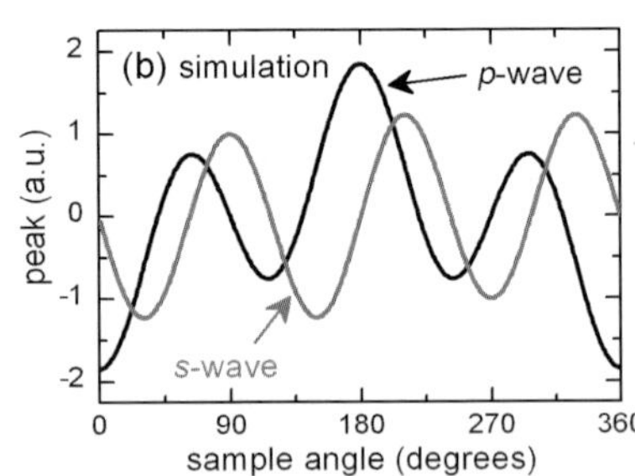

Fig. 11. (a) Plot of the peak THz signal as a function sample rotation angle. (b) Calculated sample rotation angle dependence of THz signal, assuming non-zero optical rectification. Copyright 2008 AIP. Used with permission from Ref. 19 for Fig. 11 (a).

Rotating the c-axis by 180° also rotates the direction of the built-in field by 180°, causing the photoexcited carriers to accelerate in the opposite direction. A reversal of the photoexcited carrier acceleration direction would be observed as a flip in the THz waveform polarity,[8] as seen in the THz signal from the high SF density m-GaN in Figs. 10(a) and (c). Figure 11(a) displays the sinusoidal dependence on sample rotation angle of the peak s- and corrected p-polarized emitted electric field from the high SF-density m-GaN. The 360° periodicity and 90° shift between the p- and s-polarization curves indicate that the linearly polarized THz emission associated with in-plane carrier transport rotates with sample rotation and is characteristic of real carrier transport in an in-plane electric field parallel to the c-axis.

Optical rectification is ruled out as the main THz generation mechanism, since evidence of nonlinear polarization would appear with the θ dependence of the THz signal shown in Fig. 11(b). Moreover, no dependence on pump polarization is found after accounting for absorption, and the THz signal vanishes for 400-nm excitation, indicating that the signal is associated with the generation and transport of real carriers.

For comparison, the *p*-polarized THz waveform from a ~330-μm-thick SF-free *m*-GaN substrate (from Mitsubishi Chemical Co., Ltd) is shown in Fig. 12(a). No dependence on sample rotation angle and no *s*-wave component are expected or observed from the SF-free sample, in which THz emission is governed by surface normal carrier transport similar to that in *c*-InN and *c*-InAs.[31] Comparison of the THz emission from the SF and SF-free samples indicates that the component attributed to in-plane transport dominates that from surface normal transport, even with high excess electron energies (~1 eV) and short absorption lengths (< 50 nm) at 266-nm excitation, favorable to diffusive transport.

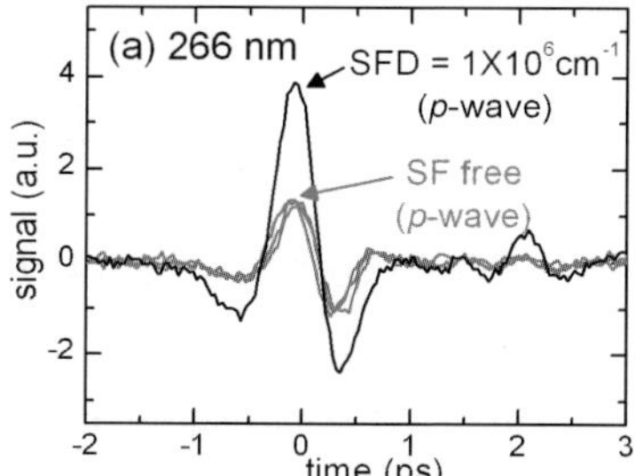

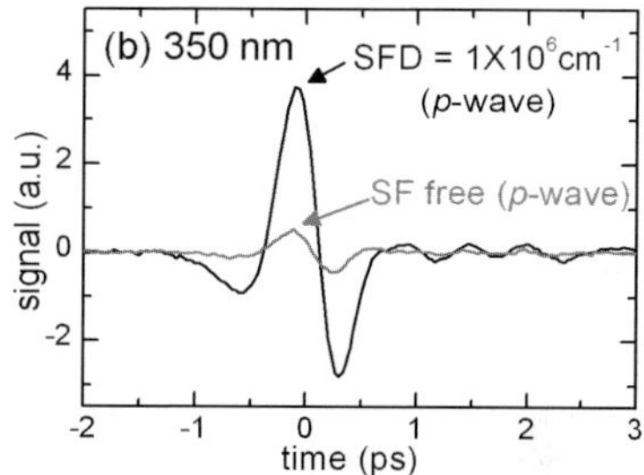

Fig. 12.   Time-resolved THz signal from a high stacking fault density and a stacking fault free m-GaN sample at (a) 266-nm and (b) 350-nm excitation wavelength.   Copyright 2008 AIP. Used with permission from Ref. 19.

The maximum *p*-polarized THz signal from the high-SF-density *m*-GaN is enhanced by nearly a factor of 3 compared to that from the SF-free sample for 266-nm excitation, and it becomes ~7 times larger for 350-nm excitation (Fig. 12(b)). The diminished signal from the SF-free *m*-GaN is primarily due to the lower excess electron energy (< 100 meV) and increased absorption depth (~100 nm) for 350-nm excitation.

Majewski *et al.*[17,18] performed ab-initio calculations of zincblende inclusions embedded in bulk wurtzite GaN and concluded that charge buildup at the wurtzite/zincblende interfaces in GaN results mainly from the spontaneous polarization P_{sp} in wurtzite material, with a negligible contribution from the piezoelectric polarization. They also concluded that the interface charge can be calculated from the bulk spontaneous polarization. Using the appropriate boundary conditions at the SF interface and Maxwell's equations, the SF-related electric field F_w in the wurtzite region can then be expressed as

$$ F_w = \frac{P_{sp}}{\varepsilon_0}\left(\frac{1}{\varepsilon_w + \varepsilon_{zb}[d_{wz}/d_{zb}]} \right), \tag{3} $$

where ε_0 is the permittivity of free space, and ε_w (ε_{zb}) is the dielectric constant of wurtzite (zincblende) GaN. The thickness of the zincblende region (in this case, the I$_1$-type SF) is $d_{zb} \approx 3c_{wz}/2 = 8$ Å, where c_{wz} is the lattice constant of wurtzite GaN, and the thickness of the wurtzite region is $d_{wz} \approx 1/n_{sf} = 100$ Å, where n_{sf} is the SF density. Using

P_{sp} = -0.034 C/m^2,[44] the maximum average in-plane electric field in the wurtzite region of an *m*-GaN sample with an I$_1$-type SF density of 1×10^6 cm^{-1} is estimated as ~290 kV/cm. This field is comparable to the bias fields applied to PC switches using low-temperature grown GaAs,[47] one of the best PC materials, but does not require electrode processing or an external bias.

5. Conclusions

Due to the strong internal polarization field in its wurtzite structure, III-Nitrides are promising novel semiconductors for efficient THz generation. With constantly improving material quality, the achievement of *p*-type doping of high-In InGaN alloys, and lower background carrier concentration, nitride-based MQWs or non-polar (or semi-polar) nitrides could soon surpass the currently employed arsenide-based THz radiation devices.

References

1. Hu, B. B. and Nuss, M. C. (1995). Imaging with terahertz waves, Opt. Lett. 20, pp. 1716–1720.
2. Mittleman, D. M., Jacobsen, R. H., and Nuss, M. C. (1996). T-Ray imaging, IEEE. J. Sel. Top. Quantum. Electron. 2 (3) pp. 679–692.
3. Grischkowsky, D., Keiding, S., van Exter, M., and Fattinger, C. (1990). Far-infrared time-domain spectroscopy with terahertz beams of dielectrics and semiconductors, J. Opt. Soc. Am. B 7, pp. 2006–2015.
4. Chuang, S. L., Schmitt-Rink, S. Greene, B. I., Saeta, P. N., and Levi, A. F. J. (1992). Optical rectification at semiconductor surfaces, Phys. Rev. Lett. 68, pp. 102–105.
5. Auston, D. H. (1975). Picosecond optoelectronic switching and gating in silicon, Appl. Phys. Lett. 26, pp. 101–103.
6. Shan, J., Weiss, C., Wallenstein, R., Beigang, R., and Heinz, T. F. (2001). Origin of magnetic field enhancement in the generation of terahertz radiation from semiconductor surfaces, Opt. Lett. 26, pp. 849–851.
7. Brown, E., Smith, F., McIntosh, K. (1993). Coherent millimeter-wave generation by heterodyne conversion in low-temperature-grown GaAs photoconductors, J. Appl. Phys. 73, pp. 1480–1484.
8. Sakai, K. (ed.) (2005). Terahertz Optoelectronics, Springer, Berlin, pp. 1–27, pp. 63–76.
9. Bernardini, F., Fiorentini, V., and Vanderbilt, D. (1997). Spontaneous polarization and piezoelectric constants of III-V nitrides, Phys. Rev. B 56, pp. R10024–R10027.
10. Planken, P. C. M., Nuss, M. C., Brener, I., Goossen, K. W., Luo, M. S. C., and Chuang, S. L. (1992). Terahertz emission in single quantum wells after coherent optical excitation of light hole and heavy hole excitons, Phys. Rev. Lett. 69, pp. 3800–3893.
11. Turchinovich, D., Uhd Jepsen, P., Monozon, B. S., Koch, M., Lahmann, S., Rossow, U., and Hangleiter, A. (2003). Ultrafast polarization dynamics in biased quantum wells under strong femtosecond optical excitation, Phys. Rev. B 68, pp. 241307(R)-1–4.
12. Chern, G. D., Shen, H., Wraback, M., Koblmüller, G., Gallinat, C., and Speck, J. (2007). Excitation Wavelength Dependence of Terahertz Emission from Indium Nitride Multiple Quantum Wells, OSA CLEO/QELS Technical Digest, CThR3.
13. Chern, G.D., Shen, H., Wraback, M., Koblmüller, G., Gallinat, C., and Speck, J., (2007b). Terahertz Emission from Indium Nitride Multiple Quantum Wells, OSA OTST Technical Digest, MA3.
14. Neumark, G. F. (1962). Theory of the anomalous photovoltaic effect of ZnS, Phys. Rev. 125, pp. 838–845.

15. Pal, U., Saha, S., Chaudhuri, A. K., and Banerjee, H. D. (1991). The anomalous photovoltaic effect in polycrystalline zinc telluride films, J. Appl. Phys. 69, pp. 6547–6555.

16. Juillaguet, S., Camassel, J., Albrecht, M., and Chassagne, T. (2007). Screening the built-in electric field in 4H silicon carbide stacking faults, Appl. Phys. Lett. 90, pp. 111902-1–3.

17. Majewski, J. A. and Vogl, P. (1998). Polarization and band offsets of stacking faults in AlN and GaN, MRS Internet J. Nitride Semicond. Res. 3, pp. 21–23.

18. Majewski, J. A., Zandler, G., and Vogl, P. (2000). Novel nitride devices based on polarization fields, Phys. Stat. Sol. A 179, pp. 285–293.

19. Metcalfe, G. D., Shen, H., Wraback, M., Hirai, A., Wu, F., and Speck, J. S., (2008). Enhanced terahertz radiation from high stacking fault density nonpolar GaN, Appl. Phys. Lett. 92, pp. 241106-1–3.

20. Metcalfe, G. D., Shen, H., Wraback, M., Hirai, A., Wu, F., and Speck, J. S. (2008b). Terahertz Emission from Nonpolar Gallium Nitride, OSA CLEO/QELS Technical Digest, CTuX5.

21. Ohtake, H., Suzuki, Y., Sarukura, N., Ono, S., Tsukamoto, T., Nakanishi, A., Nishizawa, S., Stock, M. L., Yoshida, M., and Endert, H. (2001). THz-Radiation Emitter and Receiver System Based on a 2T Permanent Magnet, 1040nm Compact Fiber Laser and Pyroelectric Thermal Receiver, Jpn. J. Appl. Phys. Part 2 40, pp. L1223–L1225.

22. Takahashi, H., Suzuki, Y., Sakai, M., Ono, S., Sarukura, N., Sugiura, T., Hirosumi, T., and Yoshida, M. (2003). Significant enhancement of terahertz radiation from InSb by use of a compact fiber laser and an external magnetic field, Appl. Phys. Lett. 82, pp. 2005–2007.

23. Nagai, M., Tanaka, K., Ohtake, H., Bessho, T., and Sugiura, T. (2004). Generation and detection of terahertz radiation by electro-optical process in GaAs using 1.56 μm fiber laser pulses, Appl. Phys. Lett. 85, pp. 3974–3976.

24. Wu, J., Walukiewicz, W., Yu, K. M., Ager III, J. W., Haller, E. E., Lu, H., Schaff, W. J., Saito, Y., and Nanishi, Y. (2002). Unusual properties of the fundamental band gap of InN, Appl. Phys. Lett. 80, pp. 3967–3969.

25. Tansley, T. and Foley, C. P. (1986). Optical band gap of indium nitride, J. Appl. Phys. 59, pp. 3241–3244.

26. Gallinat, C. S., Koblmüller, G., Brown, J. S., Bernardis, S., Speck, J. S., Chern, G. D., Readinger, E. D., Shen, H., and Wraback, M. (2006). In-polar InN grown by plasma-assisted molecular beam epitaxy, Appl. Phys. Lett. 89, pp. 032109-1–3.

27. Koblmüller, G., Gallinat, C. S., Bernardis, S., Speck, J. S., Chern, G. D., Readinger, E. D., Shen, H., and Wraback, M. (2006). Optimization of the surface and structural quality of N-face InN grown by molecular beam epitaxy, Appl. Phys. Lett. 89, pp. 071902-1–3.

28. O'Leary, S. K., Foutz, B. E., Shur, M. S., and Eastman, L. F. (2005). Steady-state and transient electron transport within bulk wurtzite indium nitride: An updated semiclassical three-valley Monte Carlo simulation analysis, Appl. Phys. Lett. 87, pp. 222103-1–3.

29. Nahata, A., Auston, D. H., and Heinz, T. F. (1996). Coherent detection of freely propagating terahertz radiation by electro-optic sampling, Appl. Phys. Lett. 68, pp. 150–152.

30. Ascazubi, R., Wilke, I., Dennison, K., Lu, H., and Schaff, W. J. (2004). Terahertz emission by InN, Appl. Phys. Lett. 84, pp. 4810–4812.

31. Chern, G. D., Readinger, E. D., Shen, H., Wraback, M, Gallinat, C. S., Koblmüller, G., and Speck, J. S. (2006). Excitation wavelength dependence of terahertz emission from InN and InAs, Appl. Phys. Lett. 89, pp. 141115-1–3.

32. Mahboob, I., Veal, T. D., McConville, C. F., Lu, H., and Schaff, W. J. (2004). Intrinsic electron accumulation at clean InN surfaces, Phys. Rev. Lett. 92, pp. 036804-1–4.

33. Dekorsy, T., Pfeifer, T., Kutt, W., and Kurz, H. (1993). Subpicosecond carrier transport in GaAs surface-space-charge fields, Phys. Rev. B 47, pp. 3842–3849.

34. Liu, K., Xu, J., Yuan, T., and Zhang, X-.C. (2006), Terahertz radiation from InAs induced by carrier diffusion and drift, Phys. Rev. B 73, pp. 155330-1–6.

35. Kane, E. O. (1957). Band structure of indium antimonide, J. Phys. Chem. Solids 1, pp. 249–261.

36. Selberherr, S. (1984). Analysis and Simulation of Semiconductor Devices, Springer, New York.

37. Mittleman, D. (ed.) (2003). Sensing with Terahertz Radiation, Springer, Berlin, pp. 158-159.
38. Takeuchi, T., Sota, S., Katsuragawa, M., Komori, M., Takeuchi, H., Amano, H., and Akasake, I. (1997). Quantum-Confined Stark Effect due to Piezoelectric Fields in GaInN Strained Quantum Wells, Jpn. J. Appl. Phys., Part 2 36, pp. L382– L385.
39. Lepkowski, S. P., Suski, T., Perlin, P., Yu Ivanov, V., Godlewski, M., Grandjean, N., and Massies, J. (2002). Study of light emission from GaN/AlGaN quantum wells under power-dependent excitation, J. Appl. Phys. 91, pp. 9622–9628.
40. Chern-Metcalfe, G. D., Readinger, E. D., Shen, H., Wraback, M., Koblmüller, G., Gallinat, C. S., and Speck, J. S. (2008). Intensity-dependent photoluminescence studies of the electric field in N-face and In-face InN/InGaN multiple quantum wells, Phys. Stat. Solidi (c) 5, pp. 1846– 1848.
41. Kurouchi, M., Naoi, H., Araki, T., Miyajma, T., and Nanishi, Y. (2005). Fabrication and Characterization of InN-Based Quantum Well Structures Grown by Radio-Frequency Plasma-Assisted Molecular-Beam Epitaxy, Jap. J. Appl. Phys. 44, pp. L230– L232.
42. Ohashi, T., Holmström, P., Kikuchi, A., and Kishino, K. (2006). High structural quality InN/In0.75Ga0.25N multiple quantum wells grown by molecular beam epitaxy, Appl. Phys. Lett. 89, pp. 041907-1–3.
43. Hirano S., Inoue, T., Shikata, G., Orihara, M., Hijikata, Y., Yaguchi, H., Yoshida S. (2007). RF-MBE growth of InN/InGaN quantum well structures on 3C–SiC substrates, J. Crystal Growth 301, pp. 513–516.
44. Ambacher, O., Majewski, J., Miskys, C., Link, A., Hermann, M., Eickhoff, M., Stutzmann, M., Bernardini, F., Fiorentini, V., Tilak, V., Schaff, B., and Eastman, L. F. (2002). Pyroelectric properties of Al(In)GaN/GaN hetero- and quantum well structures, J. Phys.: Condens. Matter 14, pp. 3399–3434.
45. Craven, M. D., Wu, F., Chakraborty, A., Imer, B., Mishra, U. K., DenBaars, S. P., and Speck, J. S. (2004). Microstructural evolution of a-plane GaN grown on a-plane SiC by metalorganic chemical vapor deposition, Appl. Phys. Lett. 84 (8), pp. 1281–1283.
46. Haskell, B. A., Chakraborty, A., Wu, F., Sasano, H., Fini, P. T., Denbaars, S. P., Speck, J. S., and Nakamura, S. (2005). Microstructure and Enhanced Morphology of Planar Nonpolar m-Plane GaN Grown by Hydride Vapor Phase Epitaxy, J. Electron. Mater. 34, pp. 357–360 and references therein.
47. Tani, M., Matsuura, S., Sakai, K., and Nakashima, S. (1997). Emission characteristics of photoconductive antennas based on low-temperature-grown GaAs and semi-insulating GaAs, Appl. Opt. 36, pp. 7853–7859.

CHAPTER 20

AN INTRAOCULAR CAMERA FOR RETINAL PROSTHESES: RESTORING SIGHT TO THE BLIND

NOELLE R. B. STILES, BENJAMIN P. McINTOSH, PATRICK J. NASIATKA,

MICHELLE C. HAUER, JAMES D. WEILAND,

MARK S. HUMAYUN, and ARMAND R. TANGUAY, JR.

Optical Materials and Devices Laboratory, Viterbi School of Engineering,
Keck School of Medicine, and Doheny Eye Institute, University of Southern California,
520 Seaver Science Center, University Park, MC-0483, Los Angeles, California, 90089-0483, USA
atanguay@usc.edu

Implantation of an intraocular retinal prosthesis represents one possible approach to the restoration of sight in those with minimal light perception due to photoreceptor degenerating diseases such as retinitis pigmentosa and age-related macular degeneration. In such an intraocular retinal prosthesis, a microstimulator array attached to the retina is used to electrically stimulate still-viable retinal ganglion cells that transmit retinotopic image information to the visual cortex by means of the optic nerve, thereby creating an image percept. We describe herein an intraocular camera that is designed to be implanted in the crystalline lens sac and connected to the microstimulator array. Replacement of an extraocular (head-mounted) camera with the intraocular camera restores the natural coupling of head and eye motion associated with foveation, thereby enhancing visual acquisition, navigation, and mobility tasks. This research is in no small part inspired by the unique scientific style and research methodologies that many of us have learned from Prof. Richard K. Chang of Yale University, and is included herein as an example of the extent and breadth of his impact and legacy.

1. Introduction

Among the leading causes of late onset blindness, both retinitis pigmentosa (RP)[1,2] and age-related macular degeneration (AMD)[3,4] cause a loss of light perception that results from degeneration of the photoreceptor cells (rods and cones) within the outer layer of the retina. Retinitis pigmentosa often begins with night blindness, which is followed by a loss of peripheral light sensitivity that then progresses toward the region of central vision, finally terminating in many cases with total loss of vision. In the case of age-related macular degeneration, dark or blurry spots begin to appear in the region of central vision, representing the onset of damage to the macular region of the retina. Progressively, the dark regions expand and combine until the entire macula exhibits visual loss. Peripheral vision is relatively unaffected in this case, but vision in the region of the macula, spanning approximately a 10° field of view, often appears blurred, dark, or black. As the macula is within the region of the retina corresponding to central (highest resolution) vision, spanning approximately a 20° field of view, this region of visual obscuration appears to those affected wherever they foveate, or direct their gaze, thereby obscuring the very features that they most desire to see.

Retinal degeneration is both widespread and prevalent, affecting large numbers of people throughout the world. Advanced age-related macular degeneration, for example, is the leading cause of irreversible central vision loss throughout the Western world in those 50 years of age or older.[4] Retinitis pigmentosa currently affects approximately 75,000 to 90,000 people in the U.S. alone, and approximately 1.7 to 1.9 million people worldwide.[5,6] Age-related macular degeneration affects an even larger population of approximately 1.75 million individuals in the U.S. currently, with the number projected to increase to approximately 3 million individuals by 2020 as the U.S. population ages.[7] Worldwide, AMD currently affects at least 20 to 25 million people, with projections of 60 to 75 million affected by 2030.[8]

Currently, there are no proven preventative measures for these two degenerative diseases, as well as no widely accepted and effective therapeutic options. As both diseases have a genetic component, one method of *a priori* intervention currently under investigation is the use of gene therapy,[9] which could potentially decrease the likelihood of expressing the disease and thereby prevent its consequences from occurring. For this approach to be generically applicable, multiple gene therapies will need to be developed in order to target the numerous mutations of each disease. Potential *a posteriori* treatments include retinal transplantation[10] and stem cell therapy,[11] both of which would be implemented after the onset of the disease and its major symptoms. These approaches show promise, although to date it has proven to be difficult to generate appropriate synaptic connections between either the transplanted retina or implanted photoreceptor-differentiated stem cells and the remaining viable cells within the host retina.[12] Recently, encouraging results have been obtained in retinal repair by transplantation of photoreceptor precursors, enabled by harvesting immature mouse donor cells at the peak of rod photoreceptor genesis.[12]

An alternative *a posteriori* approach that is potentially more viable in the near term is the development of a visual prosthesis, in which patterned electrical stimulation is applied to the retina, optic nerve, or visual cortex to achieve retinotopically-mapped percepts.[13-52] In a wide range of visual prosthesis implementations, a video camera or photodetector array captures scenes of the environment, and image-encoded signals are transmitted to a two-dimensional microstimulator array. Individual electrodes within the array excite one or more cell types that precede[13-45] or are within the visual cortex,[13,14,30,46-48] thereby giving rise to visual percepts.

The retina and the visual cortex are spatially organized with a point-to-point correspondence, so that an array of microelectrodes properly encoded with a camera-derived visual signal and proximity coupled to the retina can produce the desired retinotopic (spatially registered) mapping. In this case, advantage is taken of the fact that degeneration of the photoreceptor layer in both RP and AMD appears to leave the inner layers of the retina largely intact, such that electrical stimulation of the retinal ganglion cells, for example, can result in output signals that are transmitted from the retina to the visual cortex along dense fibers within the optic nerve. This prosthetic approach has the significant advantage that the retina is easily accessible through ocular surgery.

A visual prosthesis can also be envisioned with the microstimulator array activating some part of the optic nerve,[13,14,30,49-52] although the retinotopic organization is not so easily sorted out within the various nerve fiber bundles that comprise the optic nerve. Finally, it is also possible to generate appropriate visual percepts by implanting an electrode array directly within the visual cortex.[13,14,30,46-48] Perhaps the primary disadvantage of these two approaches is the relative inaccessibility of the optic nerve and visual cortex, requiring more invasive surgery than in the case of the retina. Even so, these approaches offer the advantage of providing possible therapeutic options for cases in which damage or disease have destroyed many or all of the functional retinal cells.

A number of other types of visual prostheses have been investigated that directly stimulate a sense other than vision, and that rely on cross-modal interactions or, with training, cross-modal plasticity to provide a sense of object identities and locations in space. One such type of visual prosthesis is based on the encoding of visual scenes in time, loudness, and frequency to provide auditory cues;[53-59] another type is based on the encoding of visual scenes with vibrating, pressure, or electrical transducers to provide spatially meaningful tactile cues.[58-62] As is the case for visual prostheses designed for the optic nerve and the visual cortex, auditory, vibrotactile, and electrotactile visual prostheses can potentially provide additional therapeutic options for all forms of blindness. Additionally, auditory and tactile visual prostheses are designed to be worn externally, and as such do not require surgical implantation procedures.[17] Although neither approach appears to restore the sensation of vision, *per se*, several cases of synthetic synaesthesia have been reported.[56,57] The extent to which these approaches induce cross-modal plasticity (with extensive training and use) that may partially interfere with the capabilities of the original sense (hearing or touch) has not yet been fully established.

The remainder of this chapter is focused on the development of an intraocular camera for retinal prostheses. Such an intraocular (rather than an extraocular) camera may also have significant advantages for visual prostheses that couple to the optic nerve or directly to the visual cortex, as well as for visual prostheses that provide auditory or tactile cues. We describe several current implementations of intraocular retinal prostheses in Section 2, the importance of foveation for the performance of human tasks in Section 3, and the concept of an intraocular camera for retinal prostheses that can provide foveation in Section 4. The concept of an eye-tracked extraocular camera for retinal prostheses that provides both foveation and an additional therapeutic option is presented in Section 5. Several key results from the study of human visual psychophysics in the low pixellation limit that imply relaxed intraocular retinal prosthesis and intraocular camera design constraints are presented in Section 6, and the design, implementation, and capabilities of a visual prosthesis simulator are described in Section 7. The design constraints that apply to the intraocular camera are then outlined in Section 8, and the intraocular camera optical system design is described in Section 9. Key image sensor array characteristics are outlined in Section 10, and integration and packaging issues are discussed in Section 11. A summary and conclusions are provided in Section 12, and future research directions are discussed in Section 13.

2. Current implementations of intraocular retinal prostheses

The human eye is shown schematically in cross section as viewed from the top of the (right) eye in Fig. 1. The temporal side of the eye is at the top of the figure, and the medial side of the eye is at the bottom, with the optic nerve angled toward mid-brain. Incoming light is refracted by the lens formed by the cornea and aqueous humor in the anterior chamber, and also by the elastic crystalline lens in the posterior chamber, thereby forming an image on the photosensitive retinal surface.

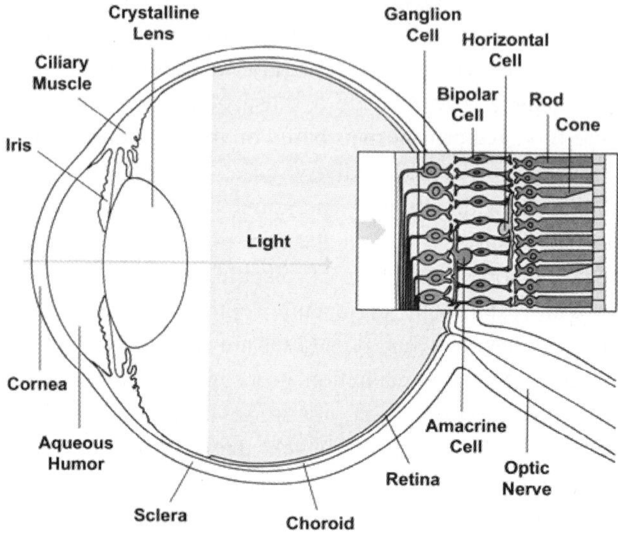

Fig. 1. Schematic diagram of the human eye and retina, illustrating the orientation of the layers of the retina with respect to the direction of optical input, as well as the cornea, aqueous humor, and crystalline lens. After David H. Hubel, Ref. 63, pg. 37.

The retina is a densely interconnected, multilayered cellular structure that is supported within the eye by the choroid, a thin nutrient-supplying and light-absorbing tissue layer. The choroid, in turn, is supported by the sclera, a thick layer of white tissue that surrounds the entire eye except within the region of the cornea. The sclera provides structural support for the eye, and is maintained in a roughly spherical shape by intraocular pressure.

Within the retina, rod and cone photoreceptors are located in close proximity to the choroid, with their cell bodies located in the external nuclear lamina (layer) of the retina, as shown schematically in Fig. 2. Densely pigmented cells within the choroidal layer absorb excess light that passes through the various layers of the retina and is not fully absorbed by the photoreceptor cells.

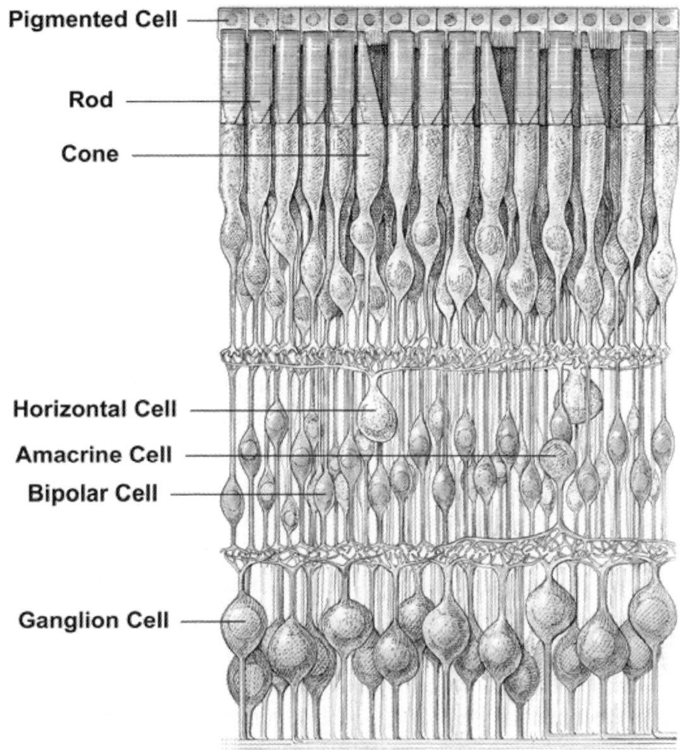

Fig. 2. Schematic diagram of the human retina, illustrating the rod and cone photoreceptor cells with their nuclei residing in the external nuclear lamina; the horizontal, amacrine, and bipolar cells in the inner nuclear lamina; and the ganglion cells in the ganglionic lamina. After David H. Hubel, Ref. 63, pg. 38.

The photoreceptor cells receive optical inputs from the environment that are imaged onto the retina by the corneal and crystalline lenses, acting in combination as an optical doublet, as described above. Outputs from the photoreceptor cells induce excitation in multiple layers of cells in response, including bipolar, horizontal, and amacrine cells within the internal nuclear lamina of the retina, and ultimately retinal ganglion cells within the ganglionic lamina of the retina. The rods and cones are distributed nonuniformly throughout the retina, with the highest density of cones occurring in the fovea and macula, and with rapidly decreasing cone density as a function of increasing eccentricity (visual angle measured from the optical axis of the eye). The density of rods is high throughout the periphery, but rods are essentially absent throughout the region of central vision.

Given the physiological architecture of the retina, multiple prosthetic approaches to replacing lost photoreceptor functionality can be envisioned, based on the fact that extracellular electrical stimulation of the remaining healthy cells in the retina can give rise to signal transmission to the brain, resulting in corresponding visual percepts.[15-17]

In the epiretinal approach,[13,14,15-30] a microstimulator (microelectrode) array is placed on the inner surface of the retina such that current injected by each electrode within the array stimulates retinal neurons, primarily the retinal ganglion cells that are in closest proximity to the electrode. In the subretinal approach,[13,14,30-38] the retina is detached and lifted in the region of the implant, and a microelectrode array is placed between the retina and the choroid, again resulting in stimulation of the neurons in closest proximity. In this case, the stimulated neurons are primarily the bipolar and horizontal cells. In the extraocular or suprachoroidal approach,[41-45] the array is either attached to the exterior of the sclera, inserted within the sclera itself, or inserted between the choroid and the sclera, thereby obviating the need to partially detach the retina as in the subretinal approach.

The advantages of the epiretinal approach to intraocular retinal prostheses include the ease of surgical implantation through a small incision in the sclera, the relative proximity of the microelectrode array to the output cells of the retina (the retinal ganglion cells) that lead directly to the optic nerve, and the relative ease of removal after implantation. The disadvantages of this approach include the possibility of exciting the axonal projections from the retinal ganglion cells themselves, the need for an attachment mechanism of the array to the retina that respects the fragility of the retinal tissue, and the difficulties inherent in transmitting power and data to the microelectrode array.

Implantation of the microelectrode array subretinally has the advantages that placement of the array between the retina and the choroid naturally forms an effective attachment mechanism, and that the subretinal location potentially allows for stimulation of the cells within the inner nuclear layer, which perform highly useful low level visual processing functions in a healthy retina. Offsetting these advantages to some degree are the difficulties inherent in any surgery requiring even partial retinal detachment, with the associated risks of further retinal detachment or hemorrhage, and similar difficulties inherent in transmitting power and data to the microelectrode array. In addition, it has recently been discovered that extensive remodeling and reprogramming of the inner nuclear and plexiform layers of the retina occurs after the onset of blindness, reducing the usefulness of stimulating the cells in the inner nuclear layer in hopes of regaining the functional processing that they would normally provide.[64-66]

An important feature of any visual prosthesis based on proximity coupling of a microelectrode array to the retina is the availability of a viable approach for removal of the array should this become necessary. With respect to both the epiretinal and subretinal approaches, preliminary reports of successful explantation of the microelectrode array following successful implantation are highly encouraging.[39,40]

Extraocular and suprachoroidal placements of the microelectrode array are advantageous in that neither the interior nor the exterior of the retina itself need be contacted by a foreign object either during or after surgery. In fact, the entire surgical procedure can occur completely extraocularly in both cases. The principal difficulty of these two approaches is the positioning of a thick dielectric membrane between the individual electrodes within the microelectrode array and the first layer of viable retinal cells, which appears to lead to additional spreading of the electric fields and currents, yielding limited resolution. One promising approach employs an array of penetrating microelectrodes that extend inwards toward the retina through the sclera and choroid, which may allow for higher resolution than purely extraocular or suprachoroidal electrode placement.[41]

In order to provide visual input to the various microstimulator arrays, two basic strategies have been employed to date. In one strategy, applicable to both epiretinal and subretinal approaches, the biological corneal and crystalline lenses are used to provide imaging with a focal plane at or near the retina, and photodetectors are integrated with the microstimulator array to function in place of the damaged photoreceptors of the eye.[13,14,30-38] This in turn involves either monolithic or hybrid integration of a two-dimensional array of silicon photodetectors with amplification and signal transformation circuitry that is capable of providing appropriate stimulation signals to each individual microelectrode within the microstimulator array. This strategy has the advantage that the natural optical elements of the eye are used for image formation, thereby providing for natural coupling of head and eye movements. In addition, the surgical procedure used for implantation is minimally complicated by the addition of photosensitive elements.

However, several key disadvantages accrue to this strategy as well, including the necessity for proximity coupling of an inflexible, planar silicon substrate to the curved and fragile retina; the difficulty of achieving a high fill factor (ratio of the photosensitive area of each pixel to the total pixel area) for high sensitivity without resorting to either multichip module integration or dual sided processing with through-substrate vias; the requirement for careful thermal management at the surface of the retina, due to the additional heat dissipation of the photodetectors and amplification circuitry above and beyond that of the power delivered to the retinal tissue by the microelectrodes themselves; and the need for transparent hermetic packaging capable of withstanding chronic implantation in the highly corrosive saline-like environment of the eye.

An alternative strategy involves the use of an extraocular camera,[13,14,15-30,37,38,41-45] typically mounted on a pair of eyeglasses as shown schematically in Fig. 3. As current implementations in clinical trials are designed for implantation of the microstimulator array in one eye only, the extraocular (video) camera may either be centrally located on the pair of eyeglasses, as shown in the figure, or offset to align with the implanted eye. The camera output signal is typically routed to a cellular-telephone-sized external visual processing unit (EVPU) that is belt mounted, and that contains a battery, power supply, camera control circuitry, signal processing circuitry, and a central processing unit (CPU) or digital signal processor (DSP) for pixellation and formatting. The EVPU also typically contains output amplification and communications circuitry for transmitting both power and the formatted video signal, either wirelessly or by wired connection, to separate microstimulator driver circuitry and the microstimulator array itself.

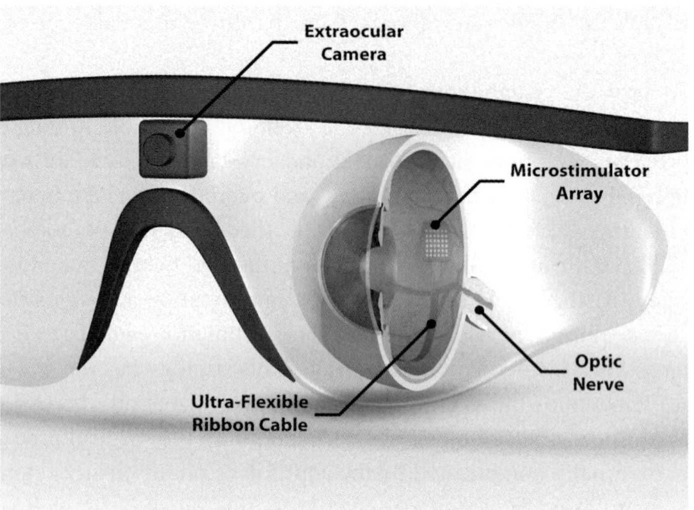

Fig. 3. Intraocular retinal prosthesis with externally mounted (extraocular) camera. The diagram of the eye has been rotated counterclockwise by 90°, and the ultra-flexible ribbon cable routed in the opposite hemisphere relative to its usual placement, for clarity of illustration.

The microstimulator driver circuitry is often placed in a thin hermetically sealed housing that is mounted to the exterior of the temporal scleral wall of the eye, and that additionally contains power conditioning electronics. In the epiretinal case, stimulation signals are transmitted from the microstimulator driver circuitry to a microstimulator array that is attached to the retina by means of a through-scleral tack.[15-21] An ultra-flexible ribbon cable carries these signals through a small incision in the sclera near the iris, as shown schematically in Figs. 3 and 4. Figure 4 also shows the relative locations of the cornea, iris, and crystalline lens, the latter enclosed within the crystalline lens sac that is suspended by thin fibrous zonules. The microstimulator array is usually placed in the macular region, nearly centered on the optical axis of the eye, and displaced from the optic disc. The optic disc is located on the surface of the retina at the head of the optic nerve, which transmits signals to the lateral geniculate nucleus (LGN) in the mid-brain. Axonal outputs from the LGN then project to area V1 within the visual cortex.

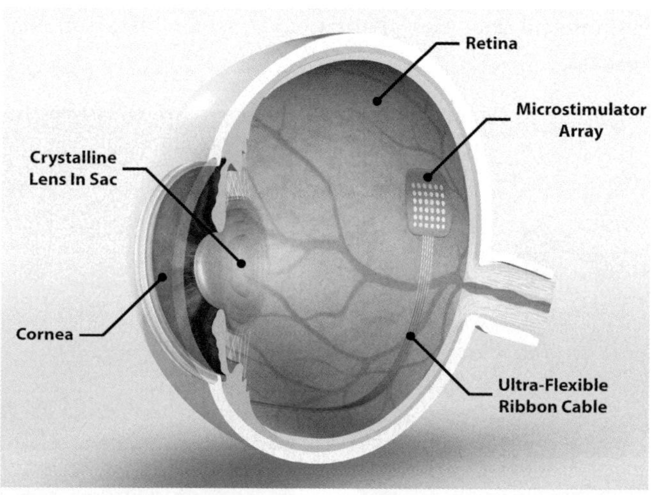

Fig. 4. Intraocular retinal prosthesis with externally mounted (extraocular) camera; enlarged view. The drawing is an illustration of a top view of the right eye, with the ultra-flexible ribbon cable routed in the opposite hemisphere relative to its usual placement for clarity of illustration.

The advantages of providing video inputs from an extraocular camera include ease of wired access to external processing circuitry within the EVPU, minimization of complexity and heat dissipation within the intraocular cavity, simple replacement of a malfunctioning camera, and capability for substitution of an upgraded camera for an earlier model. However, external placement of the image-capturing (video) source eliminates the natural coupling of head and eye movements, requiring head scanning for exploration of the environment. Rapid head motion in turn can result in disorientation, dizziness, and nausea, as described in more detail in the next section.

3. The importance of foveation

In normal foveation, or the process of aligning the highest resolution area of the retina (the fovea, within the macula) with a feature of interest in the natural environment, head and eye motions are coupled together naturally to achieve rapid target acquisition and tracking with a minimum of disorientation and dizziness. One's sense of where a particular object is within the environment relative to one's self is determined by the angular rotation of the eye with respect to the head, and the angular orientation of the head with respect to the body. This process occurs naturally, without conscious thought. In fact, physiological processes such as the vestibulo-ocular reflex (VOR) allow for stable foveation on a target of interest that is largely independent of head motion by causing the eyes to move compensatorily in the direction opposite to that of any head motion. Decoupling of these natural head and eye motions in providing input to the human visual system can thus lead to decreased performance in navigation and mobility tasks, as well as in hand-"eye" coordination. Furthermore, rapid motions of the head without this natural coupling to corresponding eye motions, as may occur quite normally in the natural environment, can also give rise to considerable disorientation, as well as dizziness and nausea.

Fig. 5. Simulation of the effects of foveation on a subject with retinitis pigmentosa (RP) that has been implanted with an intraocular retinal prosthesis. The field of view of the scene camera is shown in the upper left panel, with the central field of view outlined with a white dashed line, and the direction of foveation (gaze) outlined with a solid black line. The visual field with RP but without a visual prosthesis is shown in the upper right panel. For the head-directed case, the image from the central field of view appears in the direction of gaze, as shown in the lower left panel, while for the eye-directed case, the region of the camera field of view that is aligned with the direction of gaze is instead directed to the implant, as shown in the lower right panel.

Consider, for example, a subject with retinitis pigmentosa who has little to no remanent light sensitivity, but has been implanted with an intraocular retinal prosthesis coupled to an extraocular head-mounted (scene) camera, as shown in the simulation sequence presented in Fig. 5. The entire $32° \times 40°$ field of view of the scene camera is shown in the upper left panel in Fig. 5, with the central $10° \times 10°$ field of view outlined with a white dashed line. The same field of view as seen by the subject with retinitis pigmentosa but without the intraocular retinal prosthesis is shown in the upper right panel, assuming total blindness with no remanent light sensitivity.

A circular portion ($10°$ in diameter) of the image from the central field of view was then extracted, pixellated on a square grid with 32 pixels across the diameter of the central field of view, and then blurred with 40% Gaussian blur to simulate the effects of electric field and current spreading, as well as lateral retinal ganglion cell recruitment (see Sect. 6 below). If the pixellated image of the central field of view is coupled to the microstimulator array attached to the retina, then the resulting image percept will appear in the direction of gaze, indicated by the black solid outline in the upper left panel.

The resulting percept will be as shown in the lower left panel of Fig. 5, with the image of the face superimposed over the actual position of the stop sign. In fact, if the subject's head is motionless, but the eyes naturally foveate to different directions of gaze, the image of the face will be omnipresent, appearing wherever the subject looks. If instead the direction of gaze is tracked, and the portion of the image in the direction of gaze extracted, then the resulting percept will be as shown in the lower right panel of Fig. 5, with the perceived image of the stop sign superimposed over its actual position.

By way of contrast, consider now a subject with age-related macular degeneration (AMD), for example with little to no remanent light sensitivity in the central 15° of the visual field. In this case, the peripheral visual field will still be evident, as the subject's peripheral vision is intact. Consider such a subject to be implanted with an intraocular retinal prosthesis coupled to an extraocular head-mounted camera, as shown in the simulation sequence presented in Fig. 6.

As in the previous figure (Fig. 5), the central field of view of the scene camera is shown in the upper left panel, and the same field of view as seen by the subject with age-related macular degeneration but without the intraocular retinal prosthesis is shown in the upper right panel, assuming total blindness with no remanent light sensitivity within the central visual field. The head-directed perceived image with the intraocular retinal prosthesis is shown in Fig. 6 in the lower left panel, and the eye-directed perceived image in the lower right panel. In order to best appreciate the proper balance of central and peripheral vision, align your head to directly face the center of the photograph, and then foveate (direct your gaze) to the center of the dark circle in the upper right corner. As before, the image of the face is superimposed over the actual position of the stop sign in the head-directed case, whereas in the eye-directed case, the image of the stop sign is superimposed over its actual position in the natural environment, as desired.

Interestingly, the addition of prosthetic visual information that is incorrectly placed within the background of correctly registered peripheral vision can lead to disorientation for a subject with AMD, as can be seen in the head-directed case shown in Fig. 6 with two different spatial instantiations of the image of the face. Correct placement of prosthetic visual information within the visual field, however, allows the use of information within the peripheral field to help appreciate the information prosthetically added to the central visual field, as in the eye-directed case.

Fig. 6. Simulation of the effects of foveation on a subject with age-related macular degeneration (AMD) that has been implanted with an intraocular retinal prosthesis. As in Fig. 5, the field of view of the scene camera is shown in the upper left panel, with the central field of view outlined with a white dashed line, and the direction of foveation (gaze) outlined with a solid black line. The visual field with AMD but without a visual prosthesis is shown in the upper right panel. For the head-directed case, the image from the central field of view appears in the direction of gaze, as shown in the lower left panel, while for the eye-directed case, the region of the camera field of view that is aligned with the direction of gaze is instead directed to the implant, as shown in the lower right panel. In both cases, the remainder of the visual field is available to the subject through their still-viable peripheral vision.

The simulated images presented in Figs. 5 and 6 are derived from a single (static) frame that might represent any one of a number of frames from a video sequence recorded at 30 frames per second, as is the situation with a typical video camera. Even more dramatic effects occur with dynamic sequences, especially in navigation and mobility tasks, but even in simple human activities such as locating and shaking someone's hand. The static images in Figs. 5 and 6 were produced on a visual prosthesis simulator that we have developed to enable and support multiple visual psychophysics experiments, including performance in object recognition, navigation, and mobility tasks, as well as for evaluation of optimal image processing algorithms following scene capture. The visual prosthesis simulator is described in more detail in Sect. 7.

In order to restore the natural coupling of head and eye motions in an intraocular retinal prosthesis, placement of the image acquisition device within the eye itself is optimal, so that head and eye motions are coupled with minimal latency. To this end, the development of an intraocular camera[67-75;51,52] that naturally foveates with eye motion is described in the next section, and an eye-tracked extraocular camera that emulates the eye-directed cases shown in Figs. 5 and 6 is discussed in Sect. 5.

4. An intraocular camera for retinal prostheses

In order to provide for image acquisition within the human eye itself, several key functions must be implemented. These functions include the formation of an optical image at adequate resolution to support the envisioned retinal prosthesis, the detection of the optical image by a low power image sensor array, and the coupling of the output signal from the image sensor array to the microstimulator array. This latter interconnection also optimally involves a transformation from the native output signal levels and format of the image sensor array to that of the targeted surviving retinal cells.

The basic concept of an intraocular retinal prosthesis with an associated intraocular camera is shown in Fig. 7.[67-74] In the implementation shown, the intraocular camera is designed to fit within the crystalline lens sac, following removal of the crystalline lens by means of phacoemulsification. This procedure is commonly employed worldwide to remove the crystalline lens during cataract surgery. For those with cataracts, removal of the crystalline lens is often followed by replacement with an intraocular lens (IOL), a polymer lens supported by haptic arms that extend to the edges of the crystalline lens sac. The intraocular camera can be mounted similarly by means of haptic elements, not shown in the illustration.

The video output of the intraocular camera is transmitted either wirelessly (not shown) or by thin flexible cable to the microstimulator array driver circuitry contained within a miniature hermetically-sealed electronics housing. Output signals from the microstimulator array driver circuitry are routed by means of an ultra-flexible ribbon cable to the microstimulator array that is attached to the retina, as shown in the figure. In the implementation shown in Fig. 7, the hermetically-sealed electronics housing has been implanted within the posterior cavity of the eye (within the region normally occupied by the vitreous humor), forming a complete intraocular retinal prosthesis. Alternatively, the video signals from the intraocular camera could be transmitted outside the eye, routed to and processed within an external visual processing unit (EVPU), and then returned to microstimulator array driver circuitry located on the exterior of the eye, as in the case of the extraocular camera described in Sect. 2.

An enlarged view of one possible implementation of the intraocular camera implanted within the crystalline lens sac is shown in Fig. 8. The essential intraocular camera components are mounted within a biocompatible and hermetic housing that is sealed on the anterior end (facing the cornea) with a fused silica window. Incorporation of the window is useful in relaxing the optical system design constraints, as described in more detail in Sects. 8 and 9. An aspherical lens that is designed to be used in conjunction with the biological corneal lens forms an image of the outside world on an image sensor array placed at the posterior end of the housing. One or more additional complementary metal-oxide-semiconductor (CMOS) very-large-scale integrated (VLSI) circuits may be included to provide for image sensor array control, power conditioning, signal processing, signal transmission, and environmental monitoring.

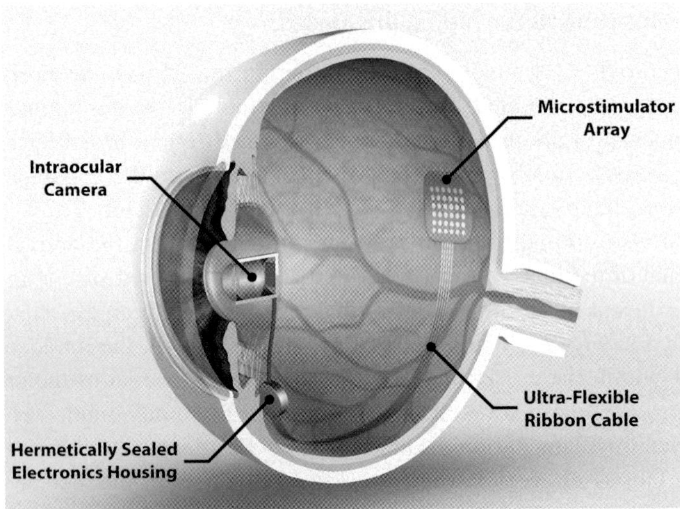

Fig. 7. Intraocular retinal prosthesis with intraocular camera (IOC).

In order to satisfy surgical constraints, as well as to fit within the crystalline lens sac, the entire intraocular camera must be ultraminiature in size. As described further in Sects. 8 and 9, the current design of the intraocular camera includes external housing dimensions of 3.18 mm in diameter and 4.5 mm in length. This in turn requires that the back focal length of the aspherical lens be on the order of 1 mm. These extreme size limitations in turn place strict design constraints on the optical system of the intraocular camera, as described in Sect. 8. As a consequence, visual psychophysics techniques have been employed extensively to determine key imaging characteristics that yield optimal percepts in the low pixellation limit. These studies, described in detail in Sect. 6, have led to significantly relaxed optical system design constraints.

Although the specific implementation of the intraocular retinal prosthesis described above comprises an intraocular camera coupled to an epiretinal microstimulator array as shown in Fig. 7, alternative implementations can easily be envisioned in which the intraocular camera provides visual input to subretinal, suprachoroidal, or extraocular microstimulator arrays.

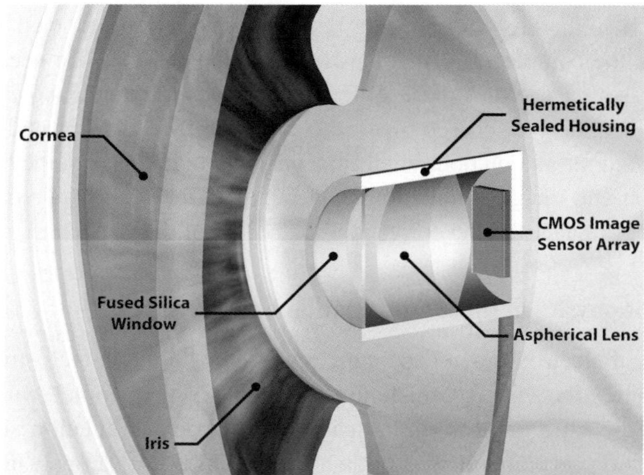

Fig. 8. Intraocular retinal prosthesis with intraocular camera (IOC); enlarged view.

5. An eye-tracked extraocular camera for retinal prostheses

The simulations of the eye-directed cases shown in Figs. 5 and 6 that were implemented in the visual prosthesis simulator suggest an alternative strategy for restoring the natural coupling of head and eye motions in an intraocular retinal prosthesis, that of an eye-tracked extraocular camera. In this case, an additional ultraminiature video camera is placed on the lower rim of the eyeglasses shown in Fig. 3, and is directed at and focused on the eye that is implanted with the intraocular retinal prosthesis. Images of the pupil of the eye are processed within the externally mounted EVPU to extract the direction of gaze in real time. The angular coordinates of the direction of gaze are then used to define the region of the image obtained by the scene camera that corresponds to the direction of gaze, and it is this region of the image that is then processed and used to drive the microstimulator array. This process restores the one-to-one correspondence between features in the natural world and their apparent location in the prosthetic image percept.

In order to accomplish this effectively, the scene camera should optimally be modified to incorporate a wide field-of-view lens, so that the full range of object space that can be acquired through foveation of the eyes is available for a given orientation of the head. This in turn requires the design of wide field-of-view lenses with very short focal lengths, which will typically exhibit barrel distortion that must be corrected in real time through an image dewarping algorithm. In addition, the EVPU must be modified to include provision for processing two video streams (one from the scene camera, and one from the eye-tracking camera) in real time, such that the extracted region of the image can be generated with minimal latency.

The eye-tracked extraocular camera has essentially the same advantages as the extraocular camera described previously in Sect. 2, without the principal disadvantage of a lack of foveation capability. Relative to the intraocular camera, however, there are additional disadvantages to the use of an eye-tracked extraocular camera, including the need for a much higher resolution wide-field-of-view camera as well as for an eye-tracking camera, the complexity of the hardware implementation, the need for additional computational resources within the EVPU, an increase in external power consumption (which affects the battery discharge time), the current difficulty of providing robust gaze tracking in both indoor and outdoor environments, and an increase in the latency between image acquisition and image stimulation resulting from the time required to extract both accurate gaze coordinates and the corresponding region of the image.

6. Visual psychophysics in the low pixellation limit

The number of electrodes that can be incorporated within the microstimulator array is limited by a number of factors, including the number of parallel wires that can be incorporated in the ultra-flexible ribbon cable that carries stimulation signals from the microstimulator driver circuitry to the microstimulator array, the number of microelectrodes that can be incorporated within the microstimulator array while retaining enough flexibility to conform to the curvature of the retina, the total power dissipated by the full set of microelectrodes when operated with appropriate levels of stimulation current and voltage at typical video frame rates (30 frames/second), and the minimal electrode size and interelectrode gaps necessary for generating distinct visual percepts without substantial crosstalk.

In the first epiretinal prosthesis chronically implanted in subjects under the auspices of a clinical trial, the array size was 4×4, resulting in a total of only 16 pixels.[19] Yet subjects implanted with this intraocular retinal prosthesis with an extraocular camera were able to differentiate among simple objects such as a plate, a cup, and a knife placed in a high contrast environment using extensive head scanning. Given the limited instantaneous field-of-view of the intraocular retinal prosthesis, head scanning allows for an expansion of the effective field-of-view by means of spatiotemporal integration, and of the resolution within a given angular field by means of spatiotemporal differentiation. With more complex scenes, of course, more pixels are required, particularly if it is desired to include tasks such as navigation and mobility with task performance that approaches that of sighted individuals.

Consider, for example, the image of a kitchen as shown in Fig. 9. The original image is 1024×1024 pixels, or approximately 1.05 million total pixels. Within the kitchen scene are a center island, a walkway to the right, a table and chairs with a flower arrangement in the foreground, a microwave oven and refrigerator on the left, two doors with glass windows to the left rear, and a stove in the right rear. A 4×4 block pixellated version of this image is shown in Fig. 10, and it is evident that one would be hard pressed to identify any of the key features of the scene from this static image alone with no further information. We have used block pixellation to demonstrate the principal features of pixellation and filtering; alternative models can be based instead on elicited percepts.

Fig. 9. Original kitchen image, 1024 × 1024 pixels, grey scale.

As the electrodes within the array are separated from one another to allow for electrical isolation, for example, it is conceivable that under certain circumstances elicited percepts will show grid lines of diminished excitation, as shown schematically in Fig. 11. As is evident from the figure, this effect also interferes with image feature identification, as it effectively adds false information not contained in the original scene.

In a more recent clinical trial, the array size is 6 × 10, comprising 60 pixels.[19] Efforts are currently underway to develop even higher resolution arrays, with an eventual goal of reaching perhaps 32 × 32 arrays with 1024 pixels. Even this resolution, of course, is far short of that characteristic of vision in normally sighted individuals. As such, it is of considerable interest to try to determine just how many pixels are required for the adequate performance of a wide range of human tasks, as well as how they should best be acquired and processed. In this section, we describe a number of key results from human psychophysical experiments that directly impact the design of intraocular retinal prostheses in general, and of an intraocular camera for retinal prostheses in particular.

Fig. 10. Pixellated kitchen image, 4 × 4 pixels, grey scale.

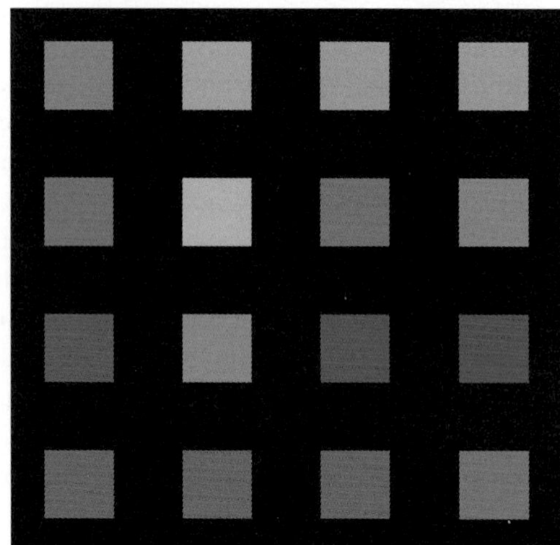

Fig. 11. Pixellated kitchen image, 4 × 4 pixels, grey scale, with grid lines at 50% duty cycle.

A wide array of human psychophysical investigations have been conducted to date that are focused on visual processing in the low pixellation limit that applies to retinal prostheses.[17,67-74,76-91] These studies have shown that remarkable capabilities can accrue to the provision of very few pixels within a given image. Furthermore, given a fixed number of pixels within the image, task performance can be increased with proper attention to the pixellation process itself, and with optimal post-pixellation processing that is implemented by a combination of digital image processing, electrode design, and the fundamental nature of the electrode/tissue interface.[17,67-74]

Consider, for example, the high resolution acquisition of an image from the natural environment, followed by a block pixellation procedure to match the number of output pixels to the number of electrodes within the microstimulator array. The value of an individual pixel in the final pixellated image can be determined in a number of ways, but perhaps the most straightforward is to simply average the values of all of the pixels in the original (input) image that fall within a given pixel of the output image. The output pixel levels may also be requantized, depending on the number of stimulation levels that can be applied to each electrode within the microstimulator array. This methodology was used to generate the 4×4 arrays of pixels shown in Figs. 10 and 11, based on the original image shown in Fig. 9.

The same procedure was applied to the original image to generate the 16×16 array of pixels shown in Fig. 12. At this degree of pixellation, several coarse features in the original image begin to be recognizable, much more so if the original image field is already well known. Post-pixellation filtering of the 16×16 image with a Gaussian convolution kernel scaled so that its standard deviation is 33% of the pixel dimension yields the surprisingly more recognizable image shown in Fig. 13. The post-pixellation filtering operation eliminates high spatial frequencies that represent false information, as they were not present in the original image.[92] Perhaps even more importantly, this filtering operation eliminates *highly-correlated* high spatial frequencies that form edges and corners, which again were not present in the original image. The presence of local associations of edges and corners will trigger higher cortical processes that tend to identify each block pixel as an individual object, effectively masking recognition of the actual objects and their natural boundaries in the scene.

Initial psychophysical testing of object recognition over a range of objects and scenes indicates not surprisingly that given block pixellation, higher resolutions yield increased object and scene recognition accuracies as well as decreased object and scene recognition times. Likewise, for a given resolution, and a given type of post-pixellation filtering (such as Gaussian convolution), optimal filtering also yields increased recognition accuracies and decreased recognition times. The optimal resolution and filtering for a given task depends strongly on the field of view spanned by the image relative to the field of view spanned by critical object and scene features. Nonetheless, pixellation at approximately 25×25 in conjunction with post-pixellation filtering appears to be sufficient for providing significantly enhanced performance for many household tasks, such as cooking, finding a dropped object, navigation, and mobility.[67-74] Interestingly, this same level of pixellation has been arrived at independently with several different fields of view, tasks, and psychophysical test procedures.[17,67-74,76-78] The original image of Fig. 9, pixellated to form a 25×25 array of pixels, is shown in Fig. 14, and both pixellated and post-pixellation filtered with a 33% Gaussian convolution in Fig. 15. The level of detail apparent in Fig. 15 is remarkable, given the fact that only 625 independent pixel values have been used to generate the image.

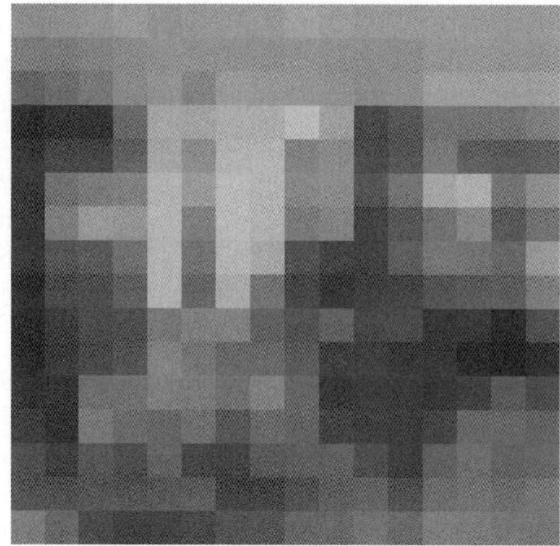

Fig. 12. Pixellated kitchen image, 16 × 16 pixels, grey scale.

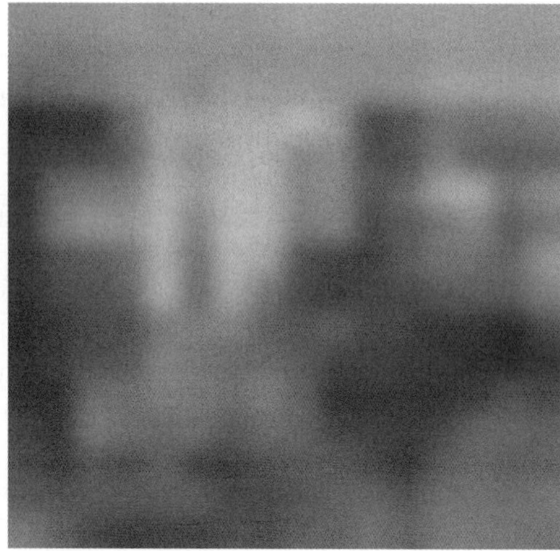

Fig. 13. Pixellated kitchen image, 16 × 16 pixels, grey scale, with 33% Gaussian post-pixellation blur.

In addition to the resolution (number of pixels within the array) and post-pixellation filtering operation, another key factor that impacts the intraocular camera is the fill factor of the image sensor array. Image acquisition with a low fill factor gives rise to undersampling, and hence to undesirable aliasing of the image.

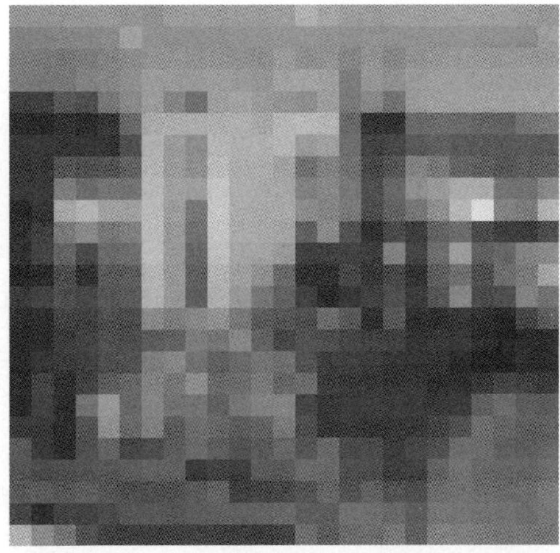

Fig. 14. Pixellated kitchen image, 25 × 25 pixels, grey scale.

Fig. 15. Pixellated kitchen image, 25 × 25 pixels, grey scale, with 33% Gaussian post-pixellation blur.

Currently available image sensor arrays are characterized by relatively low fill factors, ranging from approximately 20% to 50% due to the need for vertical and horizontal interconnection lines (implemented with metal traces) and local signal processing electronics within each geometrical pixel. In many commercially-available image sensor arrays, this effect is mitigated to some degree by the incorporation of either a microlens array or a birefringent anti-aliasing filter, or both. Nonetheless, this is an important factor to consider in the development of custom CMOS image sensor arrays specifically designed to operate in the low pixellation limit.

Fig. 16. Original bus image, 750 × 1140 pixels, grey scale.

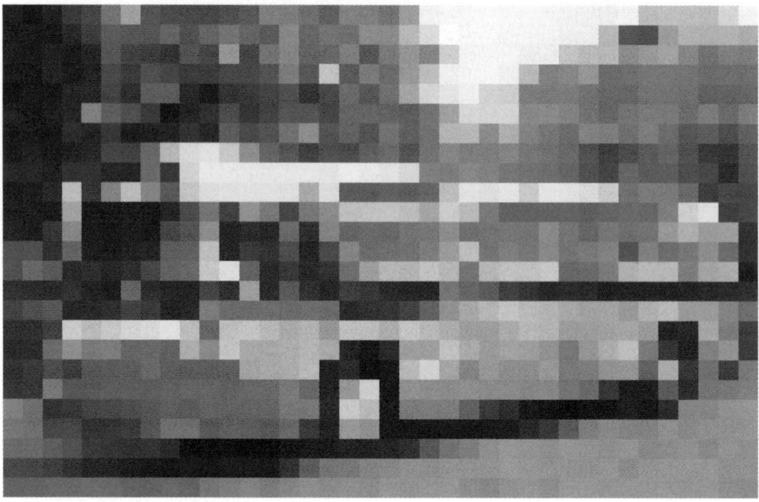

Fig. 17. Pixellated bus image, 25 × 38 pixels, grey scale, low fill factor sensor.

Fig. 18. Pixellated bus image, 25×38 pixels, grey scale, low fill factor sensor, with 40% Gaussian post-pixellation blur.

Fig. 19. Pixellated bus image, 25×38 pixels, grey scale, low fill factor sensor, with 40% Gaussian pre-pixellation blur and 40% Gaussian post-pixellation blur.

Consider, for example, the original image of a bus as shown in Fig. 16. If the original image is pixellated with a very low fill factor, in this case by dividing the 750×1140 pixel original image into 25×38 pixel regions, and filling each pixel region with the brightness value of the central pixel within that region in the original image, the resulting image is shown in Fig. 17. The effects of aliasing in this pixellated version of the image are strikingly evident. As only the central pixel is sampled within each of the 25×38 pixel regions, very small displacements of the camera during the original image acquisition will potentially yield very large changes in the value displayed in each pixel region, particularly as the central pixel crosses boundaries in the image. These effects are quite dramatic in motion video, as will be described in more detail in the next section. Post-pixellation blurring of the pixellated image of Fig. 17 yields the resultant image shown in Fig. 18, in which the effects of aliasing are still quite evident.

In order to minimize the effects of aliasing derived from pixellation with low-fill-factor image sensor arrays, pre-pixellation low pass filtering can be used to reduce the image components at high spatial frequencies. If the original image of Fig. 16 is pre-pixellation filtered with a 40% Gaussian convolution filter, then pixellated, and then post-pixellation filtered with an additional 40% Gaussian convolution filter, the resultant image is given in Fig. 19. This image shows greatly reduced aliasing, and is much closer to being a veridical lower resolution version of the original image, similar to the resultant image that would be obtained with pixellation by a high-fill-factor image sensor array sensor followed by post-pixellation filtering. In effect, high-fill-factor sensors perform a spatial average of the brightness levels within each eventual pixel of the pixellated image, thereby implementing a form of low pass filtering. Pre-pixellation filtering performs a similar function for the case of a low-fill-factor image sensor array.

The effects of grid lines of diminished excitation that may result from interelectrode gaps, as described in relation to Figs. 10 and 11 above, may also be reduced significantly by the addition of appropriate post-pixellation filtering. Consider the image in Fig. 20, which was generated from the original image of Fig. 16 by pixellation with the same methodology as used to generate Fig. 10 (equivalent to pixellation by a 100% fill factor image sensor array), followed by incorporation of grid lines at a 50% duty cycle. The presence of the grid lines greatly interferes with object recognition. If the gridded image in Fig. 20 is post-pixellation filtered with a 40% Gaussian convolution and then adjusted in brightness and contrast, the resultant image is shown in Fig. 21. In this case, a higher standard deviation Gaussian kernel was employed (as compared with the 33% Gaussian convolution used in Figs. 13 and 15) due to the presence of the relatively wide grid lines.

The visual psychophysics effects described above, taken together with physical limitations on the number of electrodes within the microstimulator array, have significant implications for the design of the intraocular camera, as they establish an acceptable resolution limit within the visual field, as well as the desirability of optimal anti-aliasing. Anti-aliasing (pre-pixellation filtering) can be implemented either with a birefringent filter, as mentioned above, or by means of optical blur, which impacts the design constraints on the intraocular camera optical system (described in more detail in Sect. 8).

Post-pixellation filtering to optimize the resultant visual percept, as described above, must be implemented at the interface between the microstimulator array and the retinal tissue, and not through an image processing algorithm; the number of electrodes within the array completely determines the number of degrees of freedom available for the generation of each image frame through the set of individual electrode stimulation levels.

Fig. 20. Pixellated bus image, 25×38 pixels, grey scale, with grid lines at 50% duty cycle.

Fig. 21. Pixellated bus image, 25×38 pixels, grey scale, with grid lines at 50% duty cycle, and with 40% Gaussian post-pixellation blur. Following the application of post-pixellation blur, the image has been adjusted in brightness and contrast to compensate for the low average intensity of the original gridded image.

The evident value of post-pixellation filtering therefore has implications for the optimal design of the electrodes within the microstimulator array (including the electrode shape), the electrode spacing, and the separation of the microelectrodes from the nearest layer of viable retinal cells (retinal ganglion cells, in the case of epiretinal stimulation).

Improved monocular depth perception represents an additional benefit that may accrue to the careful design of the overall intraocular retinal prosthesis, provided that high resolution microstimulator arrays are developed and successfully implanted, and that appropriate pre- and post-pixellation processing is implemented as described above. This effect can easily be seen by comparing Figs. 14 and 15. The block pixellation in Fig. 14 forms a two dimensional array of essentially equal-sized objects, thereby interfering with the object size cue for depth. This cue is restored to a large degree by the post-pixellation filtering process of Fig. 15. The additional monocular depth cues of occlusion, perspective (in which parallel lines converge to yield a vanishing point), lighting/shading (which should correlate to form a conception of a general lighting scheme), and parallax are all aided by appropriate pre- and post-pixellation filtering.

Of course, the effects described above have been primarily tested with simulations on normally sighted individuals, who are not subject to either gross abnormalities of the early vision system nor perturbations of a normal retinotopic map. As more and more empirical evidence emerges from ongoing clinical trials (now with more than thirty subjects implanted worldwide with intraocular retinal prostheses),[19,20,93] the conclusions drawn from human psychophysical testing will continue to be refined and applied to the design of more advanced retinal prostheses.

7. Visual prosthesis simulator

In order to extend the human psychophysical testing described above to allow for the study of the effects of motion on perception in the low pixellation limit, and particularly to allow for the study of navigation and mobility tasks, a real-time visual prosthesis simulator has been developed that allows for subject ambulation.

The visual prosthesis simulator has been designed to allow for the implementation of various simulated blindness conditions, as well as real-time image acquisition with an extraocular camera, image processing operations, eye tracking, the extraction of regions of interest based on the direction of gaze, and the formulation of a composite image stream for real-time video display. The simulator system includes a head-mounted extraocular (scene) camera, a head-mounted eye-tracking camera, dual video capture cards, a dual core PC, a head-mounted display (HMD) with a dedicated video display card, and a fifty-foot cable to allow for extended navigation and mobility experiments.

The system software includes a module for extracting the direction of gaze based on real-time processing of the images acquired by the eye-tracking camera to locate the center of mass of the pupil.[94] A separate software module extracts a region of interest in the image acquired by the scene camera that is centered on the direction of gaze and spans a given field of view. In the eye-tracked mode, this image region (rather than the central image region in the head-tracked mode) is then pixellated and placed in the field of view of the head-mounted organic light emitting diode (OLED) display in the direction of gaze. An additional software module allows for flexible pixellation, pre-pixellation filtering, and post-pixellation filtering of the region of interest, as well as establishment of array size, electrode size, electrode placement, and interelectrode spacings, all controlled through an integrated custom graphical user interface as shown in Fig. 22. The remainder of the HMD field of view is filled in with a background determined by the particular blindness condition being simulated, as in the two examples presented previously for retinitis pigmentosa and age-related macular degeneration in Figs. 5 and 6.

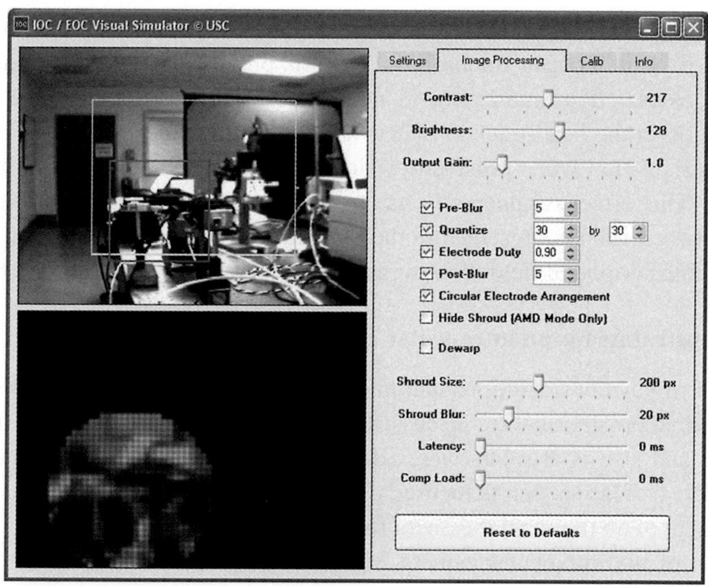

Fig. 22. Key elements of the graphical user interface of the visual prosthesis simulator, showing the full field of view of the scene camera, the field of view of the head-mounted display (light grey box), and the field of view of the region of interest (dark grey box) in the upper left panel; the image presented to the head-mounted display for the case of retinitis pigmentosa and a prosthetic visual field (with grid lines and 33% Gaussian blur) projected in the direction of gaze in the lower left panel; and the image processing control panel on the right.

One immediate conclusion that can be drawn from initial studies with the visual prosthesis simulator is that additional complications accrue to the pixellation of video sequences with motion above and beyond those outlined above for the case of static images. Striking motion artifacts appear in the low resolution limit with block pixellation if either objects move within the field of view, or the entire field of view appears to move as a consequence of head motion. For example, the individual brightness values of all of the pixels within the field of view can change dramatically with motion as the centroid of each pixel passes over various image features. The consequence is the appearance of rapidly fluctuating noise over the portion of the image field that is perceived to be moving, so much so that it distracts significantly from object recognition and tracking. A related observation is that scanning across edges (such as a doorway) produces the appearance of "jumping" borders, in which the edge appears to jump by a full pixel width at a time instead of translating smoothly. Both of these effects can be minimized by the appropriate application of pre-pixellation and post-pixellation filtering operations, even in the case of an image sensor array with a 100% fill factor.

An individual's ability to navigate through a room to locate a specific object, as well as the speed with which the individual is able to move toward a goal, are both significantly enhanced when appropriate pre-pixellation and post-pixellation filtering operations are applied to block-pixellated prosthetic images. Navigation and mobility are also significantly enhanced for operation in the eye-tracked mode over the head-tracked mode, providing additional evidence for the value of foveation in prosthetic vision. This enhancement is quite striking for both the retinitis pigmentosa and age-related macular degeneration cases. It is interesting to note that for the AMD case, the addition of prosthetic vision in the head-tracked rather than the eye-tracked mode appears to lead to increased disorientation when prosthetic vision is added, as compared with no prosthetic vision at all. This effect is apparently due to the discordance in the head-tracked mode that arises between what is presented in the prosthetic field of view and what is observed in the still-viable peripheral field of view, as described with reference to Fig. 6.

8. Design constraints for an intraocular camera

A number of key considerations that impact the provision of image acquisition for an intraocular retinal prosthesis were discussed previously in Sects. 1, 4, and 5. In particular, both surgical and size constraints were addressed briefly for the case of an intraocular camera in Sect. 4. This section is focused on the additional design constraints that impact the development of an intraocular camera for retinal prostheses.

Placement of the intraocular camera within the crystalline lens sac, as illustrated in Figs. 7 and 8, has the advantage of a natural support system for the camera that is well stabilized within the human eye. The sagittal (axial) thickness of the accommodated (fully relaxed) crystalline lens ranges from approximately 4.5 to 5 mm in adults aged 50 to 100 years old,[95] setting an upper bound on the length of the intraocular camera housing. Assuming a 5 mm scleral incision, a target for the circumference of the intraocular camera housing can be set at approximately 10 mm, assuming folding haptics and lengthwise insertion. This constraint corresponds to a housing diameter of 3.18 mm for a cylindrical housing, and a 2.5×2.5 mm cross section for a rectangular housing. The mass of the crystalline lens ranges from 200 to 300 mg,[95] thereby setting an upper bound for the intraocular camera mass.

A schematic diagram of the fourth generation prototype of the intraocular camera is presented in Fig. 23, including a hermetically-sealed camera housing, a fused silica front (anterior) window, a single element aspherical lens, a mounting spacer (that might be formed from a desiccant material to absorb moisture), a CMOS image sensor array, and a mounting adapter for the image sensor array and associated electronics. The mounting adapter will likely support several subelements in the form of a multichip module that contains one or more application-specific integrated circuits for communications and control, as well as any external components necessary to provide power conditioning and operate the CMOS image sensor array. In order to maintain the temperature at all points on the exterior of the housing within acceptable international limits for chronic implantation, the total power dissipation of all internal components should be kept at or below approximately 25 mW. Placement of the electronic components in the posterior of the intraocular camera housing creates an anisotropic thermal environment due to nonuniform heat dissipation, placing additional constraints on the overall power budget.

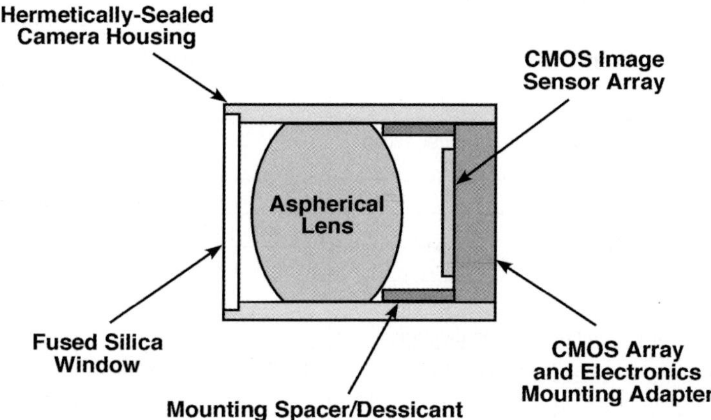

Fig. 23. Schematic diagram of the intraocular camera, fourth generation prototype.

From the results of the visual psychophysics investigations described in Sect. 6, a target resolution for the optical imaging system of 32×32 was chosen to provide for up to 1024 individual electrodes within the microstimulator array. This constraint affects the minimum thickness of the aspherical lens that must provide this resolution with only two refracting surfaces. A single element aspherical lens fabricated from a lightweight polymer material fills a large fraction of the intraocular camera housing, as shown in Fig. 23, not easily allowing for a second lens element that is traditionally employed to help reduce chromatic and other aberrations. The back focal length of the lens must be extremely short, with the image sensor array almost in proximity with the rear surface of the aspherical lens. Once the lens is designed and the effective focal length (including corneal refraction) is known, the size of the image sensor array is then determined by the desired field of view of the intraocular camera. For example, if the effective focal length is 2.5 mm, with a back focal length of 1.2 mm, and the angular field of view of the microstimulator array attached to the central retina is chosen to be $\pm 10°$, then the size of the image sensor array must be at least 0.88×0.88 mm. The minimum pixel pitch of the image sensor array for the given desired resolution is then 28 μm. This must be self-consistently determined in conjunction with the spot size of the given lens design to assure adequate overall system performance, as discussed further in Sect. 9.

Table 1. Properties of the human eye and retina.

Property	Dimension(s)	Reference(s)
Diameter of the central retina	6 mm	From Ref. 96
Angular field of view of the central retina (central vision)	20.1°; ±10°	After Ref. 97
Diameter of the macula	3 mm	From Ref. 97
Angular field of view of the macula	10°; ±5°	After Ref. 97
Diameter of the fovea	1.55 mm	After Ref. 97
Angular field of view of the fovea	5.2°	From Ref. 97
Angular field of view per unit distance on the retina	3.35°/mm	After Refs. 97, 63
Density of cone photoreceptors in the central fovea	100,000 to 300,000/mm^2	From Ref. 97
Diameter of cone photoreceptors in the fovea	1 to 4 µm	From Ref. 97
Visual acuity of human vision	1 arc min (0.016°)	After Ref. 97
Dioptric power of the cornea (diopters)	43 D	From Refs. 97, 98
Dioptric power of the crystalline lens (diopters; in air)	19 to 20 D	From Refs. 98, 97
Distance from anterior surface of the cornea to the retina	24.3 mm	From Ref. 98
Focal length of the human eye, measured from the posterior of the crystalline lens to the retina	17.1 mm	From Refs. 98, 97
Index of refraction of the cornea	1.376	From Refs. 97, 98
Index of refraction of the aqueous humor	1.336	From Ref. 98
Sagittal thickness of the crystalline lens, accommodated	4.5 to 5 mm	From Ref. 95
Mass of the crystalline lens	200 to 300 mg	From Ref. 95
Visual acuity (legal blindness) in the best (corrected) eye	> 20/200 (> 20/400)	From Refs. 99, 100
Field of view (legal blindness)	< ±10° (< ±5°)	From Refs. 99, 100

The key properties of the human eye and retina that impact the design of the intraocular camera are summarized in Table 1. Of particular note are the visual acuity and field of view limits for legal blindness in the United States and United Kingdom, listed first, as well as for the international community as defined by the UN World Health Organization.[99,100] Within the U.S. and UK, legal blindness is defined as a visual acuity of greater that 20/200 in the best (corrected) eye, or a field of view of less than ±10°.

The design of the intraocular camera is highly unusual, in that the optical system must be designed to meet these performance specifications in conjunction with the biological corneal lens and aqueous humor. As a consequence, the entire optical system is effectively a biological/inorganic hybrid. The design of a viable optical system for the intraocular camera, given these constraints, is described in detail in the next section.

9. Intraocular camera optical system design

As a biological/inorganic hybrid, the intraocular camera lens design must be undertaken in conjunction with an appropriate model of the biological cornea and aqueous humor, as shown schematically in Fig. 24.[67-74] The placement of the intraocular camera in the crystalline lens sac locates the fused silica window approximately 2 mm posterior to the innermost surface of the cornea, near the posterior surface of the iris. In order to achieve an overall housing length of 4.5 mm and leave space for the CMOS image sensor array, associated components, a layer of application specific integrated circuits, and the rear housing wall, an optical system design length of approximately 3.5 mm was chosen.

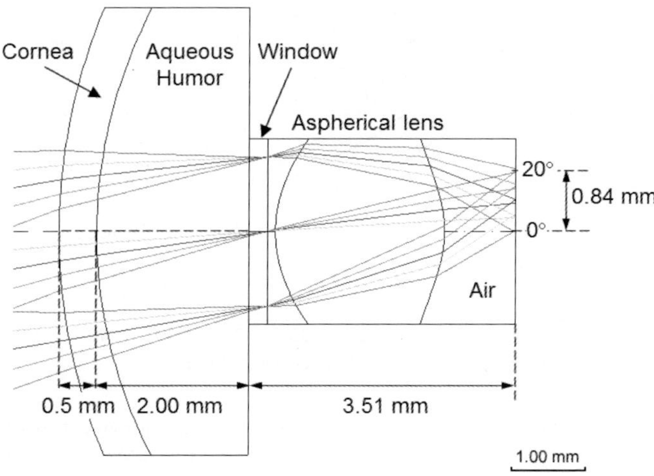

Fig. 24. Schematic diagram of the custom polymer (aspherical) lens within the intraocular camera housing, as implanted within the human eye. The rays shown are traced to and from the center and edges of a 2.0 mm aperture placed on the posterior side of the fused silica window, with entry angles of 0°, 5°, 10°, 15°, and 20°.

For the third generation intraocular camera prototype, a commercially available aspherical lens fabricated from a high index glass (Ohara PBH71, $n = 1.93306$ at 546.1 nm) with a 3.1 mm focal length was employed, as although it was designed for laser-fiber coupling in the infrared, its surface figure was very close to optimal for imaging in the visible at infinite conjugates. However, this particular optical glass has a very high density, and hence the mass of the resulting lens alone was approximately 336 mg with its external mounting flange, and approximately 100 mg without.

In order to reduce the lens mass while maintaining excellent imaging characteristics, several custom aspherical lenses were designed based on a very lightweight polymer material (Zeonex® E48R, $n = 1.5334$ at 546.1 nm). The optical properties of the biological cornea and aqueous humor were modeled by means of the Liou-Brennan eye model,[101] which includes the anatomical nonuniform thickness of the corneal lens. A fused silica optical window was added to the anterior of the housing, just before the aspherical lens, in order to provide a higher index mismatch at the first lens surface, thereby reducing the lens curvature and allowing for improved aberration control.

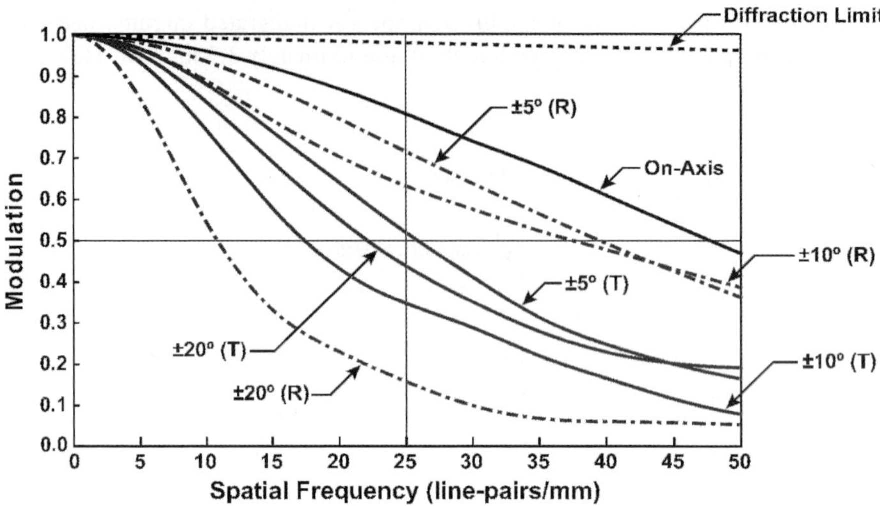

Fig. 25. Modulation transfer functions (MTFs) for the intraocular camera custom polymer lens, showing both tangential and radial rays over a 40° (±20°) field of view, with angle of incidence (field angle) as a parameter. The diffraction limit for the assumed aperture diameter (2.0 mm) is shown at the top of the graph.

In one such implementation, the lens was designed to have excellent imaging characteristics over a 20° (±10°) central field of view, with adequate imaging characteristics over a 40° (±20°) field of view. The lens design was implemented with the intraocular camera focused at a 20 cm object distance, to maximize the depth of field. The modulation transfer functions (MTFs) of this lens at a variety of field angles are shown in Fig. 25, along with the diffraction limit for the assumed aperture diameter of 2.0 mm. The MTFs over the central ±10° field of view are for the most part greater than 0.5 out to a spatial frequency of 25 line pairs/mm. The corresponding polychromatic spot diagrams at several field angles are shown in Fig. 26. On axis, the RMS spot size is about 12 μm, and the spot size is less than 30 μm over the ±10° central field of view. In fact, the RMS spot size for this lens is only 42 μm at ±20°. In both the MTF and polychromatic spot diagrams, three wavelengths spanning the visible spectrum were used, weighted (with weights w) to emphasize the central wavelength ($\lambda = 608.9$ nm, $w = 1$; $\lambda = 559.0$ nm, $w = 2$; $\lambda = 513.9$ nm, $w = 1$). A simulation of a USAF resolution target imaged through the custom polymer lens, constructed by a Monte-Carlo ray casting technique, is shown in Fig. 27. The resolution of Group 1, Element 3 is 26 line pairs/mm at the image plane, further confirming the essential features of the optical design.

Fig. 26. Spot diagrams for the custom polymer lens at several field angles over a 40° (±20°) field of view. The scale bar at the left side of the figure is 0.100 mm (100 µm) in length.

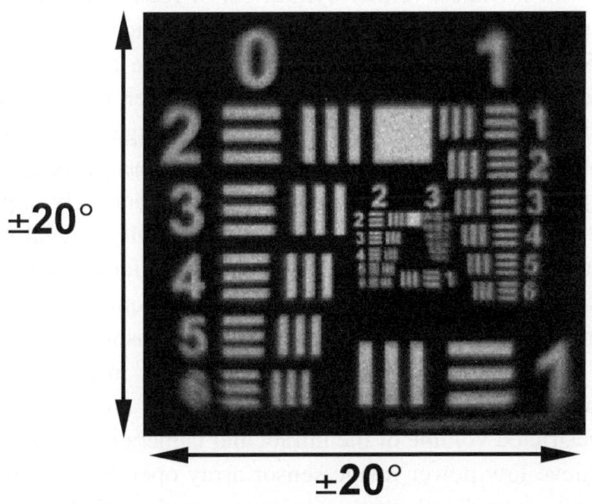

Fig. 27. Monte-Carlo ray-casting simulation of a USAF resolution target imaged through the custom polymer lens, with an image field corresponding to a 40° (±20°) field of view.

The aspherical polymer lens shown schematically in Fig. 24 was fabricated using single-point diamond turning, and antireflection-coated to provide < 1% reflectivity at normal incidence across the visible spectrum. The resulting optical system design using this lens provides for a resolution of 32 × 32 (for a ±10° field of view), and operates at a relatively fast *f*-number (*f*/0.96) with an effective focal length of 2.1 mm and a back focal length of 0.94 mm. The axial thickness of the lens is 2.22 mm, and the full ±20° field of view spans a 1.68 × 1.68 mm region of the focal plane. If the edge diameter of the lens is limited to 2.5 mm, the mass of the lens is only 9 mg. In combination with the mass of the fused silica window (3 mg), the overall mass of the IOC optical system is only 12 mg.

A key feature of this optical system design is that it exhibits an unusually wide depth of field, spanning from a few centimeters to infinity depending on where the camera is focused.[67-74] As such, it does not appear necessary to incorporate an accommodation mechanism, similar to the compression and relaxation of the crystalline lens within the human eye, that adjusts to allow objects to be in focus at either far or near distances. As the apparent magnification of an object increases with proximity to either the eye or a camera, this expanded depth of field would allow a subject implanted with the intraocular camera to bring objects very close to the eye, thereby providing natural magnification to aid in object recognition.

10. Image sensor arrays for the intraocular camera

The key constraints on the intraocular camera image sensor array are size, power dissipation, resolution, output signal format, and number of required ancillary components. The physical size of the image sensor array is limited by the interior dimensions of the intraocular camera housing. For a range of choices of housing configuration and wall thickness (either cylindrical or rectangular, and either 125 μm or 250 μm), the maximum size of the image sensor array ranges from 1.90 mm × 1.90 mm to 2.25 mm × 2.25 mm. The power dissipation should be as low as possible, perhaps below 15 mW given a total power budget of 25 mW. Given the resolution limitations of the intraocular retinal prosthesis described above, the image sensor array is required to have only moderate resolution, with an array size of at least 25 × 25 at a pixel pitch of approximately 30 μm. Smaller pixel pitches with correspondingly larger array sizes can be accommodated as well, although at a cost in sensitivity, signal-to-noise ratio, and pixellation overhead. A digital output signal format is preferable, provided that onboard analog-to-digital conversion can be performed at the required bandwidth and at low power. Most current small-form-factor CMOS image sensor arrays, however, generate NTSC compatible (analog) outputs at a considerable cost in power consumption. The number, sizes, and masses of any ancillary components are strictly limited, as space within the highly restricted volume of the intraocular camera is at a premium.

In order to achieve low power image sensor array operation, as well as to allow for integration of on-chip control and communications circuitry, CMOS image sensor arrays [particularly active pixel sensor (APS) arrays] are preferred over charge-coupled devices (CCDs). In the third generation intraocular camera prototype, an OmniVision OV5116N CMOS image sensor array was integrated with a commercial glass aspherical lens to form an intraocular camera with a reduced set of only five external passive components (required for proper image sensor array operation) and a three-wire interface. The OV5116N image sensor has an array size of 320 × 240 pixels in an image area of 3.2 mm × 2.4 mm, integrated in a chip scale package. In this reduced configuration, the total power consumption of the image sensor array was 36.6 mW. Experiments to characterize the custom polymer lens have been conducted with the OmniVision OV6920

CMOS image sensor array, which also has an array size of 320×240 pixels in an image area of $820\ \mu m \times 625\ \mu m$. The OmniVision OV6920 image sensor is integrated in a chip scale package that is only $2.3\ mm \times 2.1\ mm$ in size. Currently, we are investigating the operational characteristics of the recently released OmniVision OV6930 CMOS image sensor array, which was specifically designed for biomedical applications. The OV6930 has an array size of 400×400 pixels in an image area of $1224\ \mu m \times 1212\ \mu m$, integrated in a chip scale package that is only $1.815\ mm \times 1.815\ mm$ in size.

Although all of these image sensor arrays currently operate at power levels that are approximately twice as high as desired for this application, a custom ultra-low power APS image sensor array has already been successfully demonstrated.[102] This image sensor array has an array size of 176×144, and dissipates only $550\ \mu W$ at $1.5\ V$. In addition, the chip size is only $2\ mm \times 2\ mm$.

The current sizes of individual electrodes within the various implementations of microstimulator arrays that have been implanted to date (in the range of $100\ \mu m$ to $500\ \mu m$ in diameter or edge dimension) are quite large relative to the sizes of individual retinal ganglion cells ($5\ \mu m$ to $20\ \mu m$ in diameter).[103] With electric field and current broadening, the cell stimulation region beneath each electrode is even larger. As a consequence, each electrode is capable of stimulating a significant population of retinal ganglion cells that code for both color opponency and luminance, obviating the possibility of differentiating among individual retinal ganglion cells and thereby activating color-encoded vision at this point. Currently employed stimulation paradigms are therefore limited to grey-scale encoding. In the lowest power (and lowest bandwidth) implementations, monochrome image sensor arrays will be employed. Nonetheless, color output from the image sensor array can be quite valuable, particularly if either internal or external image processing algorithms are used for object tracking and identification, with cues provided to the subject, as these algorithms often use color to advantage.

For operation in both indoor and outdoor environments, a wide dynamic range is highly desirable. CMOS image sensor arrays with $140\ dB$ to $200\ dB$ of dynamic range have already been demonstrated.[23,104] In addition, it is of considerable interest to investigate the incorporation of biomimetic image sensor arrays that incorporate one or more features of early vision,[105] as these features will not likely be accessible through retinal stimulation.

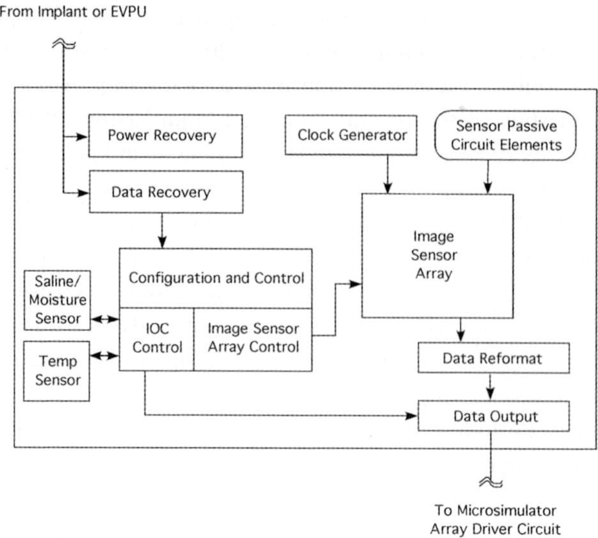

Fig. 28. Intraocular camera block diagram, including the image sensor array as well as power and data recovery, communication and control, and environmental sensing circuitry.

Within the housing of the intraocular camera, provision must be made for image sensor array control, as well as control of all of the intraocular camera functions. A block diagram of the intraocular camera electronics is given in Fig. 28, showing the input from either an implanted package or an EVPU, the image sensor array, and power and data recovery, communication and control, and environmental sensing circuitry. Also shown is the output of the intraocular camera, which can be routed directly to the microstimulator array driver circuitry, or indirectly through a wireless communication channel to the EVPU.

11. Integration and packaging of the intraocular camera

As with any chronically implanted biomedical device, the external materials of the intraocular camera, including the housing, optical window, and haptic support elements, must be fully biocompatible. In addition, the intraocular camera housing must be hermetically sealed, not only as a constituent element of overall biocompatibility, but also as a means of protection of sensitive optical elements and electronic components from the corrosive (essentially saline) environment of the human eye. Finally, it is highly desirable to design the entire intraocular camera in such a manner that it is magnetic resonance imaging (MRI) compatible, to allow for normal post-implantation diagnostic tests.

Currently investigated housing designs, with one example shown schematically in Fig. 23, include a fused silica window sealed to the anterior end of a thin-walled (125 μm to 250 μm) titanium cylinder that is either enclosed on the posterior end or sealed with a thin titanium plate. The custom polymer lens is held in place by one or more mounting spacers that may be fabricated from a desiccant material. The image sensor array will most likely be mounted in a vertically stacked multichip module (MCM) attached to a mounting adapter that terminates with either the enclosed end of the housing or the sealed titanium plate. Within the multichip module, the image sensor array as well as one or more CMOS control and communications chips will be connected to a ceramic MCM submount by flip chip bonding techniques.[106] Haptic stabilization will be provided by several poly(methyl methacrylate) (PMMA) haptic arms, similar to those used in intraocular lenses (IOLs). Mass equivalent housings (counterbalanced for the appropriate distribution of component masses) have already been integrated with haptic arms and successfully tested in saline solution under both sinusoidal and shock loading conditions at frequencies characteristic of daily human activities.[107]

In this configuration, the total mass of the intraocular camera will be in the range of 50 to 75 mg. The natural positive (upward) buoyancy of the intraocular camera housing is estimated to be approximately 28.1 mg for the rectangular housing, and 35.7 mg for the cylindrical housing, so that the net mass of the intraocular camera will range between 19.5 mg and 46.8 mg. By way of comparison, typical intraocular lenses range in mass from approximately 10 mg to 40 mg.

Power for the intraocular camera will be provided either by a wired connection to a microelectronics package attached to the side of the eye, as in the current version of the intraocular retinal prosthesis, or by inductive wireless coupling. In order to minimize electromigration effects, wired power must be transmitted within the body in AC form; as a consequence, an efficient AC to DC converter must be incorporated within the IOC housing itself irrespective of whether wired or wireless power is provided. Control data input by an ophthalmologist or transmitted from the EVPU can be combined with the power feed and then split off internally within the intraocular camera. The video signal from the intraocular camera can be similarly routed to the microelectronics package or the EVPU by either wired or inductive coupling, or even through an optical communications link established between the intraocular camera and a set of detectors mounted on the rim of a pair of eyeglasses.

In the third generation intraocular camera prototype, an image sensor array was integrated with a commercial glass aspherical lens within a carbon fiber cylinder to form an intraocular camera 7 mm in diameter and 5 mm long with a total power consumption of 36.6 mW. The power consumption in this case is due to the image sensor array and ancillary components only, without power conditioning, control, and communications circuitry.

In the fourth generation intraocular camera prototype currently in development, a much smaller image sensor array will be integrated with a fused silica window and a custom polymer lens in a cylindrical titanium housing that is 3.18 mm in diameter and 4.5 mm long, enabling critical imaging experiments at implantable camera dimensions.

12. Summary and conclusions

Intraocular retinal prostheses have shown considerable promise for restoring useful sight to those blinded by photoreceptor-degenerative diseases such as retinitis pigmentosa and age-related macular degeneration. In many current implementations of intraocular retinal prostheses, an extraocular camera is used to acquire images from the environment, requiring head motion to establish the direction of gaze and to track moving targets. Restoration of foveation with a more natural coupling of head and eye movements is possible through the use of either an eye-tracked wide-field-of-view extraocular camera, or a fully implantable intraocular camera. Human visual psychophysics experiments indicate not only the value added by the provision of foveation, but also the significant visual capabilities that can be achieved even in the low pixellation limit. Furthermore, these experiments demonstrate the importance of both pre- and post-pixellation filtering of pixellated images, with implications for the design of the intraocular camera and associated microstimulator array. The optical system of the intraocular camera can be implemented with a single refractive lens, providing for microstimulator arrays with up to 32×32 individual electrodes. The intraocular camera as currently configured is designed for chronic implantation in the crystalline lens sac by means of a surgical procedure similar to that used for cataract replacement with an intraocular lens.

13. Future research directions

Although the development of the intraocular camera is already reaching an advanced state, many research questions and challenges remain to be addressed. These future research directions include the integration and testing of the fourth generation prototype, as well as the potential incorporation of either multiple refractive elements or hybrid refractive/diffractive elements [such as diffractive optical elements (DOEs) or stratified volume diffractive optical elements (SVDOEs[108-110])] in the optical system design.[67-74]

Additional future research directions include the development of ultra-low-power wide dynamic range CMOS image sensor arrays and associated ultra-low-power analog signal chain components (such as analog-to-digital converters). The incorporation of early vision features such as lateral brightness adaptation (simultaneous contrast) and lateral chromatic adaptation[111-113] will be of considerable interest for enhancing object and scene recognizability once the microstimulator array size is increased toward the current goal of 32×32 arrays. Finally, the goal of incorporating color stimulation in an intraocular retinal prosthesis provides a worthy challenge for future research.

Acknowledgments

One of us (ART) owes an immense debt of gratitude to Prof. Richard K. Chang for his inspiration as a scientist, teacher, colleague, and friend, and for his unparalleled ability to convey novel concepts with clarity and elegance in scientific communication. We are deeply appreciative of many contributions to this effort from our colleagues, including Dr. Leon Esterowitz, Dr. Gianluca Lazzi, Dr. Gerald J. Chader, Dr. Bosco Tjan, Dr. Rajat Agrawal, Dr. Hossein Ameri, Dr. Ricardo Freda, Dr. Dilek Güven, Dr. Jaw-Chyng (Lormen) Lue, Dr. Ashish Ahuja, Satsuki Takahashi, Nutthamon Suwanmonkha, Vinit Singh, Pamela Lee, Christine Tanguay, and Morgan Leighton. We are also pleased to acknowledge Prof. David H. Hubel for permission to use two key figures from his elegant

Eye, Brain, and Vision.[63] The authors are very grateful for support from the National Science Foundation (NSF) Engineering Research Center on Biomimetic MicroElectronic Systems (EEC-0310723), from the National Science Foundation Research to Aid Persons with Disabilities Program (RAPD-0201927), and from the National Science Foundation Biophotonics Program through a Small Grant for Exploratory Research (SGER-0638469). We are also pleased to acknowledge support from the Rose Hills Foundation through several Science and Engineering Research Fellowships, and from the University of Southern California through a Provost's Undergraduate Research Fellowship and a Discovery Scholar Prize (all awarded to Noelle R. B. Stiles).

References

1. H. G. Wood-Gush, "Retinitis pigmentosa research: a review", *Journal of the Royal Society of Medicine*, **82**, 355-358, 1989.
2. G. J. Farrar, P. F. Kenna, and P. Humphries, "On the genetics of retinitis pigmentosa and on mutation-independent approaches to therapeutic intervention", *The European Molecular Biology Organization Journal*, **21**(5), 857-864, 2002.
3. R. D. Jager, W. F. Mieler, and J. W. Miller, "Age-Related Macular Degeneration", *The New England Journal of Medicine*, **358**(24), 2606-2617, 2008.
4. D. Pauleikhoff, "Neovascular Age-Related Macular Degeneration: Natural History and Treatment Outcomes", *Retina*, **25**(8), 1065-1084, 2005.
5. C. H. Bunker, E. L. Berson, W. C. Bromley, R. P. Hayes, and T. H. Roderick, "Prevalence of retinitis pigmentosa in Maine", *American Journal of Ophthalmology*, **97**(3), 357-365, 1984.
6. D. T. Hartong, E. L. Berson, and T. P. Dryja, "Retinitis Pigmentosa", *The Lancet*, **368**(9549), 1795-1809, 2006.
7. D. S. Friedman, B. J. O'Colmain, B. Muñoz, S. C. Tomany, C. McCarty, P. T. V. M. de Jong, B. Nemesure, P. Mitchell, J. Kempen, and N. Congdon, (The Eye Diseases Prevalence Research Group), "Prevalence of Age-Related Macular Degeneration in the United States", *Archives of Ophthalmology*, **122**, 564-572, 2004.
8. A. Chopdar, U. Chakravarthy, and D. Verma, "Age related macular degeneration", *British Medical Journal (BMJ)*, **326**, 485-488, 2003.
9. J. F. Wilson, "Gene Therapy, Stem Cells: Prime for Vision Restoration", *The Scientist*, **15**(18), 17-19, September 17, 2001.
10. R. B. Aramant and M. J. Seiler, "Progress in retinal sheet transplantation", *Progress in Retinal and Eye Research*, **23**, 475-494, 2004.
11. S. J. Van Hoffelen, M. J. Young, M. A. Shatos, and D. S. Sakaguchi, "Incorporation of Murine Brain Progenitor Cells into the Developing Mammalian Retina", *Investigative Ophthalmology & Visual Science*, **44**(1), 426-434, 2003.
12. R. E. MacLaren, R. A. Pearson, A. MacNeil, R. H. Douglas, T. E. Salt, M. Akimoto, A. Swaroop, J. C. Sowden, and R. R. Ali, "Retinal repair by transplantation of photoreceptor precursors", *Nature*, **444**(7116), 203-207, 2006.
13. M. S. Humayun, J. D. Weiland, G. Chader, and E. Greenbaum, Eds., *Artificial Sight: Basic Research, Biomedical Engineering, and Clinical Advances*, (Springer, New York, 2007).
14. D. Zhou and E. Greenbaum, Eds., *Implantable Neural Devices 1: Devices and Applications*, (Springer, New York, 2009), and *Implantable Neural Devices 2: Techniques and Engineering Approaches*, (Springer, New York, 2010).
15. M. S. Humayun, E. de Juan, Jr., G. Dagnelie, R. J. Greenberg, R. H. Propst, and D. H. Phillips, "Visual perception elicited by electrical stimulation of retina in blind humans", *Archives of Ophthalmology*, **114**(1), 40-46, 1996.

16. M. S. Humayun, E. de Juan, Jr., J. D. Weiland, G. Dagnelie, S. Katona, R. Greenberg, and S. Suzuki, "Pattern electrical stimulation of the human retina", *Vision Research*, **39**, 2569-2576, 1999.
17. M. S. Humayun, "Intraocular Retinal Prosthesis", *Transactions of the American Ophthalmological Society*, **99**, 271-300, 2001.
18. E. Margalit, M. Maia, J. D. Weiland, R. J. Greenberg, G. Y. Fujii, G. Torres, D. V. Piyathaisere, T. M. O'Hearn, W. Liu, G. Lazzi, G. Dagnelie, D. A. Scribner, E. de Juan, Jr., and M. S. Humayun, "Retinal Prosthesis for the Blind", *Survey of Ophthalmology*, **47**(4), 335-356, 2002.
19. M. Javaheri, D. S. Hahn, R. R. Lakhanpal, J. D. Weiland, and M. S. Humayun, "Retinal Prostheses for the Blind", *Annals, Academy of Medicine, Singapore*, **35**(3), 137-144, 2006.
20. J. D. Weiland and M. S. Humayun, "Visual Prosthesis", *Proceedings of the IEEE*, **96**(7), 1076-1084, 2008.
21. A. P. Rowley, J. J. Whalen, III, J. D. Weiland, A. R. Tanguay, Jr., and M. S. Humayun, "The Development of a Retinal Prosthesis – A Significant Biomaterials Challenge", Chapter II.5.10e in *Biomaterials Science: An Introduction to Materials in Medicine, 4th Edition*, B. D. Ratner, A. S. Hoffman, F. J. Schoen, and J. E. Lemons, Eds., (Academic Press, New York, 2010); (in press).
22. M. Schwarz, B. J. Hosticka, R. Hauschild, W. Mokwa, M. Scholles, and H. K. Trieu, "Hardware Architecture of a Neural Net Based Retina Implant for Patients Suffering from Retinitis Pigmentosa", *Proceedings of the International Conference on Neural Networks*, **2**, 653-658, 1996.
23. M. Schwarz, R. Hauschild, B. J. Hosticka, J. Huppertz, T. Kneip, S. Kolnsberg, W. Mokwa, and H. K. Trieu, "Single Chip CMOS Image Sensors for a Retina Implant System", *Proceedings of the 1998 IEEE International Symposium on Circuits and Systems*, (ISCAS '98), **6**, 645-648, 1998.
24. J. Wyatt and J. Rizzo, "Ocular implants for the blind", *IEEE Spectrum*, **33**(5), 47–53, 1996.
25. J. F. Rizzo, III, J. Wyatt, J. Loewenstein, S. Kelly, and D. Shire, "Methods and Perceptual Thresholds for Short-Term Electrical Stimulation of Human Retina with Microelectrode Arrays", *Investigative Ophthalmology & Visual Science*, **44**(12), 5355-5361, 2003.
26. J. F. Rizzo, III, J. Wyatt, J. Loewenstein, S. Kelly, and D. Shire, "Perceptual Efficacy of Electrical Stimulation of Human Retina with a Microelectrode Array during Short-Term Surgical Trials", *Investigative Ophthalmology & Visual Science*, **44**(12), 5362-5369, 2003.
27. R. Eckmiller, M. Becker, and R. Hünermann, "Dialog Concepts for Learning Retina Encoders", *Proceedings of the IEEE International Conference on Neural Networks*, **4**, 2315-2320, 1997.
28. G. Richard, R. Hornig, M. Keserü, and M. Feucht, "Chronic Epiretinal Chip Implant in Blind Patients with Retinitis Pigmentosa: Long-Term Clinical Results", *Investigative Ophthalmology & Visual Science*, **48**(5), ARVO Abstract 666, Paper B290, 2007.
29. M. Keserü, N. Post, R. Hornig, O. Zeitz, and G. Richard, "Long-Term Tolerability of the First Wireless Implant for Electrical Epiretinal Stimulation", *Investigative Ophthalmology & Visual Science*, **50**(5), ARVO Abstract 4226, Paper D759, 2009.
30. E. M. Maynard, "Visual Prostheses", *Annual Review of Biomedical Engineering*, **3**, 145-68, 2001.
31. A. Y. Chow, V. Y. Chow, M. T. Pardue, G. A. Peyman, C. Liang, J. I. Pearlman, and N. S. Peachey, "The Semiconductor-Based Microphotodiode Array Artificial Silicon Retina", *Proceedings of the IEEE International Conference on Systems, Man, and Cybernetics, 1999*, **4**, 404-408, 1999.
32. A. Y. Chow, V. Y. Chow, K. H. Packo, J. S. Pollack, G. A. Peyman, and R. Schuchard, "The Artificial Silicon Retina Microchip for the Treatment of Vision Loss from Retinitis Pigmentosa", *Archives of Ophthalmology*, **122**(4), 460-469, 2004.

33. F. Gekeler, H. Schwahn, A. Stett, K. Kohler, and E. Zrenner, "Subretinal microphotodiodes to replace photoreceptor-function: A review of the current state", *Les Seminaires Ophtalmologiques d'IPSEN*, **12**, 77-95, 2001.

34. R. Eckhorn, M. Wilms, T. Schanze, M. Eger, L. Hesse, U. T. Eysel, Z. F. Kisvárday, E. Zrenner, F. Gekeler, H. Schwahn, K. Shinoda, H. Sachs, and P. Walter, "Visual resolution with retinal implants estimated from recordings in cat visual cortex", *Vision Research*, **46**, 2675-2690, 2006.

35. F. Gekeler, P. Szurman, S. Grisanti, U. Weiler, R. Claus, T.-O. Greiner, M. Völker, K. Kohler, E. Zrenner, and K. U. Bartz-Schmidt, "Compound subretinal prostheses with extra-ocular parts designed for human trials: Successful long-term implantation in pigs", *Graefe's Archive for Clinical and Experimental Ophthalmology*, **245**, 230-241, 2006.

36. D. B. Shire, S. K. Kelly, J. Chen, P. Doyle, M. D. Gingerich, S. F. Cogan, W. A. Drohan, O. Mendoza, L. Theogarajan, J. L. Wyatt, and J. F. Rizzo, "Development and Implantation of a Minimally Invasive Wireless Subretinal Neurostimulator", *IEEE Transactions on Biomedical Engineering*, **56**(10), 2502-2511, 2009; see also "Artificial Retina", *Technology Review*, 82-86, September, 2004.

37. D. Palanker, A. Vankov, P. Huie, and S. Baccus, "Design of a high-resolution optoelectronic retinal prosthesis", *Journal of Neural Engineering*, **2**, S105-S120, 2005.

38. J. D. Loudin, D. M. Simanovskii, K. Vijayraghavan, C. K. Sramek, A. F. Butterwick, P. Huie, G. Y. McLean, and D. V. Palanker, "Optoelectronic retinal prosthesis: system design and performance", *Journal of Neural Engineering*, **4**, S72-S84, 2007.

39. G. Roessler, T. Laube, C. Brockmann, T. Kirschkamp, B. Mazinani, M. Goertz, C. Koch, I. Krisch, B. Sellhaus, H. K. Trieu, J. Weis, N. Bornfeld, H. Röthgen, A. Messner, W. Mokwa, and P. Walter, "Implantation and Explantation of a Wireless Epiretinal Retina Implant Device: Observations during the EPIRET3 Prospective Clinical Trial", *Investigative Ophthalmology & Visual Science*, **50**(6), 3003-3008, 2009.

40. J. Chen, H. A. Shaw, C. Herbert, J. I. Loewenstein, and J. F. Rizzo, III, "Extraction of a Chronically Implanted, Microfabricated, Subretinal Microelectrode Array", *Ophthalmic Research*, **42**, 128-137, 2009.

41. H. Gerding, "A new approach towards a minimal invasive retina implant", *Journal of Neural Engineering*, **4**, S30-S37, 2007.

42. Y. T. Wong, S. C. Chen, J. M. Seo, J. W. Morely, N. H. Lovell, and G. J. Suaning, "Focal activation of the feline retina via a suprachoroidal electrode array", *Vision Research*, **49**, 825-833, 2009.

43. H. Kanda, T. Morimoto, T. Fujikado, Y. Tano, Y. Fukuda, and H. Sawai, "Electrophysiological Studies of the Feasibility of Suprachoroidal-Transretinal Stimulation for Artificial Vision in Normal and RCS Rats", *Investigative Ophthalmology & Visual Science*, **45**(2), 560-566, 2004.

44. J. Ohta, T. Tokuda, K. Kagawa, T. Furumiya, A. Uehara, Y. Terasawa, M. Ozawa, T. Fujikado, and Y. Tano, "Silicon LSI-Based Smart Stimulators for Retinal Prosthesis", *IEEE Engineering in Medicine and Biology Magazine*, **25**(5), 45-59, 2006.

45. Y. Terasawa, H. Tashiro, A. Uehara, T. Saitoh, M. Ozawa, T. Tokuda, and J. Ohta, "The development of a multichannel electrode array for retinal prostheses", *Journal of Artificial Organs*, **9**, 263-266, 2006.

46. W. H. Dobelle, M. G. Mladejovsky, and J. P. Girvin, "Artificial vision for the blind: electrical stimulation of visual cortex offers hope for a functional prosthesis", *Science*, **183**(123), 440-444, 1974.

47. R. A. Normann, E. M. Maynard, P. J. Rousche, and D. J. Warren, "A neural interface for a cortical vision prosthesis", *Vision Research*, **39**, 2577-2587, 1999.

48. E. M. Schmidt, M. J. Bak, F. T. Hambrecht, C. V. Kufta, D. K. O'Rourke, and P. Vallabhanath, "Feasibility of a visual prosthesis for the blind based on intracortical microstimulation of the visual cortex", *Brain*, **119**(2), 507-522, 1996.

49. C. Veraart, C. Raftopoulos, J. T. Mortimer, J. Delbeke, D. Pins, G. Michaux, A. Vanlierde, S. Parrini, and M.-C. Wanet-Defalque, "Visual sensations produced by optic nerve stimulation using an implanted self-sizing spiral cuff electrode", *Brain Research*, **813**(1), 181-186, 1998.

50. C. Veraart, M.-C. Wanet-Defalque, B. Gerard, A. Vanlierde, and J. Delbeke, "Pattern Recognition with the Optic Nerve Visual Prosthesis", *Artificial Organs*, **27**(11), 996-1004, 2003.

51. Q. Ren, X. Chai, K. Wu, C. Zhou, and C-Sight Group, "Visual Prosthesis Based on Optic Nerve Stimulation with Penetrating Electrode Array", Chapter 10 in *Artificial Sight: Basic Research, Biomedical Engineering, and Clinical Advances*, M. S. Humayun, J. D. Weiland, G. Chader, and E. Greenbaum, Eds., (Springer, New York, 2007), 187-207.

52. X. Sui, L. Li, X. Chai, K. Wu, C. Zhou, X. Sun, X. Xu, X. Li, and Q. Ren, "Visual Prosthesis for Optic Nerve Stimulation", Chapter 2 in *Implantable Neural Devices 1: Devices and Applications*, D. Zhou and E. Greenbaum, Eds., (Springer, New York, 2009), 43-84.

53. P. B. L. Meijer, "An Experimental System for Auditory Image Representations", *IEEE Transactions on Biomedical Engineering*, **39**(2), 112-121, 1992.

54. A. Amedi, W. M. Stern, J. A. Camprodon, F. Bermpohl, L. Merabet, S. Rotman, C. Hemond, P. Meijer, and A. Pascual-Leone, "Shape conveyed by visual-to-auditory sensory substitution activates the lateral occipital complex", *Nature Neuroscience*, **10**(6), 687-689, 2007.

55. O. Collignon, P. Voss, M. Lassonde, and F. Lepore, "Cross-modal plasticity for the spatial processing of sounds in visually deprived subjects", *Experimental Brain Research*, **192**(3), 343-358, 2009.

56. J. Ward and P. Meijer, "Visual experiences in the blind induced by an auditory sensory substitution device", *Consciousness and Cognition*, **19**(1), 492-500, 2010.

57. M. J. Proulx, "Synthetic synaesthesia and sensory substitution", *Consciousness and Cognition*, **19**(1), 501-503, 2010.

58. P. Bach-y-Rita and S. W. Kercel, "Sensory substitution and the human-machine interface", *Trends in Cognitive Sciences*, **7**(12), 541-546, 2003.

59. C. Poirier, A. G. De Volder, and C. Scheiber, "What neuroimaging tells us about sensory substitution", *Neuroscience and Behavioral Reviews*, **31**(7), 1064-1070, 2007.

60. P. Bach-y-Rita, C. C. Collins, F. A. Saunders, B. White, and L. Scadden, "Vision Substitution by Tactile Image Projection", *Nature*, **221**, 963-964, 1969.

61. S. F. Frisken-Gibson, P. Bach-y-Rita, W. J. Tompkins, and J. G. Webster, "A 64-Solenoid, Four-Level Fingertip Search Display for the Blind", *IEEE Transactions on Biomedical Engineering*, **BME-34**(12), 963-965, 1987.

62. M. C. Pereira and F. Kassab, Jr., "An Electrical Stimulator for Sensory Substitution", *Proceedings of the 28th IEEE EMBS Annual International Conference*, Paper SaEP9.3, 6016-6020, 2006.

63. D. H. Hubel, *Eye, Brain, and Vision*, Volume 22 in the Scientific American Library Series, (W. H. Freeman and Company, New York), 1988.

64. B. W. Jones and R. E. Marc, "Retinal remodeling during retinal degeneration", (Review), *Experimental Eye Research*, **81**(2), 123-137, 2005.

65. B. W. Jones, C. B. Watt, and R. E. Marc, "Retinal remodeling", (Invited Review), *Clinical and Experimental Optometry*, **88**(5), 282-291, 2005.

66. R. E. Marc, B. W. Jones, J. R. Anderson, K. Kinard, D. W. Marshak, J. H. Wilson, T. Wensel, and R. J. Lucas, "Neural Reprogramming in Retinal Degeneration", *Investigative Ophthalmology & Visual Science*, **48**(7), 3364-3371, 2007.

67. P. J. Nasiatka, A. Ahuja, N. R. B. Stiles, M. C. Hauer, R. N. Agrawal, R. Freda, D. Güven, M. S. Humayun, J. D. Weiland, and A. R. Tanguay, Jr., "Intraocular Camera for Retinal Prostheses", *Investigative Ophthalmology & Visual Science*, **46**(5), ARVO Abstract 5277, Paper B480, 2005.

68. P. J. Nasiatka, A. Ahuja, N. R. B. Stiles, M. C. Hauer, R. N. Agrawal, R. Freda, D. Güven, M. S. Humayun, J. D. Weiland, and A. R. Tanguay, Jr., "Intraocular Camera Design for

Retinal Prostheses", *Annual Meeting of the Optical Society of America*, Tucson, Arizona, 2005; *FiO/LS Conference Program*, Paper FThI4, 124, 2005.

69. P. J. Nasiatka, M. C. Hauer, N. R. B. Stiles, L. Lue, S. Takahashi, R. N. Agrawal, R. Freda, M. S. Humayun, J. D. Weiland, and A. R. Tanguay, Jr., "Intraocular Camera for Retinal Prostheses", *Investigative Ophthalmology & Visual Science*, **47**(5), ARVO Abstract 3180, Paper B554, 2006.

70. P. J. Nasiatka, M. C. Hauer, N. R. B. Stiles, L. Lue, S. Takahashi, R. N. Agrawal, M. S. Humayun, J. D. Weiland, and A. R. Tanguay, Jr., "An Intraocular Camera for Retinal Prostheses", *Proceedings of the ASME 2nd Frontiers in Biomedical Devices Conference*, Irvine, California, 2007; Paper BioMed 2007-38109, 23-24, 2007.

71. N. R. B. Stiles, M. C. Hauer, P. Lee, P. J. Nasiatka, J.-C. Lue, J. D. Weiland, M. S. Humayun, and A. R. Tanguay, Jr., "Intraocular Camera for Retinal Prostheses: Design Constraints Based on Visual Psychophysics", *Annual Meeting of the Optical Society of America*, San Jose, California, 2007; *FiO/LS/OMD Conference Program*, Paper JWC46, 110, 2007.

72. M. C. Hauer, P. J. Nasiatka, N. R. B. Stiles, J.-C. Lue, R. N. Agrawal, J. D. Weiland, M. S. Humayun, and A. R. Tanguay, Jr., "Intraocular Camera for Retinal Prostheses: Optical Design", *Annual Meeting of the Optical Society of America*, San Jose, California, 2007; *FiO/LS/OMD Conference Program*, Paper FThT1, 142, 2007.

73. P. J. Nasiatka, M. C. Hauer, N. R. B. Stiles, N. Suwanmonkha, M. Leighton, J.-C. Lue, M. S. Humayun, and A. R. Tanguay, Jr., "An Intraocular Camera for Provision of Natural Foveation in Retinal Prostheses", *Biomedical Engineering Society Annual Meeting: "Engineering the Future of Biology and Medicine"*, North Hollywood, California, 2007; *BMES Conference Program*, Abstract P6.96, 2007.

74. M. C. Hauer, "Intraocular Camera for Retinal Prostheses: Refractive and Diffractive Lens Systems", Ph.D. Thesis, University of Southern California, 2009.

75. C. Zhou, Y. Liu, Q. Ren, and P. Lu, "Introduction of an implantable MOMES used in visual prosthesis", *Proceedings of the International Symposium on Biophotonics, Nanophotonics, and Metamaterials*, 183-185, 2006.

76. K. Cha, K. W. Horch, and R. A. Normann, "Simulation of a phosphene-based visual field: Visual acuity in a pixelized vision system", *Annals of Biomedical Engineering*, **20**(4), 439-449, 1992.

77. K. Cha, K. W. Horch, R. A. Normann, and D. K. Boman, "Reading speed with a pixelized vision system", *Journal of the Optical Society of America A*, **9**(5), 673-677, 1992.

78. K. Cha, K. W. Horch, and R. A. Normann, "Mobility performance with a pixelized vision system", *Vision Research*, **32**(7), 1367-1372, 1992.

79. J. Dowling, W. Boles, and A. Maeder, "Mobility assessment using simulated Artificial Human Vision", *Proceedings of the 1st IEEE Workshop on Computer Vision Applications for the Visually Impaired*, San Diego, California, 2005; in *Proceedings of the 2005 IEEE Computer Society Conference on Computer Vision and Pattern Recognition (CVPR'05)*, (IEEE Computer Society, Washington, D.C.), 32-39, 2005.

80. R. W. Thompson, Jr., G. D. Barnett, M. S. Humayun, and G. Dagnelie, "Facial Recognition Using Simulated Prosthetic Pixelized Vision", *Investigative Ophthalmology & Visual Science*, **44**(11), 5035-5042, 2003.

81. G. Dagnelie, P. Keane, V. Narla, L. Yang, J. Weiland, and M. Humayun, "Real and virtual mobility performance in simulated prosthetic vision", *Journal of Neural Engineering*, **4**, S92-S101, 2007.

82. M. Velikay-Parel, D. Ivastinovic, M. Koch, R. Hornig, G. Dagnelie, G. Richard, and A. Langmann, "Repeated mobility testing for later artificial visual function evaluation", *Journal of Neural Engineering*, **4**, S102-S107, 2007.

83. G. Dagnelie, "Psychophysical Evaluation for Visual Prosthesis", *Annual Review of Biomedical Engineering*, **10**, 339-368, 2008.

84. J. Sommerhalder, E. Oueghlani, M. Bagnoud, U. Leonards, A. B. Safran, and M. Pelizzone, "Simulation of artificial vision: I. Eccentric reading of isolated words, and perceptual learning", *Vision Research*, **43**, 269-283, 2003.

85. J. Sommerhalder, B. Rappaz, R. de Haller, A. P. Fornos, A. B. Safran, and M. Pelizzone, "Simulation of artificial vision: II. Eccentric reading of full-page text and the learning of this task", *Vision Research*, **44**, 1693-1706, 2004.

86. A. P. Fornos, J. Sommerhalder, B. Rappaz, A. B. Safran, and M. Pelizzone, "Simulation of Artificial Vision, III. Do the Spatial or Temporal Characteristics of Stimulus Pixelization Really Matter?", *Investigative Ophthalmology & Visual Science*, **46**(10), 3906-3912, 2005.

87. A. P. Fornos, J. Sommerhalder, A. Pittard, A. B. Safran, and M. Pelizzone, "Simulation of artificial vision, IV. Visual information required to achieve simple pointing and manipulation tasks", *Vision Research*, **48**, 1705-1718, 2008.

88. S. C. Chen, L. E. Hallum, N. H. Lovell, and G. J. Suaning, "Visual acuity measurement of prosthetic vision: A virtual-reality simulation study", *Journal of Neural Engineering*, **2**, S135-S145, 2005.

89. L. E. Hallum, G. J. Suaning, D. S. Taubman, and N. H. Lovell, "Simulated prosthetic visual fixation, saccade, and smooth pursuit", *Vision Research*, **45**, 775-788, 2005.

90. S. C. Chen, G. J. Suaning, J. W. Morely, and N. H. Lovell, "Simulating prosthetic vision: I. Visual models of phosphenes", *Vision Research*, **49**, 1493-1506, 2009.

91. S. C. Chen, G. J. Suaning, J. W. Morely, and N. H. Lovell, "Simulating prosthetic vision: II. Measuring functional capacity", *Vision Research*, **49**, 2329-2343, 2009.

92. L. D. Harmon and B. Julesz, "Masking in Visual Recognition: Effects of Two-Dimensional Filtered Noise", *Science*, New Series, **180**(4091), 1194-1197, 1973.

93. M. J. McMahon, A. Caspi, J. D. Dorn, K. H. McClure, M. S. Humayun, and R. J. Greenberg, "Spatial Vision in Blind Subjects Implanted with the Second Sight Retinal Prosthesis", *Investigative Ophthalmology & Visual Science*, **48**(5), ARVO Abstract 4443, 2007.

94. The visual prosthesis simulator incorporates an M-Z800 Monocular Eye-Tracking System with eMagin Z800 head mounted display (HMD) and ViewPoint EyeTracker® software, Arrington Research, Inc., Scottsdale, Arizona.

95. A. M. Rosen, D. B. Denham, V. Fernandez, D. Borja, A. Ho, F. Manns, J.-M. Parel, and R. C. Augusteyn, "In vitro dimensions and curvatures of human lenses", *Vision Research*, **46**, 1002-1009, 2006.

96. H. Kolb, E. Fernandez, and R. Nelson, "Simple Anatomy of the Retina", in *Webvision: The Organization of the Retina and Visual System*, Electronic Resource, http://www.ncbi.nlm.nih.gov/bookshelf/br.fcgi?book=webvision&part=A34, Retrieved April, 2010.

97. B. A. Wandell, *Foundations of Vision*, (Sinauer Associates, Sunderland, Massachusetts), 1995.

98. E. Hecht, *Optics, 4th Edition*, (Addison Wesley, San Francisco, California), 2002.

99. *Compilation of the Social Security Laws*, United States of America, Section 1614, http://www.ssa.gov/OP_Home/ssact/title16b/1614.htm, Retrieved April, 2010.

100. *International Classification of Diseases*, ICD-10, World Health Organization, Section H54.7, "Unspecified Visual Loss", http://apps.who.int/classifications/apps/icd/icd10online/, Retrieved April, 2010.

101. H.-L. Liou, and N. A. Brennan, "Anatomically accurate, finite model eye for optical modeling", *Journal of the Optical Society of America A*, **14**(8), 1684-1695, 1997.

102. K.-B. Cho, A. I. Krymski, and E. R. Fossum, "A 1.5-V 550-μW 176 × 144 Autonomous CMOS Active Pixel Image Sensor", *IEEE Transactions on Electron Devices*, **50**(1), 96-105, 2003.

103. R. Heber and H. Holländer, "Size and Distribution of Ganglion Cells in the Human Retina", *Anatomy and Embryology*, **168**, 125-136, 1983.

104. N. Akahane, R. Ryuzaki, S. Adachi, K. Mizobuchi, and S. Sugawa, "A 200 dB Dynamic Range Iris-less CMOS Image Sensor with Lateral Overflow Integration Capacitor using

Hybrid Voltage and Current Readout Operation", *Digest of Technical Papers, 2006 IEEE International Solid-State Circuits Conference (ISSCC 2006)*, Paper 16.7, 21-23, 2006.

105. K. A. Zaghloul and K. Boahen, "A silicon retina that reproduces signals in the optic nerve", *Journal of Neural Engineering*, **3**, 257-267, 2006.

106. A. R. Tanguay, Jr., B. K. Jenkins, C. von der Malsburg, B. Mel, G. Holt, J. O'Brien, I. Biederman, A. Madhukar, P. Nasiatka, and Y. Huang, "Vertically Integrated Photonic Multichip Module Architecture for Vision Applications", *Proceedings of the International Conference on Optics in Computing (OC 2000)*, Quebec City, Canada, 2000, R. A. Lessard and T. V. Galstian, Eds.; in *Proceedings of SPIE*, **4089**, 584-600, 2000.

107. H. G. MacDougall and S. T. Moore, "Marching to the beat of the same drummer: The spontaneous tempo of human locomotion", *Journal of Applied Physiology*, **99**, 1164-1173, 2005.

108. R. V. Johnson and A. R. Tanguay, Jr., "Stratified volume holographic optical elements", *Optics Letters*, **13**(3), 189-191, 1988.

109. G. P. Nordin, R. V. Johnson, and A. R. Tanguay, Jr., "Diffraction properties of stratified volume holographic optical elements", *Journal of the Optical Society of America A*, **9**(12), 2206-2217, 1992.

110. D. M. Chambers, G. P. Nordin, and S. Kim, "Fabrication and analysis of a three-layer stratified volume diffractive optical element high-efficiency grating", *Optics Express*, **11**(1), 27-38, 2003.

111. C. R. Tanguay, "The Fantasy of Visual Fusion", Abstract J0334, California State Science Fair, 2003; http://www.usc.edu/CSSF/History/2003/Projects/J0334.pdf, Retrieved April, 2010.

112. D. Marr, *Vision: A Computational Investigation into the Human Representation and Processing of Visual Information*, W. H. Freeman and Company, San Francisco, 1982.

113. E. H. Land, "Recent advances in retinex theory and some implications for cortical computations: Color vision and the natural image", *Proceedings of the National Academy of Sciences of the United States of America*, **80**, 5163-5169, 1983.

CHAPTER 21

MODEL FOR OPTICAL PROPAGATION IN RANDOMLY COUPLED BIREFRINGENT FIBER AND ITS IMPLEMENTATION

DIPAK CHOWDHURY and MICHAL MLEJNEK

Science and Technology
Corning Incorporated
Corning, NY 14830
USA

e-mail: chowdhurdq@corning.com

SHIVA KUMAR

Electrical and Computer Engineering
McMaster University
Communications Research Laboratory
Rm. 219, 1280 Main St. W. Hamilton Ontario L8S 4K1
Canada

e-mail: kumars@mail.ece.mcmaster.ca

This article summarizes the model for birefringent propagation in optical fibers, and describes its detail numerical implementation. Validation results against known solutions for the formulation are presented.

1. Introduction

Solution of nonlinear Schrödinger equation (NLSE) played a key role in designing optical fiber and communication systems over the last two decades. Success of the numerical scheme to solve this equation for optical fiber relies on the well known *split-step Fourier method (SSFM)*.[1,2] SSFM allowed scientists and engineers to solve the NLSE for realistic system lengths > 500Km) and for realistic number of optical channels (between 8-80) in a reasonable computational time.[3,4,5] On top of these the accuracy of SSFM for NLSE allowed simulations to predict performances of complex optical systems with practical accuracy. More work on improving the speed and accuracy has been done on SSFM for solving optical propagation through birefringent and non-birefringent fibers.[6] However, none of these articles go into the detail of the implementation of the method.

This article describes the implementation *detail* of NLSE with polarization. The coupled NLS models the propagation of polarized light in birefringent fibers. The notation used here is similar to the one given in.[1] Section 2 outlines the scalar NLSE and shows the usual split-step method of solving NLSE. Then section 3 outlines the polarized version of the NLSE which contains all the terms that couple the two orthogonally polarized field components including the four wave mixing (FWM) term. The next section (section 4) details the method used to take into account the concatenation of randomly oriented birefringent pieces. Finally section 5 outlines the initial implementation of the randomly concatenated birefringent pieces which ignores the FWM term shown in section 3.

2. Solution of NLS for the scalar case

The scalar NLS models the propagation of scalar light waves by approximating the evolution of their slowly-varying amplitudes (SVA) or wave packets having carrier waves of the form $e^{-i\omega t}$ and is:[1]

$$\frac{\partial A}{\partial z} + \beta_1 \frac{\partial A}{\partial t} + \frac{i}{2}\beta_2 \frac{\partial^2 A}{\partial t^2} - \frac{1}{6}\beta_3 \frac{\partial^3 A}{\partial t^3} + \frac{\alpha}{2} A = i\gamma |A|^2 A, \tag{1}$$

where the optical field in the guided mode is expressed as

$$E(\mathbf{r}, t) = F(x, y) A(z, t) \exp[i(\bar{\beta} z - \omega_0 t)].$$

In defining the optical field we denote the modal field distribution within the fiber cross section as $F(x, y)$ and the slowly varying envelop of the optical field as $A(z, t)$, expressed in units such that $|A(z, t)|^2 = P(z, t)$, the optical power in units of Watt. It is often more convenient for numerical solution to write this as

$$\frac{\partial A}{\partial z} = -\beta_1 \frac{\partial A}{\partial t} - \frac{i}{2}\beta_2 \frac{\partial^2 A}{\partial t^2} + \frac{1}{6}\beta_3 \frac{\partial^3 A}{\partial t^3} - \frac{\alpha}{2} A + i\gamma |A|^2 A \tag{2}$$

since we intend to evolve the entire time series along the propagation direction, z.

The scalar NLS is easily solved using the split-step approach. This is a fast, efficient method that takes advantage of the fast-Fourier transform (FFT). To approximate the solution of the NLS over a step ΔZ note that we can write

$$\frac{\partial A}{\partial z} = (\hat{D}^t + \hat{N}^t)A \tag{3}$$

where

$$\hat{D}^t = -\beta_1 \frac{\partial}{\partial t} - \frac{i}{2}\beta_2 \frac{\partial^2}{\partial t^2} + \frac{1}{6}\beta_3 \frac{\partial^3}{\partial t^3} - \frac{\alpha}{2},$$

$$\hat{D}^\omega = i(\delta\omega \, \beta_1 + \frac{1}{2}\delta\omega^2 \, \beta_2 + \frac{1}{6}\delta\omega^3 \, \beta_3) - \frac{\alpha}{2} \text{ (in frequency domain)}$$

$$\hat{N}^t = i\gamma |A|^2.$$

By evolving the field envelope by propagating using the linear operator and the nonlinear operator independently, it can be shown that after a step in the propagation direction

$$A(z + \Delta z, t) = \exp\left[\hat{D}^t \frac{\Delta z}{2}\right] \exp\left[\hat{N}^t \Delta z\right] \exp\left[\hat{D}^t \frac{\Delta z}{2}\right] A(z, t) + O\left(\Delta z^3\right). \quad (4)$$

This approximation is implemented numerically as the symmetric split-step method. It is described pictorially in Fig. 1, where the numbers indicate the sequence of steps taken.

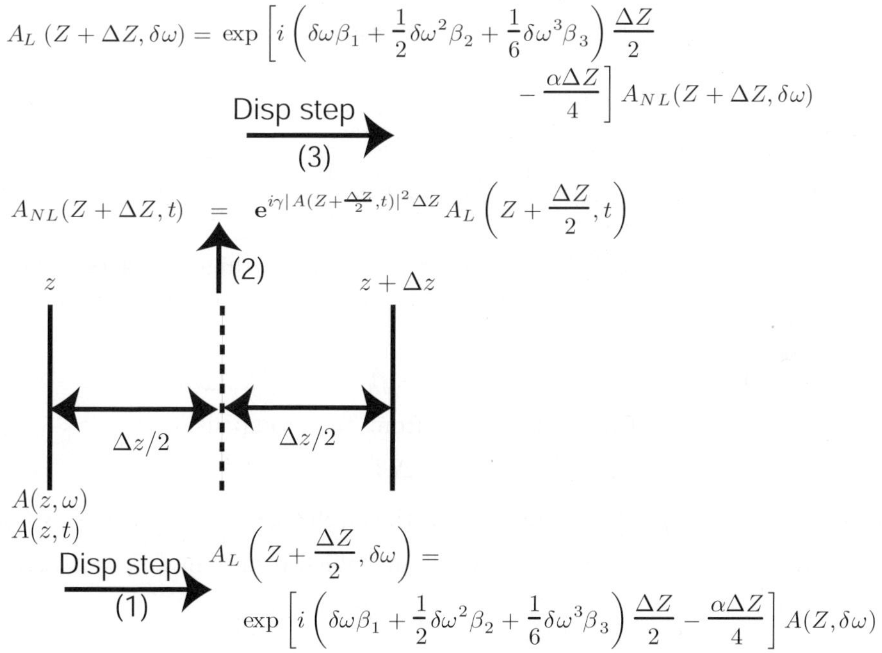

$$A_L\left(Z + \Delta Z, \delta\omega\right) = \exp\left[i\left(\delta\omega\beta_1 + \frac{1}{2}\delta\omega^2\beta_2 + \frac{1}{6}\delta\omega^3\beta_3\right)\frac{\Delta Z}{2} - \frac{\alpha\Delta Z}{4}\right] A_{NL}(Z + \Delta Z, \delta\omega)$$

Disp step (3)

$$A_{NL}(Z + \Delta Z, t) = e^{i\gamma|A(Z + \frac{\Delta Z}{2}, t)|^2 \Delta Z} A_L\left(Z + \frac{\Delta Z}{2}, t\right)$$

(2)

$z \qquad z + \Delta z$

$\Delta z/2 \qquad \Delta z/2$

$A(z, \omega)$
$A(z, t)$

Disp step (1)

$$A_L\left(Z + \frac{\Delta Z}{2}, \delta\omega\right) = \exp\left[i\left(\delta\omega\beta_1 + \frac{1}{2}\delta\omega^2\beta_2 + \frac{1}{6}\delta\omega^3\beta_3\right)\frac{\Delta Z}{2} - \frac{\alpha\Delta Z}{4}\right] A(Z, \delta\omega)$$

Fig. 1.   Solution steps for scalar NLS based on the equations shown above.

3. The coupled NLS model of polarized light propagation

In order to express the NLSE for the polarized light we start with the scalar wave equation for the slowly varying envelope:[1]

$$2i\beta_0 \frac{\partial A}{\partial z} + \left(\bar{\beta}\left(\omega\right)^2 - \beta_0^2\right) A = 0, \quad (5)$$

where the electric field is expressed as

$$\mathbf{E}(\omega, \mathbf{r}) = F(x, y)\, A(z, \delta\omega)\left[\exp(i\beta_o z)\right], \quad (6)$$

β_0 is the propagation constant at ω_0, $\delta\omega = \omega - \omega_0$, and $\bar{\beta} = (\beta(\omega) + \Delta\zeta)$ where $\beta\left(\omega\right)$ is the linear and real part of the propagation constant and $\Delta\zeta$ is the factor that takes care of the loss and the nonlinear interactions. We can rewrite Eq. 5 for the polarized light as

$$2i\beta_0 \frac{\partial A_x}{\partial z} + \left(\bar{\beta}_x(\omega)^2 - \beta_0^2\right) A_x = 0,$$

$$2i\beta_0 \frac{\partial A_y}{\partial z} + \left(\bar{\beta}_y(\omega)^2 - \beta_0^2\right) A_y = 0, \tag{7}$$

where the electric field is expressed as

$$\mathbf{E}_x(\delta\omega, \mathbf{r}) = F(x,y)\, A_x(z, \delta\omega)\, [\exp(i\beta_o z)] \quad \text{and}$$

$$\mathbf{E}_y(\delta\omega, \mathbf{r}) = F(x,y)\, A_y(z, \delta\omega)\, [\exp(i\beta_o z)],$$

$\bar{\beta}_i(\omega) = \beta_i(\omega) + \Delta\zeta_i$ for $i = x, y$, and $\beta_0 = \dfrac{\beta_{0x}(\omega_0) + \beta_{0y}(\omega_0)}{2}$. We use Taylor expansion of the propagation constant

$$\beta_i(\omega) = \beta_{0i}(\omega_0) + \beta_{1i}(\omega_0 - \omega) + \frac{1}{2}\beta_2(\omega_0 - \omega)^2 + \frac{1}{6}\beta_3(\omega_0 - \omega)^3 + \dots \quad (i = x, y) \tag{8}$$

where it is assumed that the dispersion term, β_2, and all higher orders dispersion terms are the same for both polarizations. Let us also define two other variables for later use

$$\Delta\beta_0 = \beta_x(\omega_0) - \beta_y(\omega_0) = \beta_{0x}(\omega_0) - \beta_{0y}(\omega_0),$$

$$\Delta\beta_1 = \beta_{1x} - \beta_{1y}, \quad \text{and from Taylor expansion}$$

$$\Delta\beta(\omega) = \Delta\beta_0 + \delta\omega\,\Delta\beta_1 \triangleq \Delta\beta,$$

where $\delta\omega = (\omega_0 - \omega)$. For a birefringent fiber, the birefringence is defined as $\Delta\beta(\omega) = \omega\dfrac{\delta n}{c} \approx \omega\,\Delta\tau$ where $\Delta\tau$ is expressed in ps/km (which results in $\Delta\beta_0 = \omega_0\,\Delta\tau$ and $\Delta\beta_1 = -\Delta\tau$).

Using the approximation $\left(\bar{\beta}_x^2 - \beta_0^2\right) \approx 2\beta_0\left(\bar{\beta}_x - \beta_0\right)$, the relationship

$$\beta_i - \beta_0 = \beta_i(\omega) - \beta_0 = \beta_{0i}(\omega_0) + \delta\omega\,\beta_{1i} + \frac{\delta\omega^2}{2}\beta_2 + \frac{\delta\omega^3}{6}\beta_3 - \beta_0 + \Delta\zeta_i \quad (i = x, y)$$

$$= \pm\frac{\Delta\beta_0}{2} + \delta\omega\,\beta_{1i} + \frac{\delta\omega^2}{2}\beta_2 + \frac{\delta\omega^3}{6}\beta_3 + \Delta\zeta_i \quad (+ \text{ for } i = x, - \text{ for } i = y),$$

and the formalism shown in Ref. 1 we obtain

$$\frac{\partial A_i(\omega, z)}{\partial z} = i\left(\pm\frac{\Delta\beta_0}{2} + \delta\omega\,\beta_{1i} + \frac{\delta\omega^2}{2}\beta_2 + \frac{\delta\omega^3}{6}\beta_3 + \Delta\zeta_i\right) A_i(\omega, z)$$

$$(+ \text{ for } i = x, - \text{ for } i = y). \tag{9}$$

With the current time variation ($e^{-i\omega t}$), the Fourier transform pair is defined as:

$$\mathcal{F}\{f(t)\} = F(\omega) = \int_{-\infty}^{\infty} f(t)e^{i\omega t}\,dt \quad \text{and}$$

$$\mathcal{F}^{-1}\{F(\omega)\} = f(t) = \frac{1}{2\pi}\int_{-\infty}^{\infty} F(\omega)e^{-i\omega t}\,d\omega. \tag{10}$$

With this definition of the Fourier transform and the transform of any derivative is given as

$$\mathcal{F}\{\frac{d^n f(t)}{dt^n}\} = (-i\omega)^n \, F(\omega), \quad \text{i.e.,} \quad \mathcal{F}^{-1}\{\omega^n \, F(\omega)\} = \frac{1}{(-i)^n} \frac{d^n f(t)}{dt^n}. \tag{11}$$

Applying the inverse FFT relationships to Eq. 9, we obtain (t & $\delta\omega$ are the Fourier transform pair):

$$\frac{\partial A_i(t,z)}{\partial z} = \left(\pm i\frac{\Delta\beta_0}{2} - \beta_{1i}\frac{\partial}{\partial t} - \frac{i}{2}\beta_2\frac{\partial^2}{\partial t^2} + \frac{1}{6}\beta_3\frac{\partial^3}{\partial t^3} + i\Delta\beta_i\right) A_i(t,z)$$

$$(+ \text{ for } i = x, - \text{ for } i = y). \tag{12}$$

From ref. 1 we obtain

$$\Delta\zeta_{x,y}\, A_{x,y} = \gamma\left[|A_{x,y}|^2 + \frac{2}{3}|A_{y,x}|^2\right] A_{x,y} + \frac{\gamma}{3}A^*_{x,y}A^2_{y,x} + i\frac{\alpha}{2}A_{x,y}.$$

Applying the above equation to Eq. 12 finally the coupled NLS model for polarized light waves is given as:

$$\frac{\partial A_x}{\partial z} - i\frac{\Delta\beta_0}{2}A_x + \beta_{1x}\frac{\partial A_x}{\partial t} + \frac{i}{2}\beta_2\frac{\partial^2 A_x}{\partial t^2} - \frac{1}{6}\beta_3\frac{\partial^3 A_x}{\partial t^3} + \frac{\alpha}{2}A_x =$$

$$i\gamma\left[|A_x|^2 + \frac{2}{3}|A_y|^2\right] A_x + \frac{i\gamma}{3}A^*_x A^2_y,$$

$$\frac{\partial A_y}{\partial z} + i\frac{\Delta\beta_0}{2}A_y + \beta_{1y}\frac{\partial A_y}{\partial t} + \frac{i}{2}\beta_2\frac{\partial^2 A_y}{\partial t^2} - \frac{1}{6}\beta_3\frac{\partial^3 A_y}{\partial t^3} + \frac{\alpha}{2}A_y =$$

$$i\gamma\left[|A_y|^2 + \frac{2}{3}|A_x|^2\right] A_x + \frac{i\gamma}{3}A^*_y A^2_x. \tag{13}$$

Now using $\beta_1 = (\beta_{1x}+\beta_{1y})/2$ and $T = t - \beta_1 z$ and $z = Z$, we obtain [see Appendix 1]:

$$\frac{\partial A_x}{\partial Z} - i\frac{\Delta\beta_0}{2}A_x + \frac{\Delta\beta_1}{2}\frac{\partial A_x}{\partial T} + \frac{i}{2}\beta_2\frac{\partial^2 A_x}{\partial T^2} - \frac{1}{6}\beta_3\frac{\partial^3 A_x}{\partial T^3} + \frac{\alpha}{2}A_x =$$

$$i\gamma\left[|A_x|^2 + \frac{2}{3}|A_y|^2\right] A_x + \frac{i\gamma}{3}A^*_x A^2_y,$$

$$\frac{\partial A_y}{\partial Z} + i\frac{\Delta\beta_0}{2}A_y - \frac{\Delta\beta_1}{2}\frac{\partial A_y}{\partial T} + \frac{i}{2}\beta_2\frac{\partial^2 A_y}{\partial T^2} - \frac{1}{6}\beta_3\frac{\partial^3 A_y}{\partial T^3} + \frac{\alpha}{2}A_y =$$

$$i\gamma\left[|A_y|^2 + \frac{2}{3}|A_x|^2\right] A_y + \frac{i\gamma}{3}A^*_y A^2_x. \tag{14}$$

The coupled NLS can also be solved efficiently using the split-step algorithm. Rewriting Eqs. 14 as

$$\frac{\partial A_x}{\partial Z} = (\hat{D}^T_x + \hat{N}^T_x)A_x, \text{ and}$$

$$\frac{\partial A_y}{\partial Z} = (\hat{D}^T_y + \hat{N}^T_y)A_y, \tag{15}$$

the *nonlinear* operators are

$$\hat{N}_x = i\gamma \left\{ \left[|A_x|^2 + \frac{2}{3}|A_y|^2 \right] + \frac{1}{3}\frac{A_x^*}{A_x}A_y^2 \right\}, \text{ and}$$

$$\hat{N}_y = i\gamma \left\{ \left[|A_y|^2 + \frac{2}{3}|A_x|^2 \right] + \frac{1}{3}\frac{A_y^*}{A_y}A_x^2 \right\},$$

and the *linear* operators in the frequency domain after taking Fourier transform (i.e., replacing $\frac{\partial}{\partial T}$ by $-i\delta\omega$) are[a]

$$\hat{D}_x^\omega = i\left[\frac{1}{2}(\Delta\beta_0 + \Delta\beta_1\delta\omega) + \frac{\beta_2}{2}\delta\omega^2 + \frac{\beta_3}{6}\delta\omega^3\right] - \frac{1}{2}\alpha,$$

$$= i\left[\frac{\omega\Delta\tau}{2} + \frac{\beta_2}{2}\delta\omega^2 + \frac{\beta_3}{6}\delta\omega^3\right] - \frac{1}{2}\alpha, \quad \text{and} \tag{16}$$

$$\hat{D}_y^\omega = i\left[-\frac{1}{2}(\Delta\beta_0 + \Delta\beta_1\delta\omega) + \frac{\beta_2}{2}\delta\omega^2 + \frac{\beta_3}{6}\delta\omega^3\right] - \frac{1}{2}\alpha,$$

$$= i\left[-\frac{\omega\Delta\tau}{2} + \frac{\beta_2}{2}\delta\omega^2 + \frac{\beta_3}{6}\delta\omega^3\right] - \frac{1}{2}\alpha. \tag{17}$$

The split-step solution of Eq. 15 is very similar to that used in for the scalar operator. In this case the nonlinear operator is more complicated. To implement the nonlinear step of the split step scheme for the coupled NLS in Eqs. 15 one needs to solve the following ordinary differential equations:

$$\frac{\partial A_{x,y}(Z,T_i)}{\partial Z} = i\gamma \left[|A_{x,y}(Z,T_i)|^2 + \frac{2}{3}|A_{y,x}(Z,T_i)|^2 \right] A_{x,y}(Z,T_i) +$$

$$\frac{i\gamma}{3} A_{x,y}^*(Z,T_i)\, A_{y,x}^2(Z,T_i)$$

for each T_i. Though they are integrable for many cases of interest, their solutions are in terms of elliptic functions. For most practical purposes it is sufficient to evolve the nonlinear operator using the fourth-order Runge-Kutta method.

The linear step is a simple extension of the scalar implementation. The solution after a linear propagation step in z is

$$A_x(Z+\Delta Z,T) = \exp\left[\phi_x(\delta\omega)\Delta z\right] A_x(Z,T), \tag{18}$$

$$A_y(Z+\Delta Z,T) = \exp\left[\phi_y(\delta\omega)\Delta z\right] A_y(Z,T), \tag{19}$$

where $\phi_{x,y}(\delta\omega) = i\left[\pm(\omega\Delta\tau/2) + (\beta_2/2)\delta\omega^2 + (\beta_3/6)\delta\omega^3\right] - \alpha/2$. As in the scalar case, the sequence of operations that move the solution of the coupled NLS from Z to $Z+\Delta Z$ is shown in Fig. 1. In the following, we represent the symmetric split

---

[a]Note the $\omega = \frac{2\pi c}{\lambda}$ in the equation, where λ is the wavelength where the value of ω is evaluated. In all other cases $\delta\omega$ refers to the frequency shift from the center of the simulation, where $\delta\omega = 0$.

step approximation,

$$A_{x,y}(Z + \Delta Z, T) = \exp\left[\frac{\Delta Z}{2}\hat{D}_{x,y}^T\right] \exp\left[\int_Z^{Z+\Delta Z} \hat{N}^T(Z')\,dZ'\right]$$

$$\exp\left[\frac{\Delta Z}{2}\hat{D}_{x,y}^T\right] A_{x,y}(Z,T)$$

using the notation

$$\begin{pmatrix} A_x(Z + \Delta Z, T) \\ A_y(Z + \Delta Z, T) \end{pmatrix} = \mathbf{L} \begin{pmatrix} A_x(Z, T) \\ A_y(Z, T) \end{pmatrix}. \tag{20}$$

Note that all the above equations are written and solved so that the principal axes of the birefringent fiber coincides with the principal axes of the equations. This choice of axes basically eliminates the evaluation of any extra cross-coupling terms. In the following the couplings will arise from the rotation matrices that will be introduced to orient the fields along the principal axes of the birefringent fiber.

4. Implementation of the random mode coupling model

So far we talked about NLS in a birefringent fiber. Now we are going to describe how the random mode coupling is accounted for in this system. Figure 2 illustrates two birefringent fiber pieces with their major axes oriented at θ_1 and θ_2 with respect to the lab coordinates x-y. With the local coordinates of the birefringent fiber sections designated as x' & y' respectively, we can write for propagation through the first fiber section using the local coordinate system

$$\begin{pmatrix} A_{x'}(Z + \Delta Z_1, T) \\ A_{y'}(Z + \Delta Z_1, T) \end{pmatrix} = \mathbf{L}_1 \begin{pmatrix} A_{x'}(Z, T) \\ A_{y'}(Z, T) \end{pmatrix}$$

$$\mathbf{A}'(Z + \Delta Z_1, T) = \mathbf{L}_1 \mathbf{A}'(z, T) \qquad \text{or, in the lab frame}$$

$$\mathbf{R}(\theta_1) \begin{pmatrix} A_x(Z + \Delta Z_1, T) \\ A_y(Z + \Delta Z_1, T) \end{pmatrix} = \mathbf{L}_1 \mathbf{R}(\theta_1) \begin{pmatrix} A_x(Z, T) \\ A_y(Z, T) \end{pmatrix},$$

$$\begin{pmatrix} A_x(Z + \Delta Z_1, T) \\ A_y(Z + \Delta Z_1, T) \end{pmatrix} = \mathbf{R}(-\theta_1)\, \mathbf{L}_1\, \mathbf{R}(\theta_1) \begin{pmatrix} A_x(Z, T) \\ A_y(Z, T) \end{pmatrix},$$

$$\mathbf{A}(Z + \Delta Z, T) = \mathbf{R}(-\theta_1)\, \mathbf{L}_1\, \mathbf{R}(\theta_1)\, \mathbf{A}(Z, T), \tag{21}$$

where $\mathbf{R}(\theta) = \begin{pmatrix} \cos\theta & \sin\theta \\ -\sin\theta & \cos\theta \end{pmatrix}$ is the rotation matrix.

Let us evaluate the transformation in terms of fields. From the definition of the field in Eq. 6 we can write in the laboratory coordinate system

$$\mathbf{E}(Z, T) = \mathbf{F}(x, y) \exp[i\beta_0 z] \mathbf{I} \mathbf{A}(Z, T),$$

where $\mathbf{I}$ is the unity matrix and β_0 is the average propagation constant for the fiber section. Going forward we will omitt $\mathbf{F}(x, y)$ since it is only the spatial mode

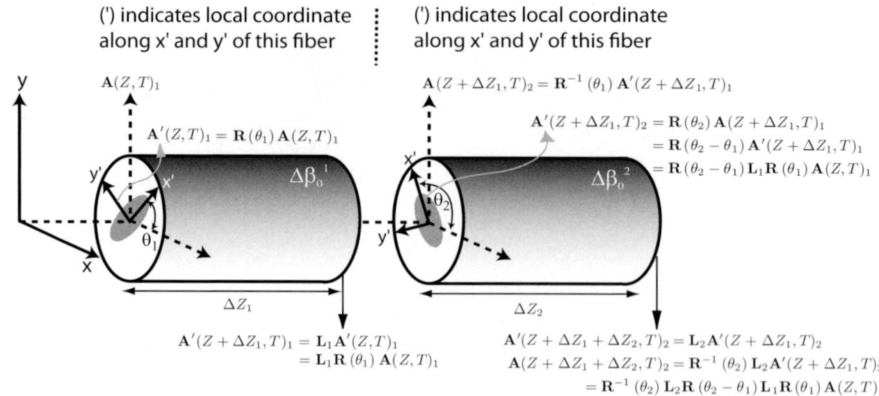

Fig. 2.   Schematic showing two birefringent fibers with two different orientations with respect to the laboratory frame. Relevant fields are also shown. The syntax is as described below in this section.

distribution of the propagating wave and does not concern us at this point for understanding the propagation properties. Using this equation we can start with the first section that we show in Fig. 2 as

$$\mathbf{E}(Z+\Delta Z_1,T) = \exp[i\left(\phi + \Delta\beta_0^{(1)}\,\Delta Z_1\right)]\,\mathbf{I}\,\mathbf{A}(Z+\Delta Z_1,T)$$

$$= \exp[i\left(\phi + \Delta\beta_0^{(1)}\,\Delta Z_1\right)]\mathbf{R}(-\theta_1)\,\mathbf{L}_1\,\mathbf{R}(\theta_1)\,\mathbf{A}(Z,T),$$

where we assume that the phase of the field at Z, i.e., beginning of the first section in the figure is ϕ. The propagation of the field through the next section results in

$$\mathbf{E}(Z+\Delta Z_1+\Delta Z_2,T) = \exp\left[i\left(\phi + \Delta\beta_0^{(1)}\,\Delta Z_1 + \Delta\beta_0^{(2)}\,\Delta Z_2\right)\right]$$
$$\mathbf{I}\,\mathbf{A}(Z+\Delta Z_1+\Delta z_2,T)$$
$$= \exp\left[i\left(\phi + \Delta\beta_0^{(1)}\,\Delta Z_1 + \Delta\beta_0^{(2)}\,\Delta Z_2\right)\right]\mathbf{R}(-\theta_2)\,\mathbf{L}_2\,\mathbf{R}(\theta_2)$$
$$\mathbf{A}(Z+\Delta Z_1,T),$$
$$= \exp\left[i\left(\phi + \Delta\beta_0^{(1)}\,\Delta Z_1 + \Delta\beta_0^{(2)}\,\Delta Z_2\right)\right] \times \mathbf{R}(-\theta_2)\,\mathbf{L}_2$$
$$\mathbf{R}(\theta_2)\,\mathbf{R}(-\theta_1)\,\mathbf{L}_1\,\mathbf{R}(\theta_1)\,\mathbf{A}(Z,T).$$

In most implimentations one ignores the phase factor that is in front of the above equation and solve only for the complex field amplitude $\mathbf{A}(z,T)$. As mentioned at the bottom of the previous section, the solution of the NLSE for each section of birefringent fiber is obtained by rotating the fields in the principal axes of the fiber sections by the rotation matrix $\mathbf{R}(\theta)$.

5. FWM term ignored

The model represented by Eqs. 21 gives a complete description of the propagation of polarized light in fibers. However, if we consider typical numbers for transmission fibers for Eqs. 17 we notice that the phase given by the mismatch term $\Delta\beta_0$ varies much more rapidly than any other phase term in the equation. All the other phase terms are comparable to $\delta\omega$ instead of ω. This fast phase variation implies that the contribution from the FWM term in Eq. 14 will be insignificant for even the smallest value of $\delta\tau$ for the fiber. Consequently, we ignore the FWM term here.

In the absence of the FWM term in 13 we have, from Eqs. 14,

$$\frac{\partial A_x}{\partial Z} - i\frac{\Delta\beta_0}{2}A_x + \frac{\Delta\beta_1}{2}\frac{\partial A_x}{\partial T} + \frac{i}{2}\beta_2\frac{\partial^2 A_x}{\partial T^2} - \frac{1}{6}\beta_3\frac{\partial^3 A_x}{\partial T^3} + \frac{\alpha}{2}A_x$$
$$= i\gamma\left[|A_x|^2 + \frac{2}{3}|A_y|^2\right]A_x,$$

$$\frac{\partial A_y}{\partial Z} + i\frac{\Delta\beta_0}{2}A_y - \frac{\Delta\beta_1}{2}\frac{\partial A_y}{\partial T} + \frac{i}{2}\beta_2\frac{\partial^2 A_y}{\partial T^2} - \frac{1}{6}\beta_3\frac{\partial^3 A_y}{\partial T^3} + \frac{\alpha}{2}A_y$$
$$= i\gamma\left[|A_y|^2 + \frac{2}{3}|A_x|^2\right]A_y. \tag{22}$$

Using these equations, we write

$$\frac{\partial A_x}{\partial Z} = (\hat{D}_x^T + \hat{N}_x^T)A_x, \text{ and}$$
$$\frac{\partial A_y}{\partial Z} = (\hat{D}_y^T + \hat{N}_y^T)A_y, \tag{23}$$

where the *nonlinear* operator is

$$\hat{N}_x = i\gamma\left\{\left[|A_x|^2 + \frac{2}{3}|A_y|^2\right]\right\}, \text{ and}$$
$$\hat{N}_y = i\gamma\left\{\left[|A_y|^2 + \frac{2}{3}|A_x|^2\right]\right\},$$

and the *linear* operator is

$$\hat{D}_x^\omega = i\left[\frac{1}{2}\omega\Delta\tau + \frac{\beta_2}{2}\delta\omega^2 + \frac{\beta_3}{6}\delta\omega^3\right] - \frac{1}{2}\alpha, \text{ and}$$
$$\hat{D}_y^\omega = i\left[-\frac{1}{2}\omega\Delta\tau + \frac{\beta_2}{2}\delta\omega^2 + \frac{\beta_3}{6}\delta\omega^3\right] - \frac{1}{2}\alpha.$$

Here, the nonlinear operator is much simpler and can be computed in a similar manner as the nonlinear operator in the scalar case. Thus, for the nonlinear operator we solve

$$\frac{\partial A_{x,y}(Z,T_i)}{\partial Z} = i\gamma\left[|A_{x,y}(Z,T_i)|^2 + \frac{2}{3}|A_{y,x}(Z,T_i)|^2\right]A_{x,y}(Z,T_i)$$

directly (see Appendix B) for each T_i. The solution for the linear operator is not changed and the algorithm is also as before, but the phase mismatch terms (FWM term) need not be retained.

6. Numerical validation

In the following we will validate the numerical scheme by comparing analytic data for N = 1 soliton that is transmitted through a length of birefringent fiber. Results are compared at the input and 3 different locations along the fiber. Before comparing the data we will describe the analytical equations for soliton.

6.1. *Analytic expression*

Let us start with the assumption that the solution of the NLS is given by

$$A(z,T) = a \operatorname{sech}(\eta T) \exp\left[-i\left(cT + bZ\right)\right]. \tag{24}$$

Inserting the above definition into the x-polarized expression of NLS (eq. 13) with the assumption that loss $\alpha = 0$, and $\beta_3 = 0$, we obtain values (with $e^{-i\omega t}$ variation)

$$a = \eta\sqrt{\frac{|\beta_2|}{\gamma}}.$$

$$b = -\frac{\omega_0 \Delta\tau}{2} - \frac{\Delta\tau^2}{8|\beta_2|} - \frac{|\beta_2|}{2}\eta^2,$$

$$c = -\frac{\Delta\tau}{2|\beta_2|},$$

where β_2 is the chromatic dispersion $\left(\frac{d^2\beta}{d\omega^2}\right)$, η is related to the soliton width and power as $\eta = \frac{1.762747174039086}{T_{FWHM}}$, T_{FWHM} is the full-width at half maxima of the soliton intensity, $\Delta\tau$ is the birefringence expressed in delay per unit length, γ is the nonlinear coefficent. The expression for the field $A(Z,T)$ in eq. 24 is valid for scalar as well as vector soliton with some scaling as we will define in the next subsections. In all the following the fiber and simulation parameters are given in Table 1.

Table 1.   All simulation parameters used in the validation.

Parameter	value	unit	Comment
T_{FWHM}	12.5	ps	
Dispersion D_2	1	ps/nm.km	$\beta_2 = -\frac{\lambda^2 D_2}{2\pi C}$, C is speed of light
wavelength λ	1550	nm	$\omega_0 = \frac{2\pi C}{\lambda}$
nonlinear index n_2	2.3×10^{-20}	m^2/W	$\gamma = \frac{2\pi n_2}{\lambda A_{eff}}$
effective area A_{eff}	50	μm^2	
birefringence $\Delta\tau$	0.5	ps/km	set to zero for scalar
fiber loss α	0	1/km	
third order disp. β_3	0	ps^3/m	

6.2. *Scalar soliton validation*

For scalar soliton, $\Delta\tau = 0$ in eq. 24. So, we obtain the soliton field in time and length as

$$A(z,T) = \eta\sqrt{\frac{|\beta_2|}{\gamma}}\operatorname{sech}(\eta T)\exp\left[-i\frac{|\beta_2|}{2}\eta^2 Z\right].$$

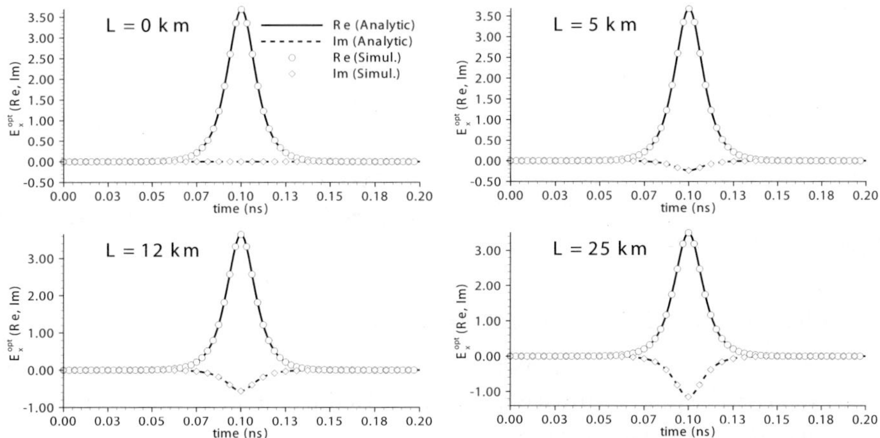

Fig. 3. Comparison of analytical and numerical propagation for a scalar soliton over a length of 25 Km. The fiber parameters are given in Table 1. Data for four different locations along the fiber are shown.

Figure 3 shows the real and imaginary part of the scalar soliton field as a function of time at four different locations. As can be seen from above figure the numerical solution agrees exactly with analytic solution for a scalar soliton case.

6.3. *Vector soliton: only x-polarization*

From eq. 24 we obtain that the expression for the soliton field in time and length is given as:

$$A_x(z,T) = \eta\sqrt{\frac{|\beta_2|}{\gamma}}\operatorname{sech}(\eta T)\exp\left[-i\left\{\frac{\Delta\tau}{2|\beta_2|}T + \left(\frac{\omega_0\Delta\tau}{2} + \frac{\Delta\tau^2}{8|\beta_2|} + \frac{|\beta_2|}{2}\eta^2\right)Z\right\}\right].$$

Figure 4 shows the real and imaginary part of the x-polarized vector soliton field as a function of time at four different locations. Only the x-polarized field is shown, since for a vector soliton with polarization aligned with one of the principal axes of polarization will remail polarized in the same direction throughout its propagation. As can be seen from above figure the numerical solution agrees exactly with analytic solution for a x-polarized vector soliton case.

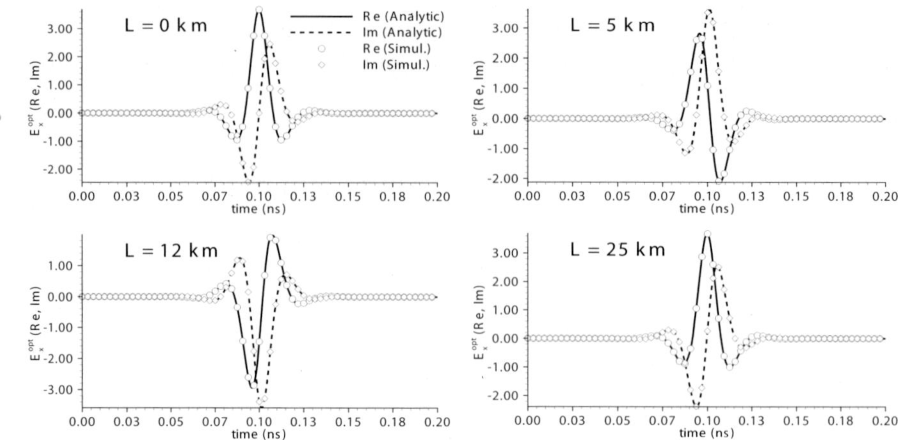

Fig. 4.   Comparison of analytical and numerical propagation for a x-polarized vector soliton over a length of 25 Km. The fiber parameters are given in Table 1. Data for four different locations along the fiber are shown.

6.4. *Vector soliton: polarized 45^0 to x-axis*

The soliton field for a 45^0 to x-axis polarized soliton is similar to a x-polarized soliton, except for a power adjustment factor. Taking the power adjustment factor into account, the fields are given as:

$$A_x(z,T) = \sqrt{\frac{3}{5}}\eta\sqrt{\frac{|\beta_2|}{\gamma}}\operatorname{sech}(\eta T)$$

$$\exp\left[-i\left\{\frac{\Delta\tau}{2|\beta_2|}T + \left(\frac{\omega_0\Delta\tau}{2} + \frac{\Delta\tau^2}{8|\beta_2|} + \frac{|\beta_2|}{2}\eta^2\right)z\right\}\right],$$

$$A_y(z,T) = \sqrt{\frac{3}{5}}\eta\sqrt{\frac{|\beta_2|}{\gamma}}\operatorname{sech}(\eta T)$$

$$\exp\left[-i\left\{-\frac{\Delta\tau}{2|\beta_2|}T + \left(-\frac{\omega_0\Delta\tau}{2} + \frac{\Delta\tau^2}{8|\beta_2|} + \frac{|\beta_2|}{2}\eta^2\right)z\right\}\right].$$

In the following two figures (5 and 6) the numerical results for the solitons are compared with the analytical results (for the x- and the y-polarized fields respectively). As can be seen that the agreement between the numerical and the analytical data is good.

7. Conclusion

In this article we presented the detail of vector NLS implementation with detail for the random mode coupling needed for a realistic simulation of polarization mode dispersion (PMD). In addition we provided some of the numerical validation for the method and its implementation for various cases of soliton propagation.

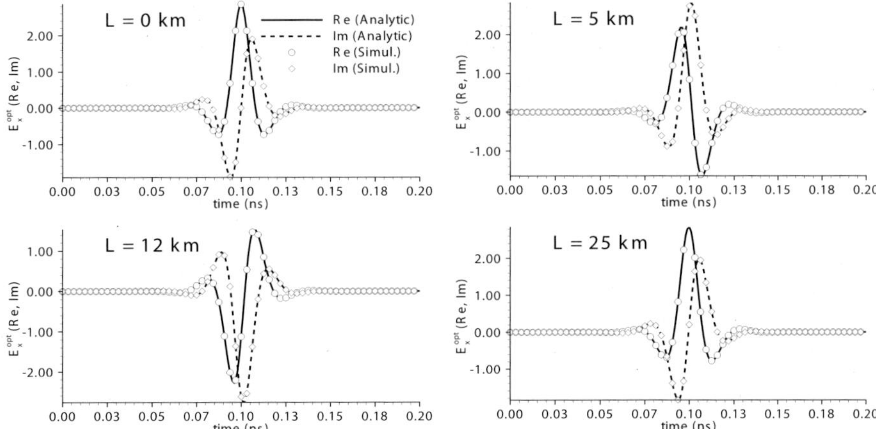

Fig. 5. Comparison of analytical and numerical propagation for a vector soliton polarized at 45^0 to the x-axis over a length of 25 Km. The fiber parameters are given in Table 1. Data for **only the x component** of the soliton at different locations along the fiber are shown.

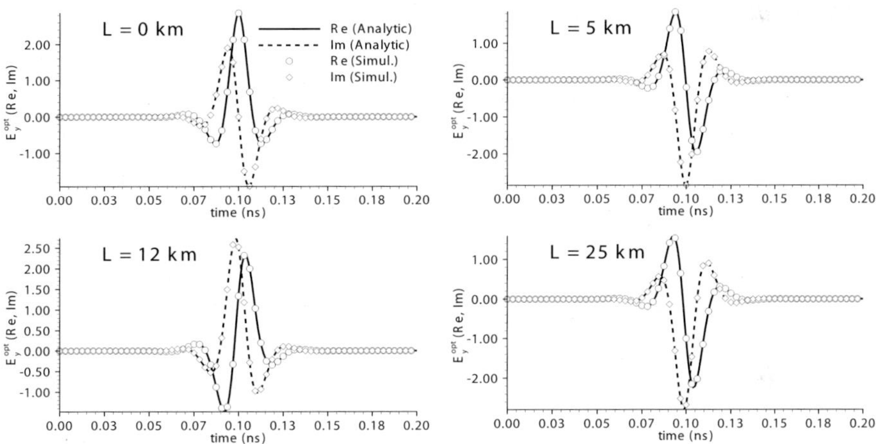

Fig. 6. Comparison of analytical and numerical propagation for a vector soliton polarized at 45^0 to the x-axis over a length of 25 Km. The fiber parameters are given in Table 1. Data for **only the y component** of the soliton at different locations along the fiber are shown.

The rationale for using soliton propagation are twofold: (1) The solution of soliton propagation in birefringent fiber is known, and (2) Soliton is particularly useful for validating numerical solutions since any numerical dispersion or accumulated error will destroy the soliton quickly. Since we demonstrated propagation over lengths longer than soliton period and did not see any instability, we are confident about the accuracy of the numerical implementation. In all the simulation and implementation we neglected the effect of the FWM term between the two polarizations. However, the FWM term between various frequency components are kept.

Application of this method for system simulation and specifically for polarization mode dispersion (PMD) impacts are given in Refs. 4 and 5, respectively.

Reference

1. G. P. Agarwal, *Nonlinear Fiber Optics* (Academic Press, New York, 1989).
2. C. R. Menyuk, "Pulse propagation in elliptically birefringent kerr medium," *IEEE J. of Quantum Electr.* **25**, 2674 (1989).
3. P. K. A. Wai and C. R. Menyuk, "Polarization mode dispersion, Decorrolation, and diffusion in optical fibers with randomly varying birefringence," *IEEE J. Lightwave Technol.* **14**, 148 (1996).
4. A. Belahlou *et. al.,* "Fiber Design Considerations for 40 Gb/s Systems," *IEEE J. Lightwave Technol.* **20**, 2290- (2002).
5. D. Q. Chowdhury, M. Mlejnek, and Y. Mauro, " Numerical Modeling of PMD," *J. Opt. Fiber Commun. Rep.* **1**, 141–149 (2004).
6. O. V. Sinkin, R. Holzlöhner, J. Zweck, and C. R. Menyuk, "Optimization of the Split-Step Fourier Method in Modeling Optical-Fiber Communications," *IEEE J. Lightwave Technol.* **21**, 61 (2003).

Appendix A. Moving coordinate with the group velocity

Using the substitution $\beta_1 = (\beta_{1x} + \beta_{1y})/2$, $T = t - \beta_1 z$, and $z = Z$, we obtain

$$\frac{\partial A_i}{\partial t} = \frac{\partial A_i}{\partial T}\frac{\partial T}{\partial t} + \frac{\partial A_i}{\partial Z}\frac{\partial Z}{\partial t} = \frac{\partial A_i}{\partial T}, \quad \text{and}$$

$$\frac{\partial A_i}{\partial z} = \frac{\partial A_i}{\partial T}\frac{\partial T}{\partial z} + \frac{\partial A_i}{\partial Z}\frac{\partial Z}{\partial z} = \frac{\partial A_i}{\partial Z} - \beta_1\frac{\partial A_i}{\partial T}.$$

Appendix B. Solution of the nonlinear part of scalar and polarized NLS

The nonlinear part of NLS is

$$\partial_z A = i\gamma |A|^2 A.$$

From this we obtain two equations:

$$A^* \partial_z A = i\gamma |A|^4 \quad \text{and}$$

$$A \partial_z A^* = -i\gamma |A|^4.$$

These leads to the power conservation relation $\partial_z |A|^2 = 0$ or $|A|^2 = A_o^2 =$ constant. So the solution for finding the the field at $Z + \Delta Z$ is

$$A(Z + \Delta Z, T) = e^{i\gamma |A(Z,T)|^2 \Delta Z} A(Z, T).$$

Similar arguments would allow the evaluation of the equation

$$\frac{\partial A_{x,y}(Z,T)}{\partial Z} = i\gamma \left[|A_{x,y}(Z,T)|^2 + \frac{2}{3}|A_{y,x}(Z,T)|^2 \right] A_{x,y}(Z,T)$$

as:

$$A_{x,y}(Z + \Delta Z, T) = e^{i\gamma \left[|A_{x,y}(Z,T)|^2 + \frac{2}{3}|A_{y,x}(Z,T)|^2 \right] \Delta z} A_{x,y}(Z, T).$$

CHAPTER 22

APPLICATIONS OF QUASI-NORMAL MODE EXPANSION: FROM MICRODROPLETS TO COMPACT STELLAR OBJECTS

PUI TANG LEUNG

Physics Department and Institute of Theoretical Physics,
The Chinese University of Hong Kong,
Shatin, Hong Kong SAR, China
ptleung@phy.cuhk.edu.hk

JUN WU

Department of Physics and Astronomy,
University of Pittsburgh, Pennsylvania 15260, USA
juw31@pitt.edu

Physical phenomena occurring in open wave systems can be analyzed in terms of relevant quasi-normal modes (QNMs), which are metastable states characterized by complex eigenfrequencies, in a way paralleling the normal mode expansion in closed systems. In this review, we discuss the application of quasi-normal mode expansion (QNME) and the properties of QNMs for various open systems, including microdroplets, black holes and neutron stars. These systems and related wave phenomena are of physical importance in their own right, and QNME is manifestly a powerful tool to describe them.

1. Introduction

Physical phenomena occurring in open wave systems, from which energy and matter are lost continuously, are ubiquitous in nature. Optical emission from excited atoms in an imperfect optical cavity (e.g., a laser resonator) and radiation of gravitational waves from relativistic stellar objects, such as black holes (BHs) and neutron stars (NSs), are two widely studied examples of these phenomena. Despite that these two kinds of systems differ greatly in their physical dimensions and nature, both of them can be analyzed in terms of relevant quasi-normal modes (QNMs) that satisfy the outgoing wave boundary wave condition at infinity.[1,2] Instead of artificially isolating an open system from its environment and considering interactions between them as minor perturbations, QNMs properly take the leakage of energy (or matter) into account from the outset. Thus, it is more natural to use quasinormal-mode expansion (QNME) to analyze the dynamics of open systems.

However, QNME should be exercised with due caution. Upon imposing the outgoing boundary condition on an open system, the system is no longer described by a hermitian operator. Hence, QNMs are characterized by complex eigenfrequencies, with the imaginary part measuring the decay rate (or growth rate for unstable systems). For leaky systems, the spatial wave function of QNMs grows exponentially at large distances. In addition, QNMs are not guaranteed to form a complete set for expansion and they are not orthogonal to each other, at least in the usual sense.[1,2]

In this review, we discuss the application of QNME in individual open systems, including dielectric microspheres, BHs and NSs. In addition to addressing the physics of these systems, special emphases will be placed on the issues of completeness, perturbation schemes and asymptotic behaviors of QNMs in each case. We will show that QNMs can still form a complete set for open systems satisfying certain conditions. Different methods to develop perturbation expansion for QNMs will also be introduced.

2. Morphology-dependent resonances of dielectric microspheres

2.1. Introduction

The interactions of light waves with dielectric spheres have attracted extensive attentions for many years.[3,4,5] Since 1980s, with the availability of intense laser light sources, many interesting optical phenomena have been observed in the scattering of electromagnetic waves from dielectric microspheres. In particular, micrometer-sized microdroplets have served as a natural example of microspheres. For example, when such droplets are illuminated with strong laser light, various optical phenomena such as resonant fluorescence,[6] stimulated Raman scattering,[7] stimulated Brillouin scattering,[8] lasing,[9] and enhanced energy transfer[10] have been observed.

A common salient feature observed in all these optical processes in microspheres is the strong enhancement of reaction rates at certain frequencies owing to the resonant effect of morphology-dependent resonances (MDRs), whose formation can be understood heuristically as follows.[4,5] Light waves propagating in a dielectric sphere may be reflected internally around its spherical surface at near glancing angle. If their wavelengths satisfy certain quantization conditions for resonances, they interfere constructively and hence form metastable states with finite lifetimes, limited by the diffractive loss at the spherical surface.

Mathematically speaking, MDRs are in fact QNMs of vectorial electromagnetic waves, which are described by the Maxwell equations. They can be observed as sharp spikes in the elastic scattering cross section,[4,5,11] whose widths are proportional to the imaginary parts of their respective complex eigenfrequencies.

2.2. Scalar wave formulation

Before introducing the full vectorial treatment of MDRs in dielectric spheres, we briefly discuss a scalar wave formulation for them to illustrate the situation for QNMs of scalar waves.[12]

Consider a non-magnetic dielectric sphere with dielectric constant $\epsilon(r)$ depending only on the radius r. In the absence of any free charge and current, it is straight-forward to show from the Maxwell equations that a time-harmonic magnetic field $\mathbf{B}(\mathbf{r}, \mathbf{t}) = \mathbf{b}(\mathbf{r}) \exp(-i\omega t)$ satisfies

$$\nabla \times \left[\frac{1}{\epsilon(r)} \nabla \times \mathbf{b} \right] - \omega^2 \mathbf{b} = 0 \,. \tag{1}$$

Hereafter gaussian units with the convention $c = 1$ and standard notations are adopted. Due to the symmetry of the system, the divergence-free (transverse) field $\mathbf{b}$ can be expressed as a linear combination of TE and TM modes:

$$\mathbf{b} = \sum_{lm} \nabla \times [\phi_{lm}(r)\mathbf{X}_{lm}] + \psi_{lm}(r)\mathbf{X}_{lm} \,, \tag{2}$$

where $\mathbf{X}_{lm}$ are vector spherical harmonics with angular momentum indices $\{lm\}$,[13] and

$$\left[-\frac{d}{dr}\rho(r)\frac{d}{dr} + \rho(r)\frac{l(l+1)}{r^2} - \rho(r)\epsilon(r)\,\omega^2 \right] \varphi = 0 \,, \tag{3}$$

with $\rho(r) = 1$, $\varphi = r\phi_{lm}$ ($\rho(r) = \epsilon(r)^{-1}$, $\varphi = r\psi_{lm}$) for TE (TM) modes.

Scalar QNMs of the sphere satisfy (3), the regular boundary condition $\varphi(r = 0) = 0$, and the outgoing wave boundary condition $\lim_{r\to\infty} [\varphi(r) \exp(-in_0\omega r)] = $ constant, where $n_0 = \lim_{r\to\infty} \sqrt{\epsilon(r)}$.[12] For stable systems the wave amplitude decays with time due to leakage, the imaginary part of the eigenfrequency ω is always negative, implying exponential divergence at large distances.

The Green's function $\tilde{D}(r, r'; \omega)$ of this scalar wave equation (TE or TM) is defined as the solution of the equation

$$\left[-\frac{d}{dr}\rho(r)\frac{d}{dr} + \rho(r)\frac{l(l+1)}{r^2} - \rho(r)\epsilon(r)\,\omega^2 \right] \tilde{D}(r, r'; \omega) = \delta(r - r') \,. \tag{4}$$

By analyzing the analytic structure of $\tilde{D}(r, r'; \omega)$, it can be proved that the QNM wave functions $f_j(r)$ form a complete (overcomplete) set:[12]

$$\sum_j \frac{\rho(r)\epsilon(r)f_j(r)f_j(r')}{2} = \delta(r - r') \,, \tag{5}$$

and the Green's function $\tilde{D}(r, r'; \omega)$ can be expanded in terms of $f_j(r)$ and the associated eigenfrequency ω_j:[12]

$$\tilde{D}(r, r'; \omega) = -\frac{1}{2} \sum_j \frac{f_j(r) f_j(r')}{\omega_j(\omega - \omega_j)} , \qquad (6)$$

for $r, r' < a$, provided that $\epsilon(r)$ has a discontinuity at $r = a$ and is constant for $r > a$. In addition, the inner product $\langle\langle f_j | f_k \rangle\rangle$ is defined as

$$\lim_{X \to \infty} \int_0^X \mathrm{d}r\, \rho(r)\epsilon(r)f_j(r)f_k(r) + \frac{i}{\omega_j + \omega_k}\rho(X)\epsilon(X)^{1/2}f_j(X)f_k(X) , \qquad (7)$$

and is proportional to δ_{jk}. Hence, the orthogonality relation of QNMs of the scalar wave equation is established.[12]

2.3. Completeness of electromagnetic MDRs

The completeness relation discussed so far concerns only scalar waves in one dimension. Its generalization to vectorial electromagnetic waves propagating in three dimensions is outlined here. First of all, the magnetic field basis vectors for TE and TM modes are defined respectively by:

$$\mathbf{b}_{1jlm} \equiv \frac{1}{i\omega_{1jl}} \nabla \times \left[\frac{f_{1jl}(r)}{r} \mathbf{X}_{lm} \right] , \qquad (8)$$

$$\mathbf{b}_{2jlm} \equiv \frac{f_{2jl}(r)}{r} \mathbf{X}_{lm} , \qquad (9)$$

where $f_{\nu jl}(r)$ $(\omega_{\nu jl})$ is the j-th normalized scalar QNM radial function (eigenfrequency) with angular momentum indices $\{lm\}$ and polarization $\nu = 1, 2$. The corresponding conjugate vector $\mathbf{b}^\dagger_{\nu jlm}$ is obtained by replacing $\mathbf{X}_{lm}$ with its complex conjugate $\mathbf{X}^*_{lm}$ in these two equation. Then, the inner product between two MDRs is given by:

$$\langle\langle \mathbf{b}_{\nu jlm} | \mathbf{b}_{\nu' j'l'm'} \rangle\rangle$$

$$\equiv \lim_{X \to \infty} \int_{r<X} \mathrm{d}^3\mathbf{r}\, \mathbf{b}^\dagger_{\nu jlm} \cdot \mathbf{b}_{\nu' j'l'm'}$$

$$+ \frac{i}{\omega_{\nu jl} + \omega_{\nu' j'l'}} \epsilon(X)^{-1/2} \int \mathrm{d}^3\mathbf{r}\, \delta(r - X)\mathbf{b}^\dagger_{\nu jlm} \cdot \mathbf{b}_{\nu' j'l'm'} . \qquad (10)$$

It is readily shown that the MDRs are orthogonal to each other.

As the functions $f_{\nu jl}(r)$ form a complete set for $r < a$, $\mathbf{b}_{\nu jlm}$ also form a (over)complete set for *transverse* vector functions there.[14,15] In general, for any function $\mathbf{V}(\mathbf{r})$ vanishing at infinity, there exists a unique decomposition separating it into a divergence-free component $\mathbf{V}_t(\mathbf{r})$ and a curl-free component $\mathbf{V}_l(\mathbf{r})$.[16] The transverse dyadic δ-functionx $\bar{\mathbf{I}}_t(\mathbf{r}, \mathbf{r}')$ is defined by:

$$\int \bar{\mathbf{I}}_t(\mathbf{r}, \mathbf{r}') \cdot \mathbf{V}(\mathbf{r}') \, \mathrm{d}^3\mathbf{r}' = \mathbf{V}_t(\mathbf{r}) . \qquad (11)$$

Specifically, if $\mathbf{V}(\mathbf{r})$ vanishes for $r \geq a$, $\bar{\mathbf{I}}_t(\mathbf{r}, \mathbf{r}')$ can be expanded as:

$$\bar{\mathbf{I}}_t(\mathbf{r}, \mathbf{r}') = \sum_{\nu j l m} \frac{\mathbf{b}_{\nu j l m}(\mathbf{r}) \, \mathbf{b}^{\dagger}_{\nu j l m}(\mathbf{r}')}{2} , \qquad \text{for } r, r' < a , \tag{12}$$

which is the completeness relation for these vectorial MDRs.[14,15]

2.4. Dyadic Green's function of Maxwell's equations

The magnetic field generated by a harmonic current source $\mathbf{J}(\mathbf{r}, t) = \mathbf{j}(\mathbf{r}) e^{-i \omega t}$ is governed by:

$$\nabla \times \left[\epsilon^{-1}(\mathbf{r}) \nabla \times \mathbf{b}(\mathbf{r}) \right] - \omega^2 \mathbf{b}(\mathbf{r}) = \nabla \times \left[4\pi \epsilon^{-1}(\mathbf{r}) \mathbf{j} \right] . \tag{13}$$

A transverse dyadic Green's function $\bar{\mathbf{G}}^b_t(\mathbf{r}, \mathbf{r}'; \omega)$ is accordingly defined as the magnetic field generated by a transverse dyadic δ-function:

$$\nabla \times \left[\epsilon^{-1}(\mathbf{r}) \nabla \times \bar{\mathbf{G}}^b_t(\mathbf{r}, \mathbf{r}'; \omega) \right] - \omega^2 \bar{\mathbf{G}}^b_t(\mathbf{r}, \mathbf{r}'; \omega) = \bar{\mathbf{I}}_t(\mathbf{r}, \mathbf{r}') . \tag{14}$$

Thus, the magnetic field generated by $\mathbf{j}(\mathbf{r})$ is given by

$$\mathbf{b}(\mathbf{r}) = \int \bar{\mathbf{G}}^b_t(\mathbf{r}, \mathbf{r}'; \omega) \cdot \nabla_{r'} \times \left[4\pi \epsilon(\mathbf{r}')^{-1} \mathbf{j}(\mathbf{r}') \right] \mathrm{d}^3 r' . \tag{15}$$

Following directly from (1), (12), (14) and the retarded nature of the Green's function, the dyadic Green's function can be expanded in terms of MDRs:[14,15]

$$\bar{\mathbf{G}}^b_t(\mathbf{r}, \mathbf{r}'; \omega) = -\frac{1}{2} \sum_{\nu j l m} \frac{\mathbf{b}_{\nu j l m}(\mathbf{r}) \, \mathbf{b}^{\dagger}_{\nu j l m}(\mathbf{r}')}{\omega_{\nu j l m}(\omega - \omega_{\nu j l m})} \qquad \text{for } r, r' < a. \tag{16}$$

The physical meaning of this expansion is of interest. Each term in the sum represents the contribution from an MDR. As the transverse dyadic Green's function is a measure of the field generated by a unit current source, MDRs with larger wave function inside the dielectric sphere or frequency closer to ω can lead to stronger field. Thus, the MDR expansion (16) successfully interprets the enhanced emission rates observed in various experiments.[6−10]

2.5. MDRs of uniform dielectric spheres

As a simple yet useful example, we consider the MDRs of a uniform dielectric sphere with refractive index n and radius a. This simple system is important because its MDRs could be solved exactly and hence could serve as the zeroth order approximation of other more complicated systems. The MDRs are solutions of the following equation:

$$\frac{\rho}{r j_l(n \omega r)} \frac{\mathrm{d}[r j_l(n \omega r)]}{\mathrm{d}r} \bigg|_{r=a} = \frac{1}{r h_l^{(1)}(\omega r)} \frac{\mathrm{d}[r h_l^{(1)}(\omega r)]}{\mathrm{d}r} \bigg|_{r=a} , \tag{17}$$

where j_l (h_l) is the spherical bessel (hankel) function of the l-th order.[12] Figure 1 shows the eigenfrequencies of $l = 10$ MDRs. For each polarization, there are two series of MDRs. While the first series consists of MDRs with low frequencies and large damping rates, the second one is characterized by high frequencies and small damping rates. The leading MDRs in the latter one are usually the dominating modes in experimental situations and approximate formulas for the frequencies and damping rates of these modes have been worked out.[17] Besides, for high frequency MDRs in the second series, the frequency difference between two adjacent MDRs and the damping rate tend to constant values asymptotically.

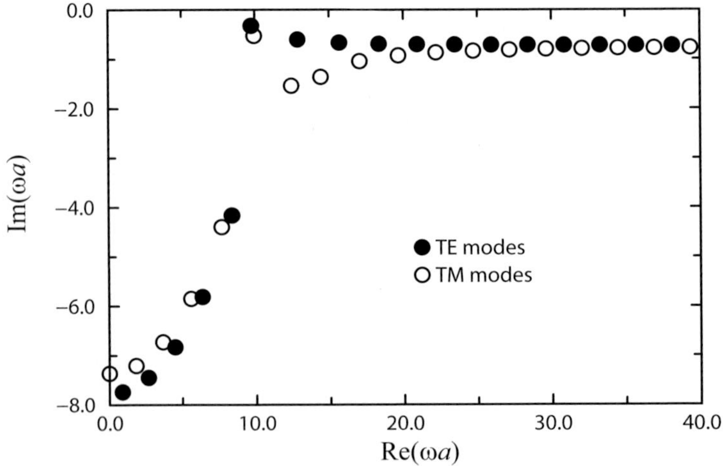

Fig. 1. Locations of the leading TE (filled circles) and TM (open circles) MDRs for $l = 10$ of a uniform dielectric sphere with $n = 1.33$ in the complex ωa plane (from Ref. 12).

2.6. Perturbation scheme

As MDRs are not orthogonal to each other in the region where they do form a complete set, the standard Rayleigh-Schrödinger perturbation theory does not apply. Instead, the Green's function method is used to establish the perturbation theory for MDRs.[15,18] As shown above, $\bar{\mathbf{G}}_t^b(\mathbf{r}, \mathbf{r}'; \omega)$ has complex frequency poles at the MDR eigenfrequencies. Thus, MDR frequencies can be found by locating the poles of $\bar{\mathbf{G}}_t^b(\mathbf{r}, \mathbf{r}'; \omega)$ perturbatively.[15,18]

Consider a system specified by dielectric constant $\epsilon(\mathbf{r})$. Its MDRs and dyadic Green's function $\bar{\mathbf{G}}_t^b$ cannot be obtained analytically. Construct a comparing system characterized by a spherically symmetric dielectric constant $\epsilon^{(0)}(r)$, whose MDRs are exactly solvable and form a complete set for expansion for $r < a$, with $r = a$ being the radius of the outermost surface of discontinuity.[12,14] Therefore, the corresponding dyadic Green's function $\bar{\mathbf{D}}_t^b(\mathbf{r}, \mathbf{r}'; \omega)$ for this system, obeying

$$\nabla \times \left\{ \left[\epsilon^{(0)}(\mathbf{r}) \right]^{-1} \nabla \times \bar{\mathbf{D}}_t^b(\mathbf{r}, \mathbf{r}'; \omega) \right\} - \omega^2 \bar{\mathbf{D}}_t^b(\mathbf{r}, \mathbf{r}'; \omega) = \bar{\mathbf{I}}_t(\mathbf{r}, \mathbf{r}') , \tag{18}$$

can be expressed as an MDR expansion for $r < a$:[14]

$$\bar{\mathbf{D}}_t^b(\mathbf{r}, \mathbf{r}'; \omega) = -\frac{1}{2} \sum_\alpha \frac{\mathbf{b}_\alpha^{(0)}(\mathbf{r}) \mathbf{b}_\alpha^{(0)\,\dagger}(\mathbf{r}')}{\omega_\alpha^{(0)} (\omega - \omega_\alpha^{(0)})} , \tag{19}$$

where $\mathbf{b}_\alpha^{(0)}(\mathbf{r})$ and $\omega_\alpha^{(0)}$ are the normalized magnetic eigenfunctions and eigenfrequencies of the MDRs of this system respectively.[14]

The difference between the original (perturbed) system and the comparing (unperturbed) system is measured by

$$\delta\rho(\mathbf{r}) \equiv \left[\epsilon^{(0)}(r) \right]^{-1} - \epsilon^{-1}(\mathbf{r}) . \tag{20}$$

If $\delta\rho(\mathbf{r})$ vanishes identically for $r > a$, a formal solution of $\bar{\mathbf{G}}_t^b$ can be written down readily:[12,15]

$$\bar{\mathbf{G}}_t^b = \bar{\mathbf{D}}_t^b + \bar{\mathbf{D}}_t^b \cdot \Delta\bar{\mathbf{D}}_t^b + \bar{\mathbf{D}}_t^b \cdot \Delta\bar{\mathbf{D}}_t^b \cdot \Delta\bar{\mathbf{D}}_t^b + \cdots , \tag{21}$$

where, for example,

$$\bar{\mathbf{D}}_t^b \cdot \Delta\bar{\mathbf{D}}_t^b = \int_{s<a} d^3s \, \bar{\mathbf{D}}_t^b(\mathbf{r}, \mathbf{s}; \omega) \cdot \nabla \times \left[\delta\rho(\mathbf{s}) \nabla \times \bar{\mathbf{D}}_t^b(\mathbf{s}, \mathbf{r}'; \omega) \right] . \tag{22}$$

Since the unperturbed MDRs $\mathbf{b}_\alpha^{(0)}$ are complete for $r < a$, the perturbed Green's function can be expanded as:

$$\bar{\mathbf{G}}_t^b(\mathbf{r}, \mathbf{r}'; \omega) = \sum_{\alpha\beta} \mathbf{b}_\alpha^{(0)}(\mathbf{r}) \, G_{\alpha\beta} \, \mathbf{b}_\beta^{(0)\,\dagger}(\mathbf{r}') . \tag{23}$$

Direct substitution of this into (21) leads to a series expansion for the matrix G:

$$G = D + D\Delta D + D\Delta D\Delta D + \cdots , \tag{24}$$

where the matrices D and Δ are given by:

$$D_{\alpha\beta} = -\frac{1}{2} \frac{1}{\omega_\alpha^{(0)} \left[\omega - \omega_\alpha^{(0)} \right]} \delta_{\alpha\beta} , \tag{25}$$

$$\Delta_{\alpha\beta} = \int_{s<a} d^3s \, \mathbf{b}_\alpha^{(0)\,\dagger}(\mathbf{s}) \cdot \nabla \times \left[\delta\rho(\mathbf{s}) \nabla \times \mathbf{b}_\beta^{(0)}(\mathbf{s}) \right] . \tag{26}$$

For non-degenerate cases where $\Delta_{\alpha\gamma} = 0$ if $\omega_\alpha^{(0)} = \omega_\gamma^{(0)}$, the self-energy matrix $W_{\alpha\beta}^{\{\gamma\}}(\omega)$ of the γ-th state is defined by the infinite series:

$$\Delta_{\alpha\beta} + \sum_{\zeta\neq\gamma} \Delta_{\alpha\zeta} D_{\zeta\zeta} \Delta_{\zeta\beta} + \sum_{\zeta\neq\gamma} \sum_{\xi\neq\gamma} \Delta_{\alpha\zeta} D_{\zeta\zeta} \Delta_{\zeta\xi} D_{\xi\xi} \Delta_{\xi\beta} + \cdots . \tag{27}$$

As the γ-th MDR is excluded in all summations, each term in these sums is regular at frequencies $\omega \simeq \omega_\gamma^{(0)}$. Hence, the infinite series can be truncated at finite orders to get an approximate expression. The matrix G is now expressible in terms of the self-energy,

$$G_{\alpha\beta} = D_{\alpha\beta} + D_{\alpha\alpha}W_{\alpha\beta}^{\{\gamma\}}D_{\beta\beta} - \frac{D_{\alpha\alpha}W_{\alpha\gamma}^{\{\gamma\}}W_{\gamma\beta}^{\{\gamma\}}D_{\beta\beta}}{2\omega_\gamma^{(0)}(\omega - \omega_\gamma^{(0)}) + W_{\gamma\gamma}^{\{\gamma\}}} , \qquad (28)$$

which is singular at the zeros of $2\omega_\gamma^{(0)}(\omega - \omega_\gamma^{(0)}) + W_{\gamma\gamma}^{\{\gamma\}}(\omega)$. Thus, the perturbed MDR frequency $\omega_\gamma = \omega_\gamma^{(0)} + \sum_{s=1}^{\infty} \omega_\gamma^{(s)}$ could be found, with the leading two orders given explicitly by:

$$\omega_\gamma^{(1)} = -\frac{\omega_\gamma^{(0)}}{2}V_{2\gamma\gamma} ,$$

$$\omega_\gamma^{(2)} = \frac{\omega_\gamma^{(0)}}{4}\sum_{\beta\neq\gamma}V_{2\gamma\beta}\frac{\omega_\beta^{(0)}}{\left[\omega_\gamma^{(0)} - \omega_\beta^{(0)}\right]}V_{2\beta\gamma} . \qquad (29)$$

Here

$$V_{2\gamma\beta} \equiv \Delta_{\gamma\beta}/(\omega_\gamma^{(0)}\omega_\beta^{(0)}) = \int_{s<a} \mathrm{d}s\, \delta\rho(\mathbf{s})\epsilon^{(0)}\mathbf{e}_\gamma^{(0)\dagger}(\mathbf{s}) \cdot \epsilon^{(0)}\mathbf{e}_\beta^{(0)}(\mathbf{s}) , \qquad (30)$$

with $\mathbf{e}_\beta^{(0)} \equiv (i/\epsilon^{(0)}\omega_\beta^{(0)})\nabla \times \mathbf{b}_\beta^{(0)}(\mathbf{s})$ being the electric field eigenfunctions of the MDRs of the unperturbed system.[14] The perturbation series outlined above is generic and applies to both TE and TM modes. After suitable revision, it can even handle degenerate cases.[18] In Ref. 18, the first-order shifts in eigenfrequencies of degenerate MDRs are obtained from the eigenvalues of a matrix measuring the coupling between degenerate MDRs.

As an example, we apply the perturbation theory developed above to locate MDR's of a dielectric sphere (with radius a and refractive index n_I) containing a smaller off-centered dielectric spherical inclusion (with radius b and refractive index n_II). The separation between their centers is d. As shown in Fig. 2, the perturbative results agree nicely with the exact values obtained from a diagonalization scheme based on the the T-matrix method,[11] demonstrating that the perturbation expansion in fact works very well.

2.7. MDRs of doped microdroplets

The perturbation scheme outlined above is a powerful method to study how MDRs would be affected by various kinds of perturbations, such as the presence of bubbles, latex spheres and shape distortions.[19–21] Specifically, the MDRs in microdroplets doped with many tiny inclusions are considered here. These systems have been studied intensively by several experimental groups and numerous interesting observations, such as anomalous lasing behaviors and spectral line broadening, have been reported.[20,22–25] Although the scattering of light waves from a sphere containing one (or a few) inclusion can be solved numerically,[11,19,26] it is still a formidable

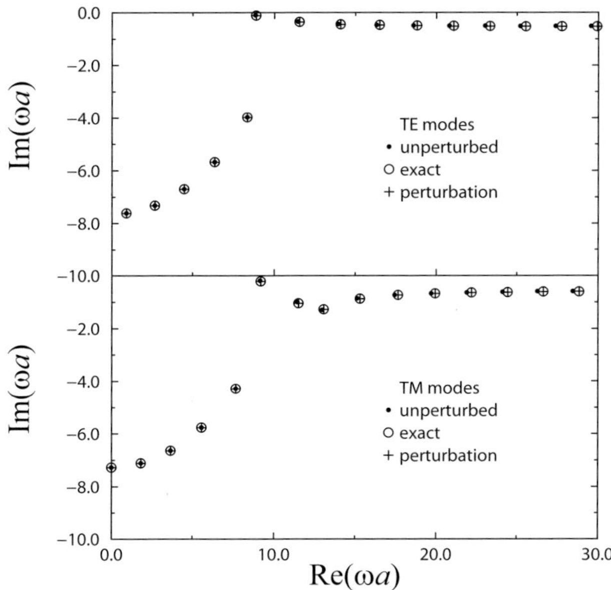

Fig. 2. The plot of the MDRs of an uniform dielectric sphere with an inclusion ($n_{\rm I} = 1.5$, $n_{\rm II} = 1.43$, $b = 0.5a$ and $d = 0.45a$). These are the $l = 10$ and $m = 2$ modes. The dots show the MDR's of the unperturbed system, the centers of the open circles show the exact positions and the crosses are the results of the second-order perturbation theory (from Ref. 15).

task to handle cases with large numbers of inclusions. In addition, usually the inclusions are randomly distributed inside the sphere, one has to perform many different simulations in order to obtain the configurational average.

Employing the degenerate perturbation method developed in Ref. 18, Leung *et al.* (2002) studied the MDRs in a dielectric sphere (with radius a and refractive index $n_{\rm I}$) containing N randomly disturbed identical tiny inclusions (with radius b and refractive index $n_{\rm II}$).[27] In the absence of the inclusions, an MDR of the host sphere can be labeled by the quantum numbers $\{\nu j l m\}$. Its eigenfrequency, $\omega_{\nu j l m}^{(0)}$, is independent of m and is $(2l + 1)$-fold degenerate. After introducing the inclusions into the host sphere, the spherical symmetry of the system is broken and the $2l + 1$ MDRs in a degenerate family thus acquire different frequencies $\omega_{\gamma p}$, with $\gamma \equiv \{\nu j l\}$ and $p = 1, 2, \ldots, 2l+1$. According to the degenerate perturbation scheme established in Ref. 18, the shifts in the MDR frequencies, $\delta\omega_{\gamma p} \equiv \omega_{\gamma p} - \omega_{\gamma}^{(0)}$, to first order in $Nb^3(n_{\rm II}^2 - n_{\rm I}^2)$, are given by the eigenvalues of the matrix T defined by:

$$\mathrm{T}_{mn} = -\frac{\omega_\gamma^{(0)}}{2}(n_{\rm II}^2 - n_{\rm I}^2)\sum_{i=1}^{N}\int_{v_i} \mathrm{d}^3\mathbf{r}\,[\mathbf{e}_{\gamma m}^{(0)}]^\dagger(\mathbf{r})\cdot\mathbf{e}_{\gamma n}^{(0)}(\mathbf{r})\,, \tag{31}$$

where v_i is the space volume occupied by the i-th inclusion.[14,15] While the positions of the MDRs depend sensitively on the exact configuration of the inclusions, the averaged data for many different configurations do show some statistical behaviors for $N \gg 1$. First, the average frequency shift $\langle \delta\omega_\gamma \rangle \equiv \langle \sum_p \delta\omega_{\gamma p} \rangle /(2l+1)$, where $\langle \rangle$ denotes configurational averaging, is proportional to Nb^3 and its real part is usually much larger than its imaginary part. Second, the normalized distribution function as a function of the real part of the MDR frequency, $\rho(x)$, yielding the probability of finding an MDR within a real frequency interval dx, where $x \equiv \mathrm{Re}(\delta\omega_{\gamma p} - \langle \delta\omega_\gamma \rangle)a$, takes on a universal form as a function of $x/(\sqrt{N}b^3)$ as shown in Fig. 3.

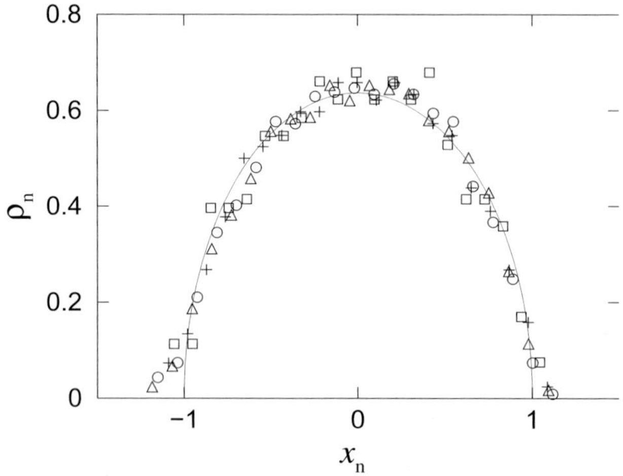

Fig. 3.   The normalized distribution function $\rho_{\mathrm{n}} \equiv 2\sigma_c\rho(x)$ is plotted against $x_{\mathrm{n}} \equiv x/(2\sigma_c)$ for TE MDRs with $l = 50$, $N = 10^5$ (circle); $l = 50$, $N = 10^6$ (square); $l = 75$, $N = 10^5$ (triangle); $l = 75$, and $N = 10^6$ (plus). $n_{\mathrm{I}} = 1.33$, $n_{\mathrm{II}} = 1.50$, and the volume ratio of the inclusions is 3% for all cases. The data all lie on a universal curve (solid line) as predicted by (33) (from Ref. 27).

Leung *et al.* (2002) found that these behaviors are analogous to those of gaussian random matrices (GRM) (see, e.g. Ref. 28), and showed that

$$\langle \delta\omega_\gamma \rangle = \langle \mathrm{T}_{mm} \rangle = -\frac{\omega_\gamma^{(0)}}{2} \frac{Nb^3(n_{\mathrm{II}}^2 - n_{\mathrm{I}}^2)I_2}{a^3 N_{1lj}^2}, \tag{32}$$

where $I_2 \equiv \int_0^a j_l^2(n_{\mathrm{I}}\omega_\gamma^{(0)}r)r^2 \, dr$.[27] Moreover, for narrow resonances with $l \gg 1$, they found that the normalized distribution function $\rho(x)$ obeys the Wigner's semi-circular distribution:[28]

$$\rho(x) = \begin{cases} (2\pi\sigma_{\mathrm{c}}^2)^{-1}\sqrt{4\sigma_{\mathrm{c}}^2 - x^2} & \text{, if } |x| < 2\sigma_{\mathrm{c}}, \\ 0 & \text{, } \quad\quad\text{otherwise.} \end{cases} \tag{33}$$

In particular, from the theory of GRMs[28] and the completeness relation of the Clebsch-Gordon coefficients they showed that:[27]

$$\frac{\sigma_{\mathrm{c}}}{|\langle \mathrm{T}_{mm} \rangle a|} = \sqrt{\frac{1}{N} \left[\frac{(2l+1)a^3 I_4}{6|I_2|^2} - 1 \right]^{1/2}}, \tag{34}$$

where $I_4 \equiv \int_0^a |j_l(n_{\mathrm{I}}\omega_\gamma^{(0)}r)|^4 r^2 \, \mathrm{d}r$. Thus, σ_{c} is proportional to $\sqrt{N}b^3$, explaining the universality observed in Fig. 3.

Besides, they also evaluated perturbatively the peaks in the total cross-section.[27] They found that each scattering peak, which originally corresponds to $2l+1$ degenerate MDRs, is split into multiple peaks. Although each of these peaks has a width close to that of the unperturbed MDR, they are slightly displaced with respect to one another and the gross spectral line is thus broadened. Following from the universality demonstrated in Fig. 3, this broadening effect is proportional to $\sqrt{N}b^3$. Their results agree qualitatively with the experimental observations of the broadening and the emergence of multi-peak structure of spectral lines upon the addition of inclusions in a droplet.[25]

3. QNMs of gravitational waves

3.1. *Introduction to gravitational waves*

In general relativity, spacetime is a dynamical entity that is measured by the metric tensor and can be distorted due to the presence of matters. In particular, gravitational waves are spacetime oscillations resulting from coherent accelerated motion of massive objects. A prominent feature of gravitational waves is that their interaction with matter is extremely weak and hence can travel from their sources to the Earth with almost zero absorption. Therefore, it is possible to probe astrophysical processes occurring far away from the Earth with gravitational waves, including the coalescence and merger of BHs and NSs, core collapse of stars, and even the dynamics of the early universe. In order to measure these waves, ultra-sensitive gravitational wave detectors of various designs have been proposed (see e.g. Ref. 29 and references therein). It is generally believed that gravitational waves can be observed and analyzed within one or two decades.

Being compact stellar objects, BHs and NSs are two kinds of promising gravitational wave emitter. When they are perturbed (or formed) in an asymmetric way, they can emit gravitational waves and in turn lose energy. In this sense, these compact stellar objects form open wave systems. Despite that the evaluation of gravitational waves emitted in violent stellar activities such as binary mergers and asymmetric core collapse has indeed posed a grand challenge to researchers in numerical relativity. Linearized theory of pulsating BHs and NSs can still provide useful insight into such complex situations. As in other open wave systems, linearized gravitational waves emitted from such compact stellar objects are analyzed in terms of QNMs.[30,31,32] In the following, we review the properties of the QNMs of Schwarzschild BHs and non-rotating NSs, with emphases on the similarities and differences between these two kinds of QNMs.

3.2. Perturbations in spacetime

The spacetime outside a spherically symmetric static compact stellar object (BH or NS) is described by the Schwarzschild metric $g_{\alpha\beta}^{(0)}$.[33] If the stellar object is perturbed dynamically, the metric $g_{\alpha\beta} = g_{\alpha\beta}^{(0)} + h_{\alpha\beta}$ in its surrounding region varies with time accordingly. The perturbation $h_{\alpha\beta}$ is a measure of the strength of the gravitational wave emitted in the process and obeys the linearized perturbed Einstein equation.[32,33]

The ten components of $h_{\alpha\beta}$ are expressible in series of vector and tensor harmonics, which can be classified into axial (a) and polar (p) parts according to their polarities under inversion. They are derivable from the solution of a Klein-Gordon-type equation:[34,35]

$$\left[\frac{d^2}{dr_*^2} + \omega^2 - V_\nu(r_*)\right] \psi_{\nu lm}(r_*) = 0, \tag{35}$$

where the tortoise coordinate r_* is given by

$$r_* = r + 2M \ln\left(\frac{r}{2M} - 1\right), \tag{36}$$

with M being the mass of the stellar object. Hereafter geometric units with $G = c = 1$ are adopted. The effective potential V_ν is given by the Regge-Wheeler potential V_{a} and the Zerilli potential V_{p} for axial and polar waves respectively, with

$$V_{\mathrm{a}}(r_*) = (r - 2M)\left[\frac{l(l+1)}{r^3} - \frac{6M}{r^4}\right], \tag{37}$$

$$V_{\mathrm{p}}(r_*) = (r - 2M)\frac{2n^2(n+1)r^3 + 6n^2Mr^2 + 18nM^2r + 18M^3}{r^4(nr + 3M)^2}, \tag{38}$$

and $n \equiv (l-1)(l+2)/2$.[34,35]

As $r_* \to \infty$, the potential tends to the standard centrifugal barrier $l(l+1)/r_*^2$, plus some correction terms of the form $(\ln r_*)^\alpha / r_*^\beta$ ($\alpha = 0, 1, \ldots$; $\beta = 3, 4, \ldots$). Despite the smallness of these correction terms, they could reflect gravitational waves at spatial infinity. Hence, a small fraction of wave energy could reside in the neighborhood of the source for a long period of time, leading to *late time tails* of gravitational waves.[36] It is obvious that such tails are not expandable in terms of discrete QNMs. Consequently, QNMs of BHs and NSs do not form complete sets for expansion.[37]

3.3. QNMs of BHs

QNMs of BHs were first observed in numerical solutions of the linearized perturbed Einstein equation.[30,38,39] Interestingly, gravitational radiation observed in many violent and nonlinear processes, such as the head-on collision of two BHs, is still dominated by the QNM ringings.[40] Rigorously speaking, QNMs of BHs are defined by the solutions of (35) which satisfy both the outgoing boundary condition at spatial infinity and the ingoing boundary condition at the event horizon $r = 2M$. From Fig. 4 showing the $l = 2$ QNMs of a Schwarzschild BH, one can readily observe several peculiarities of these QNMs.

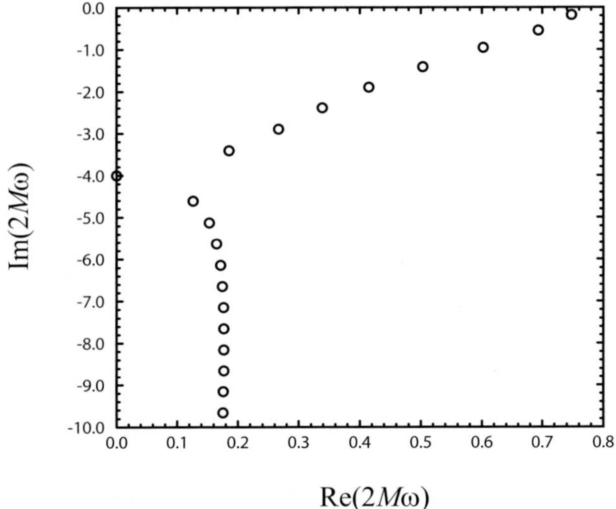

Fig. 4. The distribution of QNMs ($l = 2$) of a Schwarzschild BH. The axial and polar QNMs are identical except the algebraically special mode lying on the imaginary axis (from Ref. 41).

(1) The QNM frequency is inversely proportional to M. Thus, the mass of a BH can be inferred by its QNM signals.

(2) The number of modes for a fixed l is infinite. While $|\mathrm{Re}\,\omega|$ is bounded, $|\mathrm{Im}\,\omega|$ can increase indefinitely. One can label these modes with mode index $j = 1, 2, 3, \ldots$ in order of increasing $|\mathrm{Im}\,\omega|$. Except for a few leading modes, $|\mathrm{Im}\,\omega|$ is greater than $|\mathrm{Re}\,\omega|$, indicating large decay rates of BH QNMs. This is understandable since the potential in (35) is smooth everywhere for $\infty > r > 2M$ and gravitational waves can escape rapidly.

(3) For highly damped modes the QNM frequency is given by:

$$8\pi M\omega_j \approx \ln 3 - 2\pi i(j + 1/2) + O(1/\sqrt{j}) . \tag{39}$$

Interest in this asymptotic behavior is surging, as it is believed to be related to a calculation of the Bekenstein entropy in loop quantum gravity and to the quantum of area (see, e.g. Ref. 42 and references therein). Besides, QNMs of BHs are asymptotically similar to those of dielectric spheres with the roles of $|\mathrm{Re}\,\omega|$ and $|\mathrm{Im}\,\omega|$ interchanged.

(4) For each l there is an *algebraically special mode* lying on the imaginary axis with $2M\omega = -i(l + 2)!/[6(l - 2)!]$.[43] Whether this mode should be considered as a QNM or a total transmission mode is often an issue of debates.[44] Leung *et al.* (2001) also claimed that there are a pair of *unconventional damped modes* close

to the algebraically special mode,[45] adding more controversies to this enigmatic mode.

(5) With the exception of the algebraically special mode, axial and polar QNMs are identical. This feature can be understood as the Regge-Wheeler potential and the Zerilli potential are related by a supersymmetric transformation.[33,45,46]

As mentioned previously, QNMs of BHs do not form complete sets for expansion of spatial functions and it seems difficult to develop perturbation series for them. To this end, a logarithmic perturbative theory (LPT) has been developed to evaluate the shifts in the QNM frequencies due to a quasi-static perturbation of the BH spacetime without using the information of other QNMs.[47]

3.4. QNMs of NSs

As possible remnants of supernova explosions, NSs are built of matters with subnuclear and supranuclear densities. Their equilibrium configuration is likely to provide a direct test for theories of matters with extremely high densities (see e.g. Refs. 48, 49 and references therein). For example, Lindblom has proposed a scheme to estimate the equation of state (EOS) of nuclear matter from the mass and the radius of a NS.[50]

On the other hand, NSs undergoing non-radial oscillations are expected to be promising sources of gravitational waves (see e.g. Ref. 29 and references therein). Therefore, it is possible to unveil the internal structure of NSs and hence relevant EOS from QNMs of relevant gravitational waves. Studies along such direction, coined as *gravitational-wave asteroseismology*, have become the focus of various groups of researchers.[51−54]

The relativistic theory for non-radial oscillations of NSs was pioneered by Thorne and Campolattaro.[55] As mentioned above, such oscillations can be classified into axial and polar cases. In axial oscillations the wave field ψ_{alm}, a measure of spacetime deformation, still satisfies Eq. (35) with the Regge-Wheeler potential modified for $r < R$:

$$V_{\mathrm{a}}(r_*) = \frac{e^{\nu(r)}}{r^3} \left\{ l(l+1)r + 4\pi r^3 \left[\rho(r) - p(r) \right] - 6m(r) \right\} , \tag{40}$$

where R is the radius of the star, $-e^{\nu(r)}$ is the tt component of the unperturbed metric, $\rho(r)$, $p(r)$ are the mass density and pressure at radius r, and $m(r)$ is the mass enclosed by the radius there. It is worthy of remark that fluid elements of the star are merely spectators in axial oscillations. Hence, axial oscillations are classified as spacetime (w) mode.[56]

On the other hand, in polar oscillations of a NS there may exist close interplay between the motion of matter and the variation in spacetime. Such oscillations are usually described by a fourth-order differential system.[57] Polar QNMs fall into two categories, namely fluid and spacetime (w) modes. While the former exist even in the Newtonian limit, including the fundamental (f), pressure (p) and gravity (g) modes,[58] the latter is a feature unique to general relativity.[59] It is worthy of remark that in the polar w mode fluid elements of the star also oscillate with the spacetime. However, such fluid motion does not introduce density variations. Wu and Leung (2007) have recently made use of this point to describe polar w mode QNMs with a second-order differential system.[60]

The w modes are usually typified by high frequencies (around 5-12 kHz) and short damping times (a few tenths of a millisecond). Meanwhile, the damping times for the fluid modes are usually much longer. For example, the typical frequencies (damping times) for f, p, g modes are $1.5 - 3$ kHz ($0.1 - 0.5$ sec), $4 - 7$ kHz (several seconds), less than one hundred Hz (a few days to a few years), respectively.[32]

Fig. 5 shows the axial and polar w modes of a typical NS (APR1 EOS,[48] $M/R = 0.2$). It is interesting to note that the asymptotic behaviors of the QNMs of BHs (see Fig. 4) and the w modes of NSs are drastically different, despite that both of them are associated with oscillations in spacetime. However, the distribution of the w mode QNMs of NSs are quite similar to that of the MDRs of dielectric spheres, except that $\mathrm{Im}\,\omega$ is asymptotically proportional to $\ln(\mathrm{Re}\,\omega)$ and does not tend to a constant. The similarity is conceivable as both of them have clear surfaces delineating their borders.

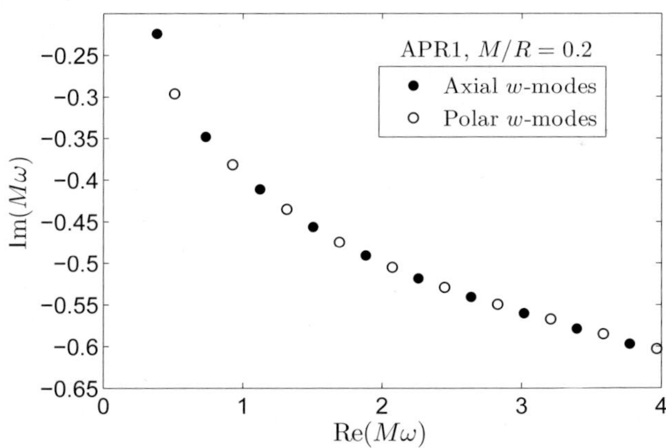

Fig. 5.   Axial and polar w mode QNMs of an APR1 NS with compactness 0.2.

3.5. Gravitational-wave asteroseismology

As the mass-radius curve for NSs reveals marked EOS dependence, QNMs of NSs are expected to behave correspondingly. It is indeed the case for p and g modes. Several schemes have been proposed accordingly to infer stellar parameters (e.g. mass, radius and EOS) from these QNMs.[52] Yet, the scaled frequency $M\omega$ of w and f modes, besides their dependence on the compactness M/R, is rather insensitive to changes in EOS.[52,53] In the light of such universal behavior, the radius and the mass of a NS can be estimated using the f mode and the first polar w mode.[52]

On the other hand, Tsui and Leung (2005)[61] attributed the universality in these modes to the observation that the mass distribution functions of most realistic NSs deviate only slightly from a simple model[62]

$$m_{\mathrm{c}}(r) = (M/2)[5(r/R)^3 - 3(r/R)^5] \,. \tag{41}$$

Furthermore, they showed that minor deviations from $m_{\mathrm{c}}(r)$ can be found from a few leading axial w modes by iteratively applying the inverse scheme of the LPT developed for NSs.[63,64] Hence, accurate EOS of nuclear matter is in turn derivable from such inversion scheme.

4. Conclusion

In this review, we have discussed the application of QNME and associated physical phenomena in dielectric spheres and compact stellar objects. These two different kinds of physical systems are used to illustrate the intriguing properties of QNMs. While QNMs of the former form a complete set for expansion, those of the latter do not owing to the long tail in the effective potential, which also leads to the late time tail in gravitational waves emitted from BHs and NSs.[36,37] On the other hand, the asymptotic behaviors of the QNMs of dielectric spheres and NSs are pretty similar because both systems possess a sharp surface to confine waves. More interestingly, asymptotic behavior of QNMs of BHs is found to be related to issues in quantum gravity.[42] All these confirm that QNME is an important tool to study open systems and deserves the attention of physicists and mathematicians as well.

Acknowledgment

We thank RK Chang and his group at Yale University for introducing us to the field of open systems and many enlightening discussions over the years. The major development of our study on QNMs was performed in collaboration with K Young, who has been the mentor of PT Leung for more than thirty years. We are deeply indebted to him for his guidance and advice. We have benefited from collaborations and discussions with HM Lai, WM Suen and ESC Ching. We are also grateful to the support of many associates and students at CUHK. Our work has been supported in part by several grants from the Hong Kong Research Grants Council [CUHK4282/00P, 401905 and 401807].

References

1. E. S. C. Ching, P. T. Leung, and K. Young, *Optical Processes in Microcavities* (World Scientific, Singapore, 1996), chap. 1, pp. 1–77.
2. E. S. C. Ching, P. T. Leung, A. M. van den Brink, W. M. Suen, S. S. Tong, and K. Young, Rev. Mod. Phys. **70**, 1545 (1998).
3. M. Kerker, ed., *Selected Papers on Light Scattering*, vol. 951 of *Proceedings of SPIE* (Bellingham, Washington, USA, 1988).
4. P. W. Barber and R. K. Chang, eds., *Optical Effects Associated with Small Particles* (World Scientific, Singapore, 1988).
5. R. K. Chang and A. J. Campillo, eds., *Optical Processes in Microcavities* (World Scientific, Singapore, 1996).
6. R. E. Benner, P. W. Barber, J. F. Owen, and R. K. Chang, Phys. Rev. Lett. **44**, 475 (1980).
7. J. B. Snow, S.-X. Qian, and R. K. Chang, Opt. Lett. **10**, 37 (1985).
8. J.-Z. Zhang and R. K. Chang, Opt. Soc. Am. B **6**, 151 (1989).
9. H. M. Tzeng, K. F. Wall, M. B. Long, and R. K. Chang, Opt. Lett. **9**, 499 (1984).
10. L. M. Folan, S. Arnold, and D. Druger, Chem. Phys. Lett. **118**, 322 (1985).
11. P. W. Barber and S. C. Hill, *Light Scattering by Particles: Computational Methods* (World Scientific, Singapore, 1990).
12. P. T. Leung and K. M. Pang, J. Opt. Soc. Am. B **13**, 805 (1996).
13. J. D. Jackson, *Classical Electrodynamics* (Wiley, New York, 1975), 2nd ed.
14. K. M. Lee, P. T. Leung, and K. M. Pang, J. Opt. Soc. Am. B **16**, 1409 (1999).
15. K. M. Lee, P. T. Leung, and K. M. Pang, J. Opt. Soc. Am. B **16**, 1418 (1999).
16. O. D. Kellogg, *Foundations of Potential Theory* (Dover Publications, New York, USA, 1953).
17. C. C. Lam, P. T. Leung, and K. Young, J. Opt. Soc. Am. B **9**, 1585 (1992).
18. S. W. Ng, P. T. Leung, and K. M. Lee, J. Opt. Soc. Am. B **19**, 154 (2002).
19. F. Borghese, P. Denti, R. Saija, and O. I. Sindoni, J. Opt. Soc. Am. A **9**, 1327 (1992).
20. R. L. Armstrong *et al.*, Opt. Lett. **18**, 119 (1993).
21. M. L. Gorodetsky, A. A. Savchenkov, and V. S. Ilchenko, Opt. Lett. **21**, 453 (1996).
22. H.-B. Lin, A. L. Huston, J. D. Eversole, A. J. Campillo, and P. Chýlek, Opt. Lett. **17**, 960 (1992).
23. J. Gu, T. E. Ruekgauer, J. G. Xie, and R. L. Armstrong, Opt. Lett. **18**, 1293 (1993).
24. S. T. H. Taniguchi, M. Nishiya and H. Inaba, Opt. Lett. **21**, 263 (1996).
25. D. Ngo and R. G. Pinnick, J. Opt. Soc. Am. B **11**, 1352 (1994).
26. G. Gouesbet and G. Gréhan, J. Mod. Opt. **47**, 821 (2000).
27. P. T. Leung, S. W. Ng, K. M. Pang, and K. M. Lee, Opt. Lett. **27**, 1749 (2002).
28. M. L. Mehta, *Random Matrices* (Academic Press, San Diego, USA, 1999), 2nd ed.
29. S. Hughes, Ann. Phys. **303**, 142 (2003).
30. W. Press, Astrophys. J. **170**, L105 (1971).
31. E. W. Leaver, Phys. Rev. D **34**, 384 (1986).
32. K. D. Kokkotas and B. G. Schmidt, Living Rev. Rel. **2**, 2 (1999).
33. S. Chandrasekhar, *The Mathematical Theory of Black Holes* (Oxford University Press, New York, 1983).
34. T. Regge and J. A. Wheeler, Phys. Rev. **108**, 1063 (1957).
35. F. J. Zerilli, Phys. Rev. Lett. **24**, 737 (1970).
36. E. S. C. Ching, P. T. Leung, W. M. Suen, and K. Young, Phys. Rev. Lett. **74**, 2414 (1995).
37. E. S. C. Ching, P. T. Leung, W. M. Suen, and K. Young, Phys. Rev. Lett. **74**, 4588 (1995).

38. C. Vishveshwara, Nature **227**, 936 (1970).

39. M. Davis, R. Rufinni, W. Press, and R. Price, Phys. Rev. Lett. **27**, 1466 (1971).

40. P. Anninos, D. Hobill, E. Seidel, L. Smarr, and W. M. Suen, Phys. Rev. D **52**, 2044 (1995).

41. P. T. Leung, Y. T. Liu, W. M. Suen, C. Y. Tam, and K. Young, Phys. Rev. D **59**, 044034 (1999).

42. M. Maggiore, Phy. Rev. Lett. **100**, 141301 (2008).

43. S. Chandrasekhar, Proc. R. Soc. London Ser. A **392**, 1 (1984).

44. A. Maassen van den Brink, Phys. Rev. D **62**, 064009 (2000).

45. P. T. Leung, A. M. van den Brink, W. M. Suen, C. W. Wong, and K. Young, J. Math. Phys. **42**, 4802 (2001).

46. A. Anderson and R. Price, Phys. Rev. D **43**, 3147 (1991).

47. P. T. Leung, Y. T. Liu, W. M. Suen, C. Y. Tam, and K. Young, Phys. Rev. Lett. **78**, 2894 (1997).

48. A. Akmal, V. R. Pandharipande, and D. G. Ravenhall, Phys. Rev. C **58**, 1804 (1998).

49. J. M. Lattimer and M. Prakash, Astrophys. J. **550**, 426 (2001).

50. L. Lindblom, Astrophys. J. **398**, 569 (1992).

51. N. Andersson and K. D. Kokkotas, Phys. Rev. Lett **77**, 4134 (1996).

52. N. Andersson and K. D. Kokkotas, MNRAS **299**, 1059 (1998).

53. O. Benhar, E. Berti, and V. Ferrari, MNRAS **310**, 797 (1999).

54. K. D. Kokkotas, T. A. Apostolatos, and N. Andersson, MNRAS **320**, 307 (2001).

55. K. S. Thorne and A. Campolattaro, Astrophys. J. **149**, 591 (1967).

56. K. D. Kokkotas, MNRAS **268**, 1015 (1994).

57. L. Lindblom and S. L. Detweiler, Astrophys. J. **53**, 73 (1983).

58. T. G. Cowling, MNRAS **101**, 367 (1941).

59. K. D. Kokkotas and B. F. Schutz, MNRAS **255**, 199 (1992).

60. J. Wu and P. T. Leung, MNRAS **381**, 151 (2007).

61. L. K. Tsui and P. T. Leung, MNRAS **357**, 1029 (2005).

62. R. C. Tolman, Phys. Rev. **55**, 364 (1939).

63. L. K. Tsui and P. T. Leung, Astrophys. J. **631**, 495 (2005).

64. L. K. Tsui and P. T. Leung, Phys. Rev. Lett. **95**, 151101 (2005).

INDEX

     Optical Processes in Microparticles and Nanostructures